T0134732

Mathematics for Industry

Volume 29

Aims & Scope

The meaning of "Mathematics for Industry" (sometimes abbreviated as MI or MfI) is different from that of "Mathematics in Industry" (or of "Industrial Mathematics"). The latter is restrictive: it tends to be identified with the actual mathematics that specifically arises in the daily management and operation of manufacturing. The former, however, denotes a new research field in mathematics that may serve as a foundation for creating future technologies. This concept was born from the integration and reorganization of pure and applied mathematics in the present day into a fluid and versatile form capable of stimulating awareness of the importance of mathematics in industry, as well as responding to the needs of industrial technologies. The history of this integration and reorganization indicates that this basic idea will someday find increasing utility. Mathematics can be a key technology in modern society.

The series aims to promote this trend by (1) providing comprehensive content on applications of mathematics, especially to industry technologies via various types of scientific research, (2) introducing basic, useful, necessary and crucial knowledge for several applications through concrete subjects, and (3) introducing new research results and developments for applications of mathematics in the real world. These points may provide the basis for opening a new mathematics-oriented technological world and even new research fields of mathematics.

More information about this series at http://www.springer.com/series/13254

Tsuyoshi Takagi · Masato Wakayama
Keisuke Tanaka · Noboru Kunihiro
Kazufumi Kimoto · Dung Hoang Duong
Editors

Mathematical Modelling for Next-Generation Cryptography

CREST Crypto-Math Project

Editors
Tsuyoshi Takagi
Kyushu University
Fukuoka
Japan

Noboru Kunihiro
The University of Tokyo
Kashiwa
Japan

Masato Wakayama
Kyushu University
Fukuoka
Japan

Kazufumi Kimoto
University of the Ryukyus
Nakagami-gun
Japan

Keisuke Tanaka
Tokyo Institute of Technology
Tokyo
Japan

Dung Hoang Duong
Institute of Mathematics for Industry
Kyushu University
Fukuoka
Japan

ISSN 2198-350X ISSN 2198-3518 (electronic)
Mathematics for Industry
ISBN 978-981-13-5309-3 ISBN 978-981-10-5065-7 (eBook)
DOI 10.1007/978-981-10-5065-7

Softcover re-print of the Hardcover 1st edition 2018

Printed on acid-free paper

This Springer imprint is published by Springer Nature
The registered company is Springer Nature Singapore Pte Ltd.
The registered company address is: 152 Beach Road, #21-01/04 Gateway East, Singapore 189721, Singapore

Preface

The CREST Crypto-Math Project: “Mathematical Modelling for Next-Generation Cryptography” supported by the Japan Science and Technology Agency (JST) aims at constructing mathematical modelling of next-generation cryptography using a wide range of mathematical theories. The goal of the book is to present mathematical background underlying a security modelling of the next-generation cryptography. The book introduces new mathematical results towards strengthening information security, simultaneously making fresh insights and developing the respective areas of mathematics. This project is supported by CREST—a funding program, which is run by the Japan Science and Technology Agency (https://cryptomath-crest.jp/english).

There were 19 papers selected for publication. The book is categorized into four parts. Part I is about mathematical cryptography. It covers both topics in post-quantum cryptography, such as multivariate public-key cryptography, code-based cryptography, hash functions based on expander graphs, isogeny-based cryptography and topics in hyperelliptic curve cryptography. Selected areas in mathematical foundation for cryptography including Ramanujan Caley graphs, quantum Rabi models and spectra of group–subgroup pair graphs are discussed in Part II. Part III is devoted to lattices and cryptography with topics ranging from security analysis for post-quantum cryptosystems based on lattices to lattice attacks on RSA cryptosystems. The last part surveys several important cryptographic protocols such as identity-based encryption and fully homomorphic encryption.

The book is suitable for graduate students and researchers. We hope that this book and its individual articles will prove useful for promoting the research on mathematical modelling for post-quantum cryptography.

Fukuoka, Japan
July 2017

Tsuyoshi Takagi
Masato Wakayama
Keisuke Tanaka
Noboru Kunihiro
Kazufumi Kimoto
Dung Hoang Duong

Contents

Introduction to CREST Crypto-Math Project

Tsuyoshi Takagi

Abstract In this article we introduce the research project "Mathematical Modelling for Prevention of Future Security Compromises (Crypto-Math)" funded by CREST, Japan Science and Technology Agency.

Keywords Security modeling · Post-quantum cryptography · Quantum Rabi model · Zeta functions · Lattice-based cryptography · Multivariate public key cryptography · Graph theory · RSA key recovery attacks

1 The Goal of CREST Crypto-Math Project

Classical cryptography has been used for enciphering techniques in the military and for diplomacy. However, contemporary cryptography has many applications in daily life such as for smartphones, DVDs, e-money, passports, electronic vehicles, and smart grids. Thus, cryptography is a fundamental technology in our society.

There are two cryptosystems that are currently in wide use: RSA Cryptosystem [1] and Elliptic Curve Cryptography (ECC) [2, 3]. Interestingly, these cryptosystems can be constructed using number theory, which has previously been thought to have no real application. However, these cryptosystems are no longer secure in the quantum computing model because the underlying mathematical problems, i.e., the integer factorization problem and discrete logarithm problem, can be solved efficiently by using quantum computers [4]. Therefore, the cryptography research community is investigating the post-quantum cryptography, which ensures the long-term security even in the era of quantum computers. The goal of our research project "CREST: Mathematical Modelling for Next-Generation Cryptography" supported by Japanese Science and Technology Agency is to eventually construct mathematical modeling of next-generation cryptography using wide-range mathematical theories and mathematical analysis of various quantum interaction models which are considered as a

T. Takagi (✉)
Institute of Mathematics for Industry, Kyushu University, 744 Motooka, Nishi-ku, Fukuoka 819-0395, Japan
e-mail: takagi@imi.kyushu-u.ac.jp

T. Takagi et al. (eds.), *Mathematical Modelling for Next-Generation Cryptography*, Mathematics for Industry 29, DOI 10.1007/978-981-10-5065-7_1

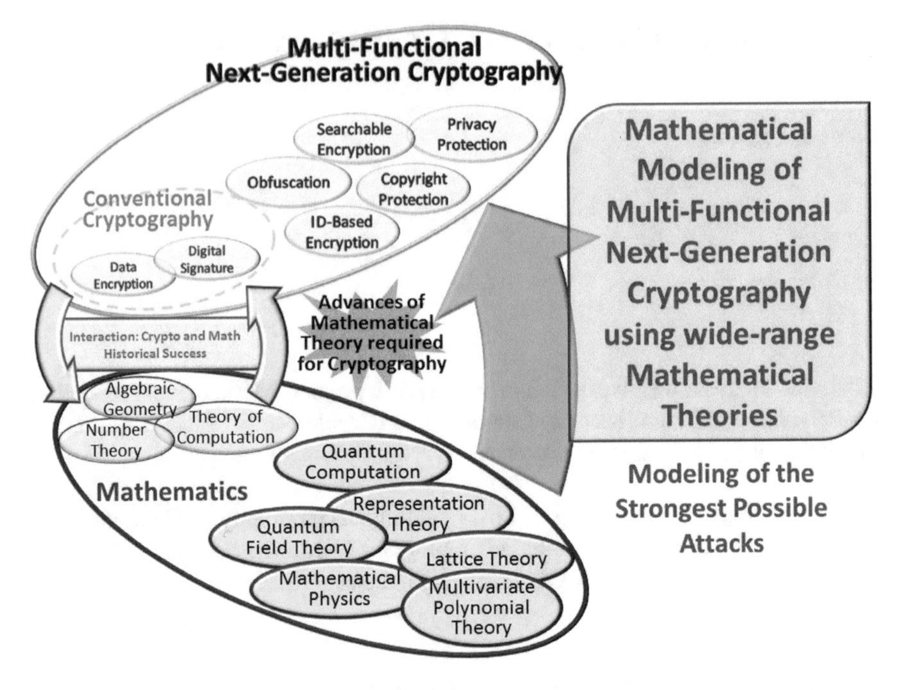

Fig. 1 Research topics in the CREST Crypto-Math Project

theoretical foundation of quantum technology including quantum information theory (Fig. 1).

Recent advances in cryptanalysis, due in particular to quantum computation and physical attacks on cryptographic devices (such as side channel attacks or power analysis), introduced increasing security risks regarding state-of-the-art cryptographic schemes. This project will focus on developing foundations for the mathematical modeling of next-generation cryptographic systems; therefore, addressing the above-mentioned risks.

To achieve this goal, a new mathematical approach will be used that will draw ideas from beyond number theory and theory of computation, which have historically proven to provide a good interchange with cryptography. Specifically, the focus will be in areas that have not yet been fully exploited for cryptographic applications such as representation theory and mathematical physics. Specifically, this project will create a platform for involving mathematicians in research focused on the promotion of a safe society, while at the same time stimulating the development of the respective branches of mathematics.

1.1 Our Research Events in 2015 and 2016

The CREST Crypto-Math Project started in November 2014, and it is a 5.5-year research project. On January 19–20, 2015, we held the first kick-off meeting, where all project members presented their expertise. To promote interaction among the project members of mathematics and cryptography, we also held 12 tutorial talks in three workshops on mathematical cryptography such as provably security techniques in cryptography, basic mathematics in quantum computing, and Ramanujan graphs. In 2015 and 2016, we organized one-day CREST workshops on the main research topics in the CREST Crypto-Math project: “Ramanujan Graphs and Cryptography”, “Geometry and Cryptography”, “L-functions and Cryptography”, “Photons and Lattices”, and “Computational Number Theory and Cryptography”.

A turning point in mathematical cryptography is that the National Security Agency (NSA) announcing a preliminary plan for transitioning to quantum-resistant algorithms in August 2015. On February 24–26, 2016, we organized the 7th International Conference on Post-Quantum Cryptography (PQCrypto 2016) [5] at Kyushu University co-organized by CREST, JST. At PQCrypto 2016, Dustin Moody gave at talk on “Post-Quantum Cryptography: NIST’s Plan for the Future”, and we intensively discussed the security analysis and efficiency estimation of post-quantum cryptography. Moreover, the National Institute of Standards and Technology (NIST) started a standardization process of post-quantum cryptography in 2016 (see their homepage at `http://www.nist.gov/pqcrypto`).

2 Recent Developments of Mathematical Cryptography

Modern cryptography has been used for not only the narrow purposes of preventing eavesdropping over telecommunications but also wide-range security applications such as protecting intellectual property and privacy-preserving computation on encrypted data. In the 1980s and 90s, public key cryptography based on the difficulty of factoring large integers started to be used for enciphering data or digital signatures. From the 1990s to the early 2000s, ID-based encryption based on elliptic curves and bilinear pairing has been used. Recently, the use of cryptography have been expanded to virtual currency, program obfuscation, privacy-protecting technology, etc. To construct such high-functional cryptography and analyze its security, we need novel mathematical theories such as representation theory, mathematical physics, multivariate polynomial theory, and lattice theory as well as advanced number theory. Therefore, mathematical theories required for cryptography have markedly progressed due to the expansion of cryptographic applications.

We now consider the criteria necessary for the mathematical modeling of modern cryptography by listing the historical developments of cryptanalysis (see Fig. 2).

(1) In the 1980s, the integer factorization algorithm and discrete logarithm problem were focused as mathematical problems that underpin the security of public key

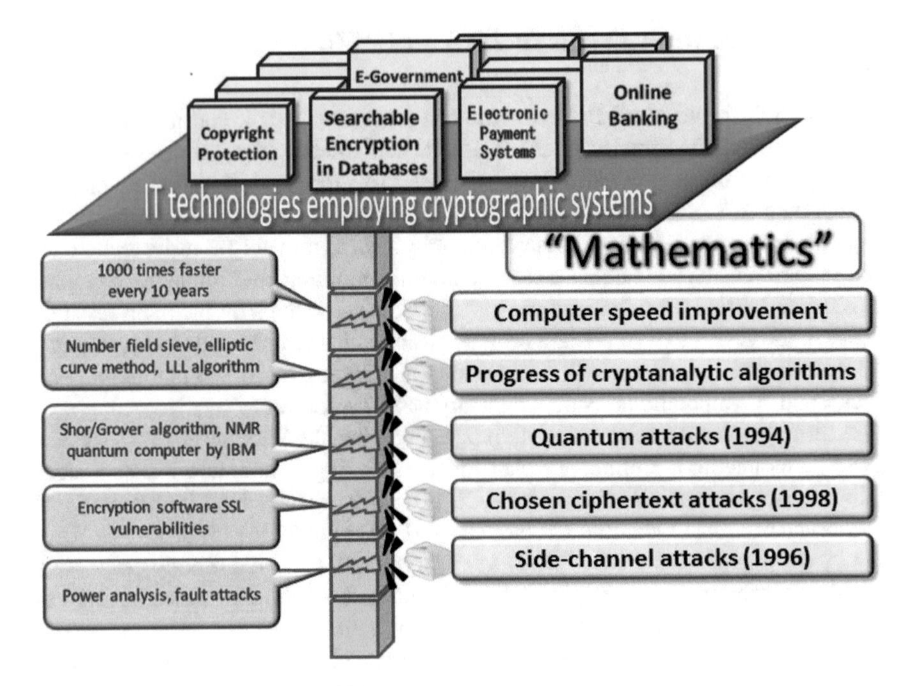

Fig. 2 Historical attacks on cryptography

cryptography. As a result, the number field sieve [6], elliptic curve method [7], and lattice basis reduction algorithm [8] have been developed, and computational number theory was established as a new subject of mathematics.
(2) In 1994, Shor proposed a polynomial-time algorithm for factoring integers using quantum computers [4], and in 2001 IBM conducted the first experiment on factoring an integer using a nuclear magnetic resonance (NMR) quantum computer. Recently, many basic experiments on enlarging quantum computers have been discussed.
(3) In 1996, physical attacks called side channel attacks, such as power analysis or fault attacks, were proposed [9]. The cold boot attack was used for deriving the secret key of an RSA cryptosystem as a type of side channel attack.
In 1998, Bleichenbacher proposed an adaptive chosen ciphertext attack on the encryption software Secure Sockets Layer (SSL) [10]. Afterwards, the indistinguishability against the adaptive chosen ciphertext attack (IND-CCA) became the standard model in cryptography. Recently, the Heartbleed attack against OpenSSL has appeared, and the forward security model has been reconsidered.
(5) The NSA scandal due to the actions of Edward Snowden has caused suspicion regarding the backdoor in the pseudorandom number generator "Dual_EC_DRBG" using the NIST elliptic curve, and reconsideration of generating safe elliptic curves has been discussed.

Shor's algorithm, which was successfully implemented by IBM in 2001, is based on the quantum phenomenon of NMR by Rabi [11] (Nobel Prize in Physics 1994).

The quantum experiment groundbreaking experimental methods that enable measuring and manipulation of individual quantum systems by Haroche and Wineland (Nobel Prize in Physics 1992) is a crucial basis for the quantum technology including quantum information theory such as quantum computers and quantum cryptography (see e.g., [12]). We notice that one of the theoretical background of the Haroche work is the Rabi oscillation [13, 14] and Jaynes–Cumming models [15], where the "quantum" Rabi model was also introduced. The quantum Rabi model is a simplest model used in quantum optics to describe interaction and matter beyond the harmonic oscillator, but only recently could this model be declared solved by Braak [16] (in 2011). It is now pointed out [17] that as physicists gain intuition for Braak's mathematical solution, it is very much expected that the result could have implications for further theoretical and experimental work that explores the interactions between light and matter from weak to extremely strong interactions. In 1996, Grover proposed a quantum-search algorithm of complexity $O(N^{1/2})$ for a function domain of size N [18], then research on a third efficient quantum algorithm proceeded. Moreover, the international conference on post-quantum cryptography started in 2006, and TU Darmstadt started the computational challenge problems of lattices in 2008, which are aimed at achieving cryptosystems secure against quantum computers. From this time, the research on post-quantum cryptography began, and research groups on the topic in governmental organization (e.g., NIST in USA, CRYPTREC in Japan) was established. Therefore, research and development in post-quantum cryptography started in collaboration with academia, industry, and government, and we need the knowledge of mathematics, such as representation theory, mathematical physic, topology (deeply interacting algebra, geometry and analysis including probability theory), which have not been studied in conventional cryptography. From this history of cryptography, we can analogically expect substantial progress from initial research in NMR in 1944 to quantum algorithms by exchanging cryptography and mathematicians of various subjects in our project.

3 Research Groups and Their Activities

There are four groups in this research project. We explain the main research activities of each group. All project members in the CREST Crypto-Math project are shown in Fig. 3. The principal investigator is Tsuyoshi Takagi from the Institute of Mathematics for Industry, Kyushu University. The co-principal investigators are Masato Wakayama (Institute of Mathematics for Industry, Kyushu University), Keisuke Tanaka (Graduate School of Information Science and Engineering, Tokyo Institute of Technology), and Noboru Kunihiro (Graduate School of Frontier Sciences, University of Tokyo). In this CREST Crypto-Math project, 25 mathematicians including 4 postdocs are working on the new mathematical problems arisen from post-quantum cryptography. See the Fig. 4 for overview of research topics in each group.

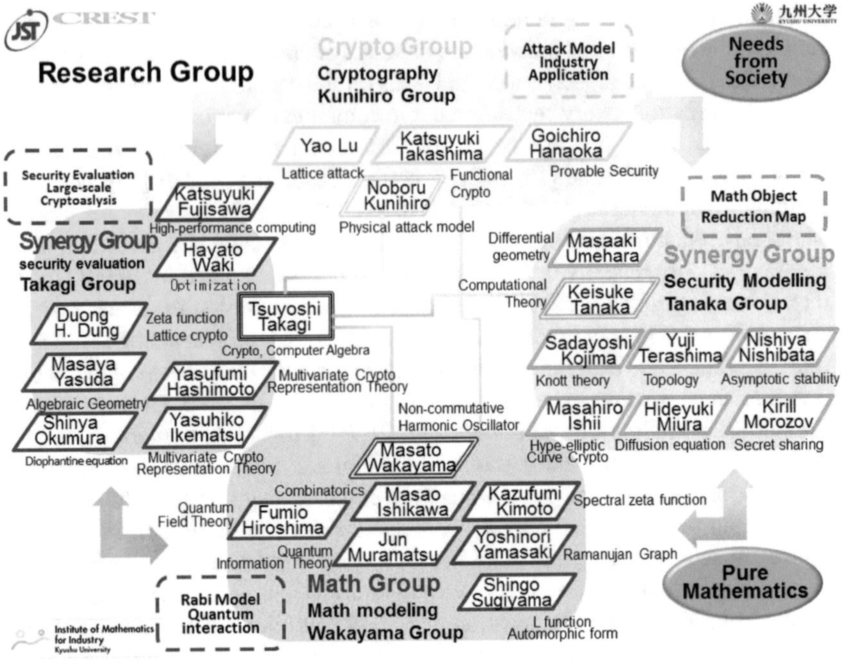

Fig. 3 Members in the CREST Crypto-Math Project

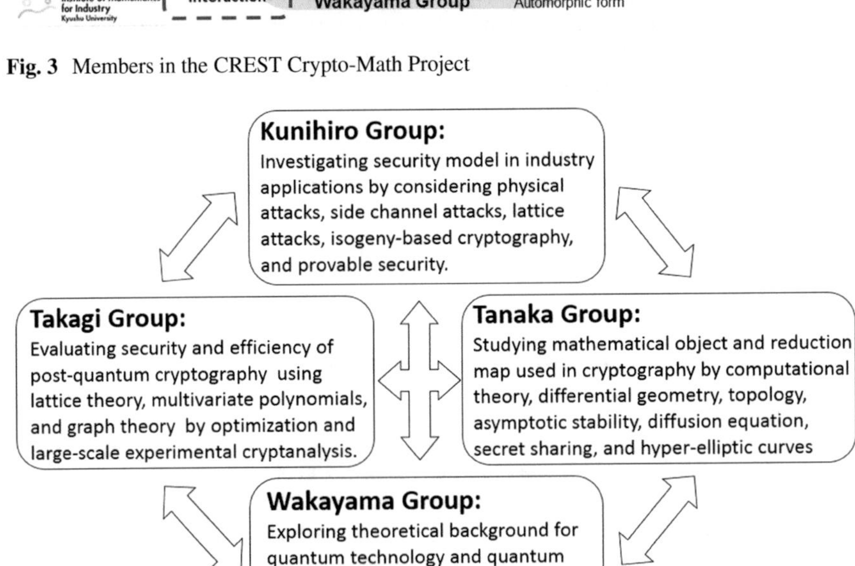

Fig. 4 Research topics in each group

3.1 Takagi Group

The Takagi group is focused on development and security evaluation of next-generation cryptographic systems, which will be resistant against attacks using quantum computers. In particular, the group will study algorithms for solving the mathematical problems underlying such systems, including the shortest vector problem on lattices (SVP) and solving systems of multivariate quadratic equations over a finite field (MQ problem). The group will also study the impact of attackers possessing massive computational resources by conducting corresponding cryptanalytic experiments with major mathematical problems underlying the above-mentioned cryptographic systems. Finally, the group will determine the possibility of using next-generation high-performance cryptographic systems in a real-world environment by building their software implementations and evaluating their performances.

Lattice-Based Cryptography: Yuan et al. presented efficient implementations of lattice-based cryptography using JavaScript, particularly, the learning with errors (LWE)-based encryptions such as Regev05 and LPR11 [19]. This paper received the Outstanding Paper in the Third International Symposium on Computing and Networking (CANDAR'15). Kudo et al. then analyzed the hardness of the LWE problem by the key recovery attack when the modulus was relatively large [20]. We also participated in the lattice challenge contest from TU Darmstadt and solved the shortest vector problem of 625 dimensions in $2^{24.0}$ s using a single CPU core [21]. As a joint study with the Wakayama group, Okumura et al. investigated the security of lattice-based encryption proposed by Garg–Gentry–Halevi [22].

Multivariate Public Key Cryptography and ECC: Hashimoto presented a rank attack on Quaternion Rainbow, which is a digital signature based on the difficulty of solving the MQ problem [23].There are some public key encryption based on the MQ problem such as SRP and ZHFE. Duong et al. reduced the key size of the SRP encryption scheme by addressing the cyclic structure of the public key [24]. Ikematsu et al. presented an efficient key generation algorithm for ZHFE [25]. Moreover, Duong et al. proposed an efficient digital signature based on the cubic UOV signature scheme [26].

Huang et al. showed improvements in the FPPR attack, which is an efficient algorithm for solving the multivariate polynomial from elliptic curve cryptography [27]. We successfully solved the discrete logarithm problem over an elliptic curve defined by a finite field of characteristic two of 29 degrees in about 34 days on AMD Opteron 6276 using the computational algebra system MAGMA.

Other Post-Quantum Cryptosystem: Morozov et al. published two papers on secret sharing and code-based encryption at IEICE Transaction [28, 29]. We also analyzed a public key cryptosystem based on Diophantine equations and showed a polynomial-time algorithm via the weighted LLL reduction [30]. Okumura et al. then discussed the post-quantum cryptosystem based on the difficulty of the section-finding problem on algebraic surfaces [31]. This paper received the Outstanding Paper in the Fourth International Symposium on Computing and Networking (CANDAR'16). As a joint study with the Kunihiro group, Tachibana et al. constructed an efficient hash function based on 3-isogeny graph of supersingular elliptic curves [32]. Finally, Jo et al. proposed a full cryptanalysis of hash functions based on cubic Ramanujan graphs [33].

3.2 Wakayama Group

The safety of RSA encryption, which is based on the computational intractability of the prime factorization, is no longer ensured if a large-scale quantum computers become a possibility. Quantum interaction models, such as the quantum Rabi model, are used in a basic element of quantum computers. The Wakayama group will study the mathematical structure of such models. Among them, noncommutative harmonic oscillators (NcHOs [34, 35]) are thought to be universal models. The group will focus on extending the existing theory and methodology on NcHOs and clarifying the structure of models treated in quantum optics from various viewpoints—representation theory, number theory, functional analysis, and dynamical systems. The group will also develop an efficient method of conducting extensive numerical experiments by using systems of orthogonal functions to verify the deep Riemann hypothesis (DRH [36–38]) for various types of zeta and L-functions. Furthermore, the group will study the DRH and its relation to post-quantum cryptography along with new constructions of Ramanujan graphs through L-functions by using probability theory and combinatorial theory.

Spectral Problem of NcHO: The Wakayama group studied NcHOs using the methodology of number theory, representation theory, analytic differential equations, and investigated the spectrum of NcHOs, the general Rabi model, and their rotation wave approximation model via representation theory. The group obtained the following results. (1) Hiroshima and Sasaki showed the simplicity of the ground state of the NcHO [39]. (2) Wakayama described the Heun differential equation of the spectrum problem of NcHOs for the even eigenvalue function [40]. (3) Employing the representation theoretical method developed in [41], Wakayama recently proved [42] the spectral degeneracies for the asymmetric quantum Rabi model demonstrated numerically by Li-Bachelor [43]. (4) Sugiyama obtained the meromorphic continuation of the spectral zeta functions for quantum Rabi models as the first step of the number theoretic approaches for deep understanding of the spectrum of these models [44].

Cayley Graph over Groups: We proved that the Wreath determinants for group–subgroup pair, which are a generalization of the group determinant defined by pairs of a group and subgroup, can be decomposed into the multiplication of binomial polynomials under proper ordering of the elements in the underlying groups [45]. Reyes-Bustos presented a sufficient condition on a Cayley-type graph for group–subgroup pairs (G, H) and certain subsets S of G that result in bipartite Ramanujan graphs and proposed the use of group–subgroup pair graphs to model linear error-correcting codes [46]. Kimoto investigated the relationship between Alon–Tarsi conjecture on the Latin square appearing in the graph-coloring problem and the Wreath-determinant spherical functions on symmetric groups [47]. Moreover, Hirano et al. determines the bound of the valency of Cayley graphs of Frobenius groups with respect to normal Cayley subsets which guarantees to be Ramanujan [48].

3.3 Tanaka Group

The Tanaka group will mainly study the following two themes on the theory of cryptography. The first theme involves investigating mathematical objects, which can be used in concrete constructions for important cryptographic primitives such as PKE, digital signature, or trapdoor one-way function. The group will focus on those that also appear in advanced topics of mathematics but not that far in the field of cryptography. The second theme involves studying reductions that are typically used in the security proofs in many cryptographic schemes. To develop the techniques applicable to the real world, the group will focus on those that originally come from advanced studies in mathematics.

Mathematical Objects: The Tanaka group constructed cryptosystems that satisfy the functions of our mathematical objects. In particular, Kitagawa et al. showed that single-bit-projection key-dependent message (KDM) security is also complete in the CCA setting, namely one can construct a PKE scheme that is KDM-CCA secure without using additional assumptions [49]. Wang et al. proposed a transformation that converts weakly existentially unforgeable signature schemes into strongly existentially unforgeable ones in the continual leakage model [50]. Wang et al. also presented a fully leakage-resilient signature scheme in the selective auxiliary input model, which captures an extremely wide class of side channel attacks that are based on physical implementations of algorithms rather than public parameters chosen [51]. Moreover, the security model of watermarking has been considered. To achieve the robustness of the embedding method, Thanh et al. embed scrambled watermark information into the low-band frequency of the q-logarithm frequency domain by using the quantization index modulation technique [52]. Thanh et al. also presented a performance analysis of robust watermarking using linear and nonlinear features secure against geometric attacks and signal processing attacks [53]. Group signatures are a class of digital signatures with enhanced privacy. Ishida et al. proposed the notion of a deniable group signature, in which an authority can issue a proof showing that the specified user is not the signer of the signature without revealing the actual signer [54]. In addition, in order to investigate the possibility to employ mathematical objects as cryptographic primitives, Umehara [55], Kojima [56], and Terashima [57] studied multiple algebraic objects such as three dimensional manifolds, knots, links and their properties including multiple invariants.

Security Reduction: Ishida et al. proposed the notion of disavowable PKE with non-interactive opening (disavowable PKENO) where, for a ciphertext and a message, the receiver of the ciphertext can issue a proof that the plaintext of the ciphertext is NOT the message and give a fairly practical construction [58]. Moreover, there are two conversion techniques that convert any PKE scheme secure against chosen plaintext attacks to a PKE scheme that is secure against CCAs. Kitagawa et al. clarified whether these two constructions are also secure in the sense of KDM-CCA security [59]. The k-wise almost independent permutations are important primitives for cryptographic schemes and combinatorial constructions, and Kawachi et al. showed lower bounds of key length for k-wise almost independent permutations and multi-message approximate secrecy [60]. Wang et al. also presented generic transformations, which allow us to convert any signature scheme satisfying the weak existential unforgeability property into one satisfying the strong existential unforgeability property [61]. In addition, in order to study the possibility to employ mathematical techniques as security proofs with reduction mappings, Nishibata [62], Miura [63] studied multiple equations including Navier–Stokes equations for fluid dynamics and their basic properties.

3.4 Kunihiro Group

The Kunihiro group will study the security model that reflects the social needs and the limit model of strongest attackers. In the real world, it has been found that an attack cannot be captured using a conventional security model, and this has caused a serious security hole due to inadequate implementation. Furthermore, it is necessary to assume a more sophisticated attack than before with high functionality of the encryption technology. This group's research target is to properly model these attacks and establish a security model that withstands them. The group will study security under an environment in which attackers reveal the private key by physical observation. The group will also conduct an analysis using more precise noise models than in previous studies and conduct security analysis in an actual environment. Through such studies, the group will propose highly secure cryptosystems and design a secure implementation. The group will also give feedback to the real world.

Attacks on RSA using Lattices: The Kunihiro group studied how to recover RSA secret keys using lattice techniques. Takayasu and Kunihiro proposed improved attacks when attackers know the most/least significant bits of secret keys and public RSA modulus can be factored even when an encryption exponent is full size or the sizes of unknown bits are less than $N^{1/3}$ [64]. Takayasu and Kunihiro also analyzed the small secret exponent attacks on Multi-Prime RSA [65] and partial key exposure attacks on the prime power RSA [66], RSA with multiple exponent pairs [67], and small CRT-exponent RSA [68]. The paper [67] received the best student paper award at 21st Australasian conference on information security and privacy (ACISP 2016). Finally, Lu et al. proposed several improvements on RSA key recovery attacks by introducing an efficient algorithm for solving the approximate common divisor problem (ACDP), which is used to attempt to find a hidden integer for two given integers that are near-multiples of a hidden integer [69, 70].

RSA Key Recovery Attacks from Noisy Version: The Kunihiro group studied how to recover RSA secret keys from their noisy versions observed by side channel attacks. Kunihiro reduced the computational cost by introducing tighter inequalities than the Hoeffding bound and gave a provable bound for crossover probabilities [71]. Next Tanigaki and Kunihiro proposed an algorithm based on the maximum likelihood approach, which can recover a secret key of symmetric key encryption in an imperfect asymmetric decay model, i.e., where bit flipping occurs in both directions [72].

The Kunihiro group also discussed algorithms for recovering secret keys of RSA cryptosystems from noisy analog data. Kunihiro and Takahashi discussed secret key recovery algorithms in accordance with a fixed probability distribution depending on the corresponding correct secret key bit [73]. They also proposed an efficient algorithm (V-based algorithm) and score function (variance-based score) by modifying the differential power analysis (DPA)-like score function to compensate for imbalanced noise. They then proved that the variance-based score is optimal in the weighted variant of a DPA-like score, and verified that their algorithm is superior to previous ones through both theoretical analysis and numerical experiments for various noise distributions.

4 Conclusion

We introduced activities of the CREST Crypto-Math project supported by the Japan Science and Technology Agency. The project's goal is to investigate secure and efficient post-quantum cryptography, which provides risk assessment of next-generation security systems, industry security applications, and security standardization. We will also contribute to new mathematical theory in cryptography as new applications of contemporary mathematics, which could result in a new career path for mathematicians in particular, post graduate students in mathematics. Finally, we will establish a research hub of mathematical cryptography via international collaboration in cryptography.

Acknowledgements I would like to thank the co-investigators of the CREST Crypto-Math Project, Masato Wakayama, Keisuke Tanaka, and Noboru Kunihiro for their valuable comments and discussions on the activities of their research groups.

References

1. R. Rivest, A. Shamir, L. Adleman, A method for obtaining digital signatures and public-key cryptosystems. Commun. ACM **21**(2), 120–126 (1978)
2. N. Koblitz, Elliptic curve cryptosystems. Math. Comput. **48**(177), 203–209 (1987)

3. V. Miller, Use of elliptic curves in cryptography, in *CRYPTO'85*. LNCS, vol. 218 (Springer, Berlin, 1985)
4. P. Shor, Polynomial-time algorithms for prime factorization and discrete logarithms on a quantum computer. SIAM J. Comput. **26**(5), 1484–1509 (1997)
5. T. Takagi (ed.), *7th International Workshop on Post-Quantum Cryptography - PQCrypto 2016*. LNCS, vol. 9606 (Springer, 2016)
6. A. Lenstra, H.W. Lenstra (eds.), *The Development of the Number Field Sieve*. Lecture Notes in Math, vol. 1554 (Springer, Berlin, 1993)
7. H. Lenstra, Factoring integers with elliptic curves. Ann. Math. **126**(3), 649–673 (1987)
8. A. Lenstra, H. Lenstra, L. Lovász, Factoring polynomials with rational coefficients. Math. Ann. **261**(4), 515–534 (1982)
9. P. Kocher, Timing attacks on implementations of Diffie-Hellman, RSA, DSS, and other systems, in *CRYPTO'96*. LNCS, vol. 1109 (Springer, 1996), pp. 104–113
10. D. Bleichenbacher, Chosen ciphertext attacks against protocols based on the RSA encryption standard PKCS #1, in *CRYPTO'98*. LNCS, vol. 1462 (Springer, 1998), pp. 1–12
11. I.I. Rabi, J.R. Zacharias, S. Millman, P. Kusch, A new method of measuring nuclear magnetic moment. Phys. Rev. **53**(4), 318–327 (1938)
12. S. Haroche, J.M. Raimond, *Exploring the Quantum, Atoms, Cavities and Photons* (Oxford University Press, Oxford, 2008)
13. I. Rabi, On the process of space quantization. Phys. Rev. **49**, 324–328 (1936)
14. I. Rabi, Space quantization in a gyrating magnetic field. Phys. Rev. **51**, 652–654 (1937)
15. E.T. Jaynes, F.W. Cummings, Comparison of quantum and semiclassical radiation theories with application to the beam maser. Proc. IEEE **51**, 89–109 (1963)
16. D. Braak, Integrability of the Rabi model. Phys. Rev. Lett. **107**, 100401–100404 (2011)
17. E. Solano, Viewpoint: the dialogue between quantum light and matter. Physics **4**, 52–68 (2011)
18. L. Grover, A fast quantum mechanical algorithm for database search, in *STOC'96* (1996), pp. 212–219
19. Y. Yuan, C.-M. Cheng, S. Kiyomoto, Y. Miyake, T. Takagi, Portable implementation of lattice-based cryptography using JavaScript. Int. J. Netw. Comput. **6**(2), 309–327 (2016)
20. M. Kudo, J. Yamaguchi, Y. Guo, M. Yasuda, Practical analysis of key recovery attack against search-LWE problem, in *IWSEC 2016*. LNCS, vol. 9836 (Springer, 2016), pp. 164–181
21. Y. Aono, Y. Wang, T. Hayashi, T. Takagi, Improved progressive BKZ algorithms and their precise cost estimation by sharp simulator, in *Eurocrypt 2016*. LNCS, vol. 9665 (Springer, 2016), pp. 789–819
22. S. Okumura, S. Sugiyama, M. Yasuda, T. Takagi, Security analysis of cryptosystems using short generators over ideal lattices. Cryptology ePrint Archive: Report 2015/1004
23. Y. Hashimoto, Cryptanalysis of the quaternion rainbow. IEICE Trans. **E98–A**(1), 144–152 (2015)
24. D.H. Duong, A. Petzoldt, T. Takagi, Reducing the key size of the SRP encryption scheme, in *ACISP 2016*. LNCS, vol. 9723 (Springer, 2016), pp. 427–434
25. Y. Ikematsu, D.H. Duong, A. Petzoldt, T. Takagi, Revisiting the efficient key generation of ZHFE, in *C2SI 2017*. LNCS, vol. 10194 (Springer, 2017)
26. D.H. Duong, A. Petzoldt, Y. Wang, T. Takagi, Revisiting the cubic UOV signature scheme, in *ICISC 2016*. LNCS, vol. 10157 (Springer, 2017), pp. 223–238
27. Y.-J. Huang, C. Petit, N. Shinohara, T. Takagi, Improvement of FPPR method to solve ECDLP. Pac. J. Math. Ind. **7**(1), 1–9 (2015)
28. R. Xu, K. Morozov, T. Takagi, Note on some recent cheater identifiable secret sharing schemes. IEICE Trans. **98–A**(8), 1814–1819 (2015)
29. R. Hu, K. Morozov, T. Takagi, Zero-knowledge protocols for code-based public-key encryption. IEICE Trans. **98–A**(10), 2139–2151 (2015)
30. J. Ding, M. Kudo, S. Okumura, T. Takagi, C. Tao, Cryptanalysis of a public key cryptosystem based on diophantine equations via weighted LLL reduction, in *IWSEC 2016*. LNCS, vol. 9836 (Springer, 2016), pp. 305–315

31. S. Okumura, K. Akiyama, T. Takagi, An estimate of the complexity of the section finding problem on algebraic surfaces, in *The Fourth International Symposium on Computing and Networking, CANDAR* vol. 2016 (2016), pp. 28–36
32. H. Tachibana, K. Takashima, T. Takagi, Constructing an efficient hash function from 3-isogenies. JSIAM Lett. (to appear)
33. H. Jo, C. Petit, T. Takagi, Full cryptanalysis of hash functions based on cubic ramanujan graphs. IEICE Trans. (to appear)
34. A. Parmeggiani, M. Wakayama, Oscillator representations and systems of ordinary differential equations. Proc. Natl. Acad. Sci. **98**, 26–30 (2001)
35. A. Parmeggiani, *Spectral Theory of Non-Commutative Harmonic Oscillators: An Introduction*, vol. 1992, Lecture Notes in Mathematics (Springer, Berlin, 2010)
36. D. Goldfeld, Sur les produitd partiels eulerians attache aux courbes elliptiques. Comptes Rendus de l'Académie des Sciences, Series I Mathematics **294**, 471–474 (1982)
37. K. Conrad, Partial Euler products on the critical line. Can. J. Math. **57**, 328–337 (2005)
38. T. Kimura, S. Koyama, N. Kurokawa, Euler products beyond the boundary. Lett. Math. Phys. **104**, 1–19 (2014)
39. F. Hiroshima, I. Sasaki, Spectral analysis of non-commutative harmonic oscillators: the lowest eigenvalue and no crossing. J. Math. Anal. Appl. **105**, 595–609 (2014)
40. M. Wakayama, Equivalence between the eigenvalue problem of non-commutative harmonic oscillators and existence of holomorphic solutions of heun differential equations, eigenstates degeneration, and the Rabi model. Int. Math. Res. Not. **3**, 759–794 (2016)
41. M. Wakayama, T. Yamasaki, The quantum Rabi model and lie algebra representations of $\mathfrak{sl}_2$. J. Phys. A: Math. Theor. **47**(33), 335203 (2014)
42. M. Wakayama, Symmetry of Asymmetric Quantum Rabi Models, arXiv:1701.03888v1 [math-ph, quant-ph]
43. Z.-M. Li, M.T. Batchelor, Algebraic equations for the exceptional eigenspectrum of the generalized Rabi model. J. Phys. A: Math. Theor. **48**, 454005 (2015)
44. S. Sugiyama, Spectral zeta functions for the quantum Rabi models. Nagoya Math. J. pp. 1-47 (2016). doi:10.1017/nmj.2016.62
45. K. Hamamoto, K. Kimoto, K. Tachibana, M. Wakayama, Wreath determinants for group-subgroup pairs. J. Comb. Theory Ser. A **133**, 76–96 (2015)
46. C. Reyes-Bustos, Cayley-type graphs for group-subgroup pairs. Linear Algebra Appl. **488**, 320–349 (2016)
47. K. Kimoto, Wreath Determinants, Spherical Functions on Symmetric Groups and the Alon-Tarsi Conjecture. Preprint
48. M. Hirano, K. Katata, Y. Yamasaki, Ramanujan cayley graphs of frobenius groups. Bull. Aust. Math. Soc. **94**(3), 373–383 (2016)
49. F. Kitagawa, T. Matsuda, G. Hanaoka, K. Tanaka, Completeness of single-bit projection-KDM security for public key encryption, in *CT-RSA 2015*. LNCS, vol. 9048 (Springer, 2015), pp. 201–219
50. Y. Wang, K. Tanaka, Generic transformation to strongly existentially unforgeable signature schemes with continuous leakage resiliency, in *ACISP 2015*. LNCS, vol. 9144 (Springer, 2015), pp. 213–229
51. Y. Wang, T. Matsuda, G. Hanaoka, K. Tanaka, Signatures resilient to uninvertible leakage, in *SCN 2016*. LNCS, vol. 9841 (Springer, 2016), pp. 372–390
52. T.M. Thanh, K. Tanaka, The novel and robust watermarking method based on q-logarithm frequency domain. Multimed. Tools Appl. pp. 1-29 (2015)
53. T.M. Thanh, K. Tanaka, Comparison of watermarking schemes using linear and nonlinear feature matching, in *KSE 2015*, (IEEE, 2015), pp. 262–267
54. A. Ishida, K. Emura, G. Hanaoka, Y. Sakai, K. Tanaka, Group signature with deniability: how to disavow a signature, in *CANS 2016*. LNCS, vol. 1052 (Springer, 2016), pp. 228–244
55. M. Hasegawa, A. Honda, K. Naokawa, K. Saji, M. Umehara, K. Yamada, Intrinsic properties of singularities of surfaces. Int. J. Math. **26**(4), 1540008 (34 pages) (2015)

56. S. Kojima, Normalized entropy versus volume for pseudo-anosovs, in *Proceedings of 62nd Symposium on Topology* (Nagoya Institute of Technology, 2015), pp. 1–10
57. T. Kitayama, Y. Terashima, Torsion functions on moduli spaces in view of the cluster algebra. Geom. Dedicata. **175**(1), 125–143 (2015)
58. A. Ishida, K. Emura, G. Hanaoka, Y. Sakai, K. Tanaka, Disavowable public key encryption with non-interactive opening. IEICE Trans. **E98–A**(12), 2446–2455 (2015)
59. F. Kitagawa, T. Matsuda, G. Hanaoka, K. Tanaka, On the key dependent message security of the Fujisaki-Okamoto constructions, in *PKC 2016*. LNCS, vol. 9615 (Springer, 2016), pp. 99–129
60. A. Kawachi, H. Takebe, K. Tanaka, Lower bounds for key length of k-wise almost independent permutations and certain symmetric-key encryption schemes, in *IWSEC 2016*. LNCS, vol. 9836 (Springer, 2016), pp. 195–211
61. Y. Wang, K. Tanaka, Generic transformations for existentially unforgeable signature schemes in the bounded leakage model. Secur. Commun. Netw. **9**(12), 1829–1842 (2016)
62. T. Nakamura, S. Nishibata, Boundary layer solution to system of viscous conservation laws in half line. Bull. Braz. Math. Soc. **47**(2), 619–630 (2016)
63. Y. Maekawa, H. Miura, On poisson operators and Dirichlet-Neumann maps in hs for divergence form elliptic operators with Lipschitz coefficients. Trans. Am. Math. Soc. **368**(9), 6227–6252 (2016)
64. A. Takayasu, N. Kunihiro, Partial key exposure attacks on CRT-RSA: better cryptanalysis to full size encryption exponents, in *ACNS 2015*. LNCS, vol. 9092 (Springer, 2015), pp. 518–537
65. A. Takayasu, N. Kunihiro, General bounds for small inverse problems and its applications to multi-prime RSA. IEICE Trans. **E100–A**(1), 50–61 (2017)
66. A. Takayasu, N. Kunihiro, How to generalize RSA cryptanalyses, in *PKC 2016*. LNCS, vol. 9615 (Springer, 2016), pp. 67–97
67. A. Takayasu, N. Kunihiro, Partial key exposure attacks on RSA with multiple exponent pairs, in *ACISP 2016*. LNCS, vol. 9723 (Springer, 2016), pp. 243–257
68. A. Takayasu, Y. Lu, L. Peng, Small CRT-exponent RSA revisited, in *Eurocrypt 2017*. LNCS (Springer, to appear)
69. Y. Lu, R. Zhang, L. Peng, D. Lin, Solving linear equations modulo unknown divisors: revisited, *Asiacrypt 2015*. LNCS, vol. 9452 (Springer, 2015), pp. 189–213
70. Y. Lu, L. Peng, R. Zhang, D. Lin, Towards optimal bounds for implicit factorization problem, in *SAC 2015*. LNCS, vol. 9566 (Springer , 2015), pp. 462–476
71. N. Kunihiro, An improved attack for recovering noisy RSA secret keys and its countermeasure, in *ProvSec 2015*. LNCS, vol. 9451 (Springer, 2015), pp. 61–81
72. T. Tanigaki, N. Kunihiro, Maximum likelihood-based key recovery algorithm from decayed key schedules, in *ICISC 2015*. LNCS, vol. 9558 (Springer, 2015), pp. 314–328
73. N. Kunihiro, Y. Takahashi, Improved key recovery algorithms from noisy RSA secret keys with analog noise, in *CT-RSA 2017*. LNCS, vol. 10159 (Springer, 2017), pp. 328–346

Part I
Mathematical Cryptography

Multivariate Public Key Cryptosystems

Yasufumi Hashimoto

Abstract This paper presents a survey on the multivariate public key cryptosystem (MPKC), which is a public key cryptosystem whose public key is a set of multivariate quadratic forms over a finite field.

Keywords Multivariate public key cryptosystem (MPKC) · Post-quantum cryptology

1 Introduction

A *Multivariate Public Key Cryptosystem (MPKC)* is a public key cryptosystem whose public key is a set of multivariate quadratic forms

$$\begin{aligned} f_1(x_1,\ldots,x_n) &= \sum_{1\le i\le j\le n} a_{ij}^{(1)} x_i x_j + \sum_{1\le i\le n} b_i^{(1)} x_i + c^{(1)}, \\ &\vdots \\ f_m(x_1,\ldots,x_n) &= \sum_{1\le i\le j\le n} a_{ij}^{(m)} x_i x_j + \sum_{1\le i\le n} b_i^{(m)} x_i + c^{(m)} \end{aligned} \tag{1}$$

over a finite field. It is known that MPKCs have advantage that the encryption (or signature verification) is faster than RSA and ECC [22]. Furthermore, since the problem of solving a system of multivariate nonlinear polynomial equations over a finite field of order 2 is NP-hard [48, 49], it has been expected that a secure cryptosystem can be constructed by a set of multivariate polynomials. Especially, after Shor [95] proposed polynomial time quantum algorithms for factoring integers and solving discrete logarithm problems, MPKCs have been considered as one of leading candidates of *Post-Quantum Cryptography* as well as the lattice-based cryptography,

---

Y. Hashimoto (✉)
Department of Mathematical Sciences, University of the Ryukyus, Nishihara-cho, Okinawa 903-0213, Japan
e-mail: hashimoto@math.u-ryukyu.ac.jp

T. Takagi et al. (eds.), *Mathematical Modelling for Next-Generation Cryptography*, Mathematics for Industry 29, DOI 10.1007/978-981-10-5065-7_2

the code-based cryptography and the isogeny-based cryptography. In fact, MPKC is included in NIST's proposals of standardization of post-quantum cryptography [24, 72, 76].

This paper presents a survey on MPKC. In Sect. 2, we describe two early MPKCs called the *Matsumoto–Imai cryptosystem (MI, C^*)* [69] and the *Moon Letter cryptosystem (ML, TsuKIFM)* [105] proposed in 1980s and the general construction of MPKCs. While these early MPKCs were already broken [29, 52, 79], the construction of maps

$$F = T \circ G \circ S \tag{2}$$

has been used in most MPKCs, where S, T are secret invertible affine maps, G is a quadratic map to be feasibly inverted and F is a public quadratic map. The central map G essentially characterizes the corresponding MPKC. The security and the speed of decryption highly depend on G. Unfortunately, at the present time, there are few works on the security proof of MPKCs. On the other hand, there are various attacks on proposed MPKCs. Such works greatly help to build secure MPKCs by pointing out which properties of G yield vulnerabilities. In Sect. 3, we give outlines of major attacks on MPKCs to explain which properties of G yield vulnerabilities of the corresponding MPKC. In Sect. 4, we describe several famous MPKCs and discuss their security based on the descriptions in Sect. 3. Finally in Sect. 5, we conclude this paper by listing open problems on MPKCs for future developments.

2 Early MPKCs and General Construction

2.1 Early MPKCs

In this subsection, we describe two early MPKCs, the Matsumoto–Imai cryptosystem [69] and the Moon Letter cryptosystem [105].

Matsumoto–Imai's cryptosystem (MI, C^* [69]).

Let $n \geq 1$ be an integer, k a finite field of even characteristic, $q := \#k$, K an n-extension of k and $\{\theta_1, \dots, \theta_n\} \subset K$ a basis of K over k. Choose an integer $i \geq 1$ such that $\gcd(q^n - 1, q^i + 1) = 1$ and define the map $\mathcal{G} : K \to K$ by

$$\mathcal{G}(X) = X^{1+q^i}. \tag{3}$$

The *secret key* of this scheme is a pair of two invertible affine maps $S, T : k^n \to k^n$ and the *public key* is

$$F := T \circ \phi^{-1} \circ \mathcal{G} \circ \phi \circ S : k^n \to k^n, \tag{4}$$

where $\phi : k^n \to K$ is a one-to-one map, e.g., $\phi(x_1, \ldots, x_n) = x_1\theta_1 + \cdots + x_n\theta_n$ for $(x_1, \ldots, x_n)^t \in k^n$. Since it holds

$$X^{q^i} = (x_1\theta_1 + \cdots + x_n\theta_n)^{q^i} = x_1\theta_1^{q^i} + \cdots + x_n\theta_n^{q^i}$$

for $X := x_1\theta_1 + \cdots + x_n\theta_n \in K, \phi^{-1}(X^{q^i})$ is a set of linear forms of $x_1, \ldots, x_n$ over k and then the public key F is quadratic over k. For a given plaintext $x \in k^n$, the *ciphertext* is $y = F(x) \in k^n$. To *decrypt* y, first calculate $Z := \phi(T^{-1}(y)) \in K$ and compute $W := Z^l \in K$ where the integer l satisfies $(1 + q^i)l \equiv 1 \bmod q^n - 1$. The plaintext is $x = S^{-1}(\phi^{-1}(W))$.

Moon Letter cryptosystem (ML, TsuKIFM [105]**).**

Let $n \geq 1$ be an integer, k a finite field and $g_1(x), \ldots, g_n(x)$ the quadratic forms of $x = (x_1, \ldots, x_n)^t$ over k given by

$$\begin{aligned}
g_1(x) &= (\text{linear form of } x_1),\\
g_2(x) &= x_2 \cdot (\text{linear form of } x_1) + (\text{quadratic form of } x_1),\\
g_3(x) &= x_3 \cdot (\text{linear form of } x_1, x_2) + (\text{quadratic form of } x_1, x_2),\\
&\ \vdots\\
g_n(x) &= x_n \cdot (\text{linear form of } x_1, \ldots, x_{n-1}) + (\text{quadratic form of } x_1, \ldots, x_{n-1}).
\end{aligned} \tag{5}$$

The *secret key* is a pair of two invertible affine maps $S, T : k^n \to k^n$ and the quadratic map $G : k^n \to k^n$ given by $G(x) = (g_1(x), \ldots, g_n(x))^t$. The *public key* is the quadratic map

$$F := T \circ G \circ S : k^n \to k^n. \tag{6}$$

For a given plaintext $x' \in k^n$, the *ciphertext* is $y = F(x') \in k^n$. To *decrypt* the cipher $y \in k^n$, first compute $z = (z_1, \ldots, z_n)^t := T^{-1}(y)$ and find $x_1 \in k$ such that $g_1(x) = z_1$. Since $g_1(x)$ is a linear form of x_1, x_1' is recovered easily. Next find $x_2 \in k$ such that $g_2(x) = z_2$. Since $g_2(x)$ is a linear form of x_2 for a fixed x_1, x_2 is recovered easily. Similarly, we can recover $x_3, \ldots, x_n \in k$ such that $g_3(x) = z_3, \ldots, g_n(x) = z_n$ recursively. The plaintext is $x' = S^{-1}(x_1, \ldots, x_n)^t$.

Unfortunately, ML had not been known well since it was proposed on the paper [105] written in Japanese at 1986. Instead, Shamir's birational signature scheme [93] presented at Crypto 1993 has been well known. These two schemes are quite similar. In fact, the map G in Shamir's scheme is given by $m = n - 1$ and $G(x) = (g_2(x), \ldots, g_n(x))^t$.

2.2 General Construction of MPKCs

Similar to MI and ML, most MPKCs have the structure $F := T \circ G \circ S$. We describe the general construction of MPKCs in this subsection.

Let $n, m \geq 1$ be integers, k a finite field and $q := \#k$. The *secret key* is a tuple of three maps (S, G, T), where $S : k^n \to k^n$, $T : k^m \to k^m$ are invertible affine maps and $G : k^n \to k^m$ is a quadratic map that is *inverted feasibly*. The *public key* F is the convolution of these three maps S, G, T:

$$F : k^n \xrightarrow{S} k^n \xrightarrow{G} k^m \xrightarrow{T} k^m.$$

For a given plaintext $x \in k^n$, the *ciphertext* $y \in k^m$ is computed by $y = F(x)$. To *decrypt* y, find $z \in k^n$ such that $G(z) = T^{-1}(y)$. Then the plaintext is $x = S^{-1}(z)$. Since G is *inverted feasibly*, one can decrypt y efficiently.

Efficiency.

One of remarkable advantage of MPKCs is the speed of encryption (or signature verification). Under the naive implementation, the ciphertext $y = (y_1, \ldots, y_m)^t \in k^m$ of a plaintext $x = (x_1, \ldots, x_n)^t \in k^n$ is computed by

$$\begin{aligned} y_i = f_i(x) = x_1 \cdot &\left(a_{11}^{(i)} \cdot x_1 + a_{12}^{(i)} \cdot x_2 + \cdots + a_{1n}^{(i)} \cdot x_n + b_1^{(i)}\right) \\ &+ x_2 \cdot \left(a_{22}^{(i)} \cdot x_2 + \cdots + a_{2n}^{(i)} \cdot x_n + b_2^{(i)}\right) \\ &+ \cdots \\ &+ x_n \cdot \left(a_{nn}^{(i)} \cdot x_n + b_n^{(i)}\right) + c_i, \qquad (1 \leq i \leq m). \end{aligned}$$

It is clear that the numbers of multiplications and additions in this computation for each $1 \leq i \leq m$ are $\ll \frac{1}{2}n^2$. Such a computation is not best possible. In fact, there have been ideas to reduce the number of operations for several MPKCs by reducing the number of parameters in the public key [43, 85, 86]. Furthermore, the average speed of encryption can be improved if several plaintexts are encrypted simultaneously. As an example, we now study the situation that one encrypts $n + 1$ plaintexts $p_1, \ldots, p_{n+1} \in k^n$. For $x = (x_1, \ldots, x_n)^t \in k^n$, let $\bar{x} := (x_1, \ldots, x_n, 1)^t \in k^{n+1}$ and denote by A_i an $(n + 1) \times (n + 1)$ matrix with

$$f_i(x) = \bar{x}^t A_i \bar{x},$$

for $1 \leq i \leq m$. Then we see that

$$\begin{pmatrix} f_i(p_1) \\ \vdots \\ f_i(p_{n+1}) \end{pmatrix} = \begin{pmatrix} \bar{p}_1^t \cdot (A_i \cdot P)_1 \\ \vdots \\ \bar{p}_{n+1}^t \cdot (A_i \cdot P)_{n+1} \end{pmatrix} \tag{7}$$

where $P := (\bar{p}_1, \ldots, \bar{p}_{n+1})$ is the $(n+1) \times (n+1)$ matrix and $(*)_j$ is the j-th column vector. The Eq. (7) means that $(f_i(p_1), \ldots, f_i(p_{n+1}))^t$ is computed by one multiplication $A_i \cdot P$ of $(n+1) \times (n+1)$ matrices and $n+1$ inner products of $(n+1)$-vectors. Thus the number of operations for encrypting $n+1$ plaintexts is $\ll (n+1)^w m$, where $2 \leq w \leq 3$ is the exponent of the matrix multiplication algorithms (see e.g., [14, 28, 66, 98], $w = 2.3728\cdots$ is the presently best estimate [66]). It is smaller than the number of operations $O(n^3 m)$ by the naive computations.

On the other hand, the size of a public key of MPKC is, in general, relatively larger than other cryptosystems. In fact, the number of coefficients of the quadratic forms in F is about $\frac{1}{2}n^2 m$, which means that, if n, m are around one hundred, the key size of public key is over several hundreds kilo bites under naive implementations. Then reducing key size is an important problem for MPKCs. Note that approaches to reduce the key size for several MPKCs are given in [43, 85, 86].

Security.

Since $F = T \circ G \circ S$, the quadratic forms in the public key F are given by

$$F(x) = \begin{pmatrix} f_1(x) \\ \vdots \\ f_m(x) \end{pmatrix} = T \begin{pmatrix} g_1(S(x)) \\ \vdots \\ g_m(S(x)) \end{pmatrix}. \tag{8}$$

The roles of the secret affine maps S, T are to transform the map G inverted feasibly into the map F not inverted feasibly. Remark that, for most MPKCs, there are nontrivial S, T such that F can be inverted efficiently. For example, on ML, if $S = \begin{pmatrix} * & & * \\ & \ddots & \\ 0 & & * \end{pmatrix}$ and $T = \begin{pmatrix} * & & 0 \\ & \ddots & \\ * & & * \end{pmatrix}$, the quadratic forms $f_1(x), \ldots, f_m(x)$ are also in the form (5), which are inverted recursively. We call such a bad pair (S, T) a *weak key*, and call a pair (S_1, T_1) an *equivalent key* if $(SS_1^{-1}, T_1^{-1}T)$ is a weak key. It is important to study which kind of (S, T) is weak, not to choose such weak keys as a secret key.

We also remark that several MPKCs are known to be insecure at all for arbitrary (S, T). In fact, two early MPKCs were already broken [29, 52, 79]. We describe how to break them in the next subsection.

2.3 *Attacks on Early MPKCs*

Patarin's attack on MI [79].

For a plaintext $x = (x_1, \ldots, x_n)^t \in k^n$ and the corresponding ciphertext $y = (y_1, \ldots, y_n)^t \in k^n$, let $X := \phi(S^{-1}(x))$ and $Y := \phi(T^{-1}(y)) = X^{1+q^i}$. It is easy to see that

$$YX^{q^{2i}} = Y^{q^i}X \left(= X^{1+q^i+q^{2i}}\right). \tag{9}$$

Since $\phi^{-1}(Y), \phi^{-1}(Y^{q^i})$ are sets of linear forms of y and $\phi^{-1}(X), \phi^{-1}(X^{q^{2i}})$ are those of x, there exist polynomials over k in the form

$$L(x, y) := \sum_{1 \le i, j \le n} \alpha_{ij} x_i y_j + \sum_{1 \le i \le n} \beta_i x_i + \sum_{1 \le j \le n} \gamma_j y_j + \delta \tag{10}$$

such that $L(x, y) = 0$ holds for arbitrary plaintext–ciphertext pairs (x, y). To determine the coefficients $\alpha_{ij}, \beta_i, \gamma_j, \delta \in k$, prepare sufficiently many pairs (x, y) of the plaintext and ciphertext, substitute them into (10) to generate a system of linear equations of variables $\alpha_{ij}, \beta_i, \gamma_j, \delta$ and solve its system. Once the attacker finds polynomials in the form (10), he/she can get candidates of the plaintext $x = (x_1, \ldots, x_n)^t$ by solving a system of linear equations derived from (10). It is known that the number of candidates of x given by this attack is $q^{\gcd(i,n)} \le q^{n/3}$, which is much smaller than $\#k^n = q^n$. □

Hasegawa–Kaneko's attack on ML [29, 52].

Let $G_1, \ldots, G_n$ be the coefficient matrices of $g_1(x), \ldots, g_n(x)$, namely $g_i(x) = x^t G_i x +$ (linear form). By the construction of g_i's, we see that

$$G_n = \begin{pmatrix} *_{n-1} & * \\ * & 0 \end{pmatrix}, \quad G_{n-1} = \begin{pmatrix} *_{n-1} & 0 \\ 0 & 0 \end{pmatrix}, \quad \ldots .$$

Since the coefficient matrices $F_1, \ldots, F_n$ of the public polynomials $f_1(x), \ldots, f_n(x)$ are given by

$$\begin{pmatrix} F_1 \\ \vdots \\ F_n \end{pmatrix} = T \begin{pmatrix} S^t G_1 S \\ \vdots \\ S^t G_n S \end{pmatrix},$$

there exist constants $\alpha_1, \ldots, \alpha_{n-1} \in k$ such that

$$\operatorname{rank}(F_i - \alpha_i F_n) \le n - 1, \quad \text{i.e.} \quad \det(F_i - \alpha_i F_n) = 0$$

for $1 \le i \le n - 1$. Then the attacker can find such α_i's by solving univariate polynomial equations. It is easy to see that such α_i's are partial information of T, which means that, once α's are recovered, the attacker can recover partial information of S easily. Further information of S, T can be recovered recursively. □

3 Major Attacks

Section 2.3 describes attacks on the early MPKCs based on the property of G. Now we want to know *what kinds of G construct secure MPKCs*. Unfortunately, we do not have complete answers; there are no MPKCs with security proofs at the present time. On the other hand, there have been various works on cryptanalysis against proposed

MPKCs. These works give answers for *what kinds of G construct insecure MPKCs,* which are quite helpful to build secure MPKCs. In this section, we describe outlines of major attacks on MPKCs.

3.1 Direct Attacks

The *direct attack* is to find a common solution of multivariate quadratic equations

$$f_1(x_1,\ldots,x_n)=y_1,\quad \ldots,\quad f_m(x_1,\ldots,x_n)=y_m \tag{11}$$

to recover the plaintext $x=(x_1,\ldots,x_n)^t\in k^n$ of a ciphertext $y=(y_1,\ldots,y_m)^t\in k^m$. The most naive approach is the *exhaustive search*, whose complexity is heuristically $O(q^{\min(m,n)}\cdot(\text{polyn.}))$. It is too heavy in general, and then the attacker requires better algorithms. Note that a faster algorithm was proposed in [16] for $q=2$.

One of standard approaches for direct attacks is by computing the Gröbner basis of the polynomial system $\{f_1(x)-y_1,\ldots,f_m(x)-y_m\}$. While the complexity of the original Gröbner basis algorithm by Buchberger [17] is $O\left(2^{2^n}\right)$, there have been improved algorithms such like the F_4- and F_5-algorithms [5, 10, 44, 45]. It is known that the complexities of these algorithms depend on the *degree of regularity* d_{reg} of the corresponding polynomial system, in fact, the complexity of F_5 algorithm is $\ll m\binom{n+d_{\mathrm{reg}}-1}{d_{\mathrm{reg}}}^w$. When the polynomial system is over-defined ($m>n$) and is *semi-regular*, d_{reg} coincides with the smallest degree of the non-positive coefficients of the polynomial $\frac{(1-t^2)^m}{(1-t)^n}$ [5]. This means that, if m is sufficiently larger than n, d_{reg} is small enough. Especially, when $m\gg\frac{1}{2}n^2$, this algorithm is in polynomial time. When the difference $m-n$ is small, one can reduce the complexity by mixing the exhaustive search with the Gröbner basis algorithm. This approach is called the *hybrid* approach. According to [10], its complexity is $O(2^{m(3.31-3.62/\log_2 q)})$ for $n=m$.

For under-defined systems ($n>m$), there are improved algorithms. When $n\geq\frac{1}{2}m(m+1)$, Cheng et al. [25] (see also [53, 65, 70]) proposed a polynomial time algorithm to find a solution x by reducing the problem of solving $\{f_1(x)=y_1,\ldots,f_m(x)=y_m\}$ to the problem of finding a solution of

$$\begin{aligned}(\text{quadratic form of } x_1)&=0,\\(\text{quadratic form of } x_1,x_2)&=0,\\&\vdots\\(\text{quadratic form of } x_1,\ldots,x_m)&=0.\end{aligned} \tag{12}$$

It is clear that (12) can be solved recursively. Even if $n<\frac{1}{2}m(m+1)$, relatively efficient algorithms are proposed in [25, 104]. For example, if $n\geq\frac{1}{2}m(m+1)-\frac{1}{2}l(l+1)$ $(1\leq l<m)$, one can reduce the corresponding problem to the problem of

finding a solution of the system of l quadratic equations of l variables, which can be solved by the Gröbner basis algorithm more efficiently than the original system of quadratic equations.

In April 2015, the *MQ Challenge* [113] started. It is a challenge to solve a given system of multivariate quadratic equations chosen randomly for $m = 2n, n \sim 1.5m$ and $q = 2, 2^8, 31$. The records of this challenge have been renewed frequently, which shows that the algorithms for the direct attack have been developing quickly.

3.2 Rank Attacks

The *rank attack* is to recover T partially when the coefficient matrices of quadratic forms have special conditions on their ranks. Let $G_1, \dots, G_m$ be the coefficient matrices of the quadratic forms $g_1(x), \dots, g_m(x)$ in the central map G, and $F_1, \dots, F_m$ those of $f_1(x), \dots, f_m(x)$ in the public key F. Due to (8), we have

$$F_j = S^t \left(\sum_{1 \le i \le m} t_{ij} G_i \right) S, \qquad (1 \le j \le m), \tag{13}$$

where t_{ij}'s are the entries in T. Then the rank of F_j coincides with the rank of $\sum_i t_{ij} G_i$. This means that, if there exist constants $c_1, \dots, c_m \in k$ such that the rank of $\sum_i c_i G_i$ is $r(< n)$, there exist constants $c'_1, \dots, c'_m \in k$ such that the rank of $\sum_i c'_i F_i$ is (at most) r. The rank attack recovers such constants $c'_1, \dots, c'_m$.

For example, on ML (Sects. 2.1, 2.3), G_n is of rank n and arbitrary linear sums of $G_1, \dots, G_{n-1}$ are of rank $n-1$. Then there exist $\alpha_1, \dots, \alpha_{n-1} \in k$ such that the rank of $F_i - \alpha_i F_n$ $(1 \le i \le n-1)$ is at most $n-1$ and the attacker can recover T partially by these constants $\alpha_1, \dots, \alpha_{n-1}$.

There are two kinds of rank attacks. One is the *min-rank attack*, which is available if there exist $\beta_1, \dots, \beta_m \in k$ such that the rank r of $\sum_{1 \le i \le m} \beta_i G_i$ is small. The other is the *high-rank attack*, which is available if there exists a small integer $1 \le l < m$, elements $\beta_1, \dots, \beta_{m-l} \in k$ and a series $\{i_1, \dots, i_{m-l}\} \subset \{1, \dots, m\}$ such that the rank of $\sum_{1 \le j \le m-l} \beta_j G_{i_j}$ is smaller than n.

When q is small, the rank attacks include exhaustive searches and their complexities are known to be $O(q^{r\lfloor \frac{m}{n} \rfloor} \cdot \text{(polyn.)})$ for the min-rank attack and $O(q^l \cdot \text{(polyn.)})$ for the high-rank attack [108]. On the other hand, when q is large, the attacker of the min-rank attack tries to find a solution of the system of polynomial equations of $(\alpha_1, \dots, \alpha_m)$ derived from the condition $\text{rank}(\sum_{1 \le i \le m} \alpha_i F_i) \le r$. Since $\text{rank}(A) \le r$ is equivalent that the determinants of all $(r+1) \times (r+1)$ minor matrices in A are zero, the corresponding equations are of degree (at most) $r+1$. While solving a system of high degree equations is difficult in general, it can be done effectively by the Gröbner basis algorithm if r is small enough since the number of equations are much larger than the number of equations, In fact, its complexity is known to be $O(\binom{n+r+1}{r+1}^w)$ for $n = m$ under several good conditions (see [11, 64]).

3.3 Conjugation Attacks

Let H_1, H_2 be linear sums of $F_1, \ldots, F_m$. Due to (13), we see that

$$H_1^{-1}H_2 = S^{-1}(Q_1^{-1}Q_2)S,$$

where Q_1, Q_2 are linear sums of $G_1, \ldots, G_m$. If $Q_1^{-1}Q_2$ has special properties for conjugation, the attacker can recover S partially.

For example, the coefficient matrices $G_1, \ldots, G_m$ on the *oil and vinegar signature scheme (OV)* (Sect. 4.1.1, [81]) are expressed by $\begin{pmatrix} 0_m & * \\ * & *_m \end{pmatrix}$, which means

$$H_1^{-1}H_2 = S^{-1}\begin{pmatrix} 0_m & * \\ * & *_m \end{pmatrix}\begin{pmatrix} *_m & * \\ * & 0_m \end{pmatrix}S = S^{-1}\begin{pmatrix} *_m & * \\ 0 & *_m \end{pmatrix}S.$$

By using the equation above, Kipnis and Shamir [63] proposed a polynomial time algorithm to recover S_1 such that $SS_1 = \begin{pmatrix} *_m & * \\ 0_m & *_m \end{pmatrix}$. Since $S = \begin{pmatrix} *_m & * \\ 0 & *_m \end{pmatrix}$ is a weak key, Kipnis–Shamir's attack breaks OV.

This attack is also available on the signature scheme YTS (Sect. 4.3.2, [55, 111]) and on MPKCs derived from a quadratic map over an extension field (Sect. 4.2.4, [23, 59, 107]), since the coefficient matrices F_i's are respectively expressed by $S^t(G_i' \otimes I_r)S$ with smaller matrix G_i' and $\tilde{S}^t\begin{pmatrix} *_N & & \\ & \ddots & \\ & & *_N \end{pmatrix}\tilde{S}$ with a divisor $N \mid n$ and a matrix $\tilde{S}$ over an extension field including the secret key S.

Remark that this attack cannot be used directly when the field is of even characteristic. When k is of even characteristic, the coefficient matrix H cannot be symmetric. Then, instead of H, the attacker will use the matrix $\hat{H} := H + H^t$. Since $\hat{H}$ is skew-symmetric ($\hat{H} + \hat{H}^t = 0$), $\hat{H}$ is not invertible when n is odd and the characteristic polynomial of $\hat{H}_1^{-1}\hat{H}_2$ is a square of a smaller degree polynomial when n is even (see e.g., [20, 40, 101]). Thus more delicate discussions are required for even characteristic cases.

3.4 Linearization Attacks

Recall that Patarin's attack on MI (Sect. 2.3, [79]) recovers polynomials in the form

$$L(x, y) := \sum_{1\le i,j\le n} \alpha_{ij}x_iy_j + \sum_{1\le i\le n} \beta_i x_i + \sum_{1\le j\le n} \gamma_j y_j + \delta \tag{14}$$

satisfying $L(x, y) = 0$ for arbitrary plaintext–ciphertext pairs (x, y). The *linearization attack* is to recover such polynomials if there exist. Once the attacker obtains such polynomials, he/she will get (candidates of) the plaintexts x of given ciphertexts y.

The basic approach to determine L is as follows. First, prepare sufficiently many plaintext–ciphertext pairs (x, y). Next, generate a system of linear equations of the coefficients in L by the pairs (x, y). Finally, solve the linear equations to determine the coefficients of L. The complexity of this attack depends on the number of monomials in L. For example, on MI, the complexity of the linearization attack is $O(n^{2w})$ since the number of monomials in (14) is $(n+1)^2$.

Such an attack is extended to MFE [37, 107] and the simple matrix encryption scheme [99]. On MFE, there exist quadratic polynomials $h_1(y), \ldots, h_{n+1}(y)$ such that

$$L(x, y) := \sum_{1 \le i \le n} x_i \cdot h_i(y) + h_{n+1}(y) \tag{15}$$

satisfies $L(x, y) = 0$ for any plaintext–ciphertext pairs (x, y). Then the complexity of the linearization attack on MFE is not large. On the simple matrix encryption scheme, the complexity is sub-exponential time since the degrees of the polynomials corresponding to $h_1(y), \ldots, h_{n+1}(y)$ are $\sqrt{n}$.

Remark that there are two ways to generate plaintext–ciphertext pairs (x, y). One is by encrypting chosen plaintexts $x \in k^n$, and the other is by decrypting chosen ciphertexts $y \in k^m$. The former is a *chosen plaintext attack (CPA)* and the latter is a *chosen ciphertext attack (CCA)*. On MI and MFE, CPA is available. On the other hand, if the decryption map $\Psi : k^m \to k^n$ is not an inversion of F, namely $\Psi(F(x)) = x$ for $x \in k^n$ but $F(\Psi(y)) \neq y$ for sufficiently many $y \in k^m$, there is a possibility that CCA is available. In fact, CCA helps to recover the decryption map of ZHFE (Sect. 4.2.4, [60, 89]). Furthermore, if hidden information is used in the decryption algorithm, CCA might be able to recover such hidden information. Actually, the additional polynomials in Zhang–Tan's variant [114, 115] can be recovered by CCA [57].

3.5 *Differential Attacks*

The *differential attack* is based on a *symmetric property* of the difference

$$Df(x, a) := f(x + a) - f(x) - f(a) + f(0),$$

for the polynomial map f associated with the corresponding MPKC. For example, Dubois et al. [41] proposed the differential attack on Sflash [1] (a variant of MI) by using the following symmetric relation:

$$D\mathcal{G}(\alpha X, a) + D\mathcal{G}(X, \alpha a) = (\alpha^{q^i} + \alpha) D\mathcal{G}(X, a), \tag{16}$$

where $\mathscr{G}(X) := X^{q^i+1}$ is the central map of MI. It is known that the differential attack is also available on l-IC and the internal perturbations of MI, HFE [42, 46, 47]. On the other hand, the security of HFE and its variations have been studied in [19, 32] and it was proved that HFEv- is secure against the differential attack.

3.6 Physical Attacks

Physical attacks, e.g., the *side channel attacks* and the *fault attacks*, have been studied for RSA [15, 62], ECC [13, 27, 30], Pairing [78], lattice- and code-based cryptosystems [2, 21, 71]. Also for MPKCs, there are several works on physical attacks.

Okeya et al. [77] proposed a *side channel attack* on Sflash [1] before Sflash was broken by the differential attack [41]. This attack can recover the random seed used for hashing in the process of signature generation. Furthermore, *fault attacks* on MPKCs was proposed at PQCryoto 2011 [61]. By comparing plaintext–ciphertext pairs given by a faulty map and those by the unfaulty (original) map, the attacker can recover the secret key S, T partially. It is known that the fault attack is available on most MPKCs under naive implementations. These works imply that MPKCs must be implemented carefully, not to be broken by physical attacks.

4 Proposed MPKCs

Until now, various MPKCs have been proposed. In this section, we describe famous MPKCs and discuss their security based on the descriptions of Sect. 3.

4.1 Stepwise Triangular Type

Recall that the central map G of ML (Sect. 2.1, [105]) is inverted recursively. While ML itself was already broken [52], the idea *decrypting step-by-step* is used in several MPKCs. We now give examples of MPKCs having the step-by-step structure.

4.1.1 Oil and Vinegar Signature Scheme

In the *Oil and Vinegar signature scheme (OV)* proposed by Patarin [81], $n = 2m$ and the quadratic map G is defined by

$$g_j(x) = \sum_{1 \le i \le m} x_i \cdot (\text{linear form of } x_{m+1}, \ldots, x_n) + (\text{quadratic form of } x_{m+1}, \ldots, x_n), \tag{17}$$

for $1 \le j \le m$. Remark that the affine map T is not necessary in OV since the polynomials in $T \circ G$ is also in the form (17). This scheme signs a message $y \in k^m$ as follows. First, choose $u_1, \ldots, u_m \in k$ randomly and find $z_1, \ldots, z_m \in k$ such that

$$\begin{aligned} g_1(z_1, \ldots, z_m, u_1, \ldots, u_m) &= y_1, \\ &\vdots \\ g_m(z_1, \ldots, z_m, u_1, \ldots, u_m) &= y_m. \end{aligned} \tag{18}$$

The signature of y is $x = S^{-1}(z_1, \ldots, z_m, u_1, \ldots, u_m)^t \in k^n$. By the definition of G, we see that $(z_1, \ldots, z_m)$ is given as a solution of m linear equations of m variables.

As already given in Sect. 3.3, an equivalent secret key of OV is recovered in polynomial time by Kipnis–Shamir's attack [63] since the coefficient matrices of $g_1, \ldots, g_m$ are in the form $\begin{pmatrix} 0_m & * \\ * & *_m \end{pmatrix}$ and $\begin{pmatrix} 0_m & * \\ * & *_m \end{pmatrix}^{-1} \begin{pmatrix} 0_m & * \\ * & *_m \end{pmatrix} = \begin{pmatrix} *_m & * \\ 0 & *_m \end{pmatrix}$. To enhance its security, Kipnis–Patarin–Goubin [65] proposed an arrangement of OV called the *Unbalanced Oil and Vinegar signature scheme (UOV)*. On this scheme, $n > 2m$ $(v := n - 2m)$ and G is given as (17) for $1 \le j \le m$. The signature generation is almost same to the original OV. It is easy to see that $S = \begin{pmatrix} *_m & * \\ 0 & *_{m+v} \end{pmatrix}$ is a *weak key* of UOV.

Different to the original OV, Kipnis–Shamir's attack is not available on UOV directly since the coefficient matrices are in the form $\begin{pmatrix} 0_m & * \\ * & *_{m+v} \end{pmatrix}$ but $\begin{pmatrix} 0_m & * \\ * & *_{m+v} \end{pmatrix}^{-1} \cdot \begin{pmatrix} 0_m & * \\ * & *_{m+v} \end{pmatrix} \ne \begin{pmatrix} * & * \\ 0 & * \end{pmatrix}$. Kipnis–Patarin–Goubin [65] also arrange Kipnis–Shamir's attack to be available on UOV with the complexity $O(q^v \cdot (\text{polyn.}))$.

The advantage of UOV is that the signature generation is elementary and the security seems enough for $v \sim m$ $(n \sim 3m)$. However, the key size is relatively larger than other MPKCs. We then need to reduce the key size of this scheme.

4.1.2 Rainbow

Rainbow [35] is the multi-layer version of UOV. For integers $o_1, \ldots, o_l, v \ge 1$, put $m := o_1 + \cdots + o_l$, $n := m + v$ and define G as follows:

$$
\begin{aligned}
g_1(x), \dots, g_{o_1}(x) &= \sum_{1 \le i \le o_1} x_i \cdot (\text{linear form of } x_{o_1+1}, \dots, x_n) \\
&+ (\text{quadratic form of } x_{o_1+1}, \dots, x_n), \\
g_{o_1+1}(x), \dots, g_{o_1+o_2}(x) &= \sum_{o_1+1 \le i \le o_1+o_2} x_i \cdot (\text{linear form of } x_{o_1+o_2+1}, \dots, x_n) \\
&+ (\text{quadratic form of } x_{o_1+o_2+1}, \dots, x_n), \\
&\vdots \\
g_{o_1+\cdots+o_{l-1}+1}(x), \dots, g_m(x) &= \sum_{o_1+\cdots+o_{l-1}+1 \le i \le m} x_i \cdot (\text{linear form of } x_{m+1}, \dots, x_n) \\
&+ (\text{quadratic form of } x_{m+1}, \dots, x_n).
\end{aligned}
$$

It is a generalization of ML and (U)OV. In fact, Rainbow with $l = n$, $o_1 = \cdots = o_n = 1$ and $v = 0$ is almost same to ML and Rainbow with $l = 1$ is just the (U)OV.

To generate a signature, first choose $u_1, \dots, u_v \in k$ randomly and find $x_{o_1+\cdots+o_{l-1}+1}, \dots, x_m \in k$ such that

$$
\begin{aligned}
g_{o_1+\cdots+o_{l-1}+1}(x_1, \dots, x_m, u_1, \dots, u_v) &= y_{o_1+\cdots+o_{l-1}+1}, \\
&\vdots \\
g_m(x_1, \dots, x_m, u_1, \dots, u_v) &= y_m.
\end{aligned} \tag{19}
$$

By the definition of G, the elements $x_{o_1+\cdots+o_{l-1}+1}, \dots, x_m$ are given as a solution of a system of o_l linear equations of o_l variables. Other parameters $x_1, \dots, x_{o_1+\cdots+o_{l-1}+1}$ can be found recursively.

The coefficient matrices of $g_1(x), \dots, g_m(x)$ are expressed by

$$
\begin{aligned}
G_1, \dots, G_{o_1} &= \begin{pmatrix} 0_{o_1} & * \\ * & *_{n-o_1} \end{pmatrix}, \\
G_{o_1+1}, \dots, G_{o_1+o_2} &= \begin{pmatrix} 0_{o_1} & 0 & 0 \\ 0 & 0_{o_2} & * \\ 0 & * & *_{n-o_1-o_2} \end{pmatrix}, \\
&\vdots \\
G_{o_1+\cdots+o_{l-1}+1}, \dots, G_m &= \begin{pmatrix} 0_{o_1+\cdots+o_{l-1}} & 0 & 0 \\ 0 & 0_{o_l} & * \\ 0 & * & *_v \end{pmatrix}.
\end{aligned} \tag{20}
$$

Then we see that a pair of $S = \begin{pmatrix} *_{o_1} & & * \\ & \ddots & \\ 0 & & *_v \end{pmatrix}$ and $T = \begin{pmatrix} *_{o_1} & & * \\ & \ddots & \\ 0 & & *_{o_l} \end{pmatrix}$ is a *weak key* of Rainbow. Due to (20), we see that the security against the high-rank attack is $O(q^{o_1} \cdot (\text{polyn.}))$, and the security against the min-rank attack is $O(q^{o_l+v} \cdot (\text{polyn.}))$.

Furthermore, since arbitrary linear sums of $G_1, \ldots, G_m$ are in the form $\begin{pmatrix} 0_{o_1} & * \\ * & *_{n-o_1} \end{pmatrix}$, (arranged) Kipnis–Shamir's attack is available also on Rainbow and its complexity is $O(q^{n-2o_1} \cdot \text{(polyn.)})$.

The parameters of Rainbow are usually chosen by $l = 2$ and $o_1 \sim o_2 \sim v$. In this case, $n \sim 1.5m$ and the security against the rank attacks and Kipnis–Shamir's attack is about $O(q^{o_1} \cdot \text{(polyn.)})$. Then Rainbow is considered to be secure enough under a suitable parameter selection and the key size is much less than UOV.

Note that there have been arrangements of Rainbow to reduce the key size. In TTS [108] and NC-Rainbow [109, 110, 112], the number of parameters in G is less than the original Rainbow and the signature generation is faster. In Cyclic Rainbow [85, 86], the number of parameters in F is less than the original Rainbow and the signature verification is faster. However, we should study the security of such arrangements carefully. For example, it is known that the security of Quaternion Rainbow over even characteristic field is almost $1/4$ of the original Rainbow of similar size [54].

4.2 *Extension Field Type*

The central map of MI (Sect. 2.1, [69]) is constructed by a univariate monomial over an extension field. While MI was already broken, the idea *generating G over an extension field* is used for several MPKCs. The central map $G : k^n \to k^m$ of such an MPKC is generally described as follows.

Let $r \geq 1$ be a common divisor of n and m, $N := n/r$, $M := m/r$, K an r-extension of k and $\{\theta_1, \ldots, \theta_r\} \subset K$ is a basis of K over k. Denote by $\phi_N : k^n \to K^N$ is a one-to-one map, e.g. $\phi_N(x_1, \ldots, x_n) = (x_1\theta_1 + \cdots + x_r\theta_r, \ldots, x_{n-r+1}\theta_1 + \cdots + x_n\theta_r)$ for $x_1, \ldots, x_n \in k$, and define a polynomial map $\mathcal{G} : K^N \to K^M$ to be inverted feasibly. The central map G is constructed by $G := \phi_M^{-1} \circ \mathcal{G} \circ \phi_N$.

$$G : k^n \xrightarrow{\phi_N} K^N \xrightarrow{\mathcal{G}} K^M \xrightarrow{\phi_M^{-1}} k^m.$$

It is known that the polynomials $g_1(x), \ldots, g_m(x)$ in $G(x)$ are quadratic forms of $x = (x_1, \ldots, x_n)^t \in k^n$ over k if and only if the polynomials $\mathcal{G}_1(X), \ldots, \mathcal{G}_M(X)$ in $\mathcal{G}(X)$ are quadratic forms of $\bar{X} := \left(X_1, \ldots, X_N, X_1^q, \ldots, \ldots, X_N^{q^{r-1}}\right)^t$ over K. It is because the one-to-one map ϕ_N is given by the matrix $\Theta_N := \left(\theta_j^{q^{i-1}} \cdot I_N\right)_{1 \leq i, j \leq r}$ where I_N is the identity matrix of size N. In fact, if $X = (X_1, \ldots, X_N)^t := (x_1\theta_1 + \cdots + x_r\theta_r, \ldots, x_{n-r+1}\theta_1 + \cdots + x_n\theta_r)^t$, it holds

$$\Theta_N x = \bar{X}.$$

Then F and G have the relation

$$F(x) = (T \circ \Theta_M^{-1}) \cdot \Big(\mathscr{G}_1(\phi_N(S(x))), \ldots, \mathscr{G}_N(\phi_N(S(x))),$$
$$\mathscr{G}_1(\phi_N(S(x)))^q, \ldots, \mathscr{G}_N(\phi_N(S(x)))^{q^{r-1}}\Big)^t,$$

and $\mathscr{G}_i(\phi_N(S(x)))^{q^j}$ is written by

$$\mathscr{G}_i(\phi_N(S(x)))^{q^j} = \bar{X}^t(\Theta_N S \Theta_N^{-1})^t G_i^{(q^j)}(\Theta_N S \Theta_N^{-1})\bar{X} + (\text{linear form of } \bar{X})$$

for some $n \times n$ matrix $G_i^{(q^j)}$ with K-entries. The matrix $G_i^{(q^j)}$ is important for the security of the extension field type MPKCs.

In this subsection, we describe several examples of such MPKCs.

4.2.1 Hidden Field Equation (HFE)

Hidden Field Equation (HFE) proposed by Patarin [79] is constructed with $n = m = r$ (namely $N = M = 1$) and

$$\mathscr{G}(X) = \sum_{0 \le i \le j \le d} \alpha_{ij} X^{q^i + q^j} + \sum_{0 \le i \le d} \beta_i X^{q^i} + \gamma,$$

where $1 \le d \ll n$ is an integer and $\alpha_{ij}, \beta_i, \gamma \in K$. The decryption of HFE is obtained by solving a univariate polynomial equation $\mathscr{G}(X) - Y = 0$ of degree $D \le 2q^d$. Its complexity is $O(D^3 + nD^2 \log q)$ by the Berlekamp algorithm [8, 9].

For the security of HFE, it has been reported that F of HFE with small d is inverted efficiently by the Gröbner basis attack [45]. It is known that the degree of regularity of the corresponding polynomial system is bounded by $\frac{1}{2}(q-1)\lfloor \log_q(2q^d - 1) + 1 \rfloor + 2$ [34, 50]. Furthermore, since the coefficient matrix of $\mathscr{G}$ as a quadratic form of $\bar{X}$ is in the form $\begin{pmatrix} *_{d+1} & \end{pmatrix}$, the min-rank attack is also available on HFE and its complexity is $\binom{n+d+2}{d+2}^w \ll n^{(d+2)w}$ [11, 64].

From these facts, we see that both the decryption speed and the security of HFE are exponential of d, namely HFE has a serious trade-off between efficiency and security. Thus HFE itself has been considered to be impractical. In Sect. 4.2.2, we describe arrangements of MI and HFE to enhance the security.

4.2.2 Variants of MI and HFE

The *minus* $(-)$ variant is to hide several polynomials in G, namely, for $G : k^n \to k^m$ with $G(x) = (g_1(x), \ldots, g_m(x))^t$, the minus $G_- : k^n \to k^{m-l}$ $(1 \le l < m)$ is defined by $G_-(x) := (g_1(x), \ldots, g_{m-l}(x))$. This is mainly used for signature schemes. To generate the signature, choose $u_1, \ldots, u_l \in k$ randomly and find $x \in k^n$ such that $G(x) = (y_1, \ldots, y_{m-l}, u_1, \ldots, u_l)^t$. Sflash [1], selected by NESSIE project

[90], is a minus variant of MI. Unfortunately, the differential attack can recover the hidden polynomials of Sflash [41, 46].

The *plus (+)* variant is to add several polynomials, namely the central map of plus is $G_+ = (g_1(x), \ldots, g_m(x), h_1(x), \ldots, h_l(x))$ where $l \geq 1$ is an integer and h_i's are quadratic forms chosen randomly. To decrypt $\tilde{y} = (y_1, \ldots, y_{m+l})^t \in k^{m+l}$, one finds $x \in k^n$ such that $G(x) = y = (y_1, \ldots, y_m)^t$ and verifies whether $(h_1(x), \ldots, h_l(x)) = (y_{m+1}, \ldots, y_{m+l})$. When $m \geq n$, the decryption of the plus variant is (probably) not too slower than the original scheme since the number of solutions of $G(x) = y$ is not many. On the other hand, when $n > m$, it is much slower since the equation $G(x) = y$ will have many solutions. See [82] for the security of MI$\pm$ (the plus variant of MI$-$).

The *vinegar (v)* variant is to add several variables. When the quadratic forms in $G(x) := (g_1(x), \ldots, g_m(x))^t$ are given by

$$g_l(x) := \sum_{1 \leq i \leq j \leq n} a_{ij}^{(l)} x_i x_j + \sum_{1 \leq i \leq n} b_i^{(l)} x_i + c^{(l)},$$

the quadratic forms in a vinegar variant $G_v(x) := (\hat{g}_1(\tilde{x}), \ldots, \hat{g}_m(\tilde{x}))$ are defined by

$$\hat{g}_l(\tilde{x}) := \sum_{1 \leq i \leq j \leq n} a_{ij}^{(l)} x_i x_j + \sum_{1 \leq i \leq n} v_i^{(l)}(x_{n+1}, \ldots, x_{n+l}) x_i + w^{(l)}(x_{n+1}, \ldots, x_{n+r}),$$

where $r \geq 1$ is an integer, $x_{n+1}, \ldots, x_{n+r}$ are additional variables, $\tilde{x} := (x_1, \ldots, x_{n+r})^t$, $v_i^{(l)}$ is a linear form and $w^{(l)}$ is a quadratic form such that for any $(u_1, \ldots, u_r) \in k^r$, $\{\hat{g}_l(x_1, \ldots, x_n, u_1, \ldots, u_r)\}_{1 \leq l \leq m}$ is equivalent to the original G. HFEv- is the vinegar variant of HFE-, and has been considered to be secure enough under suitable parameter selections (e.g., Quartz, Gui [31, 65, 80, 83, 87, 102]). Recently, Zhang and Tan proposed a new variant similar to the vinegar [114, 115]. However, the vinegar terms can be recovered easily by a chosen ciphertext attack [57].

The *projection (p)* is to reduce several variables from the quadratic forms. When the original G is given by $\{g_l(x_1, \ldots, x_n)\}_{1 \leq l \leq m}$, the projection G_p is $\{g_l(x_1, \ldots, x_{n-r}, u_1, \ldots, u_r)\}_{1 \leq l \leq m}$ with constants $u_1, \ldots, u_r \in k$. It is known that the differential attack is not available on the projection of MI- (called *PFLASH*) [96], despite the signature generation is slower than Sflash.

The *internal perturbation (IP)* [33] and the *piece in hand (PH)* [106] are randomizations of G. It is known that these variants improve the security against the Gröbner basis attack [36, 106]. However, their security should be studied carefully. In fact, the differential attack removes the perturbation on PMI (IP of MI) [46] and the linearization attack recovers additional polynomials in the 2-layer version of PH [74].

4.2.3 ZHFE

ZHFE [89] is an encryption scheme with $(N, M) = (1, 2)$. The simplest version is as follows. Let $D \geq 1$ be an integer, $\mathscr{G}_1(X), \mathscr{G}_2(X)$ quadratic forms of $\bar{X}$ such that

$$\Psi(X) := X \cdot \mathscr{G}_1(X) + X^q \cdot \mathscr{G}_2(X)$$

is of degree at most D and $\mathscr{G} : K \to K^2$ the map with $\mathscr{G}(X) := (\mathscr{G}_1(X), \mathscr{G}_2(X))$. To find $X \in K$ with $\mathscr{G}_1(X) = Y_1$ and $\mathscr{G}_2(X) = Y_2$ in the decryption process, one solves the univariate equation

$$\begin{aligned} \Psi(X, (Y_1, Y_2)) :=& \Psi(X) - XY_1 - X^q Y_2 \\ =& X \cdot (\mathscr{G}_1(X) - Y_1) + X^q \cdot (\mathscr{G}_2(X) - Y_2) = 0 \end{aligned} \tag{21}$$

of degree at most D. Similar to HFE, the complexity of the decryption is $O(D^3 + nD^2 \log q)$ by the Berlekamp algorithm.

The security of ZHFE against the attacks available on HFE has been studied in [84, 89, 116]. These works claimed that ZHFE is more secure than HFE against the direct attacks, the min-rank attacks and the differential attack. However, a *chosen ciphertext attack* can reduce the security of ZHFE to the security of HFE against the min-rank attack [60]. This means that, similar to HFE, ZHFE has a serious trade-off between efficiency and security.

4.2.4 Multivariate ($N > 1$) Version

The maps $\mathscr{G}$ for MI, HFE, and ZHFE are given by univariate polynomials, namely $N = 1$. Other than these scheme, there are MPKCs with $N > 1$. For example, MFE [107] is an extension field type MPKC with $(N, M) = (12, 15)$ and $\mathscr{G}$ has a special structure to be inverted feasibly. In multi-HFE [23], $N(= M)$ is small and the polynomials in $\mathscr{G}$ are quadratic forms chosen randomly. The map $\mathscr{G}$ for l-IC [38] is a set of multivariate higher degree monomials similar to MI.

Unfortunately, these MPKCs are known to be impractical. In fact, MFE was broken by the linearization attack [37] and the l-IC was broken by the differential attack [47]. For multi-HFE, since $\mathscr{G}_i(X) = \bar{X}^t \left(*_N \right) \bar{X} +$ (linear form), the min-rank attack [11] is available and its complexity is $O\left(\binom{n+N+1}{N+1}^{\omega}\right) = O\left(r^{(N+1)w}\right)$. Furthermore, the extension field type MPKCs with quadratic $\mathscr{G}$ and odd q are broken by the conjugation attack [59] since the public quadratic forms are in the form $f_i(x) = x^t (\Theta_N S)^t \begin{pmatrix} *_N & & \\ & \ddots & \\ & & *_N \end{pmatrix} (\Theta_N S) x +$ (linear form).

4.2.5 Noncommutative Version

In Sects. 3.2.1–4, we describe MPKCs whose central maps are derived from polynomial maps over extension *fields*. Such constructions can be generalized to *rings*, not necessarily fields. In fact, there have been several MPKCs constructed on noncommutative rings [54, 103, 109, 110, 112]. However, we cannot recommend such constructions strongly since the following theorem is well-known (see e.g. [6]).

The Artin–Wedderburn theorem. A ring $\mathscr{R}$ is a semi-simple if and only if there exist integers $n_1, \ldots, n_l \geq 1$ and division rings $K_1, \ldots, K_l$ such that

$$\mathscr{R} \simeq \mathrm{M}_{n_1}(K_1) \oplus \cdots \oplus \mathrm{M}_{n_l}(K_l),$$

where $\mathrm{M}_n(K)$ is the ring of $n \times n$ matrices of K-entries.

Furthermore, due to Wedderburn's theorem, we see that, if a semi-simple ring $\mathscr{R}$ is finite, then the rings $K_1, \ldots, K_l$ are commutative. For example, let

$$\mathscr{R} := \{a_1\sigma_1 + \cdots + a_5\sigma_5 \mid a_1, \ldots, a_5 \in k\}$$

be a ring over k with $q \equiv 1 \bmod 3$ and $\sigma_1 := \begin{pmatrix} 1 & & \\ & 1 & \\ & & 1 \end{pmatrix}$, $\sigma_2 := \begin{pmatrix} & 1 & \\ & & 1 \\ 1 & & \end{pmatrix}$, $\sigma_3 := \begin{pmatrix} & & 1 \\ 1 & & \\ & 1 & \end{pmatrix}$, $\sigma_4 := \begin{pmatrix} 1 & & \\ & & 1 \\ & 1 & \end{pmatrix}$, $\sigma_5 := \begin{pmatrix} & & 1 \\ & 1 & \\ 1 & & \end{pmatrix}$. Define $\delta_1, \ldots, \delta_5 \in \mathscr{R}$ by

$$\begin{pmatrix} \delta_1 \\ \delta_2 \\ \delta_3 \\ \delta_4 \\ \delta_5 \end{pmatrix} := 3^{-1} \begin{pmatrix} 1 & 1 & 1 & & \\ 1 & \alpha & \alpha^2 & & \\ 1 & \alpha^2 & \alpha & & \\ -1 & -1 & -1 & \alpha - 1 & \alpha^2 - 1 \\ -1 & -1 & -1 & \alpha^2 - 1 & \alpha - 1 \end{pmatrix} \begin{pmatrix} \sigma_1 \\ \sigma_2 \\ \sigma_3 \\ \sigma_4 \\ \sigma_5 \end{pmatrix},$$

where $\alpha \in \mathscr{R}$ satisfies $\alpha \neq 1$, $\alpha^3 = 1$. It is easy to see that the elements $\delta_1, \ldots, \delta_5$ have the following multiplicative relations:

$$\begin{array}{lllll} \delta_1\delta_1 = \delta_1, & \delta_1\delta_2 = 0, & \delta_1\delta_3 = 0, & \delta_1\delta_4 = 0, & \delta_1\delta_5 = 0, \\ \delta_2\delta_1 = 0, & \delta_2\delta_2 = \delta_3, & \delta_2\delta_3 = 0, & \delta_2\delta_4 = \delta_4, & \delta_2\delta_5 = 0, \\ \delta_3\delta_1 = 0, & \delta_3\delta_2 = 0, & \delta_3\delta_3 = \delta_2, & \delta_3\delta_4 = 0, & \delta_3\delta_5 = \delta_5, \\ \delta_4\delta_1 = 0, & \delta_4\delta_2 = 0, & \delta_4\delta_3 = \delta_5, & \delta_4\delta_4 = 0, & \delta_4\delta_5 = \delta_2, \\ \delta_5\delta_1 = 0, & \delta_5\delta_2 = \delta_4, & \delta_5\delta_3 = 0, & \delta_5\delta_4 = \delta_3, & \delta_5\delta_5 = 0. \end{array}$$

This means that $\mathscr{R} \simeq k \oplus \mathrm{M}_2(k)$ and, if one generates G from quadratic forms over $\mathscr{R}$, the corresponding MPKC has a risk to be broken by rank attacks or conjugation attacks.

4.3 Other MPKCs

In this subsection, we describe two MPKCs as examples not classified in neither the stepwise type nor the extension field type.

4.3.1 ABC Encryption Scheme

In the *ABC (or Simple Matrix) encryption scheme* proposed by Tao et al. [99], the central map G is generated by products among three matrices A, B, C. It is generalized as follows. Let $n, m \geq 1$ be integers with $m := 2n$, $\mathscr{R}$ a ring over k with $[\mathscr{R} : k] = n$ and $\{\xi_1, \ldots, \xi_n\} \subset \mathscr{R}$ is a basis of $\mathscr{R}$ over k. Denote by $\phi : k^n \to \mathscr{R}, \phi_2 : k^m \to \mathscr{R}^2$ one-to-one maps, e.g. $\phi(x_1, \ldots, x_n) = x_1\xi_1 + \cdots + x_n\xi_n$ and $\phi_2(y_1, \ldots, y_m) = (y_1\xi_1 + \cdots + y_n\xi_n, y_{n+1}\xi_1 + \cdots + y_m\xi_n)$ for $x_1, \ldots, x_n, y_1, \ldots, y_m \in k$, and $\mathscr{B}, \mathscr{C} : k^n \to k^n$ linear maps. For $x \in k^n$, put $A = A(x) := \phi(x)$, $B = B(x) := \phi(\mathscr{B}(x)), C = C(x) := \phi(\mathscr{C}(x)), E_1 = E_1(x) := A \cdot B, E_2 = E_2(x) := A \cdot C$ and $E(x) := (E_1(x), E_2(x))$. The central map $G : k^n \to k^m$ is defined by

$$G := \phi_2^{-1} \circ E \circ \phi.$$

For $Y_1, Y_2 \in \mathscr{R}$, one finds $x \in k^n$ with $E_1(x) = Y_1$ and $E_2(x) = Y_2$ by solving a system of linear equations derived from $C(x) = B(x)Y_1^{-1}Y_2$ or $B(x) = C(x)Y_2^{-1}Y_1$.

It is easy to see that the original ABC encryption scheme [99] is just same to the case that $\mathscr{R} = \mathrm{M}_r(k)$ with $r^2 = n$, and the *extension field cancelation (EFC)* [97] is essentially expressed as an ABC encryption scheme in the case that $\mathscr{R}$ is an n extension field of k.

The decryption of this scheme is simple and quite efficient. However, the decryption fails when A is not invertible. Especially, the probability of decryption failure for the original ABC encryption scheme [99] is about q^{-1}, which is not negligible. To reduce the probability of decryption failure, several arrangements have been proposed, e.g., taking q large, using rectangular matrices instead of A, B, C et al. [100], using a tensor type matrix as S [88]. However, the security for such arrangements should be studied carefully. It was shown that the tensor type S is a weak key [56].

For the security, it is known that the min-rank attack and the linearization attack are available on this encryption scheme. For the original ABC [99], the complexities of these attacks are $O(q^{2r} \cdot \text{(polyn.)})$ and $O((m\binom{n+r}{r})^w)$ respectively. Furthermore, Moody et al. [73] proposed another attack on this scheme with the complexity $O(q^{r+4} \cdot \text{(polyn.)})$. Then this encryption scheme (presently) has a sub-exponential time security of n. For EFC, it is known that the linearization attack can recover plaintexts easily. To prevent it, the authors of [97] recommended to use the minus and the projection of EFC. In [39], the cubic version of ABC was proposed; the polynomials in A are quadratic and then those in F, G are cubic. Though the security against the direct attack is improved, the security against the linearization attack is almost same to the original ABC.

4.3.2 YTS

YTS is a signature scheme proposed by Yasuda–Takagi–Sakurai [111] over a finite field of odd characteristic and by Zhang–Tan [115] over a field of even characteristic. We now describe the odd characteristic version.

Let $r \geq 1$ be an integer, $n := r^2$ and $m := r(r+1)/2$. Denote by $\phi : k^n \to \mathrm{M}_r(k)$, $\psi : k^m \to \mathrm{SM}_r(k)$ one-to-one maps, where $\mathrm{SM}_r(k)$ is the set of $r \times r$ symmetric matrices over k. Define two maps $\mathcal{G}_1, \mathcal{G}_2 : \mathrm{M}_r(k) \to \mathrm{SM}_r(k)$ by $\mathcal{G}_1(X) := X^t X$ and $\mathcal{G}_2(X) := X^t B^t \begin{pmatrix} I_{r-1} & \\ & \delta \end{pmatrix} BX$, where $\delta \in k$ is not a square of any elements in k and $B \in \mathrm{M}_r(k)$ is an invertible matrix. The central maps $G_1, G_2 : k^n \to k^m$ are given by

$$G_i := \psi^{-1} \circ \mathcal{G}_i \circ \phi, \qquad (i = 1, 2).$$

The public key is two maps $F_1, F_2 : k^n \to k^m$ with $F_i := T \circ G_i \circ S$ and the signature $x \in k^n$ for a message $y \in k^m$ is verified if either $F_1(x) = y_1$ or $F_2(x) = y_2$ holds. It is known that, for any $Y \in \mathrm{SM}_r(k)$, there exists $X \in \mathrm{M}_r(k)$ such that either $X^t X = Y$, $X^t \begin{pmatrix} I_{r-1} & \\ & \delta \end{pmatrix} X = Y$ holds and such X can be found feasibly [68]. This fact is used for signature generation. While the signature generation is fast, the security is not enough. Since the quadratic forms in G_i are quite sparse, an equivalent secret key can be recovered in sub-exponential time by the min-rank attack [111] and in polynomial time by the conjugation attack [55].

5 Open Problems

We conclude this paper by giving several open problems on MPKC.

1. Are there MPKCs with security proofs?

There have been several works on provable security of MPKCs [18, 92]. However, they seem still far from the security proof of proposed MPKCs. We expect that, if such an MPKC would be proposed, it could help future developments of MPKCs.

2. Which schemes are polynomial systems suitable for?

It has been considered that there are good multivariate *signature schemes*, which seem secure and efficient enough under suitable parameter selections. For example, Rainbow is one of them despite the key size is relatively large. On the other hand, there seem to be few good *encryption schemes*, except the schemes proposed recently and not yet analyzed enough. That is (maybe) because constructing a good one-to-one map by nonlinear polynomial systems is not easy. Other than signature schemes and encryption schemes, a multi-receiver signcryption scheme [67], an identity-based signature scheme [94], a public key identification schemes [91] and a stream cipher [7] were proposed. We consider that we should analyze more to use them in practice.

3. Why quadratic? How about higher (≥ 3) *degree polynomials?*
For most MPKCs, F and G are sets of quadratic forms. One of the reasons that quadratic forms are mainly used in MPKCs is that higher degree polynomials have much more coefficients, which lacks efficiency. On the other hand, there have been several MPKCs with cubic F and G (e.g., [39, 65, 75]). It has been considered that one of the advantage of "cubic" construction is to avoid attacks based on the properties of coefficient matrices. However, the attacker can get a quadratic map by taking a difference $\Delta_C F(x) := F(x+C) - F(x) = T \circ \Delta_C G \circ S$, and he/she may be able to find vulnerabilities in the coefficient matrices in $\Delta_C F(x)$. For example, in the cubic version of UOV described in Sect. 9 of [65], we can easily check that $\Delta_C F$ is equivalent to a public key of the original UOV, which means that the security of cubic version UOV is almost same to the security against the key recovery attack on the original UOV. Furthermore, another cubic version of UOV [75] was broken easily [58]. We thus consider that, to construct a cubic version of MPKC, one should study the security and efficiency carefully.
4. Are MPKCs really "Post-Quantum"?
MPKCs have been expected to be secure against quantum attacks. However, the proposed attacks on MPKCs are only by the classical computers and there are few works on the security against quantum attacks. The complexities of the proposed attacks might be improved if the attacker could implement such attacks on the quantum computers. For example, by using Grover's algorithm [51], the attacker will reduce the complexities of the attacks including the exhaustive search, e.g., the high-rank attacks on small fields. Furthermore, it is known that, on the isogeny-based cryptosystem, the security against the quantum attacks is less than the security against the attacks by the classical computers [12, 26]. We consider that (the possibility of) quantum attacks on MPKCs must be studied in near future.
5. How about relations with other NP-complete/-hard problems.
It is known that the problem of finding a solution of a system of multivariate nonlinear polynomial equations over a finite field of order 2 is NP-hard, and the correspondence between this problem and the SAT problem is given by $xy \leftrightarrow x \wedge y$, $xy + x + y \leftrightarrow x \vee y$ and $x + 1 \leftrightarrow \neg x$ [48, 49]. By using this correspondence, Bard et al. [3, 4] proposed an algorithm to solve a system of multivariate quadratic equations by the SAT-solver. We consider that studying the security of MPKCs in the view of other NP-complete/-hard problems is quite interesting.

Acknowledgements The author would like to thank the anonymous reviewer for reading the previous draft of this paper carefully and giving helpful comments to improve it. He was supported by JSPS Grant-in-Aid for Young Scientists (B) no. 26800020.

References

1. M.L. Akkar, N. Courtois, L. Goubin, R. Duteuil, A fast and secure implementation of Sflash, in *PKC'03*. LNCS, vol. 2567 (2003), pp. 267–278

2. R.M. Avanzi, S. Hoerder, D. Page, M. Tunstall, Side-channel attacks on the McEliece and Niederreiter public-ky cryptosystems. J. Crypt. Eng. **1**, 271–281 (2011)
3. G.V. Bard, *Algebraic Cryptanalysis* (Springer, Dordrecht, 2009)
4. G.V. Bard, N.T. Courtois, C. Jefferson, Efficient methods for conversion and solution of sparse systems of low-degree multivariate polynomials over $GF(2)$ via SAT-Solvers, https://eprint.iacr.org/2007/024.pdf
5. M. Bardet, J.C. Faugère, B. Salvy, B.Y. Yang, Asymptotic expansion of the degree of regularity for semi-regular systems of equations, in *MEGA'05* (2005)
6. J.A. Beachy, *Introductory Lectures on Rings and Modules* (Cambridge University Press, Cambridge, 1999)
7. C. Berbain, H. Gilbert, J. Patarin, QUAD: a practical stream cipher with provable security, in *Eurocrypt'06*. LNCS, vol. 4004 (2006), pp. 109–128
8. E.R. Berlekamp, Factoring polynomials over finite fields. Bell Syst. Tech. J. **46**, 1853–1859 (1967)
9. E.R. Berlekamp, Factoring polynomials over large finite fields. Math. Comput. **24**, 713–735 (1970)
10. L. Bettale, J.C. Faugère, L. Perret, Solving polynomial systems over finite fields: Improved analysis of the hybrid approach. ISSAC **2012**, 67–74 (2012)
11. L. Bettale, J.C. Faugere, L. Perret, Cryptanalysis of HFE, multi-HFE and variants for odd and even characteristic. Des. Codes Crypt. **69**, 1–52 (2013)
12. J.F. Biasse, D. Jao, A. Sankar, A quantum algorithm for computing isogenies between supersingular elliptic curves, in *Indocrypt'14*. LNCS, vol. 8885 (2014), pp. 428–442
13. I. Biehl, B. Meyer, V. Müller, Differential fault attacks on elliptic curve cryptosystems, in *Crypto'00*. LNCS, vol. 2000 (1880), pp. 131–146
14. D. Bini, M. Capovani, F. Romani, G. Lotti, $O(n^{2.7799})$ complexity for $n \times n$ approximate matrix multiplication. Inf. Process. Lett. **8**, 234–235 (1979)
15. D. Boneh, R.A. DeMillo, R.J. Lipton, On the importance of checking cryptographic protocols for faults, in *Eurocrypt'97*. LNCS, vol. 1233 (1997), pp. 37–51
16. C. Bouillaguet, H.C. Chen, C.M. Cheng, T. Chou, R. Niederhagen, A. Shamir, B.Y. Yang, Fast exhaustive search for polynomial systems in F_2, in *CHES'10*. LNCS, vol. 6225 (2010), pp. 203–218
17. B. Buchberger, A theoretical basis for the reduction of polynomials to canonical forms. ACM SIGSAM Bull. **10**, 19–29 (1976)
18. S. Bulygin, A. Petzoldt, J. Buchmann, Towards provable security of the unbalanced oil and vinegar signature scheme under direct attacks, in *Indocrypto'10*. LNCS, vol. 6498 (2010), pp. 17–32
19. R. Cartor, R. Gipson, D. Smith-Tone, J. Vates, On the differential security of the HFEv-signature primitive, in *PQCrypto'16*. LNCS, vol. 9606 (2016), pp. 162–181
20. A. Cayley, Sur les determinants gauches (On skew determinants). Crelle's J. **38**, 93–96 (1847)
21. P.L. Cayrel, P. Dusart, Fault injection's sensitivity of the McEliece PKC, in *Proceedings of 5th International Conference on Future Information Technology* (2010), pp. 1–6
22. A.I.T. Chen, M.S. Chen, T.R. Chen, C.M. Chen, J. Ding, E.L.H. Kuo, F.Y.S. Lee, B.Y. Yang, "SSE implementation of multivariate PKCs on modern x86 CPUs, in *CHES'09*. LNCS, vol. 5747 (2009), pp. 33–48
23. C.H.O. Chen, M.S. Chen, J. Ding, F. Werner, B.Y. Yang, Odd-char multivariate hidden field equations, http://eprint.iacr.org/2008/543
24. L. Chen, S. Jordan, Y.K. Liu, D. Moody, R. Reralta, R. Perlner, D. Smith-Tone, Report on post-quantum cryptography, in *National Institute of Standards and Technology Internal Report*, vol. 8105 (2016), http://csrc.nist.gov/publications/drafts/nistir-8105/nistir_8105_draft.pdf
25. C.M. Cheng, Y. Hashimoto, H. Miura, T. Takagi, A polynomial-time algorithm for solving a class of underdetermined multivariate quadratic equations over fields of odd characteristics, in *PQCrypto'14*. LNCS, vol. 8772 (2014), pp. 40–58
26. A. Childs, D. Jao, V. Soukharev, Constructing elliptic curve isogenies in quantum subexponential time. J. Math. Cryptol. **8**, 1–29 (2014)

27. M. Ciet, M. Joye, Elliptic curve cryptosystems in the presence of permanent and transient faults. Des. Codes Crypt. **36**, 33–43 (2005)
28. D. Coppersmith, S. Winograd, Matrix multiplication via arithmetic progressions. J. Symb. Comput. **9**, 251–280 (1990)
29. D. Coppersmith, J. Stern, S. Vaudenay, Attacks on the birational permutation signature schemes, in *Crypto'93*. LNCS, vol. 773 (1994), pp. 435–443
30. J.S. Coron, Resistance against differential power analysis for elliptic curve cryptosystems, in *CHES'99*. LNCS, vol. 1717 (1999), pp. 292–302
31. N.T. Courtois, M. Daum, P. Felke, On the security of HFE, HFEv- and Quartz, in *PKC'03*. LNCS, vol. 2567 (2003), pp. 337–350
32. T. Daniels, D. Smith-Tone, Differential properties of the HFE cryptosystem, in *PQCrypto'14*. LNCS, vol. 8772 (2014), pp. 59–75
33. J. Ding, A new variant of the Matsumoto-Imai cryptosystem through perturbation, in *PKC'04*. LNCS, vol. 2947 (2004), pp. 305–318
34. J. Ding, T.J. Hodges, Inverting HFE systems is quasi-polynomial for all fields, in *Crypto'11*. LNCS, vol. 6841 (2011), pp. 724–742
35. J. Ding, D. Schmidt, Rainbow, a new multivariate polynomial signature scheme, in *ACNS'05*. LNCS, vol. 3531 (2005), pp. 164–175
36. J. Ding, J.E. Gower, D. Schmidt, C. Wolf, Z. Yin, Complexity estimates for the F_4 attack on the perturbed Matsumoto-Imai cryptosystem, in *10th IMA International Conference on Cryptography and coding*. LNCS, vol. 3796 (2005), pp. 262–277
37. J. Ding, L. Hu, X. Nie, J. Li, J. Wagner, High order linearization equation (HOLE) attack on multivariate public key cryptosystems, in *PKC'07*. LNCS, vol. 4450 (2007), pp. 233–248
38. J. Ding, C. Wolf, B.Y. Yang, $l-$invertible cycles for multivariate quadratic (MQ) public key cryptography, in *PKC'07*. LNCS, vol. 4450 (2007), pp. 266–281
39. J. Ding, A. Petzoldt, L.C. Wang, The cubic simple matrix encryption scheme, in *PQC'14*. LNCS, vol. 8772 (2014), pp. 76–87
40. D.Z. Doković, On the product of two alternating matrices. Amer. Math. Monthly **98**, 935–936 (1991)
41. V. Dubois, P.A. Fouque, A. Shamir, J. Stern, Practical cryptanalysis of SFLASH, in *Crypto'07*. LNCS, vol. 4622 (2007), pp. 1–12
42. V. Dubois, L. Granboulan, J. Stern, Cryptanalysis of HFE with internal prturbation, in *PKC'07*. LNCS, vol. 4450 (2007), pp. 249–265
43. D.H. Duong, A. Petzoldt, T. Takagi, Reducing the key size of the SRP encryption scheme, in *ACISP'16*. LNCS, vol. 9723 (2016), pp. 427–434
44. J.C. Faugère, A new efficient algorithm for computing Grobner bases (F_4). J. Pure Appl. Algebra **139**, 61–88 (1999)
45. J.C. Faugère, A. Joux, Algebraic cryptanalysis of Hidden Field Equations (HFE) using Gröbner bases, in *Crypto'03*. LNCS, vol. 2729 (2003), pp. 44–60
46. P.A. Fouque, L. Granboulan, J. Stern, Differential cryptanalysis for multivariate schemes, in *Eurocrypt'05*. LNCS, vol. 3494 (2005), pp. 341–353
47. P.A. Fouque, G. Macario-Rat, L. Perret, J. Stern, Total break of the l-IC signature scheme, in *PKC'08*. LNCS, vol. 4939 (2008), pp. 1–17
48. A.S. Fraenkel, Y. Yesha, Complexity of problems in games, graphs and algebraic equations. Discret. Appl. Math. **1**, 15–30 (1979)
49. M.R. Garey, D.S. Johnson, *Computers and Intractability, A Guide to the Theory of NP-completeness* (W.H. Freeman, New York, 1979)
50. L. Granboulan, A. Joux, J. Stern, Inverting HFE is quasipolynomial, in *Crypto'06*, LNCS. vol. 4117 (2006), pp. 345–356
51. L.K. Grover, A fast quantum mechanical algorithm for database search, in *Proceedings 28th Annual ACM Symposium on the Theory of Computing* (1996) pp. 212–219
52. S. Hasegawa, T. Kaneko, An attacking method for a public-key cryptosystem based on the difficulty of solving a system of non-linear equations (in Japanese), in *Proceedings of 10th SITA*, vol. JA5-3 (1987)

53. Y. Hashimoto, Algorithms to solve massively under-defined systems of multivariate quadratic equations. IEICE Trans. Fundam. **E94–A**, 1257–1262 (2011)
54. Y. Hashimoto, Cryptanalysis of the quaternion rainbow, in *IWSEC'13*. LNCS, vol. 8231 (2013), pp. 244–257
55. Y. Hashimoto, Cryptanalysis of the multivariate sigature scheme proposed in PQCrypto 2013, in *PQCrypto'14*, LNCS, vol. 8772 (2014), pp. 108–125. IEICE Trans. Fundam. **99-A**, 58–65 (2016)
56. Y. Hashimoto, A note on tensor simple matrix encryption scheme, http://eprint.iacr.org/2016/065
57. Y. Hashimoto, On the security of new vinegar-like variant of multivariate signature scheme, http://eprint.iacr.org/2016/787
58. Y. Hashimoto, On the security of cubic UOV, http://eprint.iacr.org/2016/788
59. Y. Hashimoto, Key recovery attacks on multivariate public key cryptosystems derived from quadratic forms over an extension field. IEICE Tans. Fundam. **100–A**, 18–25 (2017)
60. Y. Hashimoto, Chosen ciphertext attack on ZHFE. JSIAM Lett. (2017). To appear
61. Y. Hashimoto, T. Takagi, K. Sakurai, General fault attacks on multivariate public key cryptosystems, in *PQC'11*. LNCS, vol. 7071 (2011), pp. 1–18
62. M. Joye, A.K. Lenstra, J.J. Quisquater, Chinese remaindering based cryptosystems in the presence of faults. J. Cryptol. **12**, 241–245 (1999)
63. A. Kipnis, A. Shamir, Cryptanalysis of the oil and vinegar signature scheme, in *Crypto'98*. LNCS, vol. 1462 (1998), pp. 257–267
64. A. Kipnis, A. Shamir, Cryptanalysis of the HFE public key cryptosystem by relinearization, in *Crypto'99*. LNCS, vol. 1666 (1999), pp. 19–30
65. A. Kipnis, J. Patarin, L. Goubin, Unbalanced oil and vinegar signature schemes, in *Eurocrypt'99*. LNCS, vol. 1592 (1999), pp. 206–222, extended in www.citeseer/231623.html, 2003-06-11
66. F. Le Gall, Powers of tensors and fast matrix multiplication, in *ISSAC'14, Proceedings of the 39th ISSAC* (2014), pp. 296–303
67. H. Li, X. Chen, L. Pang, W. Shi, Quantum attack-resistent certificateless multi-receiver signcryption scheme. PLoS ONE **8**(6), e49141 (2013)
68. R. Lidl, H. Niederreiter, *Finite Fields* (Addison-Wesley, London, 1983)
69. T. Matsumoto, H. Imai, Public quadratic polynomial-tuples for efficient signature-verification and message-encryption, in *Eurocrypt'88*. LNCS, vol. 330 (1988), pp. 419–453
70. H. Miura, Y. Hashimoto, T. Takagi, Extended algorithm for solving underdefined multivariate quadratic equations, in *PQCryoto'13*, LNCS, vol. 7932 (2013), pp. 118–135. IEICE Trans. Fundam. **E97-A**, 1418–1425 (2014)
71. H.G. Molter, R. Overbeck, A. Shoufan, F. Strenzke, E. Tews, Side channels in the McEliece PKC, in *PQC'08*. LNCS, vol. 5299 (2008), pp. 216–229
72. D. Moody, Post-quantum cryptography: NIST's plan for the future, in *NIST Announcement in PQCrypto'16* (2016), https://pqcrypto2016.jp/data/pqc2016_nist_announcement.pdf
73. D. Moody, R. Perlner, D. Smith-Tone, An asymptotically optimal structural attack on the ABC multivariate encryption scheme, in *PQC'14*. LNCS, vol. 8772 (2014), pp. 180–196
74. X. Nie, A. Petzoldt, J. Buchmann, Cryptanalysis of 2-layer nonlinear piece in hand method, in *CD-ARES'13*. LNCS, vol. 8128 (2013), pp. 91–104
75. X. Nie, B. Liu, H. Xiong, G. Lu, Cubic unbalance oil and vinegar signature scheme, in *Inscrypt'15*. LNCS, vol. 9589 (2015), pp. 47–56
76. NIST, Submission requirements and evaluation criteria for the Post-Quantum Cryptography standardization process (2016), http://csrc.nist.gov/groups/ST/post-quantum-crypto/documents/call-for-proposals-final-dec-2016.pdf
77. K. Okeya, T. Takagi, C. Vuillaume, On the importance of protecting Δ in SFLASH against side channel attacks. IEICE Trans. **88-A**, 123–131 (2005)
78. D. Page, F. Vercauteren, A fault attack on pairing-based cryptography. IEEE Trans. Comput. **55**, 1075–1080 (2006)

79. J. Patarin, Cryptoanalysis of the Matsumoto and Imai Public Key Scheme of Eurocrypt'88, in *Crypto'95*. LNCS, vol. 963 (1995), pp. 248–261
80. J. Patarin, Hidden fields equations (HFE) and isomorphisms of polynomials (IP): two new families of asymmetric algorithms, *Eurocrypt'96*. LNCS, vol. 1070 (1996), pp. 33–48
81. J. Patarin, The oil and vinegar signature scheme, in *The Dagstuhl Workshop on Cryptography* (1997)
82. J. Patarin, L. Goubin, N.T. Courtois, $C * -+$ and HM: variations around two schemes of T. Matsumoto and H. Imai, in *Asiacrypt'98*. LNCS, vol. 1514 (1998), pp. 35–49
83. J. Patarin, N. Courtois, L. Goubin, QUARTZ, 128-bit long digital signatures, in *CT-RSA'01*. LNCS, vol. 2020 (2001), pp. 282–297
84. R. Perlner, D. Smith-Tone, Security analysis and key modification for ZHFE, in *PQCrypto'16*. LNCS, vol. 9606 (2016), pp. 197–212
85. A. Petzoldt, S. Bulygin, J.A. Buchmann, CyclicRainbow - a multivariate signature scheme with a partially cyclic public key, in *IndoCrypt'10*. LNCS, vol. 6498 (2010), pp. 33–48
86. A. Petzoldt, S. Bulygin, J.A. Buchmann, Fast verification for improved versions of the UOV and Rainbow signature schemes, in *PQC'13*. LNCS, vol. 7932 (2013), pp. 188–202
87. A. Petzoldt, M.S. Chen, B.Y. Yang, C. Tao, J. Ding, Design principles for HFEv- based multivariate signature schemes, in *Asiacrypt'15*. LNCS, vol. 9452 (2015), pp. 311–334
88. A. Petzoldt, J. Ding, L.C. Wang, Eliminating decryption failures from the simple matrix encryption scheme (2016), http://eprint.iacr.org/2016/010
89. J. Porras, J. Baena, J. Ding, ZHFE, a new multivariate public key encryption scheme, in *PQCrypto'14*. LNCS, vol. 8772 (2014), pp. 229–245
90. B. Preneel, NESSIE Project Announces Final Selection of Crypto Algorithms, https://www.cosic.esat.kuleuven.be/nessie/deliverables/press_release_feb27.pdf
91. K. Sakumoto, T. Shirai, H. Hiwatari, Public-key identification schemes based on multivariate quadratic polynomials, in *Crypto'11*. LNCS, vol. 6841 (2011), pp. 706–723
92. K. Sakumoto, T. Shirai, H. Hiwatari, On provable security of UOV and HFE signature schemes against Chosen-Message Attack, in *PQCrypto'11*. LNCS, vol. 7071 (2011), pp. 68–82
93. A. Shamir, Efficient signature schemes based on birational permutations, in *Crypto '93*. LNCS, vol. 773 (1983), pp. 1–12
94. W. Shen, S. Tang, L. Xu, IBUOV, A provably secure Identity-Based UOV Signature Scheme, in *Proceeding CSE'13* (2013), pp. 388–395
95. P.W. Shor, Polynomial-time algorithms for prime factorization and discrete logarithms on a quantum computer. SIAM J. Comput. **26**, 1484–1509 (1997)
96. D. Smith-Tone, M.-S. Chen, B.-Y. Yang, PFLASH - secure asymmetric signatures on smart cards, in *Lightweight Cryptography Workshop* (2015), http://csrc.nist.gov/groups/ST/lwc-workshop2015/papers/session3-smith-tone-paper.pdf
97. A. Szepieniec, J. Ding, B. Preneel, Extension field cancellation: a new central trapdoor for multivariate quadratic systems, in *PQC'16*. LNCS, vol. 9606 (2016), pp. 182–196
98. V. Strassen, Gaussian elimination is not optimal. Numer. Math. **13**, 354–356 (1969)
99. C. Tao, A. Diene, S. Tang, J. Ding, Simple matrix scheme for encryption, in *PQCrypto 2013*. LNCS, vol. 7932 (2013), pp. 231–242
100. C. Tao, H. Xiang, A. Petzoldt, J. Ding, Simple Matrix - a multivariate public key cryptosystem (MPKC) for encryption. Finite Fields Appl. **35**, 352–368 (2015)
101. O. Taussky, H. Zassenhaus, On the similarity transformation between a matirx and its transpose. Pac. J. Math. **9**, 893–896 (1959)
102. R. Terada, E.R. Andrade, Comparison of two signatrue schemes based on the MQ problem and Quartz. IEICE Trans. Fundam. **99-A**, 2527–2538 (2016)
103. E. Tomae, Quo vadis quaternion? Cryptanalysis of Rainbow over non-commutative rings, in *SCN'12*. LNCS, vol. 7485 (2012), pp. 361–373
104. E. Thomae, C. Wolf, Solving underdetermined systems of multivariate quadratic equations revisited, in *PKC'12*. LNCS, vol. 7293 (2012), pp. 156–171
105. S. Tsujii, K. Kurosawa, T. Itoh, A. Fujioka, T. Matsumoto, A public-key cryptosystem based on the difficulty of solving a system of non-linear equations. IEICE Trans. Inf. Syst. (Japanese Edition), **J69-D**, pp. 1963–1970 (1986)

106. S. Tsujii, K. Tadaki, R. Fujita, Proposal for Piece in Hand Matrix: general concept for enhancing security of multivariate public key cryptosystems. IEICE Trans. **90-A**, 992–999 (2007)
107. L.C. Wang, B.Y. Yang, Y.H. Hu, F. Lai, A "medium-field" multivariate public-key encryption scheme, in *CT-RSA'06*. LNCS, vol. 3860 (2006), pp. 132–149
108. B.Y. Yang, J.M. Chen, Building secure tame-like multivariate public-key cryptosystems: the new TTS, in *ACISP'05*. LNCS, vol. 3574 (2005), pp. 518–531
109. T. Yasuda, K. Sakurai, A security analysis of uniformly-layered rainbow defined over non-commutative rings. Pac. J. Math. Ind. **6**, 81–89 (2014)
110. T. Yasuda, K. Sakurai, T. Takagi, Reducing the key size of Rainbow using non-commutative rings, in *CT-RSA'12*. LNCS, vol. 7178 (2012), pp. 68–83
111. T. Yasuda, T. Takagi, K. Sakurai, Multivariate signature scheme using quadratic forms. in *PQCrypto'13*. LNCS, vol. 7932 (2013), pp. 243–258
112. T. Yasuda, T. Takagi, K. Sakurai, Security of multivariate signature scheme using non-commutative rings. IEICE Trans. **97-A**, 245–252 (2014)
113. T. Yasuda, X. Dahan, Y.-J. Huang, T. Takagi, K, Sakurai, MQ Challenge: hardness evaluation of solving multivariate quadratic problems, in *The NIST Workshop on Cybersecurity in a Post-Quantum World, Washington, D.C*, April 2–3 (2015), https://www.mqchallenge.org/
114. W. Zhang, C.H. Tan, MI-T-HFE, A new multivariate signature scheme, in *IMACC'15*. LNCS, vol. 9496 (2015), pp. 43–56
115. W. Zhang, C.H. Tan, A secure variant of Yasuda, Takagi and Sakurai's signature scheme, in *Inscryptf15*. LNCS, vol. 9589 (2015), pp. 75–89
116. W. Zhang, C.H. Tan, On the security and key generation of the ZHFE encryption scheme, in *IWSEC'16*. LNCS, vol. 9836 (2016), pp. 289–304

Code-Based Zero-Knowledge Protocols and Their Applications

Kirill Morozov

Abstract We present a survey of recent results in the area of zero-knowledge (ZK) protocols based on coding problems and the related Learning Parities with Noise (LPN) problem. First, we sketch the constructions of two ZK code-based identification schemes: the one based on general decoding by Jain et al. (Asiacrypt 2012) and the one based on syndrome decoding by Stern (Crypto 1993). Next, we show that these two systems can also be used to implement a proof of plaintext knowledge for the code-based public key encryption schemes: the one by McEliece and the one by Niederreiter, respectively. Finally, we briefly discuss verifiable encryption and digital signatures as applications.

Keywords Code-based encryption · Zero-knowledge · Identification · Proof of plaintext knowledge · Verifiable encryption · Signatures

1 Introduction

Hard problems related to coding has been the focus of cryptographers' attention ever since introduction of the public key encryption scheme (PKE) by R.J. McEliece in 1978 [32], only a year after the famous RSA PKE [40] was presented. Very recently, an interest to such systems has been greatly boosted by the announcement of NIST future plan for standardization of the post-quantum (a.k.a. quantum-resistant) cryptographic algorithms [39]. Such algorithms are designed to withstand attacks not only by classical but also by quantum computers, and code-based cryptographic schemes are believed to belong to this class.

This survey is based on the paper: Rong Hu, Kirill Morozov, Tsuyoshi Takagi: "Zero-Knowledge Protocols for Code-Based Public Key Encryption." IEICE Transactions 98-A(10): 2139–2151 (2015) [26].

K. Morozov (✉)
School of Computing, Tokyo Institute of Technology, 2-12-1 Ookayama,
Meguro-ku, Tokyo 152-8552, Japan
e-mail: morozov@c.titech.ac.jp

T. Takagi et al. (eds.), *Mathematical Modelling for Next-Generation Cryptography*,
Mathematics for Industry 29, DOI 10.1007/978-981-10-5065-7_3

As compared to their counterparts based on number-theoretic assumptions, code-based cryptographic schemes received considerably less attention of the research community. Perhaps, this can be partially explained by the fact that employment of linear codes resulted in prohibitively large memory requirements for the early days computers. However, recent advances in computing technologies allowed us to relax the memory requirements on the one hand, while on the other hand, extensive research on implementations allowed us to optimize the existing constructions in order to accommodate even the restricted environments such as smart cards.

The purpose of this article is to make the reader familiar with code-based zero-knowledge protocols. This work is organized as follows:

1. We will sketch the constructions of two ZK identification schemes based on two famous problems in coding theory: general decoding and syndrome decoding, respectively.
2. As their immediate application, we will describe the cryptographic protocols called *proof of plaintext knowledge (PPK)* for the code-based public key encryption schemes by McEliece and by Niederreiter, respectively.
3. We will show another application of these systems for constructing the verifiable McEliece encryption.
4. Finally, we will briefly discuss the applications to digital signatures and other related results.

Stern [44] introduced a zero-knowledge (ZK) proof system for the statement related to the syndrome decoding (SD) problem, and proposed to use it as a basis for the ZK identification scheme. Jain et al. [27] introduced a ZK proof and the related ZK identification scheme based on the Exact Learning Parity with Noise (xLPN) problem, a special case of the LPN problem, where the noise vector has fixed weight. In the literature on the code-based cryptography, xLPN is often referred to as the general decoding (GD) problem. There have been quite a few other attempts to introduce code-based identification schemes, but some of them were either insecure in some sense or lacked efficiency comparable to the above mentioned proposals. One famous example is the proposal by Véron [45], which was shown to lack the zero-knowledge property in [27, Full version, Appendix A]. The identification schemes by Stern and by Jain et al. will be presented in Sects. 3 and 4, respectively.

It is easy to observe that the PKE schemes by McEliece [32] and Niederreiter [35], as well as their randomized IND-CPA versions [36] have the form similar to the GD and SD problems (see Sect. 2.3), respectively. Therefore, the proof systems in Sects. 3 and 4 can be applied to prove a plaintext knowledge for the above PKE schemes, using the plaintext and the random coins as witness [26, 34]. Such cryptographic protocols are known as *proofs of plaintext knowledge (PPK)*. We will also show that in the case of the above code-based PKE's, these protocols can be used for proving that a given ciphertext is correctly formed with respect to a given public key. These protocols are discussed in Sect. 5.

We will show a further application of one of the above constructions by presenting verifiable encryption [43] based on the McEliece scheme. Here, a party, who performed encryption, is able to prove in zero-knowledge that a given ciphertext contains a given plaintext, without disclosing the randomness used for encryption.

Finally, we will briefly discuss application of the above systems for constructing code-based digital signatures via Fiat–Shamir paradigm [20].

Concluding remarks and open questions are provided in Sect. 8.

2 Background

In this section, we follow the presentation of [26].

The Hamming distance between $x, y \in \mathbb{F}_2^n$ (i.e., the number of positions where they differ) is denoted as $d_H(x, y)$. The distance of $x \in \mathbb{F}_2^n$ to the zero-vector 0^n denoted by $w_H(x) := d_H(x, 0^n)$ is called the *weight* of x. For $x, y \in \mathbb{F}_2^n$, $x + y$ will denote the bitwise exclusive-or. Let J be an ordered subset as follows: $\{j_i, \ldots, j_m\} = J \subseteq \{1, \ldots, n\}$, then we denote a vector $(x_{j_1}, \ldots, x_{j_m}) \in \mathbb{F}_2^m$ by x_J. Similarly, we denote by $M_J \in \mathbb{F}_2^{k\times|J|}$ a restriction of the matrix $M \in \mathbb{F}_2^{k\times n}$ to the columns with indices in J. A concatenation of matrices $X \in \mathbb{F}_2^{k\times n_0}$ and $Y \in \mathbb{F}_2^{k\times n_1}$ is written as $(X|Y) \in \mathbb{F}_2^{k\times(n_0+n_1)}$, and for $k = 1$ this will denote concatenation of vectors. We denote by $x \xleftarrow{\$} \mathscr{X}$ a uniformly random sampling of an element from its domain $\mathscr{X}$. A set of $(n \times n)$ permutation matrices is denoted by $\mathscr{S}_n$. By $\mathsf{Binary}(k, n)$, we denote a set of the matrices in $\mathbb{F}_2^{k\times n}$ of rank k. All the logarithms are to the base 2, unless otherwise stated. We abbreviate *probabilistic polynomial time* as PPT. If A is a PPT algorithm which on inputs $(x_1, \ldots, x_n)$ computes the outputs $(y_1, \ldots, y_n)$, we write it as $(y_1, \ldots, y_n) \xleftarrow{\$} \mathsf{A}(x_1, \ldots, x_n)$.

We model a party taking part in an interactive two-party protocol as an interactive Turing machine. We denote by $\langle A(a), B(b)\rangle(c)$ a random variable representing the output of a party B following an execution of an interactive two-party protocol between a party A with private input a and B with private input b on a joint input c, where A and B have uniformly distributed random tapes. If a party, say A, has no input, then we omit it by writing just A (instead of $A(a)$) in the above notation. In our two-party protocols, we will denote an honest prover by P and an honest verifier by V, while a dishonest party will be denoted by $\widetilde{\mathsf{P}}$ and $\widetilde{\mathsf{V}}$, respectively.

We let κ be the security parameter. When using it as an input to algorithms, we will employ its unary representation denoted as 1^κ. We call a function $\varepsilon(\kappa)$ *negligible in* κ, if $\varepsilon(\kappa) = 2^{-\omega(\log\kappa)}$, where ω denotes the Little Omega asymptotic notation. We call a probability $1 - \varepsilon(\kappa)$ *overwhelming*, when $\varepsilon(\kappa)$ is negligible. Occasionally, we may omit mentioning the security parameter, when it is clear from the context. In these cases, by saying that a quantity is negligible (overwhelming), we mean that it is negligible (overwhelming) in the security parameter.

2.1 Linear Codes

A binary (n, k)-code $\mathscr{C}$ is a k-dimensional subspace of the vector space $\mathbb{F}_2^n$; n and k are called the *length* and the *dimension* of the code, respectively. We call $\mathscr{C}$ an (n, k, d)-code, if its so-called *minimum distance* is

$$d := \min_{x,y\in\mathscr{C};\ x\neq y} d_H(x, y).$$

For relevant topics in coding theory, we refer the reader to [31, 41].

We denote as $\mathscr{C}(G)$ the code defined by $G \in \mathbb{F}_2^{k\times n}$ as a generator matrix. It is easy to check, if a vector $v \in \mathbb{F}_2^n$ belongs to $\mathscr{C}(G)$ by computing its syndrome $syn(v)$ as explained next [41]. One can compute the parity-check matrix $H \in \mathbb{F}_2^{(n-k)\times n}$ corresponding to G, i.e., such that $HG^T = 0$, and then $syn(v) = Hv^T$. If $syn(v) = 0$, then v belongs to $\mathscr{C}(G)$, and does not belong to it otherwise.

It is known that (n, k)-codes chosen uniformly at random meet [38] the Gilbert–Varshamov bound with overwhelming probability [41]. This implies that their minimal distance is known almost surely.

2.2 Code-Based Public Key Encryption

In this subsection, which is adapted from [26], we introduce the Niederreiter and the McEliece PKE. More information on this topic, can be found, e.g., in the surveys [16, 33, 37].

2.2.1 Niederreiter PKE

The Niederreiter PKE consists of the following triplet of algorithms $(\mathscr{K}, \mathscr{E}, \mathscr{D})$ with the system parameters $n, t \in \mathbb{N}$:

• Key generation algorithm $\mathscr{K}$: On input 1^κ, choose the appropriate (n, k, t), and generate the following matrices:

– $H \in \mathbb{F}_2^{(n-k)\times n}$—the parity-check matrix of an irreducible (n, k) binary Goppa code which can correct at most t errors, where $k \geq n - t \cdot \log n$. The decoding algorithm of this code is denoted as $\mathsf{Dec}_{\mathscr{H}}$, it takes on input an $(n-k)$-bit syndrome and returns the corresponding n-bit error vector.[1]

– $M \in \mathbb{F}_2^{(n-k)\times(n-k)}$—a random non-singular matrix.

– $P \in \mathscr{S}_n$—a random permutation matrix.

– $H^{pub} = MHP \in \mathbb{F}_2^{(n-k)\times n}$.

[1] If such the vector does not exist, $\mathsf{Dec}_{\mathscr{H}}$ returns "failure." When the encryption algorithm is run correctly, this situation does not occur. Although this detail is important for practical implementations, it is immaterial for the following presentation, so that we omit mentioning it for the sake of simplicity.

Output the public key $pk = (H^{pub}, t)$ and the secret key $sk = (M, H, P)$.
• Encryption algorithm $\mathscr{E}$: On input a plaintext $m \in \mathbb{F}_2^n$ such that $w_H(m) = t$, and the public key pk, output the ciphertext $c = H^{pub} m^T$.
• Decryption algorithm $\mathscr{D}$: On input a ciphertext c and the secret key sk, calculate:
– $M^{-1}c = (HP)m^T$.
– Since $(HP)m^T = H(Pm^T)$, use the decoding algorithm $\mathsf{Dec}_{\mathscr{H}}$ to recover Pm^T.
–Output $m^T = P^{-1}Pm^T$.

It is easy to check correctness of the decryption algorithm: After the first step of decryption, we obtain a syndrome of the permuted plaintext Pm^T. Since the decoding algorithm $\mathsf{Dec}_{\mathscr{H}}$ is known, it is easy to recover the plaintext.

We note that the plaintext space of the Niederreiter PKE is the set of weight-t binary vectors. For representation of arbitrary binary vectors (of an appropriate length) as a valid plaintext, see the work by Cover [13] and its improvements by Sendrier [42].

2.2.2 McEliece PKE

The McEliece PKE consists of the following triplet of algorithms $(\mathscr{K}, \mathscr{E}, \mathscr{D})$ with system parameters $n, t \in \mathbb{N}$:
• Key generation algorithm $\mathscr{K}$: On input 1^κ, choose the appropriate (n, k, t), and generate the following matrices:
– $G \in \mathbb{F}_2^{k \times n}$—the generator matrix of an irreducible (n, k) binary Goppa code [24] correcting up to t errors, where $k \geq n - t \cdot \log n$. Its decoding algorithm is denoted as $\mathsf{Dec}_{\mathscr{G}}$, it takes on input an n-bit vector and returns the n-bit codeword, which is within the Hamming distance at most t from it.[2]
– $S \in \mathbb{F}_2^{k \times k}$—a random non-singular matrix.
– $P \in \mathscr{S}_n$—a random permutation matrix (of size n).
– $G^{pub} = SGP \in \mathbb{F}_2^{k \times n}$.
Output the public key $pk = (G^{pub}, t)$ and the secret key $sk = (S, G, P)$.
• Encryption algorithm $\mathscr{E}$: On input a plaintext $m \in \mathbb{F}_2^k$ and the public key pk, choose a vector $e \in \mathbb{F}_2^n$ of weight t at random, and output the ciphertext $c = mG^{pub} + e$.
• Decryption algorithm $\mathscr{D}$: On input c and the secret key sk, calculate:
– $cP^{-1} = (mS)G + eP^{-1}$.
– $mSG = \mathsf{Dec}_{\mathscr{G}}(cP^{-1})$.
– Let $J \subseteq \{1, \ldots, n\}$, $|J| = k$, be such that G_J is invertible.
Output $m = (mSG)_J(G_J)^{-1}S^{-1}$.

[2]If such the codeword does not exist, $\mathsf{Dec}_{\mathscr{G}}$ returns "failure." When the encryption algorithm is run correctly, this situation does not occur. Although this detail is important for practical implementations, it is immaterial for the following presentation, so that we omit mentioning it for the sake of simplicity.

It is easy to check that the decryption algorithm correctly recovers the plaintext: Since in the first step of decryption, the permuted error vector eP^{-1} is again of weight t, the decoding algorithm $\mathsf{Dec}_{\mathscr{G}}$ successfully corrects these errors in the next step.

2.2.3 IND-CPA Variants

Nojima et al. [36] introduced the efficient IND-CPA variants of the Niederreiter and the McEliece PKE's in the standard model, under hardness of the Syndrome Decoding and the LPN problems, respectively, plus hardness of the Goppa-IND problem (in both cases). These variants are referred to as the Randomized Niederreiter (respectively, McEliece) PKE.

They are constructed by a uniform padding of the plaintext. Specifically, the Randomized Niederreiter encryption is constructed in the same way, as described above, except that the ciphertext is computed as follows $c = H^{pub}(r|m)^T$, where $r \overset{\$}{\leftarrow} \{x \in \mathbb{F}_2^{n_0} \mid w_H(x) = t_0\}$, $m \in \mathbb{F}_2^{n_1}$, $w_H(m) = t_1$, and $n = n_0 + n_1$, $t = t_0 + t_1$.

The IND-CPA McEliece encryption is constructed in the same way as described above, except that the ciphertext $c = (r|m)G^{pub} + e$, where $r \overset{\$}{\leftarrow} \mathbb{F}_2^{k_0}$, $m \in \mathbb{F}_2^{k_1}$, $k = k_0 + k_1$.

We refer the reader to [36] for the security arguments and the recommended parameters.

2.3 Coding Problems

Definition 2.1 (*Syndrome Decoding (SD) Problem*).
Input: $H \overset{\$}{\leftarrow} \mathsf{Binary}(n-k, n)$, $s \overset{\$}{\leftarrow} \mathbb{F}_2^{n-k}$, and $0 < t \in \mathbb{N}$.
Output: $z \in \mathbb{F}_2^n$ such that $w_H(z) \leq t$, $Hz^T = s$.

This problem was shown to be NP-complete by Berlekamp et al. [8]. Its dual version can be formulated as below.

Definition 2.2 (*General Decoding (GD) Problem*).
Input: $G \overset{\$}{\leftarrow} \mathsf{Binary}(k, n)$, $y \overset{\$}{\leftarrow} \mathbb{F}_2^n$, and $0 < t \in \mathbb{N}$.
Output: $x \in \mathbb{F}_2^k$, $e \in \mathbb{F}_2^n$ such that $w_H(e) \leq t$, $xG \oplus e = y$.

The above problem is defined in [27] as the exact LPN problem (xLPN). For uniformity, we will use the notation "the GD problem" henceforth.

The following three problems use the quantities defined in the previous subsections. No polynomial-time algorithm is known for these problems [4, 9, 16, 21]. The polynomial-time distinguisher by Faugère et al. [18] works only for the "high-rate" Goppa codes, which can be avoided by the proper choice of parameters.

Definition 2.3 (*Niederreiter Problem*).
Input: A public key (H^{pub}, t) of the Niederreiter PKE, where $H^{pub} \in \mathbb{F}_2^{(n-k)\times n}$, $0 < t \in \mathbb{N}$, and the ciphertext $c \in \mathbb{F}_2^n$.
Output: $m \in \mathbb{F}_2^n$ such that $c = H^{pub}m^T$ and $w_H(m) = t$.

Definition 2.4 (*McEliece Problem*).
Input: A public key (G^{pub}, t) of the McEliece PKE, where $G^{pub} \in \mathbb{F}_2^{k\times n}$, $0 < t \in \mathbb{N}$, and the ciphertext $c \in \mathbb{F}_2^n$.
Output: $m \in \mathbb{F}_2^k$ such that $d_H(mG^{pub}, c) = t$.

Definition 2.5 (*Permuted Goppa Code Distinguishing (Goppa-IND) Problem*).
Input: $G \in \mathbb{F}_2^{k\times n}$.
Decide: Is G a generator matrix of an (n, k) irreducible Goppa code, or of a random (n, k)-code?

The equivalent formulation of the above problem can be made using the parity-check matrix, since it can be computed from the generator matrix by the elementary linear algebra.

2.4 Zero-Knowledge Identification Schemes

An identification scheme allows a party P (*prover*) to convince another party V (*verifier*) that the latter is indeed communicating with P. We will consider identification in the public key scenario. In this context, identification can be achieved using zero-knowledge proof systems [6, 19, 22]. A zero-knowledge proof system for an NP language L allows a prover P to convince a verifier V about a fact that a string $x \in L$ without releasing any additional knowledge apart from the fact itself. Usually, a statement of the problem in the language involves the public key pk (possibly, along with some public parameter) of the identification scheme, and the witness serves as the secret key. Indeed, informally speaking, a witness is the information that allows one to easily verify that a statement belongs to the language. Then, an algorithm that generates an instance of the NP problem serves as the key generation algorithm for the identification scheme Gen, while the ZK proof system naturally represents both the prover and the verifier algorithms. In the following definition, we omit mentioning of the security parameter.

Definition 2.6 Let $\Pi = (\mathsf{P}, \mathsf{V})$ be a tuple of PPT algorithms. Π is a *zero-knowledge proof system* for a language L, if the following conditions hold:
(Completeness) For any $x \in L$ and any witness w for x, we have $\langle \mathsf{P}(w), \mathsf{V}\rangle(x) = 1$.
(Soundness) For any $x \notin L$ and for any expected PPT $\widetilde{\mathsf{P}}$, we have that $\Pr[\langle \widetilde{\mathsf{P}}, \mathsf{V}\rangle(x) = 1]$ is negligible, where the probability is taken over the random coins of Gen, and the random tapes of both $\widetilde{\mathsf{P}}$ and V.
(Zero-knowledge) There exists an expected PPT simulator SIM such that, for any PPT $\widetilde{\mathsf{V}}$, any $x \in L$ and witness for x, the following distributions are indistinguishable:

$$\{x \in L : \langle \mathsf{P}(w), \widetilde{\mathsf{V}}\rangle(x)\},\ \{x \in L : \langle \mathsf{SIM}, \widetilde{\mathsf{V}}\rangle(x)\},$$

where the probability is taken over the random tapes of both P and $\widetilde{\mathsf{V}}$. In the case of *statistical*, respectively *computational* indistinguishability, we call the property *statistical*, respectively *computational* zero-knowledge (ZK).

2.5 Commitment Schemes

ZK proof systems require commitments. A commitment scheme consists of two phases: the first one is *committing*, where a sender P provides a receiver V with an evidence about an input m. The cheating receiver $\widetilde{\mathsf{V}}$ cannot learn m at this stage—this property is called *hiding*. In the second phase, called *opening*, P reveals m to V. The cheating sender $\widetilde{\mathsf{P}}$ cannot successfully open any other message apart from m—this property is called *binding*.

For details on the definition of commitments, we refer the reader, e.g., to the journal version of [26].

In the standard model, an efficient computationally hiding and statistically binding commitment scheme based on the GD problem can be found in [27].

For simplicity, we henceforth denote a commitment to a value x by $Com(x)$, and to a pair of values (x, y) by $Com(x, y)$, and omit the mentioning of the public commitment key.

3 Stern's Identification Scheme

In this section, we will describe the zero-knowledge identification scheme by Stern [44] based on the Syndrome Decoding problem with a matrix H, and vectors m and c as defined in Sect. 2.3.

Secret key (witness): $m \in \mathbb{F}_2^n$, $w_H(m) = t$, where the parameters n and t are chosen according to the Gilbert–Varshamov bound, such that the solution to the SD problem is unique, hereby making it the hardest [44].

Public key: (c, t), where $c = Hm^T$, i.e., c is the syndrome of m with respect to H.

Public parameter: H. In some literature H is designated as a part of the public key. However, we set it as a public parameter in order to emphasize that it can be safely reused for several instantiations.

Protocol 1 (Stern's Scheme [44]).

1. P computes $y \stackrel{\$}{\leftarrow} \mathbb{F}_2^n$, $\pi \stackrel{\$}{\leftarrow} \mathcal{S}_n$, and sends the following three commitments to V:
   - $C_1 = Com(\pi, Hy^T)$,
   - $C_2 = Com(y\pi)$,
   - $C_3 = Com((y + m)\pi)$.

2. V sends a challenge $b \stackrel{\$}{\leftarrow} \{0, 1, 2\}$.
3. P replies as follows, while V performs the following checks and rejects, if any check fails:
   - If $b = 0$,
     - P sends y, π, and opens C_1 and C_2.
     - V checks validity of the opened values.
   - If $b = 1$,
     - P sends $y + m$, π, and opens C_1 and C_3.
     - V checks validity by computing $H^{pub}y^T = H(y + m)^T + c$, and then verifying that the opening of C_1 is $(\pi, H(y + m)^T + c)$, and the opening of C_3 is $(y + m)\pi$.
   - If $b = 2$,
     - P sends $y\pi$, $m\pi$, and opens C_2 and C_3.
     - V checks validity of the opened values by verifying that C_2 opens to $y\pi$, C_3 opens to $y\pi + m\pi$, and that $w_H(m\pi) = t$.

Denote a protocol consisting of r independent sequential iterations of Protocol 1 by $\mathsf{ZKID}_{\mathsf{Stern}}(H, c; m)$, with some appropriately chosen r.

Theorem 3.1 *The protocol* $\mathsf{ZKID}_{\mathsf{Stern}}(H, c; m)$ *is a zero-knowledge proof for the binary vectors* m *of weight* t *such that* $Hm^T = c$ *for a given* c*, in the standard model, according to Definition 2.6 assuming that the SD problem is hard, and the underlying commitment scheme is secure.*

The proof is provided in the Appendix. Here, we emphasize one point: All the checks in the above protocol are essential to avoid the attacks by the cheating prover. For example, if checking of the weight for $m\pi$ is not performed in the last line of the protocol, then the attack can be derived from actions of the simulator—see the description in the proof of Lemma 9.4 (in the Appendix) for the case $b = 1$.

4 Jain et al.'s Identification Scheme

We now describe the ZK identification scheme by Jain et al. [27] based on the GD problem, with a matrix G and vectors m, e, and c as defined in Sect. 2.3.
Secret key (witness): $m \in \mathbb{F}_2^k$, $e \in \mathbb{F}_2^n$ such that $w_H(e) = t$, where the parameters (n, k, t) are chosen according to the Gilbert–Varshamov bound, such that the solution to the GD problem is unique.
Public key: (c, t), where $c = mG + e$.
Public parameter: G.
Protocol 2 (Jain et al.'s Scheme [27]).

1. P computes $u \stackrel{\$}{\leftarrow} \mathbb{F}_2^k$, $y \stackrel{\$}{\leftarrow} \mathbb{F}_2^n$, $\pi \stackrel{\$}{\leftarrow} \mathscr{S}_n$, and sends three commitments to V:
   $C_1 = Com(\pi, uG + y)$,
   $C_2 = Com(y\pi)$,
   $C_3 = Com((y + e)\pi)$.

2. V sends $b \xleftarrow{\$} \{0, 1, 2\}$.
3. P replies as follows, while V performs the following checks and rejects, if any check fails:
   - If $b = 0$,
     - P sends π, $uG + y$, and $y\pi$, and opens C_1 and C_2.
     - V checks validity by computing $(uG + y) + (y\pi)\pi^{-1}$ and verifying that it is in $\mathscr{C}(G)$.
   - If $b = 1$,
     - P sends π, $uG + y$, and $(y + e)\pi$, and opens C_1 and C_3.
     - V checks validity by computing $(uG + y) + ((y + e)\pi)\pi^{-1} + c$ and verifying that it is in $\mathscr{C}(G)$.
   - If $b = 2$,
     - P sends $y\pi$ and $(y + e)\pi$, and opens C_2 and C_3.
     - V checks validity by computing $w_H(y\pi + (y + e)\pi)$ and verifying that it is equal to t.

We note that Jain et al. [27] present the security proof relying on the fact that their protocol is a special variant of Σ-protocols, where soundness error is larger than 1/2. They claimed that the major results for Σ-protocols hold for their variant as well. However, we choose to use the standard security argument for the ZK proof of knowledge. The reason is that in Σ-protocols, a special honest-verifier ZK property is proved, while the security against malicious verifier is obtained via some secure transformation, e.g., by Damgård et al. [15]. However, Hu et al. [26] observed that the Jain et al. protocol is readily secure against malicious verifier, and constructed an alternative proof to reflect this fact.

Denote a protocol consisting of r independent sequential iterations of Protocol 2 by $\mathsf{ZKID}_{\mathsf{JKPT}}(G, c; m, e)$, with some appropriately chosen r.

Theorem 4.1 *The protocol $\mathsf{ZKID}_{\mathsf{JKPT}}(G, c; m, e)$ is a zero-knowledge proof for the binary vectors m and e with $w_H(e) = t$, for a given c such that $c = mG + e$, in the standard model, according to Definition 2.6 assuming that the GD problem is hard, and the underlying commitment scheme is secure.*

The proof can be adapted from the journal version of [26].

5 Proof of Plaintext Knowledge for Code-Based Cryptosystems

Proof of plaintext knowledge (PPK) protocols were first introduced by Aumann and Rabin [3]. These constructions applied to any PKE in the generic manner. Katz [28] presented efficient PPK's for RSA, Rabin, ElGamal, and Paillier public key encryption schemes. Goldwasser and Kharchenko presented PPK for the lattice-based Ajtai–Dwork PKE [23], and then other lattice-based schemes were covered in [7, 46, 47]. In particular, Xagawa et al. [47] and Xagawa and Tanaka [46] used a modification of Stern's scheme. This scheme was also used by Kobara et al. [30] for proving that the code-based encryption is well-formed in the setting of oblivious transfer.

Let us formally define PPK. The following definition adapted from [28] was used in [26]. For a PKE scheme $(\mathscr{K}, \mathscr{E}, \mathscr{D})$, denote by $c = \mathscr{E}_{pk}(m; R)$ a ciphertext of a plaintext m under the public key pk using the randomness R. We will call (m, R) a *witness* to the decryption of c under pk. Informally, in a PPK protocol, a sender P proves to a receiver V the knowledge of a witness to the decryption for some ciphertext c under the known public key pk.

Definition 5.1 Let $\Pi = (\mathsf{P}, \mathsf{V})$ be a tuple of PPT algorithms. Π is a *proof of plaintext knowledge* for an encryption scheme $(\mathscr{K}, \mathscr{E}, \mathscr{D})$ if the following conditions hold:
(Completeness) For any pk output by $\mathscr{K}(1^\kappa)$ and any c with witness w to the decryption of c under pk, we have that $\langle \mathsf{P}(w), \mathsf{V}\rangle(pk, c) = 1$ (when V outputs 1 we say it *accepts*).
(Validity) For a public key pk output by $\mathscr{K}(1^\kappa)$, a ciphertext c produced under such the pk, and for any expected PPT $\widetilde{\mathsf{P}}$, we have that $\Pr[\langle \widetilde{\mathsf{P}}, \mathsf{V}\rangle(pk, c) = 1]$ is negligible, where the probability is taken over the random coins of $\mathscr{K}$, and the random tapes of both $\widetilde{\mathsf{P}}$ and V.
(Zero-knowledge) There exists an expected PPT Turing machine SIM (called a *simulator*) such that, for any pk output by $\mathscr{K}(1^\kappa)$, any PPT $\widetilde{\mathsf{V}}$, and any w, the following distributions are indistinguishable:

$$\{c = \mathscr{E}_{pk}(m; R) : \langle \mathsf{P}(w), \widetilde{\mathsf{V}}\rangle(pk, c)\}$$

and

$$\{c = \mathscr{E}_{pk}(m; R) : \langle \mathsf{SIM}, \widetilde{\mathsf{V}}\rangle(pk, c)\},$$

where the probability is taken over the random tapes of both P and $\widetilde{\mathsf{V}}$. In the case of *statistical*, respectively *computational* indistinguishability, we call the property *statistical*, respectively *computational* zero-knowledge (ZK).

Now, we observe that the constructions in Sects. 3 and 4 can be applied to realize proof of plaintext knowledge for the Niederreiter and the McEliece PKE's, respectively, with minor adjustments.

Consider the Niederreiter PKE as described in Sect. 2.2.1. We can observe similarity of its structure to an instance of the SD problem with the difference that the parity-check matrix of a random code will be replaced by the Niederreiter public key.

Let us consider $\mathsf{ZKID}_{\mathsf{Stern}}(H^{pub}, c; m)$, where H^{pub} is generated as described in Sect. 2.2.1, then we can obtain the following result [26].

Theorem 5.2 ([26]) *The protocol* $\mathsf{ZKID}_{\mathsf{Stern}}(H^{pub}, c; m)$ *is a proof of plaintext knowledge for the Niederreiter public key encryption in the standard model, accord-*

ing to Definition 5.1 assuming that the Niederreiter problem is hard, and the underlying commitment scheme is secure.

We will omit the proof, which can be found in the journal version of [26], and only explain two important points, which must be taken into account. First, we note that the proof of Theorem 3.1 (see the Appendix) does not put any restrictions on the distribution of the parity-check matrix H as long as the related decoding problem remains hard. This allows us to safely replace with H with H^{pub}, when switching from hardness of SD to hardness of the Niederreiter Problem (see Sect. 2.3). In particular, the above security proof by itself does not require hardness of the Goppa-IND problem.

Second, we need to deal with the problem of maliciously generated instances. Note that in the case of identification schemes, the hard instance (responsible for the public/secret key pair) is assumed to be generated by the (honest) key generation algorithm. At the same time, in the PPK scenario, the ciphertext is presumably generated by the prover herself, and an assumption that she behaves honestly at that time, would be too strong. To this end, note that—similarly to [28]—the soundness in Lemma 9.3 is achieved in the strong sense. For the details, we refer the reader to the survey by Bellare and Goldreich [5], where such property is called a *strong validity*—for this reason, we call it *validity* in Definition 5.1 as well. Technically, it is first proven that if there exists no witness—or in other words, if the prover presents an invalid ciphertext—then she has only a negligible probability to be accepted. At the same time, if the prover did manage to be accepted with a non-negligible probability, then we construct a so-called witness extractor—an algorithm running in expected polynomial time, which computes the witness.

In a manner similar to the above, we consider $\mathsf{ZKID}_{\mathsf{JKPT}}(G^{pub}, c; m, e)$, where G^{pub} is generated as described in Sect. 2.2.2, and obtain the following theorem.

Theorem 5.3 ([26]) *The protocol* $\mathsf{ZKID}_{\mathsf{JKPT}}(G^{pub}, c; m, e)$ *is a proof of plaintext knowledge for the McEliece public key encryption in the standard model, according to Definition 5.1 assuming that the McEliece problem is hard, and the underlying commitment scheme is secure.*

Note that the protocol $\mathsf{ZKID}_{\mathsf{Stern}}(H^{pub}, c; r|m)$ achieves PPK for the Randomized Niederreiter PKE as well. The only difference is the structure of the error vector, however, its weight remains fixed. Therefore, if both of its components (r and m) are known to the prover, the protocol would work in the same way. In the same manner, the $\mathsf{ZKID}_{\mathsf{JKPT}}(G^{pub}, c; r|m)$ achieves PPK for the Randomized McEliece PKE.

Finally, we note that the above PPK protocols can be used as ZK proof of correctness for the respective encryptions. Indeed, the proof ensures the weight of the error vector, with respect to a given public key—in turn, this ensures correct decryption due to successful decoding.

6 Code-Based Verifiable Encryption

Verifiable encryption was first introduced by Stadler [43] with application to publicly verifiable secret sharing, and later generalized by Asokan et al. [2] and applied to fair exchange of digital signatures. The follow-up works on this topic include the papers by Camenisch and Damgård [10] and by Camenisch and Shoup [11]. In the post-quantum setting, verifiable encryption based on the Ajtai–Dwork PKE was presented by Goldwasser and Kharchenko [23].

Formally, we obtain the following definition.

Definition 6.1 Let $(\mathscr{K}, \mathscr{E}, \mathscr{D})$ be a public key encryption scheme, let R be a binary relation and let $L_R = \{x | \exists w : (x, w) \in R\}$. A *secure verifiable encryption* scheme for a relation R consists of a two-party protocol between P and V such that the following conditions hold:
(Completeness) For any pk output by $\mathscr{K}(1^\kappa)$ and any $x \in L_R$, we have that $\langle \mathsf{P}(x), \mathsf{V}\rangle(pk, c) = 1$ (when V outputs 1 we say it *accepts*).
(Validity) For any pk output by $\mathscr{K}(1^\kappa)$, any $x' \notin L_R$, and for any expected PPT cheating prover $\widetilde{\mathsf{P}}$, $\Pr[\langle \widetilde{\mathsf{P}}(x'), \mathsf{V}\rangle(pk, c) = 1]$ is negligible, where the probability is taken over the random tapes of both $\widetilde{\mathsf{P}}$ and V.
(Zero-knowledge) There exists a PPT simulator SIM such that for any pk output by $\mathscr{K}(1^\kappa)$, any PPT $\widetilde{\mathsf{V}}$, and any $x \in L_R$, the following distributions are computationally indistinguishable:

$$\{x \in L_R : \langle \mathsf{P}(x), \widetilde{\mathsf{V}}\rangle(pk, c)\}$$

and

$$\{x \in L_R : \langle \mathsf{SIM}, \widetilde{\mathsf{V}}\rangle(pk, c)\},$$

where the probability is taken over the random tapes of both P and $\widetilde{\mathsf{V}}$.

This definition captures only the properties related to verifiability. We implicitly assume that a scheme in question is indeed a public key encryption scheme.

Note that in the case of the original McEliece encryption, the problem of checking the equality relation for some $m' \in \mathbb{F}_2^k$ is trivial and does not require interactions. Specifically, given the ciphertext $c = mG^{pub} + e$ (as defined in Sect. 2.2.2), one can compute $y = m'G^{pub} + c$, and check if $w_H(y) = t$. Then, $y = (m + m')G^{pub} + e$, and if $m' = m$, we have that $y = e$. At the same time, if $m' \neq m$, then taking into account that $\mathscr{C}(G^{pub})$ by construction has a minimum distance at least $2t + 1$, we have that $w_H((m + m')G^{pub}) \geq 2t + 1$ and hence $w_H(y) \geq t + 1$.

Therefore, we assume that the IND-CPA variant of the McEliece PKE, as described in Sect. 2.2.3, is used for encryption. Specifically, the ciphertext is computed as $c = (r|m)G^{pub} + e$, where in particular, $r \in \mathbb{F}_2^{k_0}$, $m \in \mathbb{F}_2^{k_1}$, $k = k_0 + k_1$. Moreover, we assume that the ciphertext presented by the prover is correctly formed. In principle, this assumption can be removed, if we require the prover to run the Protocol 2 as $\mathsf{PPK}_M(G^{pub}, c; (r|m), e)$, i.e., the part of the witness m in Protocol 2 will be replaced with $(r|m)$. We observe that here, a PPK will play the role of the proof of correctness of the ciphertext.

Finally, we make an additional assumption that all the rows of the public key $G^{pub} \in \mathbb{F}_2^k$ are linearly independent. We note that this assumption is natural as the public key is typically constructed this way.

Witness: $r \in \mathbb{F}_2^{k_0}$, $m \in \mathbb{F}_2^{k_1}$, and $e \in \mathbb{F}_2^n$, where the parameters k_0, k_1, and n are described in Sect. 2.2.3.

Public data: $m' \in \mathbb{F}_2^k$, (G^{pub}, t) such that $G^{pub} \in \mathbb{F}_2^{k \times n}$—a public key of the IND-CPA McEliece PKE, and $c = (r|m)G^{pub} + e$—the ciphertext of the IND-CPA McEliece PKE, where $r \in \mathbb{F}_2^{k_0}$, $m \in \mathbb{F}_2^{k_1}$, $k = k_0 + k_1$.

Consider a partition of the public key as follows:

$$(G^{pub})^T = ((G_0^{pub})^T | (G_1^{pub})^T),$$

where $G_0^{pub} \in \mathbb{F}_2^{k_0 \times n}$ and $G_1^{pub} \in \mathbb{F}_2^{k_1 \times n}$ are the sub-matrices of G^{pub} corresponding to the randomness and the plaintext, respectively.

Protocol 3 (Verifiable McEliece PKE).

1. P and V both compute $c' = c + m' G_1^{pub}$.
2. P and V run the protocol $\mathsf{PPK}_M(G_0^{pub}, c'; r, e)$.

Theorem 6.2 ([26]) *The above protocol is a verifiable McEliece PKE for equality relation in the standard model, according to Definition 6.1 assuming that the variant of the McEliece PKE described in Sect. 2.2.3 is IND-CPA.*

7 Code-Based Signatures

Another immediate application of the ZK proofs systems described in Sects. 3 and 4 are code-based signatures obtained via Fiat–Shamir transform [20]. As explained in Sect. 3, Stern's identification scheme has soundness error 2/3, and hence, in order to improve the round complexity of the identification scheme, as well as the resulting signature size, Cayrel et al. [12] presented a 5-pass identification scheme based on q-ary codes with soundness error $q/2(q-1)$, that approaches 1/2 as q is increasing. A variant of the Fiat–Shamir transform for such schemes proposed by Dagdelen et al. [14] will yield a signature based on the above scheme.

It is worth noting that the Stern identification scheme [44] was proposed for the binary case, but it can be easily generalized to the q-ary case, if the scramble-permutation transformation of [12] is employed.[3] A non-asymptotic analysis by Hu et al. [25] shows that even for $q = 4$, i.e., when the soundness error of both Cayrel et al.'s and q-ary Stern's schemes is 2/3, the latter is slightly superior in terms of both

[3]In fact, this is the way, in which Stern's scheme was employed in the context of lattices by Kawachi et al. [29].

computational and communication cost. We note that the results of [25] confirm, as expected, that for $q = 3$, q-ary Stern's is superior, while for $q > 4$, Cayrel et al.'s becomes superior in the above sense.

8 Conclusion

We presented the code-based zero-knowledge identification schemes by Stern and by Jain et al., and discussed their applications to proof of plaintext knowledge for code-based PKE, verifiable code-based PKE, and code-based signatures.

The future research topics include construction of proof systems for other problems and their applications for widening the range of available code-based functionalities. Examples include a threshold ring signature by Aguilar Melchor et al. [1] and a group signature by [17].

Acknowledgements The author is supported by a Kakenhi Grant-in-Aid for Scientific Research (C) 15K00186 from Japan Society for the Promotion of Science. The author would like to thank anonymous reviewers for their helpful comments.

9 Appendix: Proof of Theorem 3.1

We adapt the proof of [26]. It generally follows the argument of [44], but for the proof of soundness it uses the argument from [45], since it is shorter. We emphasize that the gap in the proof of [45] pointed out in the full version of [27] concerned only the proof of the zero-knowledge property.

Completeness. It is easy to check that P who knows the plaintext m can answer all the three challenges correctly. This implies that $\langle \mathsf{P}(m), \mathsf{V}\rangle(pk, c) = 1$.

Soundness.

Lemma 9.1 *Protocol 1 is sound according to Definition 2.6, if the underlying commitment scheme is binding, the SD problem is hard, and* $r(\kappa) = \omega(\log \kappa)$.

The proof of this lemma follows from the two auxiliary lemmas presented next. In their proofs, we will omit mentioning the fact that the parameters r and ε depend on the security parameter κ, for simplicity.

Lemma 9.2 *If the witness does not exist, then the probability for* $\widetilde{\mathsf{P}}$ *to be accepted in the above protocol is at most* $\left(\frac{2}{3}\right)^r$, *after r rounds.*

Proof We show that if $\widetilde{\mathsf{P}}$'s replies to all the three challenges are accepted, then a (valid) witness can be computed from them. This will contradict the assumption, and imply that $\widetilde{\mathsf{P}}$ is not able to answer all the three challenges at the same time, hence his probability to be accepted is at most $\frac{2}{3}$ in every round.

Consider the following challenge–response pairs:

- $b = 0 : (y_0, \pi_0)$,
- $b = 1 : (w_1, \pi_1)$ (w_1 corresponds to $y + m$),
- $b = 2 : (z_2, t_2)$ (correspond to $y\pi$ and $m\pi$, respectively).

Since, the information in the opened commitments is consistent by assumption, we have: $(\pi_0, Hy_0^T) = Open(C_1) = (\pi_1, Hw_1^T + c)$. Since binding holds, we conclude that $\pi_0 = \pi_1$ and $Hy_0^T = Hw_1^T + c$. Similarly, by consistency of the commitments C_2 and C_3, and by the binding property, we can show that $z_2 = y_0\pi_0, z_2 + t_2 = w_1\pi_1$, and $w_H(t_2) = t$. Therefore, we have that $t_2 = z_2 + (t_2 + z_2) = (y_0 + w_1)\pi_0$ such that $w_H(y_0 + w_1) = t$. Now from $H(y_0 + w_1)^T = Hy_0^T + Hw_1^T = c$, we conclude that $y_0 + w_1$ is a valid witness.

Lemma 9.3 *If* V *accepts* $\widetilde{\mathsf{P}}$*'s proof with probability at least* $(\frac{2}{3})^r + \varepsilon$*, then there exists an expected PPT algorithm* WE *which, with overwhelming probability, computes a witness* m*.*

Proof Let $\mathscr{T}(RA)$ be an execution tree of the protocol $(\widetilde{\mathsf{P}}, \mathsf{V})$, where RA is the random tape of $\widetilde{\mathsf{P}}$. This tree is constructed as follows: A vertex will represent the commitments made by $\widetilde{\mathsf{P}}$, and the edges will be labeled by the challenges of V. An edge will be present only if $\widetilde{\mathsf{P}}$ is able to correctly reply to the challenge. Remember that V can send 3 possible challenges at each stage. First, we will argue that as long as the binding property of the commitment holds, a witness m can be computed from a vertex with 3 descendants, that is from the correct answers to three challenges. Next, we will show that a PPT WE can find such a vertex in $\mathscr{T}(RA)$ with overwhelming probability.

Let v be a vertex with three descendants. This corresponds to a situation, where three commitments C_1, C_2, and C_3 have been made and where the three challenges were correctly answered. Then, the witness can be computed from these correct answers as described in Lemma 9.2.

Next, we can use the argument from [45] to show that the probability for $\mathscr{T}(RA)$ to have a vertex with three descendants is at least ε. We give this argument here for the sake of completeness.

Let us consider the random tape RA of $\widetilde{\mathsf{P}}$ as a set of μ elements, from which $\widetilde{\mathsf{P}}$ randomly picks its values and let $Q = \{1, 2, 3\}$. These two sets are considered as probability spaces, both of them with uniform distribution.

A pair $(a, b) \in (RA \times Q)^r$ represents the commitments, challenges, and responses communicated between $\widetilde{\mathsf{P}}$ and V. This is indeed the case, since the random tape of the prover, along with the challenges, uniquely defines all the messages sent by her during the protocol. A pair (a, b) is called *valid*, if the execution of $(\widetilde{\mathsf{P}}, \mathsf{V})$ is accepted.

Let V be the subset of valid pairs in $(RA \times Q)^r$. By the hypothesis of the lemma,

$$\frac{|V|}{|(RA \times Q)^r|} \geq \left(\frac{2}{3}\right)^r + \varepsilon.$$

Let $\Omega_r \subset RA^r$ be such that:
• If $a \in \Omega_r$, then $2^r + 1 \leq |\{b : (a, b) \text{ are valid}\}| \leq 3^r$,
• If $a \in RA^r \setminus \Omega_r$, then $0 \leq |\{b : (a, b) \text{ are valid}\}| \leq 2^r$.

Then, we write $V = \{\text{valid } (a, b), a \in \Omega_r\} \cup \{\text{valid } (a, b), a \in RA^r \setminus \Omega_r\}$, therefore $|V| \leq |\Omega_r| \cdot 3^r + (\mu^r - |\Omega_r|) \cdot 2^r$. Taking into account that $|RA^r| = \mu^r$ and $|Q^r| = 3^r$, we have

$$\frac{|V|}{|(RA \times Q)^r|} \leq \left(\frac{|\Omega_r|}{|RA^r|} + 2^r\left(3^{-r} - \frac{|\Omega_r|}{|(RA \times Q)^r|}\right)\right) \leq \frac{|\Omega_r|}{|RA^r|} + \left(\frac{2}{3}\right)^r.$$

Now, it follows that $|\Omega_r|/|RA^r| \geq \varepsilon$, which shows that the probability that $\widetilde{\mathsf{P}}$ replies correctly to at least $2^r + 1$ challenges, by choosing random values from RA, is at least ε. Moreover, in this case, $\mathscr{T}(RA)$ has at least $2^r + 1$ leaves. Indeed, by construction of $\mathscr{T}(RA)$, a correctly answered challenge corresponds to an edge, and therefore, the number of leaves is lower bounded by the number of correctly answered challenges. This implies that $\mathscr{T}(RA)$ has at least one vertex with three descendants. Now, the machine WE will simply rewind the above $\widetilde{\mathsf{P}}$ polynomially many times, hereby finding an execution tree containing a vertex with three descendants with overwhelming probability, as claimed. Specifically, we can directly use the analysis by Stern from Lemma 1 in the journal version of [44] to verify that the number of necessary rewindings is $\frac{10}{\varepsilon^3}$. □

Note that the machine WE constructed in the above proof, finds a valid witness, hereby contradicting hardness of the SD problem, unless the binding property of the commitment is violated. Therefore, for a cheating prover $\widetilde{\mathsf{P}}$, we must have $\Pr[\langle\widetilde{\mathsf{P}}, V\rangle(pk, c) = 1] \leq (2/3)^r + \varepsilon$, which is negligible in κ.

Zero-knowledge. Let us denote by $\mathscr{R}$ the communication tape for P and V, that is a concatenation of all bits they exchange during the protocol. We consider the probability distributions on $\mathscr{R}$.

Lemma 9.4 *Protocol 1 is computational (respectively statistical) zero-knowledge according to Definition 5.1, if the underlying commitment scheme is computationally (respectively statistically) hiding.*

Proof We construct a simulator SIM, which generates, in expected PPT, a communication tape $\mathscr{R}_s$, whose distribution is indistinguishable from that of $\mathscr{R}$ in a computational or statistical sense (depending on the type of commitments, which are used).

Suppose that $\widetilde{\mathsf{V}}$ chose a particular strategy depending on the information received from P. Denote this strategy by $St(C_1, C_2, C_3)$.

The simulator SIM works as follows:

1. Pick a challenge $b \stackrel{\$}{\leftarrow} \{0, 1, 2\}$.

    • If $b = 0$, choose $y \stackrel{\$}{\leftarrow} \mathbb{F}_2^n$, $\pi \stackrel{\$}{\leftarrow} \mathscr{S}_n$, compute $C_1 = Com(\pi, H^{pub}y^T)$, $C_2 = Com(y\pi)$, $C_3 = Com(0)$, and $Rep = (y, \pi)$, where by Rep, we denote

the reply of the prover.
Clearly, the distributions of C_1, C_2, C_3, and Rep are identical to those from the communication tape of the actual protocol.

- If $b = 1$, choose $y \stackrel{\$}{\leftarrow} \mathbb{F}_2^n$, $\pi \stackrel{\$}{\leftarrow} \mathscr{S}_n$, and $w = y + z$, where $z \in \mathbb{F}_2^n$ is such that $H^{pub}z^T = c$, $z \neq m$, $w_H(z) \neq t$. Note that such the vector w can be computed in polynomial time as shown in [37, Proposition 1]. Then, compute $C_1 = Com(\pi, H^{pub}y^T), C_2 = Com(0), C_3 = Com(w\pi)$, and $Rep = (w, \pi)$. It is easy to check that the openings of the above commitments and Rep will pass the verification of Step 3 in Protocol 1, and also that distributions of the commitments and Rep are identical to those in the actual protocol. In particular, in the simulation, the distribution of w is uniform over $\mathbb{F}_2^n$, and hence the contents of C_3 has the distribution identical to that in Protocol 1.

- If $b = 2$, choose $y \stackrel{\$}{\leftarrow} \mathbb{F}_2^n$, $\pi \stackrel{\$}{\leftarrow} \mathscr{S}_n$, and $z \stackrel{\$}{\leftarrow} \{x \in \mathbb{F}_2^n | w_H(x) = t\}$. Then, compute $C_1 = Com(0)$, $C_2 = Com(y\pi)$, $C_3 = Com((y + z)\pi)$, and $Rep = (y\pi, z\pi)$. It is again easy to check that the values in Rep will pass the verification of Step 3 in Protocol 1, and that distributions of the commitments and Rep are identical to those in the actual protocol.

2. SIM computes $b' = St(C_1, C_2, C_3)$.
3. If $b = b'$, then SIM writes on the tape $\mathscr{R}_s$ the values H^{pub}, b, Rep, otherwise it goes to Step 1.

Note that in the above simulator, in the case of commitments to zero, we use the hiding property of the commitment to ensure that the distributions in question are identical.

We can see that in $3r$ rounds on the average, SIM produces the communication tape $\mathscr{R}_s$, which is indistinguishable from the communication tape $\mathscr{R}$ produced by the honest parties, who execute r rounds of Protocol 1.

We conclude that $\langle \mathsf{P}(m), \widetilde{\mathsf{V}} \rangle (pk, c)$, and $\langle \mathsf{SIM}, \widetilde{\mathsf{V}} \rangle (pk, c)$ are indistinguishable. Note that the simulation is perfect by itself, and the type of indistinguishability, statistical or computational—and hence the type of the ZK proof, which we obtain—depends solely on the underlying commitment scheme.

Using Lemmas 9.1 and 9.4, and the observation on the completeness, we conclude the proof of Theorem 3.1. □

References

1. C. Aguilar Melchor, P. Cayrel, P. Gaborit, F. Laguillaumie, A new efficient threshold ring signature scheme based on coding theory. IEEE Trans. Inf. Theory **57**(7), 4833–4842 (2011)
2. N. Asokan, V. Shoup, M. Waidner, Optimistic fair exchange of digital signatures (Extended Abstract), in *EUROCRYPT 1998* (1998), pp. 591–606

3. Y. Aumann, M.O. Rabin, A proof of plaintext knowledge protocol and applications. Manuscript. June, 2001. Available as slides from 1998 IACR Distinguished Lecture by M.O. Rabin: http://www.iacr.org/publications/dl/rabin98/rabin98slides.ps
4. A. Becker, A. Joux, A. May, A. Meurer, Decoding random binary linear codes in $2^{n/20}$: how $1+1=0$ improves information set decoding, in *EUROCRYPT 2012* (2012), pp. 520–536
5. M. Bellare, O. Goldreich, On defining proofs of knowledge, in *CRYPTO 1992* (1992), pp. 390–420
6. M. Bellare, M. Fischlin, S. Goldwasser, S. Micali, Identification protocols secure against reset attacks, in *EUROCRYPT 2001* (2001), pp. 495–511
7. R. Bendlin, I. Damgård, Threshold decryption and zero-knowledge proofs for lattice-based cryptosystems, *TCC 2010* (2010), pp. 201–218
8. E. Berlekamp, R. McEliece, H. van Tilborg, On the inherent intractability of certain coding problems. IEEE Trans. Inf. Theory **24**, 384–386 (1978)
9. D.J. Bernstein, T. Lange, C. Peters, Smaller decoding exponents: ball-collision decoding, in *CRYPTO 2011* (2011), pp. 743–760
10. J. Camenisch, I. Damgård, Verifiable encryption, group encryption, and their applications to separable group signatures and signature sharing schemes, in *ASIACRYPT 2000* (2000), pp. 331–345
11. J. Camenisch, V. Shoup, Practical verifiable encryption and decryption of discrete logarithms, *CRYPTO 2003* (2003), pp. 126–144
12. P. Cayrel, P. Véron, S.M. El Yousfi Alaoui, A zero-knowledge identification scheme based on the q-ary syndrome decoding problem, in *Selected Areas in Cryptography 2010* (2010), pp. 171–186
13. T. Cover, Enumerative source encoding. IEEE Trans. Inf. Theory **19**(1), 73–77 (1973)
14. Ö. Dagdelen, D. Galindo, P. Véron, S.M. El Yousfi Alaoui, P. Cayrel, Extended security arguments for signature schemes, in *AFRICACRYPT 2012* (2012), pp. 19–34. Journal version: Ö. Dagdelen, D. Galindo, P. Véron, S.M. El Yousfi Alaoui, P. Cayrel, Extended security arguments for signature schemes. Des. Codes Cryptogr. **78**(2), 441–461 (2016)
15. I. Damgård, O. Goldreich, T. Okamoto, A. Wigderson, Honest verifier vs dishonest verifier in public coin zero-knowledge proofs, in *CRYPTO 1995* (1995), pp. 325–338
16. D. Engelbert, R. Overbeck, A. Schmidt, A summary of McEliece-type cryptosystems and their security. J. Math. Cryptol. **1**, 151–199 (2007)
17. M.F. Ezerman, H.T. Lee, S. Ling, K. Nguyen, H. Wang, A provably secure group signature scheme from code-based assumptions, in *ASIACRYPT (1)* (2015), pp. 260–285
18. J. Faugére, A. Gauthier-Umana, V. Otmani, L. Perret, J. Tillich, A distinguisher for high rate McEliece cryptosystems, in *Information Theory Workshop (ITW)* (2011), pp. 282–286
19. U. Feige, A. Fiat, A. Shamir, Zero knowledge proofs of identity, in *STOC 1987* (1987), pp. 210–217. Journal version: U. Feige, A. Fiat, A. Shamir, Zero-knowledge proofs of identity. J. Cryptol. **1**(2), 77–94 (1988)
20. A. Fiat, A. Shamir, How to prove yourself: practical solutions to identification and signature problems, in *CRYPTO 1986* (1986), pp. 186–194
21. M. Finiasz, N. Sendrier, Security bounds for the design of code-based cryptosystems, in *ASIACRYPT 2009* (2009), pp. 88–105
22. O. Goldreich, *Foundations of Cryptography I: Basic Tools* (Cambridge University Press, Cambridge, 2001)
23. S. Goldwasser, D. Kharchenko, Proof of plaintext knowledge for the Ajtai–Dwork cryptosystem, in *TCC 2005* (2005), pp. 529–555
24. V. Goppa, A new class of linear error-correcting codes (in Russian). Probl. Peredachi Inf. **6**, 24–30 (1970). Russian Academy of Sciences
25. R. Hu, K. Morozov, T. Takagi, On zero-knowledge identification based on q-ary syndrome decoding, in *AsiaJCIS 2013* (2013), pp. 12–18
26. R. Hu, K. Morozov, T. Takagi, Proof of plaintext knowledge for code-based public-key encryption revisited, in *ASIACCS 2013* (ACM, 2013), pp. 535–540. Journal version: R. Hu, K. Morozov, T. Takagi, Zero-knowledge protocols for code-based public-key encryption. IEICE Trans. **98-A**(10), 2139–2151 (2015)

27. A. Jain, S. Krenn, K. Pietrzak, A. Tentes, Commitments and efficient zero-knowledge proofs from learning parity with noise, in *ASIACRYPT 2012*, LNCS, vol. 7658 (2012), pp. 663–680. Full version: A. Jain, S. Krenn, K. Pietrzak, A. Tentes, Commitments and Efficient Zero-Knowledge Proofs from Hard Learning Problems. Cryptology ePrint Archive, Report 2012/513 (2012), http://eprint.iacr.org/2012/513
28. J. Katz, Efficient and non-malleable proofs of plaintext knowledge and applications, in *EUROCRYPT 2003* (2003), pp. 211–228
29. A. Kawachi, K. Tanaka, K. Xagawa, Concurrently secure identification schemes based on the worst-case hardness of lattice problems, in *ASIACRYPT 2008* (2008), pp. 372–389
30. K. Kobara, K. Morozov, R. Overbeck, Coding-based oblivious transfer, in *MMICS 2008* (2008), pp. 142–156
31. F. MacWilliams, N.J.A. Sloane, *The Theory of Error-Correcting Codes* (North-Holland, Amsterdam, 1992)
32. R.J. McEliece, A public-key cryptosystem based on algebraic coding theory, Deep Space Network Progress Report (1978)
33. K. Morozov, Code-based public-key encryption, *A Mathematical Approach to Research Problems of Science and Technology*, Mathematics for Industry, vol. 5 (Springer, Berlin, 2014), pp. 47–55
34. K. Morozov, T. Takagi, Zero-knowledge protocols for the McEliece encryption, in *ACISP 2012* (2012), pp. 180–193
35. H. Niederreiter, Knapsack-type Cryptosystems and algebraic coding theory. Probl. Control Inf. Theory **15**(2), 159–166 (1986). Russian Academy of Sciences
36. R. Nojima, H. Imai, K. Kobara, K. Morozov, Semantic security for the McEliece cryptosystem without random oracles. Design. Codes Cryptogr. **49**(1–3), 289–305 (2008)
37. R. Overbeck, N. Sendrier, Code-based cryptography, in *Post-Quantum Cryptography*, ed. by D.J. Bernstein, J. Buchmann, E. Dahmen (Springer, Berlin, 2009), pp. 95–145
38. J.N. Pierce, Limit distributions of the minimum distance of random linear codes. IEEE Trans. Inf. Theory **13**, 595–599 (1967)
39. Request for Comments on Post-Quantum Cryptography Requirements and Evaluation Criteria: A Notice by the National Institute of Standards and Technology on 08/02/2016, http://csrc.nist.gov/groups/ST/post-quantum-crypto/rfc-july2016.html
40. R. Rivest, A. Shamir, L. Adleman, A method for obtaining digital signatures and public-key cryptosystems. Commun. ACM **21**(2), 120–126 (1978)
41. R. Roth, *Introduction to Coding Theory* (Cambridge University Press, Cambridge, 2006)
42. N. Sendrier, Encoding information into constant weight codewords, in *ISIT'2005* (2005), pp. 435–438
43. M. Stadler, Publicly verifiable secret sharing, in *EUROCRYPT 1996* (1996), pp. 190–199
44. J. Stern, A new identification scheme based on syndrome decoding, in *CRYPTO 1993* (1993), pp. 13–21. Journal version: J. Stern, A new paradigm for public key identification. IEEE Trans. Inf. Theory **42**(6), 1757–1768 (1996)
45. P. Véron, Improved identification schemes based on error-correcting codes. Appl. Algebra Eng. Commun. Comput. **8**(1), 57–69 (1996)
46. K. Xagawa, K. Tanaka, Zero-knowledge protocols for NTRU: application to identification and proof of plaintext knowledge, in *ProvSec 2009* (2009), pp. 198–213
47. K. Xagawa, A. Kawachi, K. Tanaka, Proof of plaintext knowledge for the Regev cryptosystems, Technical report C-236, Tokyo Institute of Technology (2007)

Hash Functions Based on Ramanujan Graphs

Hyungrok Jo

Abstract Cayley hash functions are a family of cryptographic hash functions constructed from Cayley graphs, with appealing properties such as a natural parallelism and a security reduction to a clean, well-defined mathematical problem. As this problem involves non-Abelian groups, it is a priori resistant to quantum period finding algorithms and Cayley hash functions may therefore be a good foundation for post-quantum cryptography. Four particular parameter sets for Cayley hash functions have been proposed in the past, and so far dedicated preimage algorithms have been found for all of them. These algorithms do however not seem to extend to generic parameters, and as a result it is still an open problem to determine the security of Cayley hash functions in general. In this chapter, we introduce how to design hash functions based on Ramanujan graphs, which can be considered as an optimal expander graphs in a sense of qualities of transmission network schemes. We introduce a polynomial time preimage attack against Cayley hash functions based on two explicit Ramanujan graphs. We suggest some possible ways to construct the Cayley hash functions that may not be affected by this type of attacks as open problems, which can contribute to a better understanding of the hard problems underlying the security of Cayley hash functions.

Keywords Expander graphs · Ramanujan graphs · LPS Ramanujan graphs · Cubic Ramanujan graphs · Cayley graphs · Cayley hash functions · Lifting attacks

1 Hash Functions

A hash function is a basic cryptographic scheme which is used ubiquitously in cryptographic applications. A *hash function* is a function that takes as inputs an arbitrarily long string of bits and outputs a bit string of a finite, fixed length. It requires that the hashed values are easy to compute, making both hardware and software

H. Jo (✉)
Graduate School of Mathematics, Kyushu University, 744 Motooka, Nishi-ku, Fukuoka 819-0395, Japan
e-mail: h-jo@math.kyushu-u.ac.jp

T. Takagi et al. (eds.), *Mathematical Modelling for Next-Generation Cryptography*, Mathematics for Industry 29, DOI 10.1007/978-981-10-5065-7_4

implementations practical. Such a function should satisfy with certain properties as collision resistant, second preimage resistant, and preimage resistant.

Let $n \in \mathbb{N}$ and let $\mathscr{H} : \{0, 1\}^* \to \{0, 1\}^n; m \mapsto h = \mathscr{H}(m)$ where $\{0, 1\}^*$ is a set of bit strings of arbitrary length and $\{0, 1\}^n$ is a set of bit strings of fixed length n. The function $\mathscr{H}$ is said to be

- **collision resistant** if it is *computationally infeasible* to find $m, m' \in \{0, 1\}^*, m \neq m'$, such that $\mathscr{H}(m) = \mathscr{H}(m')$;
- **second preimage resistant** if given $m \in \{0, 1\}^*$, it is *computationally infeasible* to find $m' \in \{0, 1\}^*, m \neq m'$, such that $\mathscr{H}(m) = \mathscr{H}(m')$;
- **preimage resistant** if given $h \in \{0, 1\}^n$, it is *computationally infeasible* to find $m \in \{0, 1\}^*$ such that $h = \mathscr{H}(m)$.

Roughly speaking, "*computationally infeasible*" can be considered as with any big cluster of computers cannot perform the task successfully. From a theoretical point of view, it means that no probabilistic algorithm that runs in time polynomial in n can succeed in performing the task for large values of the parameter n with a probability larger than the inverse of some polynomial function of n. We do not give an explanation what "*computationally infeasible*" means, precisely. It is possible to check a precise definition and a rigorous proof of it in [11].

2 Expander Graphs and Ramanujan Graphs

Most of explicit Ramanujan graphs are Cayley graphs. There is an explicit Ramanujan graph constructed by Pizer [18], which is the unique non-Cayley graph so far, based on the theory of supersingular elliptic curve isogeny. Though there are some cryptographic applications [3, 7] with Pizer's Ramanujan graphs, we focus on Ramanujan graphs constructed from Cayley graphs in this chapter.

We briefly introduce the background of expander graphs and Ramanujan graphs. We review how to construct hash functions from expander graphs which described by Charles et al. [4].

Let $X = (V, E)$ be a (undirected) finite graph with a vertex set V and an edge set E. It is said that a graph is *k-regular* if every vertex has k edges incident on it. For a vertex-subset $F \subset V$, the boundary ∂F is the set of all the edges between F and $V \setminus F$. The *expanding constant* of X is

$$h(X) = \min \left\{ \frac{|\partial F|}{|F|} : F \subseteq V \text{ such that } 0 < |F| < \frac{|V|}{2} \right\}. \tag{1}$$

If we view X as a network transmitting information, then $h(X)$ measures the "quality" of X as a network. Roughly, we can say that expander graphs are "sparse" graphs with "strong connectivity".

Definition 2.1 ([6]) Let $(X_m)_{m\geq 1}$ be a family of graphs $X_m = (V_m, E_m)$ indexed by $m \in \mathbb{N}$. Furthermore, fix $k \geq 2$. Such a family $(X_m)_{m\geq 1}$ of finite, connected, k-regular graphs is **a family of expanders** if $|V_m| \to +\infty$ for $m \to +\infty$, and if there exists $\varepsilon > 0$, such that $h(X_m) \geq \varepsilon$ for every $m \geq 1$.

The adjacency matrix of an undirected graph is symmetric, and therefore all its eigenvalues are real. For a finite, connected, k-regular graph X with N vertices, the largest eigenvalue is k, and all others are strictly smaller. Order the eigenvalues as follows:

$$\mu_0 = k > \mu_1 \geq \mu_2 \geq \cdots \geq \mu_{N-1} \geq -k. \tag{2}$$

The expanding constant can be estimated spectrally by means of a double inequality due to Alon and Milman [1] and, to Dodziuk [8] as follows:

$$\frac{k-\mu_1}{2} \leq h(X) \leq \sqrt{2k(k-\mu_1)}. \tag{3}$$

In order to have good quality expanders, the *spectral gap* $k - \mu_1(X_m)$ has to be as large as possible. However, the spectral gap cannot be too large. A theorem of Alon and Boppana [6] says that for a family $(X_m)_{m\geq 1}$ of finite, connected, k-regular graphs, with the number of vertices in the graphs tending to infinity, $\liminf \mu_1(X_m) \geq 2\sqrt{k-1}$. This motivates the definition of a *Ramanujan graph*.

Definition 2.2 ([6]) A finite, connected, k-regular graph X is **Ramanujan** if, for every eigenvalue μ of the adjacency matrix of X other than $\pm k$, one has $|\mu| \leq 2\sqrt{k-1}$. We call $2\sqrt{k-1}$ as a **Ramanujan bound**.

So, if for some $k \geq 3$ we succeed in constructing an infinite family of k-regular Ramanujan graphs, we will get the optimal ones from the spectral point of view. A random walk on an expander graph mixes very quickly. The endpoint of a random walk approximates the uniform distribution after $O(\log N)$ steps on an expander graph with N vertices [13].

3 LPS Ramanujan Graphs and Cubic Ramanujan Graphs

We introduce explicit Ramanujan graphs constructed from Cayley graphs. We give some preliminaries about a Cayley graph and a quaternion algebra.

Suppose that G is a group and S is a generating-set of G, which is symmetric and does not contain the identity. A *Cayley graph* $Cay_{G,S} = (V, E)$ over G with respect to S is a $|S|$-regular graph that constructed from as follows: V contains a vertex v_g associated to each element $g \in G$, and E contains the directed edge (v_{g_i}, v_{g_j}) iff there is some $s \in S$ such that $g_j = g_i s$.

We will describe some preliminaries of general quaternion algebras, which refer to Eichler's lecture note [10].

Let D be a quaternion algebra over $\mathbb{Q}$ defined as follows. It is generated by the basis $[1, \omega, \Omega, \omega\Omega]$ over $\mathbb{Q}$ with the relations:

$$\text{For } s, t \in \mathbb{Q}, \omega^2 = s, \Omega^2 = t \text{ and } \omega\Omega + \Omega\omega = 0.$$

Any element α of D can be represented uniquely as $\alpha = a_0 + a_1\omega + a_2\Omega + a_3\omega\Omega$ for some $a_i \in \mathbb{Q}$. The conjugate of α, denoted by $\overline{\alpha}$, is $\alpha = a_0 - a_1\omega - a_2\Omega - a_3\omega\Omega$. The trace of α is $\mathrm{T}(\alpha) = \alpha + \overline{\alpha} = 2a_0$. The norm of α is $\mathrm{N}(\alpha) = \alpha\overline{\alpha} = {a_0}^2 - s{a_1}^2 - t{a_2}^2 + st{a_3}^2$. The algebra is called definite or indefinite according to whether its norm is a definite or indefinite quadratic form.

A quaternion algebra is said to *split* over a field extension L of $\mathbb{Q}$ if $\mathrm{D} \otimes L$ is isomorphic to the matrix algebra $\mathrm{Mat}(2, L)$. Let p be a prime number and $\mathbb{Q}_p$ be the p-adic completion of $\mathbb{Q}$, then the quaternion algebra $\mathrm{D}_p = \mathrm{D} \otimes \mathbb{Q}_p$ will split over $\mathbb{Q}_p$ (or "D splits at p") for almost all primes p.

An *order* $\mathbb{T}$ in D is a ring of elements of D which is a rank 4 module over the integers consisting of 1 and elements of D whose norms and traces are integers. An order $\mathbb{T}$ in D is defined to be *maximal* if $\mathbb{T}$ is not properly contained in any other order in D.

For a given order $\mathbb{T}$, we can obtain the number of inequivalent right ideals of $\mathbb{T}$, where two right ideals M and N for $\mathbb{T}$ are equivalent if there exists an α in D such that $M = \alpha N$ and $\mathrm{N}(\alpha) \neq 0$. We can similarly obtain the number of inequivalent left ideals of $\mathbb{T}$ and we know it is equal to the number of inequivalent right ideals of $\mathbb{T}$. We consider the number of inequivalent right (or left) ideals of a maximal order of D, and it does not depend on the choice of the maximal order. It is said to be the *class number* of the quaternion algebra D and denoted by h.

3.1 Lubotzky–Phillips–Sarnak Ramanujan Graphs

Lubotzky et al. [15] constructed the explicit Ramanujan graphs based on a Hamiltonian quaternion algebra. We call these graphs as LPS Ramanujan graphs in short.

Let p and q be two distinct primes, p odd, small (≥ 5) and q relatively large ($> p^8$), both p and q are congruent to 1 modulo 4. We remark that those conditions for prime numbers p, q ensure that the desired graphs are connected.

In order to avoid bipartite graphs, we only consider the cases when p is a quadratic residue modulo q. We assume that p is a square modulo q. LPS Ramanujan graph $X^{p,q}$ can be constructed as follows. The vertices of $X^{p,q}$ are the matrices in $G = \mathrm{PSL}(2, \mathbb{Z}/q\mathbb{Z})$, i.e., the invertible 2×2 matrices with entries in $\mathbb{Z}/q\mathbb{Z}$ that have unitary determinant together with the equivalence relation $M = -M$ for any matrix M. A matrix M in G is connected to the matrices Ms where the s's are the following explicitly defined matrices.

From the Jacobi's four-square theorem [6, 12] for the number of representations of an integer as a sum of four squares, there are exactly $8(p+1)$ solutions (a_0, a_1, a_2, a_3) to the equation ${a_0}^2 + {a_1}^2 + {a_2}^2 + {a_3}^2 = p$. Among these, there are exactly $p+1$ solutions with odd $a_0 > 0$ and even a_1, a_2, a_3. To each such (a_0, a_1, a_2, a_3) we associate the matrix

$$s = \begin{pmatrix} a_0 + \mathbf{i}a_1 & a_2 + \mathbf{i}a_3 \\ -a_2 + \mathbf{i}a_3 & a_0 - \mathbf{i}a_1 \end{pmatrix}, \tag{4}$$

where $\mathbf{i}$ is an integer satisfying $\mathbf{i}^2 \equiv -1$ modulo q.

This gives us a generating-set S_{LPS} of $p+1$ matrices in $\mathrm{PGL}(2, \mathbb{Z}/q\mathbb{Z})$. The determinant p of generators is square modulo q (since p is a quadratic residue modulo q), hence they lie in $\mathrm{PSL}(2, \mathbb{Z}/q\mathbb{Z})$ which is the index 2 subgroup of $\mathrm{PGL}(2, \mathbb{Z}/q\mathbb{Z})$. The $X^{p,q} = Cay_{G,S_{LPS}}$ has $q(q^2-1)/2$ vertices and is a $p+1$-regular graph.

There is a correspondence between a quaternion algebra over $\mathbb{Z}/q\mathbb{Z}$ and a 2×2 matrix group over $\mathbb{Z}/q\mathbb{Z}$, i.e., $\mathrm{Mat}(2, \mathbb{Z}/q\mathbb{Z})$.

For $s, t \in \mathbb{Z}/q\mathbb{Z}$, let $H_{s,t}(\mathbb{Z}/q\mathbb{Z})$ be the quaternion algebra over $\mathbb{Z}/q\mathbb{Z}$ defined by

$$H_{s,t}(\mathbb{Z}/q\mathbb{Z}) = \{\alpha = a_0 + a_1\omega + a_2\Omega + a_3\omega\Omega \,|\, a_i \in \mathbb{Z}/q\mathbb{Z}\}.$$

Here $\omega^2 = s$, $\Omega^2 = t$ and $\omega\Omega + \Omega\omega = 0$. If s and $-t$ are a quadratic residue modulo q, that is, $\sqrt{s}, \sqrt{-t} \in \mathbb{Z}/q\mathbb{Z}$, we can explicitly give an isomorphism ψ as follows:

$$\psi : H_{s,t}(\mathbb{Z}/q\mathbb{Z}) \to \mathrm{Mat}(2, \mathbb{Z}/q\mathbb{Z});$$

$$a_0 + a_1\omega + a_2\Omega + a_3\omega\Omega \mapsto \begin{pmatrix} a_0 + a_1\sqrt{s} & \sqrt{-t}(a_2 + a_3\sqrt{s}) \\ -\sqrt{-t}(a_2 - a_3\sqrt{s}) & a_0 - a_1\sqrt{s} \end{pmatrix}.$$

Then if $s = -1, t = -1$, we have the *Hamiltonian quaternion algebra* over $\mathbb{Z}/q\mathbb{Z}$, which is the based quaternion of LPS Ramanujan graphs. There are more arguments to construct LPS Ramanujan graphs based on the Hamiltonian quaternion algebra in [6].

3.2 Cubic Ramanujan Graphs

We will describe the explicit Ramanujan graphs that Chiu [5] constructed in 1992. We consider the quaternion algebra D over $\mathbb{Q}$ given by $s = -2$ and $t = -13$. It is definite with norm $\mathrm{N}(\alpha) = {a_0}^2 + 2{a_1}^2 + 13{a_2}^2 + 26{a_3}^2$. Then we have D over $\mathbb{Q}$ splitting at the prime 2.

We will consider the generating-set $S_{CUB} = \{\omega, \rho, \overline{\rho}\}$ where $\rho = \frac{1}{4}(2 + \omega + \omega\Omega)$ and $\overline{\rho}$ is the conjugate of ρ. Each generator of S_{CUB} is furnished by norm 2. The cubic Ramanujan graph will be realized as the Cayley graphs of a free group generated by S_{CUB}. By Proposition 4.1 in [5], we know that D splits at the prime 2 and has class number 1 for right ideals of a maximal order.

Let $\mathbb{T} = \mathbb{Z}[1, \omega, \Omega, \rho]$, which is a maximal order of D. Let Λ' be the set of $\alpha \in \mathbb{T}$ with $\mathrm{N}(\alpha) = 2^k$ for some $k \geq 1$. Identify α and β if $\pm 2^\nu \alpha = \beta$ for some $\nu \in \mathbb{Z}$, and let Λ be the set of equivalence classes. Thus, Λ consists of elements from a maximal order $\mathbb{T} = \mathbb{Z}[1, \omega, \Omega, \rho]$ of norm 2^k for some $k \geq 1$ (up to equivalence).

We want finite graphs. For each prime $q \neq 2$ or 13, subgroups $\Lambda(q)$ of finite index in Λ is defined as the kernel of the homomorphism

$$\varphi : \Lambda \to T(\mathbb{Z}/q\mathbb{Z})^*/Z(T(\mathbb{Z}/q\mathbb{Z})^*),$$

where $T(\mathbb{Z}/q\mathbb{Z}) = \{\alpha = a_0 + a_1\omega + a_2\Omega + a_3\rho | a_i \in \mathbb{Z}/q\mathbb{Z}\}$, $Z(T(\mathbb{Z}/q\mathbb{Z})^*)$ is the central subgroup of $T(\mathbb{Z}/q\mathbb{Z})^*$, and $[\alpha] \mapsto (\alpha \bmod q)Z(T(\mathbb{Z}/q\mathbb{Z})^*)$, where $[\alpha]$ is the equivalence class of α in Λ.

Let $q \neq 2, 13$ be a prime number such that $\sqrt{-2}$ and $\sqrt{13}$ belong to $\mathbb{Z}/q\mathbb{Z}$ (i.e., $x^2 \equiv -2 \pmod q$ and $x^2 \equiv 13 \pmod q$ are solvable). Then we can identify the basis elements $1, \omega, \Omega, \omega\Omega$ with

$$\begin{pmatrix} 1 & 0 \\ 0 & 1 \end{pmatrix}, \begin{pmatrix} \sqrt{-2} & 0 \\ 0 & -\sqrt{-2} \end{pmatrix}, \begin{pmatrix} 0 & \sqrt{13} \\ -\sqrt{13} & 0 \end{pmatrix}, \begin{pmatrix} 0 & \sqrt{-26} \\ \sqrt{-26} & 0 \end{pmatrix} \tag{5}$$

We identify each element $\omega, \rho, \overline{\rho}$ with

$$\begin{pmatrix} \sqrt{-2} & 0 \\ 0 & -\sqrt{-2} \end{pmatrix}, \frac{1}{4}\begin{pmatrix} 2+\sqrt{-2} & \sqrt{-26} \\ \sqrt{-26} & 2-\sqrt{-2} \end{pmatrix}, \frac{1}{4}\begin{pmatrix} 2-\sqrt{-2} & -\sqrt{-26} \\ -\sqrt{-26} & 2+\sqrt{-2} \end{pmatrix}. \tag{6}$$

We consider the generating-set S_{CUB} for cubic Ramanujan graphs as $\{w, \rho, \overline{\rho}\}$. Similarly as LPS Ramanujan graphs, we assume that 2 is a quadratic residue modulo q and the vertices of $X^{2,q}$ are the matrices in $G = \mathrm{PSL}(2, \mathbb{Z}/q\mathbb{Z})$. Then $X^{2,q} = Cay_{G,S_{CUB}}$ has $q(q^2-1)/2$ vertices and is a 3-regular graph.

4 Cayley Hash Functions

In this section, we construct a hash function over Cayley graphs as described in [4]. We consider a Cayley graph $Cay_{G,S}$ with a given group G and its generating-set S. Let $n := |S| - 1$. In order to avoid trivial collisions, we define a function π which determines the sequence of the elements of the generating-set orderly.

Choose a function

$$\pi : \{0, 1, \ldots, n-1\} \times S \to S \tag{7}$$

such that for any $s \in S$ the set $\pi(\{0, 1, \ldots, n-1\} \times \{s\})$ is equal to $S \setminus \{s^{-1}\}$.

A *Cayley hash function* $\mathscr{H}$ is a function from $\{0, 1\}^*$ to G with respect to π. Let s_0 and g_{SV} be arbitrary fixed elements of S and G respectively (g_{SV}: *the starting vertex*). First, the input message in $\{0, 1\}^*$ is converted to a base-n number $x_1 \ldots x_k$ and the elements $s_i = \pi(x_i, s_{i-1})$ are computed recursively (Table 1).

Table 1 LPS Ramanujan graphs and cubic Ramanujan graphs for constructing Cayley hash functions

	LPS Ramanujan graph $X^{p,q}$ $(p \geq 5)$ [15]	Cubic Ramanujan graph $X^{2,q}$ $(p = 2)$ [5]
Parameters	q: large prime ($\geq p^8$), $p \equiv q \equiv 1 \pmod 4$, $\left(\frac{p}{q}\right) = 1$	$q \neq 2, 13$ be a prime such that $\left(\frac{2}{q}\right) = \left(\frac{-2}{q}\right) = \left(\frac{13}{q}\right) = 1$
Norm	$a_0{}^2 + a_1{}^2 + a_2{}^2 + a_3{}^2 = p$	$a_0{}^2 + 2a_1{}^2 + 13a_2{}^2 + 26a_3{}^2 = 2$
Based group	$\mathrm{PSL}(2, \mathbb{Z}/q\mathbb{Z})$	$\mathrm{PSL}(2, \mathbb{Z}/q\mathbb{Z})$
Generating set	$S_{LPS} = \{s_j\}_{j=0,\ldots,p}$ where $s_j = \begin{pmatrix} a_{0_j} + \mathbf{i}a_{1_j} & a_{2_j} + \mathbf{i}a_{3_j} \\ -a_{2_j} + \mathbf{i}a_{3_j} & a_{0_j} - \mathbf{i}a_{1_j} \end{pmatrix}$, $(a_{0_j}, a_{1_j}, a_{2_j}, a_{3_j})$ satisfying $a_{0_j}{}^2 + a_{1_j}{}^2 + a_{2_j}{}^2 + a_{3_j}{}^2 = p$ ($a_{0_j} > 0$ odd, $a_{1_j}, a_{2_j}, a_{3_j}$ even)	$S_{CUB} = \{\omega, \rho, \overline{\rho}\}$ where $\omega = \begin{pmatrix} \sqrt{-2} & 0 \\ 0 & -\sqrt{-2} \end{pmatrix}$ $\Omega = \begin{pmatrix} 0 & \sqrt{13} \\ \sqrt{-13} & 0 \end{pmatrix}$ $\rho = (2I + \omega + \omega\Omega)/4$
$\|S\|$	$p + 1$	3
Connectedness	Yes	Yes
Based on	Hamiltonian quaternion algebra $(s = -1, t = -1)$, which is not split at 2.	General quaternion algebra $(s = -2, t = -13)$

The hashed value of the input message is the product of group elements

$$\mathscr{H}(x_1 \ldots x_k) = g_{SV} s_1 \ldots s_k. \tag{8}$$

4.1 LPS and Cubic Hash Functions

In Charles et al.'s paper [4], they propose LPS hash functions which is based on LPS Ramanujan graphs $X^{p,q}$. Since the way to construct hash functions based on LPS Ramanujan graphs are similar to the way to construct hash functions based on cubic Ramanujan graphs, we only describe Cayley hash functions based on cubic Ramanujan graphs. We will call these hash functions based on LPS Ramanujan graphs and cubic Ramanujan graphs as *LPS hash functions* and *cubic hash functions*, respectively.

We define a choose function $\pi : \{0, 1\} \times S_{CUB} \to S_{CUB}$ such that for any $s \in S_{CUB}$ the set $\pi(\{0, 1\} \times \{s\})$ is equal to $S_{CUB} \setminus \{s^{-1}\}$. Let s_0 and g_{SV} be arbitrary fixed elements of S_{CUB} and G, respectively. Cubic hash functions receive the converted input message as a base-2 number, which is $x = x_1 \ldots x_k$.

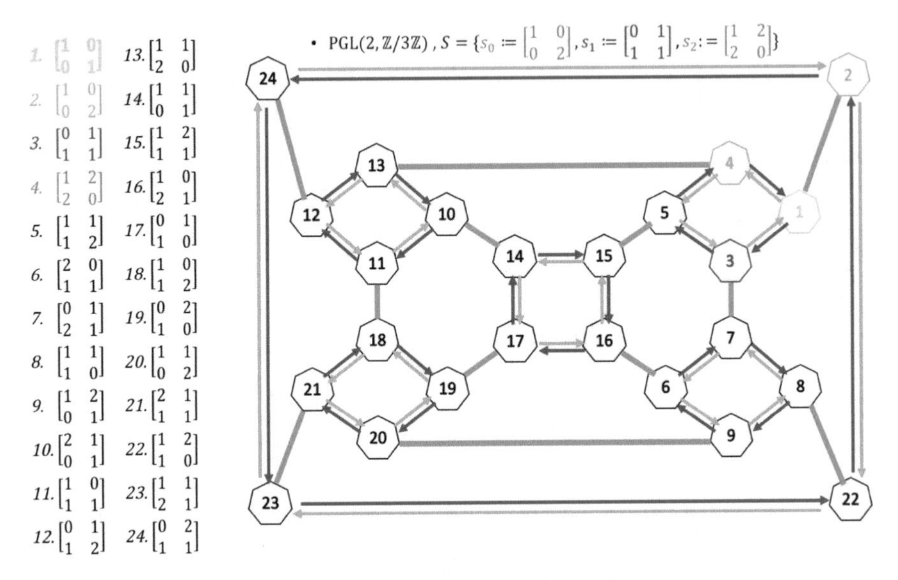

Fig. 1 Cayley graphs over PGL(2, $\mathbb{Z}/3\mathbb{Z}$) with the generating-set S

Starting at g_{SV}, multiply by an element of S_{CUB} determined by the first digit of x to get to the next vertex with choose function π which computes the elements $s_i = \pi(x_i, s_{i-1})$, recursively.

The hashed value of the input message is the product of group elements

$$\mathscr{H}(x_1 \dots x_k) = g_{SV} s_1 \dots s_k. \tag{9}$$

Here are Cayley graphs $Cay_{\mathrm{PGL}(2,\mathbb{Z}/3\mathbb{Z}),S}$ in Fig. 1 and a toy-example of Cayley hash function based on $Cay_{\mathrm{PGL}(2,\mathbb{Z}/3\mathbb{Z}),S}$ in Fig. 2.

$Cay_{\mathrm{PSL}(2,\mathbb{Z}/17\mathbb{Z}),S}$ is the smallest cubic Ramanujan graph over a projective special linear group which satisfies the conditions for a prime q. It has 2448 vertices. Although we know that a Cayley graph over a projective general linear group is a bipartite graph, the number of vertices of $Cay_{\mathrm{PGL}(2,\mathbb{Z}/3\mathbb{Z}),S}$ is only 24 which is smaller than one of $Cay_{\mathrm{PSL}(2,\mathbb{Z}/17\mathbb{Z}),S}$. Since it is easier and more convenient to check and also harmless to show how those work as a toy-example of Cayley hash functions, we will use $Cay_{\mathrm{PGL}(2,\mathbb{Z}/3\mathbb{Z}),S}$ for our toy-example of the Cayley hash function.

All the elements in PGL(2, $\mathbb{Z}/3\mathbb{Z}$) are listed in Fig. 1. We have the input message m as a base-2 number $(10110100)_2$ and the starting vertex is s_2. Along the given function π in Fig. 2, from the leftmost bit of the input message m, we transform each bit to one of elements of the generating-set S.

For example, for the first time, since we have the leftmost bit of m as 1 and the starting vertex s_2, we transform the leftmost bit 1 to s_2 according to $\pi(1, s_2) = s_2$. For the next move, since we have the second bit 0 from the leftmost and the previous transformed bit is s_2, we transform the second bit 0 to s_0 according to $\pi(0, s_0) = s_2$.

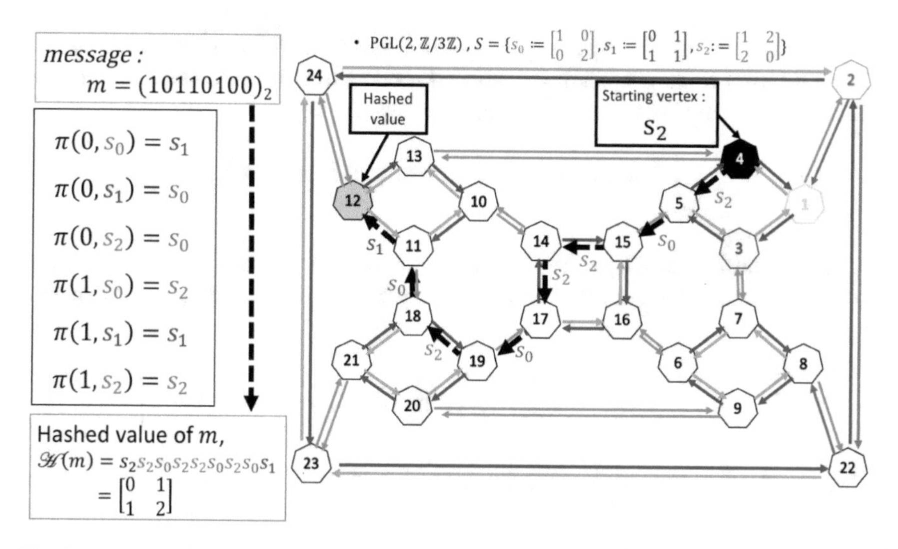

Fig. 2 A toy-example of Cayley hash functions based on $Cay_{\mathrm{PGL}(2,\mathbb{Z}/3\mathbb{Z}),S}$

If we iterate this process until the last bit of the input message m is transformed, we obtain the sequence of elements of S. The hashed value $\mathcal{H}(m)$ of m is the product of $s_2s_2s_0s_2s_2s_0s_2s_0s_1$, that is, $\begin{pmatrix} 0 & 1 \\ 1 & 2 \end{pmatrix} \in \mathrm{PGL}(2, \mathbb{Z}/3\mathbb{Z})$.

In order to use the Ramanujan graphs for cryptographic uses, we need to have a large girth, where the girth of a graph is the length of the smallest cycle in the graph. Finding a collision is the same as finding a cycle in the graphs $X^{p,q}$. In LPS case, Lubotzky et al. [15] argued that the girth is greater than or equal to $2\log_p q$. In cubic case, Chiu [5] described that the girth of $X^{2,q}$ is greater than or equal to $2\log_2 q$ in the same manner as LPS case. In other words, for finding a collision in the graph $X^{2,q}$, we need to find products of at least $2\log_2 q$ generators.

4.2 *Cryptanalysis of LPS Hash Functions*

Lifting attacks are one of the most powerful tools against Cayley hash functions so far. It is also applied to find collisions of LPS hash function by Tillich and Zémor [20]. They lift the elements of $\mathrm{PSL}(2, \mathbb{Z}/q\mathbb{Z})$ to elements in a subset of $\mathrm{Mat}(2, \mathbb{Z}[i])$ where $i^2 = -1$ (cf. $\mathbf{i}^2 = -1 \pmod q$). Remark that $\mathbb{Z}[i]$ is the ring of Gaussian integers.

In this attack, the lifts of the generators do not generate the whole $\mathrm{Mat}(2, \mathbb{Z}[i])$, but only a subset Ξ of very small density. We call this Ξ as a *lifted set*. A lifted set Ξ contains all the elements whose determinants are powers of p.

In a group theory, there is a *representation problem*, which is for a given group G and a subset S thereof, finding a reduced product of subset elements of length at most L that is equal to the unit element of the group, that is

$$\prod_{0 \le i < \ell} s_i = 1 \text{ with } s_i \in S, s_i s_{i+1} \neq 1 \text{ and } \ell \le L.$$

Tillich and Zémor [20] solve the representation problem of LPS hash functions by lifting the identity to a lifted set Ξ which is the relevant set of S_{LPS} with entries in $\mathbb{Z}[i]$.

$$\Xi = \left\{ \begin{pmatrix} a_0 + ia_1 & a_2 + ia_3 \\ -a_2 + ia_3 & a_0 - ia_1 \end{pmatrix} : (a_0, a_1, a_2, a_3) \in E_k, k > 0 \right\}$$

where E_k for some integer $k > 0$ is the set of 4-tuples $(a_0, a_1, a_2, a_3) \in \mathbb{Z}^4$ such that

$$\begin{cases} {a_0}^2 + {a_1}^2 + {a_2}^2 + {a_3}^2 = p^k \\ a_0 > 0, a_0 \equiv 1, a_1 \equiv a_2 \equiv a_3 \equiv 0 \pmod 2. \end{cases}$$

They proceed a lifting attack along three steps. The first step is the most crucial part among them. In this step, we find a matrix M in Ξ which is not of the form $p^r I$ for any r, where I is the identity matrix and such that if the complex entries is replaced by their corresponding values in $\mathbb{Z}/q\mathbb{Z}$ then the matrix of the form λI is obtained. This amounts to finding integers x, y, z satisfying the norm equation ${a_0}^2 + 4q^2(x^2 + y^2 + z^2) = p^k$ with a chosen $a_0 = p^{k'} - 2mq^2$ for some integer m and $k = 2k'$. More details about solving this norm equation can be checked in Tillich and Zémor's paper [20].

In the second step, we find the unique factorizations of M. Any $M \in \Xi$ can be uniquely expressed as a form $M = \pm p^r M_1 \cdots M_e$ where $M_i \in \Sigma$. Here, $\Sigma \subset \Xi$ is the set corresponding to S_{LPS} (and so $|\Sigma| = |S_{LPS}| = p + 1$) [6]. We find the greatest integer r such that p^r divides the 4 entries of M. And we start by finding the rightmost element in the factorization of M by computing all products of the form $\pm M M_i^{-1}$ where $M_i \in \Sigma$. If we iterate this process, we can have the unique factorization of M.

At the final step, if we recover the factorizations of M by modular reduction, the representation problem of LPS hash functions is solved.

Petit et al. [17] follow this guideline for finding a preimage of LPS hash functions. However, it was not directly applicable. The main difference appears in the first step. For a given matrix $M \in \mathrm{PSL}(2, \mathbb{Z}/q\mathbb{Z})$ which has square determinant modulo q. We can write it in the form $M = \begin{pmatrix} A_0 + \mathbf{i}A_1 & A_2 + \mathbf{i}A_3 \\ -A_2 + \mathbf{i}A_3 & A_0 - \mathbf{i}A_1 \end{pmatrix}$.

Lifting the representation problem amounts to finding integers a_0, a_1, a_2, a_3 which are not divisible by p and λ satisfying the following conditions:

$$\begin{cases} (a_0, a_1, a_2, a_3) \in E_k \\ (a_0, a_1, a_2, a_3) \equiv \lambda(A_0, A_1, A_2, A_3) \pmod{q}. \end{cases}$$

We write $a_0 = A_0\lambda + wq$, $a_1 = A_1\lambda + xq$, $a_2 = A_2\lambda + yq$, and $a_3 = A_3\lambda + zq$ with $w, x, y, z \in \mathbb{Z}$. Our norm equation can be represented by

$$(A_0\lambda + wq)^2 + (A_1\lambda + xq)^2 + (A_2\lambda + yq)^2 + (A_3\lambda + zq)^2 = p^k.$$

If at least one of A_1, A_2, A_3 are nonzero, it is at first sight not easy to solve this norm equation. When we consider this norm equation modulo q^2, the term $2q(wA_0 + xA_1 + yA_2 + zA_3)\lambda$ is left. Since those terms do not vanish, the coefficients of degree-2 terms are huge (at least q), and the equation is not easy to solve.

Therefore, they reduce the first step to only four diagonal matrices with decomposing the preimage to four diagonal matrices and up to four graph generators. Since almost same strategies are applied to against cubic hash functions with decomposing the preimage to anti-diagonal matrices and it will be described in Sect. 5, we omit the back of Petit et al.'s. More details about full strategies to find preimages of LPS hash functions can be checked in Petit et al.'s paper [17].

5 Cubic Hash Functions and Their Security

In this section, we propose the specific way to find preimages of cubic hash functions. Suppose we are given a matrix $M \in \mathrm{PSL}(2, \mathbb{Z}/q\mathbb{Z})$ which has square determinant modulo q. We are asked to find a preimage, which is a factorization of it with elements in S_{CUB}. We can write it in the form

$$M = \begin{pmatrix} A_0 + A_1\sqrt{-2} & A_2\sqrt{13} + A_3\sqrt{-26} \\ -A_2\sqrt{13} + A_3\sqrt{-26} & A_0 - A_1\sqrt{-2} \end{pmatrix}. \tag{10}$$

Lifting the representation problem amounts to finding integers A_0, A_1, A_2, A_3 and λ satisfying the following conditions:

Fix $k = \lceil 6 \log_2 q \rceil$. We write $a_0 = A_0\lambda + xq$, $a_1 = A_1\lambda + yq$, $a_2 = A_2\lambda + zq$ and $a_3 = A_3\lambda + wq$ with $x, y, z, w \in \mathbb{Z}$. For convenience, we choose q so that k is even, that is, $k = 2k'$ for some k' an integer.

Then norm equation becomes $(A_0\lambda + xq)^2 + 2(A_1\lambda + yq)^2 + 13(A_2\lambda + zq)^2 + 26(A_3\lambda + wq)^2 = 2^{2k'}$. In analogy to [17], we find preimages for anti-diagonal matrices with A_2 and/or A_3 nonzero, and then we write any matrix as a product of four anti-diagonal matrices and up to three generators. Altogether this leads to an efficient probabilistic algorithm that finds preimages of cubic hash function.

5.1 Preimages for Anti-diagonal Matrices

Now we show how to find a factorization of a matrix

$$M = \begin{pmatrix} & A_2\sqrt{13} + A_3\sqrt{-26} \\ -A_2\sqrt{13} + A_3\sqrt{-26} & \end{pmatrix}$$

such that $13{A_2}^2 + 26{A_3}^2$ is a square modulo q, where one of A_2 and A_3 is nonzero. We write $a_0 = xq$, $a_1 = yq$, $a_2 = A_2\lambda + zq$ and $a_3 = A_3\lambda + wq$.

We need to find integer solutions to

$$(xq)^2 + 2(yq)^2 + 13(A_2\lambda + zq)^2 + 26(A_3\lambda + wq)^2 = 2^{2k'}. \tag{11}$$

By modulo q on (11), we have $13\lambda^2({A_2}^2 + 2{A_3}^2) \equiv 2^{2k'} (\text{mod } q)$. Now we fix λ a square roots of $2^{2k'}({A_2}^2 + 2{A_3}^2)^{-1}13^{-1} \pmod q$.

Let $m := (2^{2k'} - 13({A_2}^2 + 2{A_3}^2)\lambda^2)/q$. By modulo q^2 on (11), we have

$$13({A_2}^2\lambda^2 + 2A_2\lambda zq) + 26({A_3}^2\lambda^2 + 2A_3\lambda wq) \equiv 2^{2k'} \pmod{q^2}.$$

From this equation above, we randomly choose z and w. We can deform the equation from (11) as

$$x^2 + 2y^2 = N \text{ where } N := (2^{2k'} - 13((A_2\lambda + zq)^2 + 26(A_3\lambda + wq)^2))/q^2.$$

Pick a random $w \in \{0, 1, \ldots, q-1\}$ and we assume that A_2 is nonzero then compute $z = \frac{m}{26A_2\lambda} - \frac{2A_3w}{A_2} \pmod q$ until getting w and z, which guarantee that $N > 0$.

If 2 and N are not coprime, we proceed to the whole algorithm from the beginning again until 2 and N are coprime.

Finally, we use the Cornacchia's Algorithm [2] to find the solutions x, y on $x^2 + 2y^2 = N$. We choose u, v in $\{1, \ldots, q-1\}$ which satisfy $\mathrm{u}^2 \equiv -2 \pmod q$ and $\mathrm{v}^2 \equiv 13 \pmod q$, respectively. Then we get the lifted matrix

$$\overline{M} = \begin{pmatrix} xq + yq\mathrm{u} & (A_2\lambda + zq)\mathrm{v} + (A_3\lambda + wq)\mathrm{uv} \\ -(A_2\lambda + zq)\mathrm{v} + (A_3\lambda + wq)\mathrm{uv} & xq - yq\mathrm{u} \end{pmatrix} \in \mathrm{Mat}(2, \mathbb{Z}).$$

The second step is to factorize the lifted element $\overline{M}$ into products of lifted elements of generators ω, ρ and $\overline{\rho}$. We try to multiply each lifted element of generators on the right side of the lifted element. If it turns out all elements in a matrix are divisible by 2, we put the inverse of that generator as a bit for products of its factorization. We iterate this procedure, recursively.

At last, we apply modular reduction to each element of the factorization of $\overline{M}$.

5.2 Cornacchia's Algorithm

In 1908, Cornacchia [2] gave an algorithm for solving the diophantine equation $x^2 + dy^2 = N$ where $1 \le d < N$, d and N are coprime, N may not be prime. The algorithm is briefly described as follows:

1. Put $r_0 = N$ and $r_1{}^2 \equiv -d \pmod{N}, 0 \le r_1 \le (N/2)$.
   (If no such r_1 exist, there can be no primitive solution to the original equation.)
2. Using the Euclidean algorithm, compute $r_{i+2}{}^2 \equiv r_i \pmod{r_{i+1}}$ recursively until we arrive at $r_k{}^2 < N$.
3. If $(N - r_k{}^2/d)$ is a square integer, say s^2, we get the solution (r_k, s).

Assuming N behaves "as a random number" then we will need $O(\log d \log N)$ trials before getting one N which can be applied to the Cornacchia's algorithm.

The securities of LPS and cubic hash functions depend on how to find the specific solutions of their norm equations. In other words, the Cornacchia's algorithm (the Euclidean algorithm) is the essential procedure, which is directly related to their securities.

5.3 Reduction to the Anti-diagonal Case

Now we show how to decompose any matrix $M \in \mathrm{PSL}(2, \mathbb{Z}/q\mathbb{Z})$ into a product of anti-diagonal matrices and generators. We may assume that all the entries of M are nonzero. We reduce the number of variables in each anti-diagonal matrix by multiplying one component's inverse of this matrix. We will show how to find $(\lambda, \beta_1, \beta_2, \gamma_1, \gamma_2)$ with the last four being squares, such that

$$\begin{pmatrix} M_1 & M_2 \\ M_3 & M_4 \end{pmatrix} = \lambda \begin{pmatrix} 0 & 1 \\ \beta_1 & 0 \end{pmatrix} \begin{pmatrix} f_1 & f_2 \\ f_3 & f_4 \end{pmatrix} \begin{pmatrix} 0 & 1 \\ \beta_2 & 0 \end{pmatrix} = \lambda \begin{pmatrix} \beta_2 f_4 & f_3 \\ \beta_1\beta_2 f_2 & \beta_1 f_1 \end{pmatrix} \tag{12}$$

and

$$\begin{pmatrix} f_1 & f_2 \\ f_3 & f_4 \end{pmatrix} = \rho \begin{pmatrix} 0 & 1 \\ \gamma_1 & 0 \end{pmatrix} \rho \begin{pmatrix} 0 & 1 \\ \gamma_2 & 0 \end{pmatrix} \rho = \frac{1}{32} \begin{pmatrix} f_1' & f_2' \\ f_3' & f_4' \end{pmatrix}$$

where $\rho = \frac{1}{4}\begin{pmatrix} 2+\sqrt{-2} & \sqrt{-26} \\ \sqrt{-26} & 2-\sqrt{-2} \end{pmatrix}$ and

$$\begin{cases} f_1' = (-26 - 13\sqrt{-2})\gamma_2\gamma_1 + (6 + 3\sqrt{-2})\gamma_2 + (-26 - 13\sqrt{-2})\gamma_1 - 26 - 13\sqrt{-2} \\ f_2' = (-13\gamma_2\gamma_1 + 3\gamma_2 + 3\gamma_1 + 3)\sqrt{-26} \\ f_3' = (3\gamma_2\gamma_2 + 3\gamma_2 + 3\gamma_1 - 13)\sqrt{-26} \\ f_4' = (-26 + 13\sqrt{-2})\gamma_2\gamma_1 + (-26 + 13\sqrt{-2})\gamma_2 + (6 - 3\sqrt{-2})\gamma_1 - 26 + 13\sqrt{-2}. \end{cases}$$

We can see the matrix Eq. (12) is equivalent to the following system:

$$\begin{cases} M_1M_4f_2f_3 - M_2M_3f_1f_4 = 0, & \beta_1M_2f_1 - M_4f_3 = 0 \\ \beta_2M_4f_2 - M_3f_1 = 0, & \lambda f_3 - M_2 = 0 \end{cases}$$

We concentrate on solving the first equation, which is quadratic in γ_1, γ_2:

$$\begin{aligned} M_1M_4f_2f_3 - M_2M_3f_1f_4 = & -3(M_1M_4 - M_2M_3)(13{\gamma_1}^2 + 10\gamma_1 - 3){\gamma_2}^2 \\ & +2((-15{\gamma_1}^2 + 78\gamma_1 - 15)M_1M_4 \\ & +15({\gamma_1}^2 + 10\gamma_1 + 1)M_2M_3)\gamma_2 \\ & +3(M_1M_4 - M_2M_3)(3{\gamma_1}^2 - 10\gamma_1 - 13). \end{aligned}$$

Finally, finding a preimage algorithm proceeds as follows

1. Choose a random γ_1 which is a square.
2. Compute the discriminant of the quadratic equation in γ_2. If it is not a square, go back to 1.
3. Solve the quadratic equation. If none of the roots is a square, go back to 1. Otherwise we assign one of the roots, which is a square, to γ_2.
4. Solve $\beta_1M_2f_1 - M_4f_3 = 0$ to get β_1. If β_1 is not a square, go back to 1.
5. Solve $\beta_2M_4f_2 - M_3f_1 = 0$ to get β_2. If β_2 is not a square, go back to 1.

This concludes the exposition of our algorithm.

6 Open Problems from Further Extensions of Lifting Attacks

The main part of preimage attacks is to apply the Cornacchia's algorithms to their norm equations obtained by the underlying quaternion algebra as mentioned. In Table 2, we describe each norm equation of Ramanujan graphs and constant term N of the deformed norm equations for using the Cornacchia's algorithm.

In the case of LPS hash functions in Sect. 4, LPS Ramanujan graph $X^{p,q}$ has norm equations ${a_0}^2 + {a_1}^2 + {a_2}^2 + {a_3}^2 = p^k$. Since $s = -1, t = -1$, their norm equation are just sums of four squares as above. Roughly, we can say that finding preimage of LPS hash functions requires finding the integer solution (a_0, a_1) of ${a_0}^2 + {a_1}^2 = N$ where $N := p^k - {a_2}^2 - {a_3}^2$ with random variables a_2 and a_3.

In the case of cubic hash functions in Sect. 5, we know that $s = -2, t = -13$ for cubic Ramanujan graphs $X^{2,q}$. Thus its norm equation is ${a_0}^2 + 2{a_1}^2 + 13{a_2}^2 + 26{a_3}^2 = 2^k$. For this case, we applied the Cornacchia's algorithm to solve the deformed norm equation ${a_0}^2 + 2{a_1}^2 = N$ where $N := 2^k - 13{a_2}^2 - 26{a_3}^2$ with random variables a_2 and a_3. However, in either of cases, since $|s|$ and $|t|$ are too small to interrupt the Cornacchia's algorithm, LPS and cubic hash functions are easily

Table 2 Norm equations and the Cornacchia's algorithm used for Ramanujan graphs

Ramanujan graphs	Norm equation and N for the Cornacchia's algorithm
LPS Ramanujan graph $X^{p,q}$ ($p \geq 5$) [15] (Preimage attack in [17])	$a_0{}^2 + a_1{}^2 + a_2{}^2 + a_3{}^2 = p^k$ $N := p^k - a_2{}^2 - a_3{}^2$
Cubic Ramanujan graph $X^{2,q}$ ($p = 2$) [5] (Proposed preimage attack)	$a_0{}^2 + 2a_1{}^2 + 13a_2{}^2 + 26a_3{}^2 = 2^k$ $N := 2^k - 13a_2{}^2 - 26a_3{}^2$
Other Ramanujan graphs (Possible attacks)	$a_0{}^2 - sa_1{}^2 - ta_2{}^2 + sta_3{}^2 = p^k$ (splitting at p) $N := p^k + ta_2{}^2 - sta_3{}^2$ (class number condition)

cryptanalyzed. Thus, we need other secure classes that $|s|$ and $|t|$ are both big to resist the Cornacchia's algorithm.

In a similar manner of Chiu's work [5], we can expect to construct collision resistant or preimage resistant Cayley hash functions based on the other Ramanujan graphs which has the norm equation $a_0{}^2 - sa_1{}^2 - ta_2{}^2 + sta_3{}^2 = p^k$. There are many possibilities to construct other Ramanujan graphs with larger class numbers than Chiu's cases or with a constant term $N := p^k + ta_2{}^2 - sta_3{}^2$ of the Cornacchia's algorithm under some special properties of the prime p. In the following, we argue in two possible ways to find the other Ramanujan graphs.

Splitting at the Prime p The structure of LPS Ramanujan graphs is based on the Hamiltonian quaternion algebra, which is not split at the prime 2. For $p = 2$ case, Chiu [5] construct cubic Ramanujan graphs, and the desired algebra for constructing those Ramanujan graphs is the definite quaternion algebra splitting at the prime 2 and having a maximal order of class number 1. It also seems possible to argue with the case the prime $p \neq 2$ with class number 1 in a way of Chiu's constructions. First of all, we should find the proper s and t which create the definite quaternion algebra splitting at the prime p and having class number 1 for right ideals of a maximal order.

Theorem 6.1 ([5]) *Let* D *be a definite quaternion algebra over* $\mathbb{Q}$ *of class number* 1 *which splits at* p*, and let* $\mathbb{T}$ *be a maximal order of* D*. Then there exist* $1 + p + \cdots + p^k$ *elements, unique up to units, of norm* p^k*.*

Then we look for a maximal order $\mathbb{T}$ containing the $p + 1$ elements (up to units) of norm p furnished by Theorem 6.1. From this part, we construct the desired Cayley graphs in a way of Chiu's cubic Ramanujan graphs [5]. Actually, the norm equation of Ramanujan graphs is determined by the structure of its maximal order. This can be much more complicate to find the integral solutions.

Maximal Order with Class Number $h \geq 1$ In Chiu's paper [5], he refereed Theorem 6.2. from Eichler's lecture note [9] as below.

Theorem 6.2 ([9]) *Let* D *be a definite quaternion algebra over* $\mathbb{Q}$, $\mathbb{T}$ *a maximal order of* D. *Then the class number of right (or left) ideals of* $\mathbb{T}$ *is*

$$h = \frac{1}{12}\prod_{p|\delta}(p-1) + \frac{1}{4}\prod_{p|\delta}\left(1-\left(\frac{-4}{p}\right)\right) + \frac{1}{3}\prod_{p|\delta}\left(1-\left(\frac{-3}{p}\right)\right),$$

where δ *is the product of all the primes* p *for which* D *does not split at* p, *and* $\left(\frac{n}{p}\right)$ *is Kronecker's extension of the Legendre symbol.*

From Theorem 6.2, we can estimate the class number of a maximal order in a given quaternion algebra. This shows that $h = 1 \iff \delta = 2, 3, 5, 7, 13$ and $h = 2 \iff \delta = 11, 17, 19, 30, 42, 70, 78$. From Ibukiyama's work [14], it gives an explicit form of a maximal order of the quaternion algebra over $\mathbb{Q}$ which does not split at given primes. It means it is possible to find the definite quaternion algebra over $\mathbb{Q}$ with a given maximal order of class number h related to a given δ.

It does not exist as a refined way to build these Ramanujan graphs with two properties above so far. Since it is hard to guess how secure those hash functions based on those Ramanujan graphs are, it is worth investigating. As open problems, we suggest that there are potential ways to find new other Ramanujan graphs splitting at the prime p with class number h greater than or equal to 1. Historically, since these schemes are based on noncommutative algebraic structure, it is also meaningful in the era of post-quantum cryptography.

References

1. N. Alon, V. Milman, λ_1, isoperimetric inequalities for graphs, and superconcentrators. J. Comb. Theory B **38**(1), 73–88 (1985)
2. J. Basilla, On the solution of $x^2 + dy^2 = m$. Proc. Jpn. Acad. A Math. **80**(5), 40–41 (2004)
3. J.F. Biasse, D. Jao, A. Sankar, A quantum algorithm for computing isogenies between supersingular elliptic curves, in *INDOCRYPT*, LNCS, vol. 8885 (2014), pp. 428–442
4. D. Charles, K. Lauter, E. Goren, Cryptographic hash functions from expander graphs. J. Cryptol. **22**(1), 93–113 (2009)
5. P. Chiu, Cubic Ramanujan graphs. Combinatorica **12**(3), 275–285 (1992)
6. G. Davidoff, P. Sarnak, A. Valette, *Elementary Number Theory, Group Theory and Ramanujan Graphs* (Cambridge University Press, Cambridge, 2003)
7. L. De Feo, D. Jao, J. Plût, Towards quantum-resistant cryptosystems from supersingular elliptic curve isogenies. J. Math. Cryptol. **8**(3), 209–247 (2014)
8. J. Dodziuk, Difference equations, isoperimetric inequality and transience of certain random walks. Trans. Am. Math. Soc. **284**(2), 787–794 (1984)
9. M. Eichler, The basis problem for modular forms and the traces of the Hecke operators, in *Modular Functions of One Variable*, vol. 320, ed. by W. Kuyk (Springer, Heidelberg, 1973), pp. 75–152
10. M. Eichler, S. Sundaravaradan, Lectures on modular correspondences. Tata Institute of Fundamental Research (1956), http://www.math.tifr.res.in/~publ/ln/tifr09.pdf
11. O. Goldreich, *Foundations of Cryptography* (Cambridge University Press, Cambridge, 2004)

12. M. Hirschhorn, A simple proof of Jacobi's four-square theorem. Proc. Am. Math. Soc. **101**(3), 436–438 (1987)
13. H. Hoory, N. Linial, A. Wigderson, Expander graphs and their applications. Bull. Am. Math. Soc. **43**(4), 439–561 (2006)
14. T. Ibukiyama, On maximal orders of division quaternion algebras over the rational number field with certain optimal embeddings. Nagoya. Math. J. **88**, 181–195 (1982)
15. A. Lubotzky, R. Phillips, P. Sarnak, Ramanujan graphs. Combinatorica **8**(3), 261–277 (1988)
16. G. Margulis, Explicit group-theoretical constructions of combinatorial schemes and their application to the design of expanders and concentrators. Probl. Peredachi Inf. **24**(1), 51–60 (1988)
17. C. Petit, K. Lauter, J. Quisquater, Full cryptanalysis of LPS and Morgenstern hash functions, in *SCN*, LNCS, vol. 5229 (2008), pp. 263–277
18. A.K. Pizer, Ramanujan graphs and Hecke operators. Bull. Am. Math. Soc. **23**(1), 127–137 (1990)
19. P. Sarnak, *Some Applications of Modular Forms* (Cambridge University Press, Cambridge, 1999)
20. J. Tillich, G. Zèmor, Collisions for the LPS expander graph hash function, in *EUROCRYPT*, LNCS, vol. 3027 (2008), pp. 254–269

Pairings on Hyperelliptic Curves with Considering Recent Progress on the NFS Algorithms

Masahiro Ishii

Abstract In this paper, we analyze and reexamine the key lengths of the pairings on the hyperelliptic curves of genus 2 and considering the estimated run time of the (special) extended tower number field sieve. Pairing-based cryptosystems have become a major research topic in cryptography and have attracted more attention because of the increasing interest in the efficient and functional cryptographic protocols, e.g., functional encryption. Recently, the algorithm of number field sieve and its variants have made progress, and it is urgently necessary to estimate key lengths of pairings taking into account of impact of the algorithms. We report the detailed computational cost of the pairings on the Kawazoe–Takahashi curves of genus 2, and give the comparison of our pairing and the pairing on the BLS24 elliptic curves at the 192-bit security level. The estimated cost of our pairing is approximately 2.5 times more than the cost of the BLS24 pairing.

Keywords Twisted ate pairing · Kawazoe–Takahashi curves · Key length · Security levels · Extended tower number field sieve · Hyperelliptic curves · Jacobians · Discrete logarithms in finite fields

1 Introduction

The recent progress on Number Field Sieve (NFS) and its variants have a direct effect on pairing-based cryptosystems since pairings are defined over finite fields $\mathbb{F}_{p^k}$ in which the NFS is the best-known method to solve discrete logarithms. Especially, the extended tower number field sieve (exTNFS) and the special exTNFS (SexTNFS) algorithms can be applied to the security parameters given by pairing-friendly curves which define the state-of-the-art elliptic curve pairings. The key lengths of any previous pairings in the literature were set without considering the run time of exTNFS and

M. Ishii (✉)
Department of Mathematical and Computing Sciences, Tokyo Institute of Technology, 2-12-1 Ookayama, Meguro-ku, Tokyo 152-8550, Japan
e-mail: mishii@c.titech.ac.jp

T. Takagi et al. (eds.), *Mathematical Modelling for Next-Generation Cryptography*, Mathematics for Industry 29, DOI 10.1007/978-981-10-5065-7_5

SexTNFS. Therefore re examination and refresh of adequate key lengths of pairings should be done immediately. However, it is hard to determine reliable key lengths from the asymptotic complexity formula of the NFS algorithms without enough tests to solve DLPs in finite fields by these algorithms.

Menezes et al. [24] analyzed the run time of NFS variants more concretely, and reported the adjusted parameters for the pairings on the elliptic curves. They offered a conservative choice to set the key lengths of pairings with the experimental results for the run time of the relation collection stage of exTNFS and SexTNFS.

In the author's PhD dissertation [17], the author comprehensively explored pairings on hyperelliptic curves of genus 2 at the 192- and 256-bit security levels and reported that the twisted Ate pairings on the Kawazoe–Takahashi curves [21] are the optimum choice and difference in the performance of a pairing between the elliptic and genus 2 cases.

In this paper, we re examine the key lengths of the pairings on Kawazoe–Takahashi curves by following the analyzing method by [24]. We also show the computational costs of the pairings which are slightly less than those in [17].

The remainder of this paper is organized as follows. Section 2 outlines the security of pairings with previous works about the NFS. We explain the selection of appropriate curves of genus 2 for effective pairings in Sect. 3. In Sect. 4 we report the adequate sizes of p for Kawazoe–Takahashi curves for each security level and estimated run time of the exTNFS and SexTNFS. We then construct the pairings with the explicit security parameters, and report the detailed computational costs in Sect. 5. Finally, we conclude the paper in Sect. 6.

2 The Security of Pairings

Let C be a hyperelliptic curve defined over $\mathbb{F}_q$ and let $\mathrm{Jac}_C(\simeq \mathrm{Pic}_C^0)$ denote the Jacobian of C. For the r-torsion subgroup of Jacobian variety $\mathrm{Jac}_C(\mathbb{F}_{q^k})[r]$ and the embedding degree k which is the smallest integer satisfying that $r|(q^k-1)$, several pairings $e\colon \mathbb{G}_1 \times \mathbb{G}_2 \to \mathbb{G}_T$ have been exploited by many researchers. More precisely, the subgroups $\mathbb{G}_1$ and $\mathbb{G}_2$ of $\mathrm{Jac}_C(\mathbb{F}_{q^k})[r]$ given by

$$\mathbb{G}_1 := \mathrm{Jac}_C(\mathbb{F}_{q^k})[r] \cap \ker(\pi - [1]),$$
$$\mathbb{G}_2 := \mathrm{Jac}_C(\mathbb{F}_{q^k})[r] \cap \ker(\pi - [q])$$

are used to construct various pairings where π denotes the q-th power Frobenius map. Then pairing values lie in $\mathbb{G}_T = \mathbb{F}_{q^k}$.

The security level of pairings are determined by setting r and q^k adequately against the complexity of the solving the ECDLP and DLP in the finite fields $\mathbb{F}_{p^k}$. Today, the bit-lengths of r and q^k should be set to at least 384 and 7936, respectively [8, NIST and ECRYPT II Recommendations], to construct a pairing at the 192-bit security level. The Table 1 shows key sizes, that are size of subgroups r and size of extension fields

Table 1 Recommended key sizes for pairings at desired security levels following [8, NIST and ECRYPT II Recommendations]

Security level	Subgroup size (r)	Extension field size (q^k)	Embedding degree (k)			
			$\rho \approx 1$	$\rho \approx 2$	$\rho \approx 3$	$\rho \approx 4$
128	256	3248	$12.7g$	$6.3g$	$4.2g$	$3.2g$
192	384	7936	$20.6g$	$10.3g$	$6.9g$	$5.2g$
256	512	15424	$30.1g$	$15.0g$	$10.0g$	$7.5g$

q^k, and embedding degrees with various ρ-values $\rho = (g \log q)/\log r$ for desired security levels. Note that genus g in the table should be 1 or 2.

The recent progress on NFS is remarkable and that heavily affect the key lengths of $\mathbb{F}_{q^k}$ for the pairings. The major theoretical and practical breakthrough in the computation of discrete logarithms in finite fields of small characteristics and also other fields was made [3, 5]. As a result, type 1 [14] (symmetric, i.e., $\mathbb{G}_1 = \mathbb{G}_2$) pairings defined on supersingular curves over finite fields of small characteristics have been vulnerable. Some type 1 pairings, however, still exist and are obtained by using supersingular curves over large characteristic fields in the elliptic case [28, 30] and in the genus 2 case [13] that are unsuitable for achieving high security levels because of their small embedding degrees.

Recently, Kim and Barbulescu [22] provided a new variant of the Number Field Sieve algorithm which is known as exTNFS (extended tower number field sieve) for discrete logarithms in $\mathbb{F}_{p^n}$ and improved the asymptotic complexity. Their proposal method improved the polynomial selection by combining the TNFS (tower number field sieve) [6] construction with the conjugation method [4]. Furthermore, the methods SNFS [19] and SexTNFS [18, 22] for the characteristic p which has a special form affect the key length of pairings on families of pairing-friendly curves. We refer the reader to [24, Sects. 3 and 4] for the detailed overview of the NFS and its variants.

For the usual L_Q-notation, we write

$$L_Q(\ell, c) = \exp\left((c + o(1))(\log Q)^{\ell}(\log\log Q)^{1-\ell}\right)$$

where $\mathbb{F}_Q = \mathbb{F}_{p^n}$. The asymptotic run time of the NFS variants are reported in the beginning of [24, Sect. 5]. The current key sizes by ECRYPT II Recommendations can be estimated by the asymptotic complexity given by $A \cdot L_Q(1/3, (64/9)^{1/3})$ where $o(1) = 0$ and $A = 2^{-14}$ as described in [24, Sect. 6.2]. There is a great gap between the asymptotic run time and practical run time of the NFS algorithms. If we use directly a complexity formula $A \cdot L_Q(1/3, (32/9)^{1/3})$ where $A = 2^{-14}$ for the pairing-friendly curves, the resulting value is much larger than the run time with more concrete analyses by [24]. Those differences are shown in Sect. 4.

3 Pairing-Friendly Hyperelliptic Curves of Genus 2

Here we show that the Kawazoe–Takahashi curves are suitable to construct a genus 2 pairing at the high security levels.

3.1 The Selection of Pairing-Friendly Families

Similar to the elliptic case [11], we should select the *pairing-friendly* curves so that we can apply several techniques to speed up pairing computations. The definitions of pairing-friendly can be found in some literature [11, Definition 2.3], [2, Sect. 3.1]. In general, the ρ-value of pairing-friendly curves should be small and close to 1 which is the theoretical lower bound.

Higher Degree Twists

To reduce the costs of arithmetic in the extension fields, we first consider the twists of C. The twists are another curves which are isomorphic to C over an algebraic closure $\overline{K}$ where C is defined over the field K. If C has a twist of higher degree d, we can perform computations for a divisor $D_2 \in \mathbb{G}_2$ in $\mathbb{F}_{q^{k/d}}$. In the genus 2 case, we can obtain a twist of degree at most 10, where $\mathrm{char}(q) > 5$ [29, Sect. 3]. For $d = 10$, the curve $C\colon y^2 = x^5 + a$ over $\mathbb{F}_q$, where $q = 1 \pmod 5$ has a twist of degree 10, that is $C_t\colon y^2 = x^5 + ac$ $(c \in \mathbb{F}_q)$, with the map

$$\psi\colon \mathrm{Jac}_{C_t} \to \mathrm{Jac}_C;\ (x, y) \mapsto (c^{-\frac{1}{5}}x, c^{-\frac{1}{2}}y).$$

The form of ψ, however, does not permit the use of the denominator elimination method to compute the pairing. Furthermore, we do not have an explicit method to generate a family of such curves which has a lower ρ-value. Hence, we should consider other curves.

Then it is natural to take a curve with degree 8. The methods by Kawazoe and Takahashi [21] provide families of curves

$$C\colon y^2 = x^5 + ax, \tag{1}$$

and these curves have a twist

$$C_t\colon y^2 = x^5 + acx\ (c \in \mathbb{F}_q) \tag{2}$$

if $q = 1 \pmod 8$ and we have that $\psi\colon \mathrm{Jac}_{C_t} \to \mathrm{Jac}_C;\ (x, y) \mapsto (c^{-\frac{1}{4}}x, c^{-\frac{1}{8}}y)$.

In this case, we can use the denominator elimination method. In addition, we can perform arithmetic in the divisor class group by using the explicit formula and the dedicated coordinate system [10].

We can adopt the following strategies for reducing the computational cost of pairing on the Kawazoe–Takahashi curves.

Field Constructions

Instead of *pairing-friendly fields* $\mathbb{F}_{p^k}$ [23] where k is of the form $2^i 3^j$ and $p = 1 \pmod{12}$, Benger and Scott proposed a *towering-friendly field* $\mathbb{F}_{q^m}$ [7, Definition 2] where q is a prime power, for which all prime divisors of m also divide $q - 1$. Then a towering-friendly field $\mathbb{F}_{q^m}$ is the tower of subextensions which can be constructed by binomials. We can therefore perform extension field arithmetic in pairing computations effectively by using a towering-friendly field.

Indeed, we take the curves whose embedding degree k is 16 and 24 for the 192- and 256-bit security levels, respectively.

Squaring in Cyclotomic Subgroups

Granger and Scott [16] extended the methods to perform a squaring in the extension field using the cyclotomic subgroups offered by [15, 27]. They showed that a squaring in the order $\Phi_6(p)$ cyclotomic subgroup of $\mathbb{F}_{p^6}$

$$G_{\Phi_6(p)} = \{a \in \mathbb{F}_{p^6} \mid a^{\Phi_6(p)} = a^{p^2-p+1} = 1\}$$

requires 3 squarings in $\mathbb{F}_{p^2}$ with $p = 1 \pmod 6$.

More generally, Karabina [20] provided a new squaring formulae and comprehensive squaring formulas in $G_{\Phi_6(p)}$ including the results by Granger and Scott by computing Gröbner bases. These squaring formulas are effective for performing the final exponentiation. If $p = 1 \pmod 6$ and k is divisible by 6, we can apply the squaring methods to compute the exponentiation $g^{\frac{\Phi_k(p)}{r}}$ (the hard part of the final exponentiation).

For the curves with $k = 24$, we can adopt these squaring formulas, and we can consider squarings in the cyclotomic subgroup $G_{\Phi_2(p)}$ for the curves with $k = 2$.

From the above advantages, we conclude that Kawazoe–Takahashi curves are the most efficient for constructing a pairing at the high security level.

3.2 *The Explicit Constructions of the Kawazoe–Takahashi Curves*

Here we describe the explicit constructions of the Kawazoe–Takahashi cyclotomic family of curves with $k = 16$ and 24 for pairings at the 192- and 256-bit security levels respectively. This family of curves is explicitly given by using the cyclotomic polynomial and the security parameters can be parametrized by the polynomials over the rational field.

For $k = 16$, it is possible to construct the cyclotomic family of type I [21, Sect. 6.1], where $(k, h) = (16, 5)$ and h is a positive integer with $(k, h) = 1$ which determines the parameters and structure of the curve. The curve C (1) is defined over $\mathbb{F}_p$, where $p = 1 \pmod 8$ such that the parameter p and r, which is a prime factor of the order of $\mathrm{Jac}_C(\mathbb{F}_p)$, are parametrized by $t \in \mathbb{Z}$ as follows:

$$
\begin{aligned}
r(t) &= \frac{\Phi_{16}(t)}{2} = \frac{t^8+1}{2}, \\
p(t) &= \frac{1}{8}(2t^{14} - 4t^{13} + 2t^{12} + t^{10} + 2t^9 + t^8 + 2t^6 + 4t^5 + 2t^4 + t^2 + 2t + 1).
\end{aligned} \tag{3}
$$

Hence, the ρ-value $\rho = g \log p / \log r$ is approximately 3.5 since $p \approx r^{14/8}$.

We can also use the method for generating the cyclotomic family of type II [21, Sect. 6.2], where $(k, h) = (24, 11)$. This is similar to the case $k = 16$ in which these curves C (1) are defined over $\mathbb{F}_p$, where $p = 1 \pmod 8$. Then $r(t)$ and $p(t)$ are given as follows:

$$
\begin{aligned}
r(t) &= \Phi_{24}(t) = t^8 - t^4 + 1, \\
p(t) &= \frac{1}{8}(2t^{12} + 4t^{11} + 3t^{10} - 2t^9 - t^8 + 4t^7 - 3t^6 + 2t^5 + t^4 - 4t^3 + 3t^2 - 2t + 1).
\end{aligned} \tag{4}
$$

We then have that $\rho \approx 3$.

4 Analyses of the Complexities of the NFS by [24] for the Kawazoe–Takahashi Pairings

We here provide the analysis of the bit-length of security parameters of the Kawazoe–Takahashi curves for the 192- and 256-bit security levels by following the method to examine the run time of the TNFS algorithms by [24]. We follow their notations to describe our analyses.

In the same manner as [24], we first determine the size of p so that pairings defined over $\mathbb{F}_{p^k}$ achieve the given security levels. More precisely, we find $\mathrm{lgp}(\eta)$ which is the minimum value of $\lg p$ for a given η where $Q = p^n$ and $n = \eta\kappa$ under the TNFS setting. Then we determine the p as the maximum of $\mathrm{lgp}(\eta)$ for any η.

After that we fix the adequate value of p and other parameters are fixed, we adjust the value of B which is the size of the factor base so that the cost of the relation collection stages balances with the cost of the linear algebra stages [24, Sect. 6.3].

Tables 2 and 3 summarize the approximate run time of exTNFS and SexTNFS for the 192- and 256-bit security levels. These estimated run time are given with considering the constants $\mathfrak{C}(\eta, 2\kappa, H)$, $\mathfrak{C}(\eta, \kappa, H)$, and $\mathfrak{C}(\eta, \kappa, H)$, $\mathfrak{C}(\eta, \kappa\lambda, H)$ appeared in analysis of the run time of the relation collection stage of the exTNFS and SexTNFS, respectively where $H = 2$.

Table 2 Approximate run time of exTNFS and SexTNFS that achieve the 192-bit security level

Kawazoe–Takahashi curves: $n = 16,\ \rho \approx 3.5,\ \lambda = 14,\ \|\|\Gamma\|\|_\infty = 1/2$						
Algorithm	Constants	η	κ	$\lg p$	$\lg Q$	lg(run time)
exTNFS	Without	4	4	672	10752	197
exTNFS	With	4	4	672	10752	213
SexTNFS	Without	8	2	672	10752	288
SexTNFS	With	8	2	672	10752	415

Table 3 Approximate run time of exTNFS and SexTNFS that achieve the 256-bit security level

Kawazoe–Takahashi curves: $n = 24,\ \rho \approx 3,\ \lambda = 12,\ \|\|\Gamma\|\|_\infty = 1/2$						
Algorithm	Constants	η	κ	$\lg p$	$\lg Q$	lg(run time)
exTNFS	Without	4	6	864	20736	256
exTNFS	With	4	6	768	18432	269
SexTNFS	Without	2	12	768	18432	264
SexTNFS	With	2	12	768	18432	426

For the 192-bit security level, the bit-length of p is minimum value for all algorithms since the bit-length of r should be at least 384 and the ρ-value is approximately 3.5. However, the bit-length of p is at least 864 for constructing a pairing on the Kawazoe–Takahashi curve at the 256-bit security level.

As described in [24, Sect. 6], the estimated run time of the SexTNFS are larger than those of exTNFS that is contrary to expectation with the asymptotic run time. If we adopt the asymptotic complexity formula $A \cdot L_Q(1/3, (32/9)^{1/3})$ where $A = 2^{-14}$ to the case $k = 16$ and 24, the bit-lengths of p should be at least 876 and 1142 to achieve the 192- and 256-bit security levels, respectively. These values seem to be oversized for the key lengths of the pairings. In [24, 6.1] the authors gave the experimental results that ensure the accuracy of their estimations.

5 Computational Costs of the Pairings on Kawazoe–Takahashi Curves

In this section, we show the computational costs of the twisted Ate pairings on the Kawazoe–Takahashi curves as described in Sect. 3 in terms of number of multiplications in the definition field $\mathbb{F}_p$. By the run-time estimations of the exTNFS and SexTNFS for the Kawazoe–Takahashi curves in Sect. 4, we take the explicit security parameter p and r of the curves. Note that the parameter where $k = 16$ is same as previously shown by [17]. Here we reexamine the computational costs of the Miller part and the final exponentiation of the pairing and offer a more accurate estimation.

For the Kawazoe–Takahashi curves C (1), we denote the degree of the twists d and let $c \in \mathbb{F}_{p^e}$ be a l-th power non-residue, where $e = k/d$ and l divides d. The twist of C can be defined as C_t (2) and we have the isomorphism

$$\begin{aligned} \psi : C_t(\mathbb{F}_{p^k}) &\to C(\mathbb{F}_{p^k}) \\ P(x, y) &\mapsto \psi(P) = (c^{-\frac{2}{d}}x, c^{-\frac{5}{d}}y). \end{aligned}$$

For a divisor D, $\varepsilon(D)$ denotes the effective part of the reduced divisor of D. For any divisor $D_2 \in \mathbb{G}_2$, we denote D_2' as a divisor in $\mathrm{Jac}_{C_t}(\mathbb{F}_p)[r]$ such that $D_2 = \psi(D_2')$. Here $\psi(D)$ is defined by $\sum \psi(P)$ where $D = \sum P$.

5.1 *The Constructions of Pairings*

In contrast to the elliptic case, the cost of arithmetic on Jac_C is considerably higher than that on elliptic curves. Hence, we should consider the twisted versions of the optimal Ate pairings so that we can perform arithmetic in the divisor class group over the definition field $\mathbb{F}_p$, not over the extension field $\mathbb{F}_{p^e}$. In this paper we focus on the twisted Ate pairings. For more details and other constructions of the pairings, see [17, Chap. 5].

A pairing value is calculated by evaluating its Miller function at input divisors with the Miller's algorithm. In general, for a divisor class $\overline{D} \in \mathrm{Jac}_C(\mathbb{F}_{q^k})[r]$ and an integer i, the Miller function $f_{i,D}$ is defined as a rational function with divisor

$$(f_{i,D}) = iD - [i]D,$$

where $[i]D$ is the reduced divisor which is equivalent to iD.

Twisted Ate Pairing with $k = 16$

Zhang [29] showed that

$$\begin{aligned} a^{\mathrm{twist}} : \mathbb{G}_1 \times \mathbb{G}_2 &\to \mu_r \\ (\overline{D_1}, \overline{D_2}) &\mapsto f_{q^{ei} \pmod r, D_1}(D_2)^{(q^k-1)/r} \end{aligned}$$

is a bilinear pairing [29, Theorem 4]. We can find the smallest $p^{ei} \pmod r$ for the parameters (3) such that $p^{10} \pmod r = t^2$. We therefore obtain the bilinear pairing

$$a^{\mathrm{twist}}(D_1, D_2) = f_{t^2, D_1}(D_2)^{(q^{16}-1)/r}. \tag{5}$$

Twisted Ate Pairing with $k = 24$

In the case of $k = 24$ with the parameters (4), the smallest $p^{ei} \pmod r$ is $p^9 = t^3 \pmod r$. Then the twisted Ate pairing is given as follows:

$$a^{\mathrm{twist}}(D_1, D_2) = f_{t^3, D_1}(D_2)^{(q^{24}-1)/r}. \tag{6}$$

Note that we cannot take the optimal Miller loop lengths t for the twisted pairings in both cases $k = 16$ and 24, since the degree of twists d is less than k.

5.2 Notations for the Cost Estimation

We denote the cost of a multiplication and a squaring in $\mathbb{F}_{p^i}$ by M_i and S_i, respectively, and I_i denotes the cost of an inversion in $\mathbb{F}_{p^i}$. Let S_c denote the cost of a cyclotomic squaring in $\mathbb{F}_k$. We suppose that the cost of a squaring is equal to that of a multiplication in $\mathbb{F}_p$, that is, $M_1 = S_1$.

Using the explicit formula and the dedicated coordinate system given by [10], we can perform a doubling and mixed addition in $\mathrm{Jac}(\mathbb{F}_{p^i})$ of C and C_t with $35M_i + 5S_i$ and $36M_i + 5S_i$, respectively.

For simplicity, we denote the cost of each part of the pairing computation by T_* described in [9, 10, 25] as follows:

- T_D : the cost of doubling a nondegenerate divisor,
- T_A : the cost of adding two nondegenerate divisors,
- T_G : the cost of evaluating the line function G at the divisor $\varepsilon(\psi(D_2'))$.

Then the cost of evaluating the Miller function $f_{\lambda, D_1}(\varepsilon(\psi(D_2')))$ (without including the final exponentiation) for a reduced divisor D_2' can be written by

$$\mathbf{D}(T_D + T_G + S_k + bM_k) + \mathbf{A}(T_A + T_G + bM_k),$$

where $\mathbf{D}$ and $\mathbf{A}$ denote the number of doublings and additions in Jac_C for the Miller loop, respectively, and $b = \deg(\varepsilon(D_2'))$.

Since D_2' is defined over the extension field $\mathbb{F}_{p^e} \setminus \mathbb{F}_p$, we can take degenerate divisors D_2' as the input to reduce the cost of pairings [9]. In this paper we suppose that $b = 1$.

5.3 Evaluating a Line Function

Here we reveal the cost of evaluating the line function T_G in the same manner in [9, Sect. 5.5]. Each step in the Miller loop, we need to compute rational functions (with applying the denominator elimination technique)

$$\begin{aligned} c_D(x, y) &= (\tilde{r} z_{11})y - ((s_1' z_{11})x^3 + l_2 x^2 + l_1 x + l_0), \\ c_A(x, y) &= (\tilde{r} z_{21})y - ((s_1' z_{21})x^3 + l_2 x^2 + l_1 x + l_0) \end{aligned} \tag{7}$$

in the dedicated coordinate system by [10], and evaluate at the input divisor $\varepsilon(\psi(D_2')) = (c^{-\frac{2}{d}}x, c^{-\frac{5}{d}}y)$ for a doubling and an addition, respectively, where the parameters are specified in [9, Tables 4, 5].

After precomputing $(c^{-\frac{2}{d}}x)^2$, $(c^{-\frac{2}{d}}x)^3$ with $S_e + M_e$, we can compute each of $c_D(\psi(D_2'))$ and $c_A(\psi(D_2'))$ with $T_G = e \cdot M_1 + 3e \cdot M_1$, where $\psi(D_2')$ is degenerate. Note that the T_G described here is different from those of [9, 17] and is revised downward since we can construct a suitable extension field $\mathbb{F}_k$ by using the parameter c which defines the twist so that we can omit the multiplication cost by the map ψ.

5.4 For the Embedding Degree k = 16

We first show the explicit security parameters of the curve with $k = 16$, and provide the costs with computing the Miller loop and the final exponentiation.

We take the following curve $C\colon y^2 = x^5 + 7x$ where

$$
\begin{aligned}
t &= 282575562145795 = 2^{48} + 2^{40} + 2^{30} + 2^{16} + 2 + 1, \\
p &= p(t) \text{ (671 bits)}, \\
r &= r(t) \text{ (384 bits)}
\end{aligned}
$$

for the 192-bit security level. Then we can construct the extension field $\mathbb{F}_{p^{16}} = \mathbb{F}_p[x]/(x^{16} - 3)$ since $p = 5 \pmod{12}$. Table 4 lists the costs of basic arithmetic in the extension fields. $p/p^3/p^5/p^7$-, p^2/p^6-, and p^4-Frobenius operations in $\mathbb{F}_{p^{16}}$ require $14M_1$, $12M_1$, and $8M_1$, respectively.

For the twisted Ate pairing (5) where D_2' is degenerate,

$$
\begin{aligned}
t^2 &= 79848948322012052144836182025 \\
&= 2^{96} + 2^{89} + 2^{80} + 2^{79} + 2^{71} + 2^{65} + 2^{60} + 2^{57} + 2^{50} + 2^{49} \\
&\quad + 2^{47} + 2^{42} + 2^{41} + 2^{33} + 2^{31} + 2^{18} + 2^{17} + 2^3 + 1
\end{aligned}
$$

and we have that $\mathbf{D} = 96$ and $\mathbf{A} = 18$.

Table 4 Costs of arithmetic in the extension fields for $\mathbb{F}_{p^{16}}$

Field	Multiplication	Squaring	Inversion
$\mathbb{F}_{p^2}$	$3M_1$	$2M_1$	I_2
$\mathbb{F}_{p^4}$	$9M_1$	$6M_1$	I_4
$\mathbb{F}_{p^8}$	$27M_1$	$18M_1$	I_8
$\mathbb{F}_{p^{16}}$	$81M_1$	$54M_1$	$I_1 + 134M_1$
$G_{\Phi_2(p^8)}$	$81M_1$	$2M_8 = 36M_1$ [16]	Conjugation

Miller loop

The cost of precomputation for evaluating line functions described in Sect. 5.3 is $S_2 + M_2 = 5M_1$. Therefore, we have that $T_G = 2 \cdot M_1 + 6 \cdot M_1 = 8M_1$ with the above precomputation cost. Hence, the computation of the Miller loop requires

$$5M_1 + 96(35M_1 + 5S_1 + 8M_1 + S_{16} + M_{16}) + 18(36M_1 + 5S_1 + 8M_1 + M_{16})$$
$$= 19913M_1.$$

Final Exponentiation

The cost of computing the hard part dominates the total cost for performing the final exponentiation. In this paper, we adopted the method by Scott et al. [26] with computing a vectorial addition chain for performing a multi-exponentiation to obtain the resulting value of the hard part.

Their method can be presented briefly as follows. For $l = \Phi_k(p)/r$, we write l as the p-adic number

$$l = l_0 + l_1 p + \cdots + l_{\varphi(k)-1} p^{\varphi(k)-1}.$$

After performing p^i-Frobenius g^{p^i}, we then compute g^l by performing a multi-exponentiation with a short vectorial addition chain derived from the coefficients l_i. Indeed, the addition chain can be computed using the set of coefficients for the polynomials $l_i(t)$.

We also consider the technique by Fuentes–Castañeda et al. [12, Sect. 3] which adopted the lattice-based method for multiple l' of l. We applied this method to our parameters, however, we did not find any short addition chain which was better than result given by the naive Scott's method.

For r and p (3), we consider $4l(t) = 4 \cdot \frac{(p(t)^8+1)}{r(t)} = \sum_{i=0}^{7} l_i(t) p(t)^i$.

We then compute a vectorial addition chain from the coefficients of $l_i(t)$

$$\{(1000), (0100), (0010), (0001), (2000), (2100), (4200), (4210), (8420), (8421)\}.$$

This implies that 6 multiplications in $\mathbb{F}_{p^{16}}$ including three cyclotomic squarings are required to obtain the final exponentiation. In [17], the author treated the parameter $p(t)$ as the polynomial over the integer ring by taking $t = 2x + 1$. This restriction causes the size of coefficients of the polynomial to increase, as a result the calculated vectorial addition chain was considerably long.

Computing g^t requires 48 cyclotomic squarings and 5 multiplications in $F_{p^{16}}$ for an element $g \in G_{\Phi_2(p^8)}$. Hence, one can compute g^{t^i} $(1 \leq i \leq 13)$ with $13(48S_c + 5M_{16}) = 624S_c + 65M_{16}$. We then compute $(g^{t^i})^{p^j}$ with $(26 \cdot 14 + 13 \cdot 12 + 7 \cdot 8)M_1 = 576M_1$ for the coefficients of $l_i(t)$.

Therefore, performing the final exponentiation using the vectorial chain requires

$$M_{16} + I_{16} + 624S_c + 65M_{16} + 576M_1 + 49M_{16} + 3S_c + 3M_{16} = 32840M_1 + I_1.$$

Consequently, the estimated cost of the twisted Ate pairing with $k = 16$ is $52753M_1 + I_1$.

5.5 *For the Embedding Degree k = 24*

Similar to the case $k = 16$, the computational cost of the pairing with $k = 24$ and the following explicit security parameters can be estimated.

We take the following curve $C\colon y^2 = x^5 + 25x$ where

$$\begin{aligned} t &= 9444734091639197466625 = 2^{73} + 2^{50} + 2^{17} + 2^{16} + 1, \\ p &= p(t)\ (875\text{ bits}), \\ r &= r(t)\ (585\text{ bits}) \end{aligned}$$

for the 256-bit security level. Then we can construct the extension field $\mathbb{F}_{p^{24}}$ as $\mathbb{F}_p[x]/(x^{25} - 5)$ since $p = 2 \pmod 5$. It also suffices that $p = 1 \pmod 6$.

Table 5 contains the costs of basic arithmetic in the extension fields. $p/p^3/p^5/p^7$-, p^2/p^6-, and p^4-Frobenius operations in $\mathbb{F}_{p^{24}}$ require $22M_1$, $12M_1$, and a conjugation, respectively.

For t^3, we have $\mathbf{D} = 220$ and $\mathbf{A} = 42$.

Miller loop

We have that $T_G = 3 \cdot M_1 + 9 \cdot M_1 = 12M_1$ with the precomputation which requires $S_3 + M_3 = 11M_1$. Hence, the computation of the Miller loop requires

$$\begin{aligned} &11M_1 + 220(35M_1 + 5S_1 + 12M_1 + S_{24} + M_{24}) + 42(36M_1 + 5S_1 + 12M_1 + M_{24}) \\ &= 79881M_1. \end{aligned}$$

Table 5 Costs of arithmetic in the extension fields for $\mathbb{F}_{p^{24}}$

Field	Multiplication	Squaring	Inversion
$\mathbb{F}_{p^2}$	$3M_1$	$2M_1$	I_2
$\mathbb{F}_{p^3}$	$6M_1$	$5M_1$	I_3
$\mathbb{F}_{p^4}$	$9M_1$	$6M_1$	I_4
$\mathbb{F}_{p^{12}}$	$54M_1$	$36M_1$	I_8
$\mathbb{F}_{p^{24}}$	$162M_1$	$108M_1$	$I_1 + 275M_1$
$G_{\Phi_4(p^6)}$	$162M_1$	$6M_4 = 54M_1$ [16]	Conjugation

Final Exponentiation

For r and p (4), we consider $8l(t) = 8 \cdot \frac{(p(t)^8 - p(t)^4 + 1)}{r(t)} = \sum_{i=0}^{7} l_i(t)p(t)^i$.

We then compute a vectorial addition chain

$$\{(10000), (01000), (00100), (00010), (00001), (20000), (21000), (21100), (21001), (21110), (42101), (42220), (84321)\}$$

from the polynomials $l_i(t)$.

In the same manner as the case $k = 16$, we can estimate the cost of computing the final exponentiation as follows:

$$M_{24} + I_{24} + 803S_c + 44M_{24} + 786M_1 + 54M_{24} + 2S_c + 6M_{24} = 61541M_1 + I_1.$$

Consequently, the estimated computational cost of the twisted Ate pairing with $k = 24$ is $141422M_1 + I_1$.

5.6 Comparisons

Here we provide comparisons of our pairings on Kawazoe–Takahashi curves with the state-of-the-art pairings on the elliptic curves. When we take into account the recent progress of the NFS algorithms and the asymptotic analysis of their run time by [24], we should compare our results with the result for BLS24 pairing described in [1, Sect. 8] since the parameters of other (BN, BLS12, and KSS) curves should be readjusted so that they achieve the 192-bit security level according to [24, Table 5].

Table 6 shows the comparison of our pairings and the elliptic curve pairing at the 192-bit security level.

As noted in [1, Sect. 8], we can consider the scaled costs in software implementations on 64-bit platforms. Indeed, we have that $m_{671} \approx 1.86m_{512}$ and the scaled cost of our pairing with $k = 16$ is approximately 2.5 times more than the cost of the BLS24 pairing.

Table 6 Comparison of the cost of the pairings at the 192-bit security level

Curve	Pairing	Phase	Operations in $\mathbb{F}_p$
BLS24 [1]	Optimal Ate	ML	$14927m_{480}$
		FE	$25412m_{480} + i_{480}$
		Total	$40339m_{480} + i_{480}$
Kawazoe–Takahashi16	Twisted Ate	ML	$19913m_{671}$
		FE	$32840m_{671} + i_{671}$
		Total	$52753m_{671} + i_{671}$

For the 192- and 256-bit security level, we should analyze security parameters for several pairing-friendly and computational costs of pairings on them comprehensively. According to the results by [24], the bit-length of p should be larger than 1000 for BN and BLS12 curves for the 192-bit security level. Hence, the BN and BLS12 pairings might not be adequate to construct pairings at the 192-bit security level. Meanwhile, it seems that KSS18 curves are suitable and competitive with the BLS24 curves for constructing efficient pairings.

6 Conclusion

In this paper we reexamined the security parameters of the pairings on the Kawazoe–Takahashi curves with considering the recent progress on the NFS algorithms. We followed the analyzing method to estimate the runtimes of the relation collection stage of the exTNFS and SexTNFS by [24].

The bit-length of p should be at least 672 and 864 to achieve the 192- and 256-bit security levels for the pairings on Kawazoe–Takahashi curves. For the 192-bit security level, the scaled cost of our pairing with $k = 16$ is approximately 2.5 times more than the cost of the BLS24 pairing which is the state-of-the-art elliptic pairing in the literature when we take into account the exTNFS and SexTNFS.

It is necessary to analyze other pairing-friendly elliptic curves to achieve a comprehensive comparison between genus 2 pairings and elliptic pairings. Further work is required to analyze the key lengths of pairings on not only Kawazoe–Takahashi curves of genus 2 but also other pairing-friendly curves including elliptic curves by evaluating norms in the NFS algorithms. Furthermore, there is a considerable difference between cost of pairings on elliptic curves and that of pairings on genus 2 curves. One of the causes comes from the discrepancy of ρ-values. The ρ-values of Kawazoe–Takahashi families in this paper take $3 \leq \rho \leq 3.5$ which are somewhat larger than those of pairing-friendly elliptic curves. This is the one of biggest disadvantage for pairings on genus 2 curves. Therefore, a method for constructing pairing-friendly genus 2 curves with $\rho < 2$ should be exploited.

References

1. D.F. Aranha, L. Fuentes-Castañeda, E. Knapp, A. Menezes, F. Rodríguez-Henríquez, Implementing pairings at the 192-bit security level, in *Pairing-Based Cryptography - Pairing 2012*, vol. 7708, Lecture Notes in Computer Science, ed. by M. Abdalla, T. Lange (Springer, Berlin, 2013), pp. 177–195
2. J. Balakrishnan, J. Belding, S. Chisholm, K. Eisenträger, K.E. Stange, E. Teske, Pairings on hyperelliptic curves, in *CoRR*, http://arxiv.org/abs/0908.3731v2 (2009)
3. R. Barbulescu, P. Gaudry, A. Guillevic, F. Morain, Improving NFS for the discrete logarithm problem in non-prime finite fields, in *Advances in Cryptology - EUROCRYPT 2015*, vol. 9056, Lecture Notes in Computer Science, ed. by E. Oswald, M. Fischlin (Springer, Berlin, 2015), pp. 129–155

4. R. Barbulescu, P. Gaudry, A. Guillevic, F. Morain, Improving NFS for the discrete logarithm problem in non-prime finite fields, in *Advances in Cryptology - EUROCRYPT 2015: 34th Annual International Conference on the Theory and Applications of Cryptographic Techniques, Sofia, Bulgaria, April 26–30, 2015, Proceedings, Part I*, ed. by E. Oswald, M. Fischlin (Springer, Berlin, 2015), pp. 129–155
5. R. Barbulescu, P. Gaudry, A. Joux, E. Thom, A heuristic quasi-polynomial algorithm for discrete logarithm in finite fields of small characteristic, in *Advances in Cryptology - EUROCRYPT 2014*, vol. 8441, Lecture Notes in Computer Science, ed. by P. Nguyen, E. Oswald (Springer, Berlin, 2014), pp. 1–16
6. R. Barbulescu, P. Gaudry, T. Kleinjung, The tower number field sieve, in *Advances in Cryptology - ASIACRYPT 2015: 21st International Conference on the Theory and Application of Cryptology and Information Security, Auckland, New Zealand, November 29 - December 3, 2015, Proceedings, Part II*, ed. by T. Iwata, H.J. Cheon (Springer, Berlin, 2015), pp. 31–55
7. N. Benger, M. Scott, Constructing tower extensions of finite fields for implementation of pairing-based cryptography, in *Arithmetic of Finite Fields: Third International Workshop*, ed. by M.A. Hasan, T. Helleseth, WAIFI 2010, Istanbul, Turkey, June 27–30, 2010. Proceedings (Springer, Berlin, 2010), pp. 180–195
8. BlueKrypt: - cryptographic key length recommendation, http://www.keylength.com (2012)
9. X. Fan, G. Gong, D. Jao, Speeding up pairing computations on genus 2 hyperelliptic curves with efficiently computable automorphisms, in *Pairing-Based Cryptography – Pairing 2008*, ed. by S. Galbraith, K. Paterson. Lecture Notes in Computer Science, vol. 5209 (Springer, Berlin, 2008), pp. 243–264. doi:10.1007/978-3-540-85538-5_17
10. X. Fan, G. Gong, D. Jao, Efficient pairing computation on genus 2 curves in projective coordinates, in *Selected Areas in Cryptography*, vol. 5381, Lecture Notes in Computer Science, ed. by R. Avanzi, L. Keliher, F. Sica (Springer, Berlin, 2009), pp. 18–34
11. D. Freeman, M. Scott, E. Teske, A taxonomy of pairing-friendly elliptic curves. J. Cryptol. **23**(2), 224–280 (2010)
12. L. Fuentes-Castañeda, E. Knapp, F. Rodríguez-Henríquez, Faster hashing to $\mathbb{G}_2$, in *Selected Areas in Cryptography: 18th International Workshop, SAC 2011, Toronto, ON, Canada, August 11–12, 2011, Revised Selected Papers*, ed. by A. Miri, S. Vaudenay (Springer, Berlin, 2012), pp. 412–430
13. S.D. Galbraith, X. Lin, D.J.M. Morales, Pairings on hyperelliptic curves with a real model, in *Pairing-Based Cryptography – Pairing 2008*, ed. by S. Galbraith, K. Paterson. Lecture Notes in Computer Science, vol. 5209 (Springer, Berlin, 2008), pp. 265–281
14. S.D. Galbraith, K.G. Paterson, N.P. Smart, Pairings for cryptographers. Discret. Appl. Math. **156**(16), 3113–3121 (2008). doi:10.1016/j.dam.2007.12.010
15. R. Granger, D. Page, N.P. Smart, High security pairing-based cryptography revisited, in *Algorithmic Number Theory: 7th International Symposium, ANTS-VII*, Berlin, Germany, July 23–28, 2006. Proceedings, ed. by F. Hess, S. Pauli, M. Pohst (Springer, Berlin, 2006), pp. 480–494
16. R. Granger, M. Scott, Faster squaring in the cyclotomic subgroup of sixth degree extensions, in *Public Key Cryptography – PKC 2010: 13th International Conference on Practice and Theory in Public Key Cryptography*, Paris, France, May 26–28, 2010. Proceedings, ed. by P.Q. Nguyen, D. Pointcheval (Springer, Berlin, 2010), pp. 209–223
17. M. Ishii, Pairings on hyperelliptic curves of genus 2 at high security levels. Ph.D. thesis, Nara Institute of Science and Technology (2016), http://library.naist.jp/dspace/handle/10061/11005
18. J. Jeong, T. Kim, Extended tower number field sieve with application to finite fields of arbitrary composite extension degree. Cryptol. ePrint Arch. Rep. 2016/526 (2016), http://eprint.iacr.org/2016/526
19. A. Joux, C. Pierrot, The special number field sieve in $\mathbb{F}_{p^n}$, application to pairing-friendly constructions, in *Pairing-Based Cryptography – Pairing 2013: 6th International Conference*, Beijing, China, November 22–24, 2013, Revised Selected Papers, ed. by Z. Cao, F. Zhang (Springer International Publishing, Berlin, 2014), pp. 45–61
20. K. Karabina, Squaring in cyclotomic subgroups. Math. Comput. **82**(281) (2013), http://dx.doi.org/10.1090/S0025-5718-2012-02625-1

21. M. Kawazoe, T. Takahashi, Pairing-friendly hyperelliptic curves with ordinary jacobians of type $y^2 = x^5 + ax$, in *Pairing-Based Cryptography - Pairing 2008*, vol. 5209, Lecture Notes in Computer Science, ed. by S. Galbraith, K. Paterson (Springer, Berlin, 2008), pp. 164–177
22. T. Kim, R. Barbulescu, Extended tower number field sieve: A new complexity for the medium prime case, in *Advances in Cryptology - CRYPTO 2016: 36th Annual International Cryptology Conference, Santa Barbara, CA, USA, August 14–18, 2016, Proceedings, Part I*, ed. by M. Robshaw, J. Katz (Springer, Berlin, 2016), pp. 543–571
23. N. Koblitz, A. Menezes, Pairing-based cryptography at high security levels, in *Cryptography and Coding: 10th IMA International Conference*, Cirencester, UK, December 19–21, 2005. Proceedings, ed. by N.P. Smart (Springer, Berlin, 2005), pp. 13–36
24. A. Menezes, P. Sarkar, S. Singh, Challenges with assessing the impact of NFS advances on the security of pairing-based cryptography. Cryptol. ePrint Arch. Rep. 2016/1102 (2016), http://eprint.iacr.org/2016/1102
25. C. Ó hÉigeartaigh, M. Scott, Pairing calculation on supersingular genus 2 curves, in *Selected Areas in Cryptography: 13th International Workshop, SAC 2006*, ed. by E. Biham, A.M. Youssef. Lecture Notes in Computer Science, vol. 4356 (Springer, Berlin, 2007), pp. 302–316
26. M. Scott, N. Benger, M. Charlemagne, L. Dominguez Perez, E. Kachisa, On the final exponentiation for calculating pairings on ordinary elliptic curves, in *Pairing-Based Cryptography - Pairing 2009*, vol. 5671, Lecture Notes in Computer Science, ed. by H. Shacham, B. Waters (Springer, Berlin, 2009), pp. 78–88
27. M. Stam, A.K. Lenstra, Efficient subgroup exponentiation in quadratic and sixth degree extensions, in *Cryptographic Hardware and Embedded Systems - CHES 2002: 4th International Workshop Redwood Shores*, CA, USA, August 13–15, 2002 Revised Papers, ed. by B.S. Kaliski, ç.K. Koç, C. Paar (Springer, Berlin, 2003), pp. 318–332
28. T. Teruya, K. Saito, N. Kanayama, Y. Kawahara, T. Kobayashi, E. Okamoto, Constructing symmetric pairings over supersingular elliptic curves with embedding degree three, in *Pairing-Based Cryptography – Pairing 2013*, ed. by Z. Cao, F. Zhang. Lecture Notes in Computer Science, vol. 8365 (Springer, Berlin, 2014), pp. 97–112
29. F. Zhang, Twisted ate pairing on hyperelliptic curves and applications. Sci. China Inf. Sci. **53**(8), 1528–1538 (2010)
30. X. Zhang, K. Wang, Fast symmetric pairing revisited, in *Pairing-Based Cryptography – Pairing 2013*, ed. by Z. Cao, F. Zhang. Lecture Notes in Computer Science, vol. 8365 (Springer, Berlin, 2014), pp. 131–148

Efficient Algorithms for Isogeny Sequences and Their Cryptographic Applications

Katsuyuki Takashima

Abstract We summarize efficient isogeny sequence computations on elliptic and genus 2 Jacobians. For cryptographic purposes, sequences of low-degree isogenies are important. Then we focus on sequences of 2- and 3-isogenies on elliptic curves and $(2, 2)$- and $(3, 3)$-isogenies on genus 2 Jacobians. Our aim is to explicitly describe the low-degree isogeny sequence computations and improve them for cryptographic applications such as post-quantum cryptosystems and random self-reducibility of discrete logarithm problem (DLP).

Keywords Isogeny · Expander graph · Post-quantum cryptography · Random self-reducibility of dlp

1 Introduction

Computing a sequence of isogenies of elliptic curves is a new cryptographic basic operation in some applications. For example, a cryptographic hash function from expander graphs, proposed in [8], consists of computing an isogeny sequence, which is based on hardness of constructing an isogeny between two isogenous curves. A Diffie–Hellman type key exchange protocol based on isogenies is given in [13, 30], which is considered as a candidate for a post-quantum public key primitive [12] called supersingular isogeny Diffie–Hellman (SIDH). Both applications employ supersingular elliptic curves and the graph consisting of them. One decisive feature for the applications is that the graph is Ramanujan, which has an *optimal* expanding property. In the ordinary elliptic curve (resp. genus 2 Jacobian) case, isogeny sequence computation is important for analyzing security (hardness) of the discrete logarithm problem (DLP) on the different isogenous elliptic curves [21] (resp. genus 2 Jacobians [6, 22]). The graphs of (same level) ordinary curves also have an expanding property (under GRH) and an (almost) volcano structure, which are crucial for our analyses.

K. Takashima (✉)
Mitsubishi Electric, 5-1-1 Ofuna, Kamakura Kanagawa 247-8501, Japan
e-mail: Takashima.Katsuyuki@aj.MitsubishiElectric.co.jp

T. Takagi et al. (eds.), *Mathematical Modelling for Next-Generation Cryptography*, Mathematics for Industry 29, DOI 10.1007/978-981-10-5065-7_6

Lubicz–Robert [26] described an efficient algorithm for the computation of separable isogenies between abelian varieties represented in the coordinate system given by algebraic theta functions. Subsequently, Cosset–Robert [10] improved the algorithm for the genus 2 case, where the kernel of isogenies are given by maximal isotropic subgroups with respect to the Weil pairing. For the above applications, we iterate several low-degree isogeny computations. In particular, in the elliptic curve case, isogenies of degree two and three are basic building blocks (see SIDH given in Sect. 2.3.1). Our main purpose is to provide *explicit descriptions of low-degree isogeny sequence computation on elliptic and genus 2 curves* for *practical* cryptography. In particular, we mainly focus on the following four cases: (1) 2-isogeny sequence on elliptic curves, (2) 3-isogeny sequence on elliptic curves, (3) (2, 2)-isogeny sequence on genus 2 Jacobians, (4) (3, 3)-isogeny sequence on genus 2 Jacobians. All the prefixes 2, 3, (2, 2) and (3, 3) indicate that used isogenies are of low degree. In the elliptic curve case, we have a useful Vélu's formula, so we improve the iterated computation of the formula below. In the genus 2 case, general Vélu type algorithms are given by Cosset and Robert [10]. However, they are based on theta coordinates, so we just focus on low-degree cases, (2, 2) and (3, 3), both cases have explicit and compact algorithms [15, 32, 33, 37] on Mumford coordinates. Before describing these algorithms, we survey two target cryptographic applications of the low-degree isogeny sequence computation.

Isogeny-Based Cryptosystems Charles et al. [8] proposed an algorithm for computing a sequence of 2-isogenies between two *supersingular* elliptic curves. Their method is based on Vélu's formulas, which computes a 2-isogeny explicitly for that purpose. Yoshida–Takashima [39] improved the computation and showed that all computations can stay in $\mathbb{F}_{p^2}$. It makes the sequence computation a reasonable cryptographic operation. Childs et al. [9] attack the isogeny computation problem, and obtained subexponential time quantum algorithm for *ordinary* elliptic curves. In contrast, the attack cannot be employed to the supersingular case (because of non-commutativity of endomorphism rings). Based on the observation, De Feo et al. [13] proposed SIDH protocol using supersingular elliptic curves. Recently, the detailed implementation of SIDH was investigated by Costello et al. [12].

Charles et al. [7] suggested higher dimensional abelian varieties for post-quantum cryptosystems. They show that some special classes of abelian varieties and isogenies form Ramanujan graphs and are suitable for cryptographic applications.

Random Self-reducibility of DLP Kohel [23] first showed that the graph consisting of ordinary elliptic curves has a special structure called volcano (see [35] for the survey). Based on the volcano structure, Jao, Miller, and Venkatesan [21] showed random self-reducibility of the discrete logarithm problem within ordinary isogenous elliptic curves. We say that two ordinary elliptic curves E_1, E_2 defined over a finite field have the same level if their rings of endomorphisms $\mathrm{End}(E_i)$ are isomorphic. An ordinary curve isogeny is classified as horizontal, ascending, or descending (on the volcano) according to the associated change of levels. For traveling in the volcano, we iterate low-degree isogeny computations.

Brooks et al. [6] showed that (ℓ, ℓ)-isogeny graphs of abelian surfaces have similar (but complicated) level structures, which can be used for computing isogenies from an arbitrary absolutely simple ordinary abelian surface toward one with maximal endomorphism ring. Since Jetchev et al. [22] proved random self-reducibility of the discrete logarithm problem within (principally polarizable) isogenous ordinary abelian surfaces with the same maximal endomorphism ring, the algorithm in [6] realizes a random self-reduction of the problems between two arbitrary isogenous ordinary abelian surfaces.

Organization In Sect. 2, background materials are summarized. In Sect. 3 (resp. 4), isogeny sequence computations on elliptic curves (resp. genus 2 Jacobians) are described. Section 5 gives some concluding remarks.

2 Elliptic Curves and Hyperelliptic Genus 2 Jacobians

Here, we show background materials for elliptic curves and genus 2 Jacobians. Section 2.1 introduces several basic facts on elliptic curves, and gives Vélu's formulas for isogeny as fundamental operations. Section 2.2 gives a basic one-way function from isogeny computation as is explained in [8, 13]. Section 2.3 gives two explicit constructions of trapdoor homomorphisms from isogenies, Algorithms 1 and 2.

2.1 Elliptic Curves

We summarize facts about elliptic curves. For details, see [31], for example.

Let p be a prime greater than 3 and $\mathbb{F}_p$ be the finite field with p elements. Let $\overline{\mathbb{F}}_p$ be its algebraic closure. An elliptic curve E over $\overline{\mathbb{F}}_p$ is given by the Weierstrass normal form

$$E : Y^2 = X^3 + AX + B \tag{1}$$

for A and $B \in \overline{\mathbb{F}}_p$ where the discriminant of the RHS of Eq. (1) is nonzero. We denote the point at infinity on E by O_E. Elliptic curves are endowed with a unique algebraic group structure, with O_E as neutral element. The j-invariant of E is $j(E) := j(A, B) := 1728\frac{4A^3}{4A^3+27B^2}$. Conversely, for $j \neq 0, 1728 \in \overline{\mathbb{F}}_p$, set $A = A(j) = \frac{3j}{1728-j}$, $\quad B = B(j) = \frac{2j}{1728-j}$. Then the obtained E in Eq. (1) has j-invariant j. Two elliptic curves over $\overline{\mathbb{F}}_p$ are isomorphic if and only if they have the same j-invariant. For a positive integer n, the set of n-torsion points of E is $E[n] = \{P \in E(\overline{\mathbb{F}}_p) \mid nP = O_E\}$.

Given two elliptic curves E and $\tilde{E}$ over $\overline{\mathbb{F}}_p$, a homomorphism $\phi : E \to \tilde{E}$ is a morphism of algebraic curves that sends O_E to $O_{\tilde{E}}$. A nonzero homomorphism is called an isogeny, and a separable isogeny with the cardinality ℓ of the kernel is called

ℓ-isogeny. We consider only *separable* isogenies in this paper, i.e., any isogeny is separable here. An elliptic curve E over $\overline{\mathbb{F}}_p$ is called supersingular if there are no points of order p, i.e., $E[p] = \{O_E\}$. The j-invariants of supersingular elliptic curves lie in $\mathbb{F}_{p^2}$ (see [31, Chap. V, Theorem 3.1]). A non-supersingular elliptic curve (over $\overline{\mathbb{F}}_p$) is called ordinary.

2.2 Basic Facts of Elliptic Curve Isogeny Graphs

We first consider a graph consisting of ℓ-isogenies between supersingular elliptic curves. The graph has an expanding property (expander graph), and is called a Pizer graph [8, 29] (or an isogeny graph [34]). We then see that a graph consisting of isogenies between (isogenous) ordinary elliptic curves also has an expanding property, which implies random self-reducibility of DLP.

Pizer Graph The Pizer graph $\mathscr{G} = \mathscr{G}_\ell(p) = (\mathscr{V}, \mathscr{E})$ with a small prime ℓ consists of isomorphism classes of supersingular elliptic curves over $\overline{\mathbb{F}}_p$ as vertex set $\mathscr{V}$, and (informally) their ℓ-isogenies as edge set $\mathscr{E}$. Precisely, the vertex set $\mathscr{V}$ is the set of supersingular j-invariants and edges $(j, j') \in \mathbb{F}_{p^2}^2$ present with multiplicity k whenever j' is a root of $\Phi_\ell(j, Y)$ with multiplicity k, where $\Phi_\ell(X, Y)$ is the classical modular polynomial (see Definition 1 in [34]). Equivalently, $\ell + 1$ edges are coming from any vertex in $\mathscr{V}$, and when the vertex is represented by an elliptic curve E, they are associated with $\ell + 1$ ℓ-torsion cyclic subgroups on E. For each edge from E, the other vertex is the quotient curve E/K where K represents the corresponding subgroup of order ℓ.

The graph is directed, in which the direction of the edge associated with (j, j') (resp. (E, K)) is defined to be from j to j' (resp. from E to $E' = E/K$). The in-degree and out-degree of any vertex are $\ell + 1$, and it has a multi-edge (j, j') when $\Phi_\ell(j, Y)$ has a multiple root $Y = j'$ with multiplicity ≥ 2. Moreover, it has a self-loop (j, j) when $\Phi_2(X, X)$ has a root $X = j$.

$\mathscr{G}$ is known to have a rapidly mixing property, which is crucial for cryptographic applications. The family of Pizer graphs $\mathscr{G}_\ell(p)$ for a fixed ℓ and various p is called (a family of) Ramanujan graphs (with a fixed degree $\ell + 1$), which are optimal expander graphs. For details, see [8, 29], for example.

One-wayness of Isogeny Computation against Quantum Computers We have a one-way function (Eq. (2)) from isogeny (sequence) computation

$$\text{Isogeny } \phi : (E, K) \underset{\text{hard}}{\overset{\text{easy}}{\rightleftarrows}} (E, E/K), \tag{2}$$

where E is a supersingular elliptic curve (EC) and $K (\subset E[\ell^\kappa])$ is an order-ℓ^κ cyclic torsion subgroup where $p = \Theta(2^\lambda)$, $\kappa = \Theta(\log p) = \Theta(\lambda)$ for the security parameter λ. First, from the expanding property (or rapidly mixing property) explained above, by walking just $\kappa = \Theta(\log p)$-times iteratively, our ending point $E' := E/K$ has almost uniform distribution in the isogeny graph. This improves efficiency of the forward direction function evaluation in Eq. (2).

Childs et al. [9] proposed a *subexponential time quantum algorithm* for the inverse direction function given in Eq. (2), i.e., for the isogeny problem between *ordinary* elliptic curves. However, there exists *no* subexponential time quantum algorithm for the isogeny problem between *supersingular* elliptic curves while Biasse et al. [3, 14] made a progress on the exponential time algorithm based on a Grover-type quantum search. Therefore, our isogeny function given in Eq. (2) is considered as one-way even against quantum adversaries at present.

Random Self-reducibility of DLP We say that two ordinary elliptic curves E_1, E_2 defined over $\mathbb{F}_q$ have the same level if their rings of endomorphisms $\mathrm{End}(E_i)$ are isomorphic. Jao et al. [21] showed random self-reducibility of the discrete logarithm problem within ordinary isogenous elliptic curves in a fixed level. The obtained graphs of ordinary elliptic curves are equivalent to the Cayley graphs of narrow ray class groups, especially, of abelian groups. Although no Cayley graphs of abelian groups give a family of expander graphs *with a constant degree* [25, 27] in general, the graphs given in [21] are shown to be expanding families *by allowing loosely increasing degrees* (under the generalized Riemann hypotheses (GRH)), which leads to Theorem 2.1 below. If we consider all the isogenous curves with different levels, the graphs have an important structure called volcano, which is also applied to an efficient deterministic supersingular elliptic curve identification algorithm [34].

Theorem 2.1 (Theorem 1.6 in [21]) *Assume GRH for the characters of narrow ray class groups of imaginary quadratic fields. If there is an algorithm $\mathscr{A}$ which solves the discrete logarithm problem on a positive fraction μ of the elliptic curves in a given level. There exists an absolute polynomial $p(x)$ such that one can probabilistically solve the discrete logarithm problem on any curve in the same level with expected runtime $\frac{p(\log q)}{\mu}$ times the maximal runtime of $\mathscr{A}$.*

2.3 Two Types of Isogeny Sequence Computation

At present, there exist two types of algorithms for computing isogeny sequences, one by Charles et al. [8] and another by De. Feo et al. [13]. The first is given in a clear manner according to Eq. (2) and the second has an advantage of less additional restriction for the system parameter p.

2.3.1 Isogeny Sequence Algorithm by De. Feo et al. [13]

Let ℓ be a small prime, for example, $\ell = 2, 3, \ldots$, and a large prime p satisfies $\ell^\kappa \mid p \pm 1$. Then, a supersingular EC has a rational subgroup $(\mathbb{Z}/\ell^\kappa\mathbb{Z})^2 \subseteq E(\mathbb{F}_{p^2}) \simeq (\mathbb{Z}/(p \pm 1)\mathbb{Z})^2$. In other words, all the ℓ^κ-torsion points are defined over $\mathbb{F}_{p^2}$. For using a point R in $E[\ell^\kappa]$, we can compute an isogeny $\phi : E \to E/\langle R\rangle$ by iteratively using Vélu's formula for ℓ-isogenies. (In [13], the algorithm is used for establishing a DH-type key exchange.)

In De Feo et al.'s algorithm, for an input point R in $E[\ell^\kappa]$, first set $E_0 := E$, $R_0 := R$ and, for $0 \le i < \kappa$, let

$$E_{i+1} := E_i/\langle \ell^{\kappa-i-1}R_i\rangle, \quad \psi_i : E_i \to E_{i+1}, \quad R_{i+1} := \psi_i(R_i),$$

where $R_i \in E_i[\ell^{\kappa-i}]$, $\ell^{\kappa-i-1}R_i$ is in $E_i[\ell]$ and then ψ_i is an ℓ-isogeny. The composition gives the desired

$$\phi := \psi_{\kappa-1}\cdots\psi_0 : E = E_0 \to E_\kappa = E/\langle R\rangle.$$

We describe the algorithm in Algorithm 1 and call it $\mathsf{Isog}^{\mathsf{djp}}_{\ell,\kappa}$ after De Feo, Jao, and Plût. A trapdoor for computing isogenies is given by the kernel generating point R.

Algorithm 1 $\mathsf{Isog}^{\mathsf{djp}}_{\ell,\kappa}$: DJP-type isogeny sequence computation of degree ℓ^κ from E_0 (given in [13])

Require: An initial elliptic curve E_0 and a nonzero point R in $E_0[\ell^\kappa]$.
Ensure: Isogenous curve E_κ such that $E_\kappa = E_0/\langle R\rangle$.
1: set $R_0 := R$,
2: **for** $0 \le i < \kappa$ **do**
3: compute $E_{i+1} := E_i/\langle \ell^{\kappa-i-1}R_i\rangle$, $\psi_i : E_i \to E_{i+1}$, and $R_{i+1} := \psi_i(R_i)$ by Vélu's formula, where $R_i \in E_i[\ell^{\kappa-i}]$, $\ell^{\kappa-i-1}R_i$ is in $E_i[\ell]$ and then ψ_i is an ℓ-isogeny.
4: **end for**
5: we set the composition $\phi := \psi_{\kappa-1}\cdots\psi_0 : E_0 \to E_\kappa = E_0/\langle R\rangle$. return E_κ (or $j(E_\kappa)$).

Supersingular Isogeny Diffie–Hellman (SIDH) De Feo et al. [13] proposed an isogeny-based Diffie–Hellman type key exchange protocol called supersingular isogeny Diffie–Hellman (SIDH), whose implementation is investigated in detail in [12] and subsequently in [1, 2, 4, 24]. Very recently, the security is analyzed in [17].

For two small primes $\ell_{\mathrm{A}}, \ell_{\mathrm{B}}$ (e.g., $\ell_{\mathrm{A}} = 2, \ell_{\mathrm{B}} = 3$), we choose a large prime p such that $p \pm 1 = f \cdot \ell_{\mathrm{A}}^{e_{\mathrm{A}}}\ell_{\mathrm{B}}^{e_{\mathrm{B}}}$. Then, we also choose a supersingular elliptic curve E over $\mathbb{F}_{p^2}$ with $E(\mathbb{F}_{p^2}) \simeq (\mathbb{Z}/(p\pm 1)\mathbb{Z})^2 \supseteq (\mathbb{Z}/\ell_{\mathrm{A}}^{e_{\mathrm{A}}}\mathbb{Z})^2 \oplus (\mathbb{Z}/\ell_{\mathrm{B}}^{e_{\mathrm{B}}}\mathbb{Z})^2$. We use isogenies $\phi_{\mathrm{A}}, \phi_{\mathrm{B}}$ with kernels of orders $\ell_{\mathrm{A}}^{e_{\mathrm{A}}}, \ell_{\mathrm{B}}^{e_{\mathrm{B}}}$, respectively. We use the following commutative diagram for the DH-type key exchange between Alice and Bob.

$$\begin{array}{ccc} E & \xrightarrow{\phi_{\mathrm{A}}} & E_{\mathrm{A}} := E/\langle R_{\mathrm{A}}\rangle \\ \downarrow{\scriptstyle \phi_{\mathrm{B}}} & & \downarrow{\scriptstyle \phi'_{\mathrm{B}}} \\ E_{\mathrm{B}} := E/\langle R_{\mathrm{B}}\rangle & \xrightarrow{\phi'_{\mathrm{A}}} & E/\langle R_{\mathrm{A}}, R_{\mathrm{B}}\rangle \end{array} \qquad \begin{aligned} \text{for } \ker\phi_{\mathrm{A}} &= \langle R_{\mathrm{A}}\rangle \subset E[\ell_{\mathrm{A}}^{e_{\mathrm{A}}}], \\ \ker\phi_{\mathrm{B}} &= \langle R_{\mathrm{B}}\rangle \subset E[\ell_{\mathrm{B}}^{e_{\mathrm{B}}}], \\ \ker\phi'_{\mathrm{A}} &= \langle \phi_{\mathrm{B}}(R_{\mathrm{A}})\rangle \subset E_{\mathrm{B}}[\ell_{\mathrm{A}}^{e_{\mathrm{A}}}], \\ \ker\phi'_{\mathrm{B}} &= \langle \phi_{\mathrm{A}}(R_{\mathrm{B}})\rangle \subset E_{\mathrm{A}}[\ell_{\mathrm{B}}^{e_{\mathrm{B}}}]. \end{aligned}$$

We first choose generators P_A, Q_A, P_B, Q_B such that $E[\ell_A^{e_A}] = \langle P_A, Q_A\rangle$, $E[\ell_B^{e_B}] = \langle P_B, Q_B\rangle$ and then set the curve $E/\mathbb{F}_p$ and the above generators as public parameters. DH-type key exchange is given as below.

Alice		**Bob**
$m_A, n_A \stackrel{U}{\leftarrow} (\mathbb{Z}/\ell_A^{e_A}\mathbb{Z})^\times$: Alice's secret key,	$\xrightarrow{E_A,\ \phi_A(P_B),\ \phi_A(Q_B)}$	$m_B, n_B \stackrel{U}{\leftarrow} (\mathbb{Z}/\ell_B^{e_B}\mathbb{Z})^\times$: Bob's secret key,
$R_A := m_A P_A + n_A Q_A$,	$\xleftarrow[E_B,\ \phi_B(P_A),\ \phi_B(Q_A)]{}$	$R_B := m_B P_B + n_B Q_B$,
$\phi_A : E \to E_A := E/\langle R_A\rangle$,		$\phi_B : E \to E_B := E/\langle R_B\rangle$,
$\mathscr{K}_{\texttt{Alice}} :=$ $E_B/\langle m_A\,\phi_B(P_A) + n_A\,\phi_B(Q_A)\rangle$		$\mathscr{K}_{\texttt{Bob}} :=$ $E_A/\langle m_B\,\phi_A(P_B) + n_B\,\phi_A(Q_B)\rangle$

Here, since $\langle m_A\,\phi_B(P_A) + n_A\,\phi_B(Q_A)\rangle = \langle\phi_B(R_A)\rangle = \ker\phi'_A$, $\langle m_B\,\phi_A(P_B) + n_B\,\phi_A(Q_B)\rangle = \langle\phi_A(R_B)\rangle = \ker\phi'_B$, it holds that $\mathscr{K}_{\texttt{Alice}} = E_B/\ker\phi'_A = E/\langle R_A, R_B\rangle = E_A/\ker\phi'_B = \mathscr{K}_{\texttt{Bob}}$, and $\mathscr{K} := \mathscr{K}_{\texttt{Alice}} = \mathscr{K}_{\texttt{Bob}}$ is a shared key between them. Alice's output includes $\phi_A(P_B), \phi_A(Q_B)$ as well as E_A, the security is based on the hardness of isogeny problem with the auxiliary inputs.

2.3.2 Isogeny Sequence Algorithm by Charles et al. [8]

Let ℓ be a small prime s.t. $\ell \mid p \pm 1$ for an odd prime $p > 3$, then a supersingular EC is defined over $\mathbb{F}_{p^2}$ with $E(\mathbb{F}_{p^2}) \simeq (\mathbb{Z}/(p \pm 1)\mathbb{Z})^2 \supseteq E[\ell] \simeq (\mathbb{Z}/\ell\mathbb{Z})^2$. In other words, all the ℓ-torsion points are defined over $\mathbb{F}_{p^2}$. Therefore, by iteratively choosing ℓ-isogenies in a random manner, we can obtain a random ℓ^κ-isogeny from any supersingular E. We consider computing a ℓ-isogeny sequence

$$E_0 \xrightarrow{\psi_0} E_1 \xrightarrow{\psi_1} \cdots \xrightarrow{\psi_{\kappa-2}} E_{\kappa-1} \xrightarrow{\psi_{\kappa-1}} E_\kappa \tag{3}$$

where E_i are supersingular without backtracking, i.e., $\psi_i \neq \widehat{\psi_{i+1}}$ for $i = 0, \ldots, \kappa - 2$ and all ψ_i are given by Vélu's formulas (Eq. (5)). The isogeny sequence starting from E_0 is determined by a walk data in $(\mathbb{Z}/\ell\mathbb{Z})^\kappa$. As is explained before, a supersingular elliptic curve E_0 and the ℓ-torsion points on E_0 are defined over $\mathbb{F}_{p^2}$.

As the out-degree of any vertex is three and the walk we consider has no backtracking, we have ℓ possibilities to proceed to the next vertex in $\mathscr{V}$ at $i \geq 1$. For $i = 0$, we fix ℓ possibilities $(\psi_{0,0}, \ldots, \psi_{0,\ell-1})$ from E_0 at the beginning. For each $i = 0, \ldots, \kappa - 1$, a next step is determined by a lexicographical order in $\mathbb{F}_{p^2}$ for choosing a next j-invariant (In fact, any deterministic selection can be used). That is, we associate a walk data $\omega = \omega_0\omega_1\cdots\omega_{\kappa-1} \in (\mathbb{Z}/\ell\mathbb{Z})^\kappa$ with a sequence (3) where $\psi_0 \in \{\psi_{0,0}, \ldots, \psi_{0,\ell-1}\}$. Our goal is to compute the j-invariant $j_\kappa = j(E_\kappa)$ from $j_0 = j(E_0)$ and a walk data $\omega \in (\mathbb{Z}/\ell\mathbb{Z})^\kappa$ that determines the next vertices. For the details, see [8, 39]. We describe the algorithm in Algorithm 2 and call it $\mathrm{Isog}_{\ell,\kappa}^{\mathsf{clg}}$ after Charles, Lauter, and Goren. A trapdoor for computing isogenies is given by the walk data ω.

We will give modified versions of Algorithm 2 for $\ell = 2$ and 3, namely, Algorithms 3, 4 for $\ell = 2$ and Algorithms 5, 6 for $\ell = 3$. They give efficiency improvements over Algorithm 2.

Algorithm 2 $\mathrm{Isog}_{\ell,\kappa}^{\mathsf{clg}}$: CLG-type isogeny sequence computation of degree ℓ^κ from E_0 (given in [8, 39])

Require: An initial elliptic curve E_0 and all the selector numbers $\omega := \{\omega_i\}_{0 \le i < \kappa} \in (\mathbb{Z}/\ell\mathbb{Z})^\kappa$.
Ensure: Isogenous curve E_κ determined by ω.
1: **for** $0 \le i < \kappa$ **do**
2: select all next kernel points R_i, which generates a order-ℓ subgroup in $E_i[\ell] \setminus \psi_{i-1}(E_{i-1}[\ell])$ if $i \neq 0$ (resp., in $K_i := \{$ some fixed ℓ candidate order-ℓ subgroups in $E_i[\ell] \setminus \{O_{E_i}\}$ $\}$ if $i = 0$).
3: R_i, is determined from ω_i by a lexicographic order in $\mathbb{F}_{p^2}^2$.
4: we set $\psi_i : E_i \to E_{i+1} := E_i/\langle R_i \rangle$ for the selected R_i.
5: **end for**
6: we set the composition $\phi := \psi_{\kappa-1} \cdots \psi_0 : E_0 \to E_\kappa$. return E_κ (or $j(E_\kappa)$).

Comparison with De Feo et al.'s Algorithm I compared the two methods for computing isogeny sequences. We call these DJP-type and CLG-type algorithms, respectively.

As is shown, the DJP-type computation is applicable to SIDH, and other public key cryptosystems, e.g., encryption and signatures [18], and so on. While the DJP-type needs one scalar multiplication for each degree-ℓ isogeny, the CLG-type extracts a next kernel point for each step. The latter operation requires quadratic (resp. cubic) root finding for the $\ell = 2$ (resp. $\ell = 3$) case as is shown in Sect. 3.2 (resp. 3.3), hence easy to execute. However, for a large ℓ, we should use a general root finding algorithm, which is more costly than low-degree ℓ cases. Therefore, since the CLG-type one has better efficiency than the DJP-type at least for the important low-degree $\ell (= 2, 3)$ cases, the CLG-type is suitable to hash function applications, but not easy to apply to SIDH.

The CLG-type with degree $\ell = 2$ can allow a wide range of parameter choices for supersingular elliptic curves since 2 divides the order of $\mathbb{F}_{p^2}$-rational point group, i.e., $(p \pm 1)^2$ for any odd prime p. Since the DJP-type needs a ℓ^κ-torsion to be defined in a low extension degree finite field, e.g., $\mathbb{F}_{p^2}$, the prime p is of restricted form such that $p \pm 1 = f \cdot \ell_{\mathrm{A}}^{e_{\mathrm{A}}} \ell_{\mathrm{B}}^{e_{\mathrm{B}}}$, i.e., can allow only a narrow range of parameters p.

2.4 Hyperelliptic Curves of Genus 2 and Their Jacobians

Let p be an odd prime > 5. Then, a hyperelliptic curve of genus 2 over $\overline{\mathbb{F}}_p$ is given by

$$C : Y^2 = f(X)$$

where $\deg(f(X)) = 5$ or 6 and $f(X)$ has no multiple zeros. Let the zeros of $f(X)$ be $(a_0, \ldots, a_4)$, or $(a_0, \ldots, a_5)$. Then, $P_m := (a_m, 0)$ for $0 \le m \le 4$ or 5 are called *Weierstrass points* (When $\deg(f) = 5$, the infinity point gives another Weierstrass point). Given a hyperelliptic curve C of genus 2, we can define a group variety J_C, the Jacobian. A point D on J_C is given by a divisor class of C of degree 0, which is a formal sum of points on C modulo linear equivalence. When $\deg(f(X)) = 5$, D is represented by a pair of polynomials, in other words, as a set,

$$J_C = J_C(\overline{\mathbb{F}}_p) = \left\{ \begin{array}{r|l} D := (u(X), v(X)) & u(X) \mid v(X)^2 - f(X), u(X) : \text{monic}, \\ \in \overline{\mathbb{F}}_p[X]^2 & \deg(v(X)) < \deg(u(X)) \le 2 \end{array} \right\} \tag{4}$$

where $\overline{\mathbb{F}}_p[X]$ is the polynomial ring whose coefficient field is $\overline{\mathbb{F}}_p$. When $\deg(f(X))$ $= 6$, a point in J_C is given by a pair $(u(X), v(X))$ s.t. $u(X) \mid v(X)^2 - f(X)$, $u(X)$: monic, and $\deg(v(X)) < \deg(u(X)) \le 2$. Such a representation is called Mumford representation (or Mumford coordinates). An addition of divisors naturally gives an algebraic addition law on J_C. For details, see [16]. Jacobian J_C is called supersingular if it is isogenous (over $\overline{\mathbb{F}}_p$) to a product of two supersingular elliptic curves, and a curve C is called supersingular if J_C is supersingular. Let J be the Jacobian and K be a maximal isotropic subgroup of $J[\ell]$ for the Weil pairing. Then, we have an isogeny from J to J/K, which is called (ℓ, ℓ)-isogeny.

Comparison with EC systems While EC public key cryptosystems based on hardness of DLP (e.g., ECDH) seem to have better efficiency than genus 2 curve based ones (see e.g., [11]), nowadays both of the genus 1 and 2 systems are seriously considered as candidates for real applications. Therefore, security assessments of the both cases are important. In particular, self-reducibility of DLP in the genus 2 ordinary case should be investigated much. Efficient isogeny sequence computation is necessary for the study.

Using genus 2 hyperelliptic curves, we can obtain larger valencies (or degrees) in isogeny graphs than elliptic curve ones for the same prime ℓ. For example, the regular graph of (2, 2)-isogenies in Sect. 4.2 (for $\ell = 2$) has degree 15, which is larger than degree 3 (for $\ell = 2$) and 4 (for $\ell = 3$) for the graphs in Sects. 3.2 and 3.3, respectively. The fact might give better throughput cryptosystems in the genus 2 (e.g., hash functions or SIDH). However, since there does not exist enough researches of genus 2 graphs in the literatures from cryptographic points of view, we need to study the genus 2 case much.

Since genus 2 isogenies has seemingly complicated formulas (compared with the genus 1 case), to reduce the computation, we try to seek efficient algorithms in the genus 2 case in this work by generalizing genus 1 case (in some sense).

3 CLG-Type Isogeny Sequence Computation on Elliptic Curves

3.1 Vélu's Formulas

We compute the ℓ-isogeny by using Vélu's formulas for a small prime $\ell = 2, 3, \ldots$. Vélu gave in [38] the explicit formulas of the isogeny $\psi : E \to \tilde{E}$ and the equation of the form (5) of $\tilde{E}$ when E is given by Eq. (1) and $K := \ker \psi$ is explicitly given. Then there exists a unique isogeny $\psi : E \to \tilde{E}$ s.t. $K := \ker \psi$, and we denote E' by E/K.

For an elliptic curve E and a cyclic group $K (\subset E)$ of order ℓ, Vélu's formula [38] gives an isogenous curve E/K and the associated isogeny $E \ni (x, y) \mapsto (\tilde{x}, \tilde{y}) \in E/K$. For computing it, for $E : Y^2 = X^3 + AX + B$ and point $Q = (x_Q, y_Q) \neq O_E \in K$, we define $g_Q^x = 3x_Q^2 + a$, $g_Q^y = -2y_Q$, and $t_Q = 2g_Q^x$ if $Q \in E[2]$, $t_Q = g_Q^x$ if $Q \notin E[2]$, $u_Q = (g_Q^y)^2$. For $S := (K - \{O_E\})/\pm 1$, let $t = \sum_{Q \in S} t_Q$, $w = \sum_{Q \in S}(u_Q + x_Q t_Q)$, $\tilde{A} = A - 5t$, $\tilde{B} = B - 7w$, then,

$$\tilde{E} = E/K : Y^2 = X^3 + \tilde{A}X + \tilde{B}, \qquad \tilde{x} = x + \sum_{Q \in S} \left(\frac{t_Q}{x - x_Q} + \frac{u_Q}{(x - x_Q)^2} \right),$$

$$\tilde{y} = y - \sum_{Q \in S} \left(\frac{2u_Q y}{(x - x_Q)^3} + \frac{t_Q(y - y_Q) - g_Q^x g_Q^y}{(x - x_Q)^2} \right) \tag{5}$$

gives the curve and isogeny. Analogs of Vélu's formulas for isogenies on alternate models of elliptic curves are given in [28].

3.2 2-Isogeny Sequences on Elliptic Curves

Charles et al. [8] proposed an algorithm for computing a sequence of 2-isogenies between supersingular elliptic curves based on Vélu's formulas [38]. In [39], we described simple algorithms based on compact expressions of 2-isogenies, without some redundancy in the description in [8].

Let p be an odd prime > 3, $\mathbb{F}_p$ the finite field of order p, and $\overline{\mathbb{F}}_p$ an algebraic closure of $\mathbb{F}_p$. For $0 \leq i \leq \kappa$, let $E_i/\overline{\mathbb{F}}_p$ be a supersingular elliptic curve given by the short Weierstrass normal form $Y^2 = f_i(X)$ with $\deg(f_i) = 3$. Let $(a_{i,0}, 0)$, $(a_{i,1}, 0)$, and $(a_{i,2}, 0)$ be 2-torsion points on E_i.

In [39], we considered the computation of the sequence (3) of 2-isogenies ψ_i associated to $(a_{i,0}, 0)$ without backtracking. Here, we denote $(a_{i,1}, 0)$ as the 2-torsion point associated with the backtracking, i.e., the dual isogeny $\hat{\psi}_{i-1}$. Then we obtained the following simple recurrence formulas between $(a_{i,0}, a_{i,1}, a_{i,2})$ and $(a_{i+1,0}, a_{i+1,1}, a_{i+1,2})$:

$$a_{i+1,1} = -2a_{i,0} \text{ and } a_{i+1,0} \in \left\{ a_{i,0} \pm 2\sqrt{(a_{i,0} - a_{i,1})(a_{i,0} - a_{i,2})} \right\}. \tag{6}$$

Here, note that there is a square root term in the RHS of the second formula of (6). Based on Eq. (6), we proposed two *simple* algorithms for a sequence (3) in [39]. Here we describe the second one (Algorithms 3 and 4). In Algorithm 3, `select` and $\texttt{select}_0$ are selector functions which determine the next step by a lexicographical order in $\mathbb{F}_{p^2}$, and σ_t is the t-th elementary symmetric polynomial of three variables for $t = 2, 3$.

Overall, Algorithm 3 is given by iterations of permutations of 2-torsion (Weierstrass) points

$$(a_{i,0}, a_{i,1}, a_{i,2}) \overset{\psi_i}{\longmapsto} (a_{i+1,0}, a_{i+1,1}, a_{i+1,2}) \tag{7}$$

for $i = 0, \ldots, \kappa - 1$. This is generalized to (2, 2)-isogeny sequence computation (Eq. (11)).

Moreover, we showed that when *appropriately* choosing a starting *supersingular* elliptic curve $E_0/\mathbb{F}_{p^2}$, all 2 torsions on E_i, i.e., $(a_{i,m}, 0)$, are defined in $\mathbb{F}_{p^2}$, and then *all* the computation of the proposed algorithms stays in $\mathbb{F}_{p^2}$.

Algorithm 3 2-isogeny sequence computation

Require: $j_0 = j(E_0)$, walk data $\omega = \omega_0\omega_1 \ldots \omega_{\kappa-1} \in \{0, 1\}^\kappa$.
Ensure: $j_\kappa = j(E_\kappa)$.
1: $(A_0, B_0) \leftarrow (A(j_0), B(j_0))$, $(a_{0,0}, a_{0,1}) \leftarrow \texttt{select}_0(A_0, B_0, \omega_0)$.
2: **for** $i \leftarrow 0$ to $\kappa - 2$ **do**
3: $\quad (a_{i+1,0}, a_{i+1,1}, a_{i+1,2}) \leftarrow \texttt{Isog2}(a_{i,0}, a_{i,1}, a_{i,2}, \omega_{i+1})$.
4: **end for**
5: $\xi \leftarrow \alpha_{\kappa-1}^2$, $A_\kappa \leftarrow -4\sigma_2(a_{\kappa-1,0}, a_{\kappa-1,1}, a_{\kappa-1,2}) - 15\xi$,
$B_\kappa \leftarrow -8\sigma_3(a_{\kappa-1,0}, a_{\kappa-1,1}, a_{\kappa-1,2}) - 14a_{\kappa-1,0}\xi$, $j_\kappa \leftarrow j(A_\kappa, B_\kappa)$.
6: **return** j_κ.

Algorithm 4 `Isog2` : 2-isogeny computation

Require: $a_{i,0}, a_{i,1}, a_{i,2}$, select bit ω_{i+1}.
Ensure: $a_{i+1,0}, a_{i+1,1}, a_{i+1,2}$.
1: $\delta \leftarrow (a_{i,1} - a_{i,0})(a_{i,2} - a_{i,0})$, $\eta \leftarrow \delta^{\frac{1}{2}}$, $\lambda_0 \leftarrow a_{i,0} + 2\eta$, $\lambda_1 \leftarrow a_{i,0} - 2\eta$,
$a_{i+1,0} \leftarrow \texttt{select}(\lambda_0, \lambda_1, \omega_{i+1})$, $a_{i+1,1} \leftarrow -2a_{i,0}$, $a_{i+1,2} \leftarrow \texttt{select}(\lambda_0, \lambda_1, 1 - \omega_{i+1})$.
2: **return** $a_{i+1,0}, a_{i+1,1}, a_{i+1,2}$.

3.3 3-Isogeny Sequences on Elliptic Curves

We generalized the above 2-isogeny sequence algorithm to 3-isogeny one (Tachibana et al. [36]).

We obtained the following simple recurrence formulas between $(\alpha_{i,0}, \alpha_{i,1}, \alpha_{i,2}, \alpha_{i,3})$ and $(\alpha_{i+1,0}, \alpha_{i+1,1}, \alpha_{i+1,2}, \alpha_{i+1,3})$. Here, as in the 2-isogeny case, we denote $\alpha_{i,1}$ as the x coordinate of the 3-torsion point associated with the backtracking, i.e., the dual isogeny $\hat{\psi}_{i-1}$.

$$\alpha_{i+1,1} = -3\alpha_{i,0} \quad \text{and}$$
$$\alpha_{i+1,0} \in \left\{ \alpha_{i,0} + \zeta^k \sqrt[3]{-s + \sqrt{s^2 + t^3}} + \zeta^{3-k} \sqrt[3]{-s - \sqrt{s^2 + t^3}} \text{ for } k = 0, 1, 2 \right\},$$

where ζ is a primitive 3rd root of unity, $t = -6(3\alpha_{i,0}^2 + A_i)$ and $s = -6(15\alpha_{i,0}^3 + 11A_i\alpha_{i,0} + 9B_i)$. Based on the above recurrence relation, we proposed an isogeny sequence algorithm for a sequence (3) by iterating computations from (A_i, B_i, α_i) to $(A_{i+1}, B_{i+1}, \alpha_{i+1})$ where $\alpha_i := \alpha_{i,0}, \alpha_{i+1} := \alpha_{i+1,0}$ (Algorithms 5 and 6). The selector functions $\texttt{select}$ and $\texttt{select}_0$ are defined in a similar manner as in Algorithms 3 and 4.

Algorithm 5 3-isogeny sequence computation

Input: $j_0 = j(E_0)$, walk data $\omega = \omega_0\omega_1...\omega_{\kappa-1} \in \{0, 1, 2\}^\kappa$
Output: $j_\kappa = j(E_\kappa)$
1: $(A_0, B_0) \leftarrow (A(j_0), B(j_0))$,
   $\alpha_0 \leftarrow \texttt{select}_0(A_0, B_0, \omega_0)$
2: **for** $i = 0$ to $\kappa - 2$ **do**
3: $\quad (A_{i+1}, B_{i+1}, \alpha_{i+1}) \leftarrow \texttt{Isog3}(A_i, B_i, \alpha_i, \omega_{i+1})$
4: **end for**
5: $\xi_1 \leftarrow \alpha_{\kappa-1}^2$, $\xi_2 \leftarrow \alpha_{\kappa-1}\xi_1$, $\xi_3 \leftarrow A_{\kappa-1}\alpha_{\kappa-1}$,
   $A_\kappa \leftarrow -(9A_{\kappa-1} + 30\xi_1)$, $B_\kappa \leftarrow -(70\xi_2 + 42\xi_3 + 27B_{\kappa-1})$, $j_\kappa \leftarrow j(A_\kappa, B_\kappa)$
6: **return** j_κ

Algorithm 6 $\texttt{Isog3}$: 3-isogeny computation

Input: $A_i, B_i, \alpha_i, \omega_{i+1} \in \{0, 1, 2\}$, two roots of unity ζ, ζ^2 of the order three
Output: $A_{i+1}, B_{i+1}, \alpha_{i+1}$
1: $\xi_1 \leftarrow \alpha_i^2, \xi_2 \leftarrow \alpha_i\xi_1$, $\xi_3 \leftarrow \xi_2 + A_i\alpha_i + B_i$
   $A_{i+1} \leftarrow -(9A_i + 30\xi_1)$, $B_{i+1} \leftarrow 15B_i - 42\xi_3 - 28\xi_2$ /* Vélu's formula */
2: /* Solve the cubic equation */
   $t \leftarrow -6(3\xi_1 + A_i), s \leftarrow -6(4\xi_2 + 11\xi_3 - 2B_i)$, $\delta \leftarrow \sqrt{s^2 + t^3}, u \leftarrow \sqrt[3]{-s + \delta}, v \leftarrow -t/u$,
   $\lambda_0 \leftarrow \alpha_i + u + v$, $\lambda_1 \leftarrow \alpha_i + \zeta u + \zeta^2 v$, $\lambda_2 \leftarrow \alpha_i + \zeta^2 u + \zeta v$
3: $\alpha_{i+1} \leftarrow \texttt{select}(\lambda_0, \lambda_1, \lambda_2, \omega_{i+1})$
4: **return** $A_{i+1}, B_{i+1}, \alpha_{i+1}$

4 CLG-Type Isogeny Sequence Computation on Genus 2 Jacobians

4.1 Generalized Vélu's Formulas in Genus 2

Cosset and Robert [10] give a generalization of Vélu's formulas in the genus 2 case.

Theorem 4.1 ([10]) *Let J be the Jacobian of an hyperelliptic curve of genus 2 over a finite field k of characteristic different from 2. Let K be a maximal isotropic subgroup of $J[\ell]$ for the Weil pairing, and K is rational. Then we can compute the isogeny $J \to J/K$ in Mumford coordinates using $\widetilde{O}(\ell^{2+r})$ arithmetic operations in k, where $r = 2$ if $\ell \equiv 1 \bmod 4$, $r = 4$ otherwise.*

However, their algorithm is given in theta coordinates so that we need computation overhead for the transformation between Mumford coordinates and theta coordinates. Therefore, hereafter, we give two compact (2, 2)- and (3, 3)-isogeny computations (on Mumford coordinates) below.

4.2 (2, 2)-Isogeny Sequences on Genus 2 Jacobians

Richelot Isogeny We explain an isogeny of a hyperelliptic curve of genus 2, called *Richelot isogeny* ([5, 32] etc.), which is a (2, 2)-isogeny.

Let $G_j(X) \in \overline{\mathbb{F}}_p[X]$ for $j = 0, 1, 2$ be 3 monic polynomials of $\deg(G_j) \le 2$ such that $\prod_{j=0}^{2} G_j(X)$ is of degree 5 or 6 and squarefree. Then

$$C : Y^2 = f(X) = d \prod_{j=0}^{2} G_j(X) \tag{8}$$

where $d \in \overline{\mathbb{F}}_p^*$ is a curve of genus 2. Using coefficients $g_{j,k}$ of $G_j(X) = \sum_{k=0}^{2} g_{j,k} X^k$, let M be the matrix $(g_{j,k})_{0 \le j,k \le 2}$. Here, note that if $\deg(G_j) = 1$, then $g_{j,2} = 0$. If $\deg(G_j) = 2$, we denote the zeros of $G_j(X)$ by a_{2j} and a_{2j+1}, i.e., $G_j(X) = (X - a_{2j})(X - a_{2j+1})$. Hereafter, we consider permutations of $(a_0, \dots, a_5)$ for the description of the Richelot isogeny. For that purpose, we use a special symbol "∞" to treat the case that $G_j(X)$ is linear, i.e., $G_j(X) = X - a$ where $a = a_{2j}$ or a_{2j+1}. Then, we consider that a and ∞ are the two zeros of $G_j(X)$, and treat permutations of six elements $(a_0, \dots, a_5)$ including ∞.

Suppose that the determinant of $M = (g_{j,k})_{0 \le j,k \le 2}$ is non-zero. Hereafter, prime "$'$" means differentiation by the variable X. We then define the bracket product $[G_{j+1}(X), G_{j+2}(X)]$ and its transform to the monic one, $\tilde{G}_j(X)$, below,

$$[G_{j+1}(X), G_{j+2}(X)] := G'_{j+1}(X)G_{j+2}(X) - G'_{j+2}(X)G_{j+1}(X),$$
$$\tilde{G}_j(X) := \tau_j^{-1}[G_{j+1}(X), G_{j+2}(X)]$$

where τ_j is the leading coefficient of $[G_{j+1}(X), G_{j+2}(X)]$. Here, and in similar places throughout this paper, we will take addition w.r.t. the index of G to mean addition modulo 3. Then the degree of $\prod_{j=0}^{2} \tilde{G}_j(X)$ is 5 or 6 [32]. Let $\tilde{f}(X) := \tilde{d} \prod_{j=0}^{2} \tilde{G}_j(X)$ where $\tilde{d} := d \cdot \tau_0\tau_1\tau_2 \cdot \det(M)^{-1}$. Using $\tilde{f}(X)$, we then obtain a curve of genus 2

$$\tilde{C} : Y^2 = \tilde{f}(X) = \tilde{d} \prod_{j=0}^{2} \tilde{G}_j(X) \ \text{ with } \ \tilde{d} := d \cdot \tau_0\tau_1\tau_2 \cdot \det(M)^{-1}.$$

The curve $\tilde{C}$ is called a *Richelot dual* of C. Here, we call the above correspondence *Richelot operator* $\mathscr{R}$ according to B. Smith [32].

$$\mathscr{R} : (G_0(X), G_1(X), G_2(X), d) \mapsto (\tilde{G}_0(X), \tilde{G}_1(X), \tilde{G}_2(X), \tilde{d}).$$

However, this $\mathscr{R}$ is slightly different from that in [32].

A Sequence of Richelot Isogenies Let C be given by Eq. (8). Richelot isogenies from J_C are determined by splitting $(G_0(X), G_1(X), G_2(X))$ of $f(X)$. This corresponds to a splitting of the zero-points of $f(X)$ into three pairs, i.e., (a_0, a_1), (a_2, a_3), and (a_4, a_5). Therefore, the number of Richelot isogenies from C is $\binom{6}{2} \cdot \binom{4}{2}/3! = 15$. We consider computing a walk consisting of Richelot isogenies

$$J_0 \xrightarrow{\psi_0} J_1 \xrightarrow{\psi_1} \cdots \xrightarrow{\psi_{\kappa-2}} J_{\kappa-1} \xrightarrow{\psi_{\kappa-1}} J_\kappa \tag{9}$$

without backtracking, i.e., ψ_{i+1} is not the dual of ψ_i for $i = 0, \ldots, \kappa - 2$. Hence, at each step, there exist $14 = 15 - 1$ possible choices to go forward. In sequence (9), J_i is the Jacobian of C_i, which is given below,

$$\begin{aligned} &C_i/\overline{\mathbb{F}}_p : Y^2 = f_i(X) = d_i \prod_{j=0}^{2} G_{i,j}(X) \ \text{ where} \\ &G_{i,j}(X) = \sum_{k=0}^{2} g_{i,j,k} X^k = \begin{cases} (X - a_{i,2j})(X - a_{i,2j+1}) & \text{if } \deg(G_{i,j}) = 2, \\ X - a_{i,2j} \ \text{ or } \ X - a_{i,2j+1} & \text{if } \deg(G_{i,j}) = 1 \end{cases} \end{aligned} \tag{10}$$

where $d_i \neq 0$ and $\det(M_i) \neq 0$ for $M_i := (g_{i,j,k})_{0 \le j,k \le 2}$. For the Richelot dual $\tilde{C}_{i+1}$ (after applying ψ_i to J_{C_i}), we use similar notation $\tilde{G}_{i+1,j}(X)$. Here, we note that, if $\det(M_i) = 0$, then J_{C_i} has an isogeny to a product of elliptic curves $E_1 \times E_2$ [32].

The goal is to compute the C_κ from C_0 and a walk data $\omega \in \{0, \ldots, 13\}^\kappa$. For $i = 1, \ldots, \kappa - 1$, the i-th step in sequence (9) for computing ψ_i consists of the following 2 procedures

1 Permutation of the zero-points $(a_{i,m})_{m=0,\ldots,5}$ of $f_i(X)$, then construct polynomials $(G_{i,j}(X))_{j=0,1,2}$ according to Eq. (10).
2 Isogeny calculation by the Richelot operator $\mathscr{R}$, i.e.,
$\tilde{G}_{i+1,j}(X) = \tau_{i,j}^{-1}[G_{i,j+1}(X), G_{i,j+2}(X)]$ for $j = 0, 1, 2$, where $\tau_{i,j}$ is the leading coefficient of $[G_{i,j+1}(X), G_{i,j+2}(X)]$.

To permute 6 zero-points of $\tilde{G}_{i,0}(X)$, $\tilde{G}_{i,1}(X)$ and $\tilde{G}_{i,2}(X)$, we must solve the quadratic equations $\tilde{G}_{i,j}(X) = 0$ for $j = 0, 1, 2$. Hence, the three square root computations are the most time-consuming as in the genus 1 case. Therefore, the algorithm is given by iterations of permutations of Weierstrass points

$$(a_{i,0}, a_{i,1}, a_{i,2}, a_{i,3}, a_{i,4}, a_{i,5}) \overset{\psi_i}{\longmapsto} (a_{i+1,0}, a_{i+1,1}, a_{i+1,2}, a_{i+1,3}, a_{i+1,4}, a_{i+1,5}) \tag{11}$$

for $i = 0, \ldots, \kappa - 1$ as in 2-isogeny sequence computation (Eq. (7)). Precisely, one more multiplicative factor d_i (Eq. (10)) is also repeatedly calculated. For detailed explanations of the algorithm, refer to [37].

Moreover, we also demonstrated all the above calculations stay in $\mathbb{F}_{p^2}$ or $\mathbb{F}_{p^4}$.

4.3 (3, 3)-Isogeny Sequences on Genus 2 Jacobians

Dolgachev–Lehavi Method Dolgachev–Lehavi [15] proposed an efficient method for computing (3, 3)-isogeny on genus 2 Jacobians, then Smith [33] described the algorithm explicitly using elementary mathematical terms. In the following, we describe only the general case for simplicity, for the details, see [33].

For an input curve C, $\rho : (x : y : z) \mapsto (w_0 : \cdots : w_6) := (z^6 : xz^5 : \cdots : x^5z : x^6)$ defines a map of C into $\mathbb{P}^6$. For a divisor $D := [P + Q - D_\infty]$ on C where D_∞ is the divisor at infinity, a line $\mathscr{L}_D := \mathscr{L}_{P,Q}$ in $\mathbb{P}^6$ is given by connecting $\rho(P)$ and $\rho(Q)$ (and well-defined). Also, a hyperplane H in $\mathbb{P}^6$ is given by $H : \sum_{i=0}^{6} c_i W_i = 0$ where $f(X) = \sum_{i=0}^{6} c_i X^i$ with $c_6 = 1$ is the defining polynomial of C.

As an input to the isogeny computation, a maximal isotropic subgroup K in the Jacobian $J_C[3]$ as well as the defining polynomial f are given. First, choose a minimal subset $S \subset K$ such that $K = S \cup (-S) \cup \{0\}$ (similarly as Velu's formula), i.e., $\sharp S = 4$. For each $D \in S$, compute intersection point $\iota_D := \mathscr{L}_D \cap H \in \mathbb{P}^6$. Set 7×4 matrix $I := (\iota_D^t)_{D \in S}$.

Compute a basis $\varphi_0, \ldots, \varphi_3$ for the kernel of the 7×4 matrix I where the last one is the coefficient vector $\varphi_3 := (c_0, \ldots, c_6)$. Set polynomials

$$\Phi_i(X) := \textstyle\sum_{j=0}^{6} \varphi_{ij} X^j \quad \text{for } i = 0, 1, 2 \quad \text{where } \varphi_i = (\varphi_{i\,0}, \ldots, \varphi_{i\,6}).$$

We compute conic and cubic curves, $\mathscr{Q}$ and $\mathscr{C}$, in $\mathbb{P}^2$ from $(\Phi_i)_{i=0,1,2}$ to obtain six Weierstrass points and defining polynomial $\tilde{f}$ for the target curve $\tilde{C}$. For each $0 \leq i \leq j \leq 2$, compute vector $r_{ij} \in \mathbb{F}_q^6$ of coefficients $r_{ij} := (r_{(ij)\ell})_{\ell=0,..,5}$ such that $(\Phi_i \Phi_j (X) \bmod f(X)) = \sum_{\ell=0}^{5} r_{(ij)\ell} X^\ell$. Set 6×6 matrix $R := (r_{(ij)\ell})_{0 \leq i \leq j \leq 2,\ \ell=0,..,5}$. Compute a generator vector (q_{ij}) for the kernel of R. Set the quadratic form and conic as

$$Q(V_0, V_1, V_2) := \sum_{0 \leq i \leq j \leq 2} q_{ij} V_i V_j \quad \text{and} \quad \mathscr{Q} := \{Q = 0\} \subset \mathbb{P}^2.$$

For each $0 \le i \le j \le k \le 2$, compute vector $t_{ijk} \in \mathbb{F}_q^6$ of coefficients $t_{ijk} := (t_{(ijk)\ell})_{\ell=0,..,5}$ such that $(\Phi_i \Phi_j \Phi_k(X) \bmod f(X)) = \sum_{\ell=0}^{5} t_{(ijk)\ell} X^\ell$. Set 10×6 matrix $T := (t_{(ijk)\ell})_{0\le i\le j\le k\le 2,\ \ell=0,..,5}$. Compute a nonzero vector (γ_{ijk}) for the kernel of T. Set the cubic form and curve as

$$\Gamma(V_0, V_1, V_2) := \textstyle\sum_{0\le i\le j\le k\le 2} \gamma_{ijk} V_i V_j V_k \quad \text{and} \quad \mathscr{C} := \{\Gamma = 0\} \subset \mathbb{P}^2.$$

We obtain our target curve $\tilde{C}$ as the double cover of $\mathscr{Q}$ ramified at six points $\mathscr{C} \cap \mathscr{Q}$, which give six Weierstrass points on $\tilde{C}$. Therefore, first compute a rational curve parametrization $(V_0 : V_1 : V_2) = (P_0(X, Z) : P_1(X, Z) : P_2(X, Z))$ of the conic $\mathscr{Q}$ and then output

$$\tilde{f}(X, Z) := \Gamma(P_0(X, Z), P_1(X, Z), P_2(X, Z)) \quad \text{and} \quad \tilde{C} : Y^2 = \tilde{f}(X, 1).$$

Dolgachev–Lehavi Isogeny Sequence Order three torsion points on a Jacobian are calculated as follows [19, 20]: Let the Mumford coordinates (u_1, u_0, v_1, v_0) with $u(X) = X^2 + u_1 X + u_0$ and $v(X) = v_1 X + v_0$ (see Eq. (4)). By solving the third division polynomial equation $\mathfrak{R}(u_1) = 0$, we determine u_1 and then other u_0, v_1, v_0 are determined using auxiliary polynomials $\mathfrak{S}$, $\mathfrak{W}$, and $\mathfrak{Z}$ as follows:

$$\mathfrak{R}(u_1) = 0,\ u_0 = \mathfrak{S}(u_1),\ v_1^2 = \mathfrak{W}(u_1),\ v_0 = v_1 \mathfrak{Z}(u_1),$$

where $\deg(\mathfrak{R}) = 40,\ \deg(\mathfrak{S}), \deg(\mathfrak{W}), \deg(\mathfrak{Z}) < 40$. First, we obtain a maximal isotropic subgroup K in $J_C[3]$ by solving the above polynomial systems. Then, we apply the above algorithm of Dolgachev–Lehavi (and division polynomial solving), iteratively.

Improving the iteration process (isogeny sequence computation) will be treated with elsewhere.

5 Concluding Remarks

Bos and Friedberger [4] investigated special primes $p := 2^{e_A}\ell^{e_B} \pm 1$ for implementing SIDH with efficient arithmetic modulo p, for searching "SIDH-friendly primes" (see also Costello et al. [12]). They focus on comparison of arithmetic operations with different primes p. The primary choice for ℓ is $\ell = 3$, but they also consider other choices $\ell = 5, 7, \ldots$ in [4]. Therefore, one research direction is to construct explicit and efficient isogeny sequence computation for other than 2 and 3, for example, $\ell = 5$ for the above comparison. Another approach for the comparison and searching SIDH-friendly primes is to explicitly determine the expansion factor of Pizer graph $\mathscr{G}_\ell(p)$, especially for the case that $\ell = 2, 3$. A large expansion factor leads to a better mixing property (efficient cryptography !), and we can choose a desirable p used for a standard parameter selection.

Similarly, also for the ordinary isogeny graph, since random self-reducibility of DLP depends on the expansion factor, the precise estimate is important.

Acknowledgements The author would like to thank Kazuto Matsuo for his valuable comments on genus 2 division polynomials given in Sect. 4.3.

References

1. R. Azarderakhsh, D. Jao, K. Kalach, B. Koziel, C. Leonardi, Key compression for isogeny-based cryptosystems. AsiaPKC **2016**, 1–10 (2016)
2. R. Azarderakhsh, B. Koziel, A. Jalali, M.M. Kermani, D. Jao, NEON-SIDH: efficient implementation of supersingular isogeny Diffie-Hellman key-exchange protocol on ARM. IACR Cryptol. ePrint Archive **2016**, 669 (2016). (To appear in CANS 2016)
3. J. Biasse, D. Jao, A. Sankar, A quantum algorithm for computing isogenies between supersingular elliptic curves. INDOCRYPT **2014**, 428–442 (2014)
4. J.W. Bos, S. Friedberger, Fast arithmetic modulo $2^x p^y \pm 1$. IACR Cryptol. ePrint Arch. **2016**, 986 (2016)
5. J.B. Bost, J.F. Mestre, Moyenne arithmético-géométrique et périodes des courbes de genre 1 et 2. Gaz. Math. Soc. France **38**, 36–64 (1988)
6. E.H. Brooks, D. Jetchev, B. Wesolowski, Isogeny graphs of ordinary abelian varieties. IACR Cryptol. ePrint Arch. **2016**, 947 (2016)
7. D. Charles, E. Goren, K. Lauter, Families of Ramanujan graphs and quaternion algebras. in *Groups and Symmetries: From Neolithic Scots to John McKay* (2009), pp. 53–80
8. D. Charles, K. Lauter, E. Goren, Cryptographic hash functions from expander graphs. J. Crypt. **22**(1), 93–113 (2009)
9. A. Childs, D. Jao, V. Soukharev, Constructing elliptic curve isogenies in quantum subexponential time. J. Math. Crypt. **8**(1), 1–29 (2014)
10. R. Cosset, D. Robert, Computing (ℓ, ℓ)-isogenies in polynomial time on jacobians of genus 2 curves. Math. Comput. **84**, 1953–1975 (2015)
11. C. Costello, P. Longa, Four||: Four-dimensional decompositions on a ||-curve over the mersenne prime, in *ASIACRYPT 2015, Part I* (2015), pp. 214–235
12. C. Costello, P. Longa, M. Naehrig, Efficient algorithms for supersingular isogeny Diffie-Hellman, in *CRYPTO 2016, Part I* (2016), pp. 572–601
13. L. De Feo, D. Jao, J. Plût, Towards quantum-resistant cryptosystems from supersingular elliptic curve isogenies. J. Math. Cryptol. **8**(3), 209–247 (2014)
14. C. Delfs, S.D. Galbraith, Computing isogenies between supersingular elliptic curves over $\mathbb{F}_p$. Des. Codes Cryptogr. **78**(2), 425–440 (2016)
15. I. Dolgachev, D. Lehavi, On isogenous principally polarized abelian surfaces, in *Curves and Abelian Varieties, Contemporary Mathematics*, vol. 465 (2008), pp. 51–69
16. S. Galbraith, *Mathematics of Public Key Cryptography* (Cambridge University Press, Cambridge, 2012)
17. S.D. Galbraith, C. Petit, B. Shani, Y.B Ti, On the security of supersingular isogeny cryptosystems, in *ASIACRYPT 2016, Part I* (2016), pp. 63–91
18. S.D. Galbraith, C. Petit, J. Silva, Signature schemes based on supersingular isogeny problems. IACR Cryptol. ePrint Arch. **2016**, 1154 (2016)
19. P. Gaudry, É. Schost, Construction of secure random curves of genus 2 over prime fields. EUROCRYPT **2004**, 239–256 (2004)
20. P. Gaudry, É. Schost, Genus 2 point counting over prime fields. J. Symb. Comput. **47**(4), 368–400 (2012)
21. D. Jao, S.D. Miller, R. Venkatesan, Expander graphs based on GRH with an application to elliptic curve cryptography. J. Number Theory **129**, 1491–1504 (2009)

22. D. Jetchev, B. Wesolowski, On graphs of isogenies of principally polarizable abelian surfaces and the discrete logarithm problem, in *CoRR* (2015), https://arxiv.org/abs/1506.00522
23. D. Kohel, Endomorphism rings of elliptic curves over finite fields. Ph.D. thesis, University of California at Berkeley (1996)
24. B. Koziel, R. Azarderakhsh, S.H.F. Langroudi, M.M. Kermani, Post-quantum cryptography on FPGA based on isogenies on elliptic curves. IACR Cryptol. ePrint Arch. **2016**, 672 (2016). (To appear in IEEE Transactions on Circuits and Systems (TCAS-I))
25. M. Krebs, A. Shaheen, *Expander Families and Cayley Graphs: A Beginner's Guide* (Oxford University Press, Oxford, 2011)
26. D. Lubicz, D. Robert, Computing isogenies between abelian varieties. Compos. Math. **148**, 1483–1515 (2012)
27. A. Lubotzky, B. Weiss, Groups and expanders, in *Expanding Graphs, Proceedings of a DIMACS Workshop*, vol. 1992 (1992), pp. 95–110
28. D. Moody, D. Shumow, Analogues of Vélu's formulas for isogenies on alternate models of elliptic curves. Math. Comput. **85**, 1929–1951 (2016)
29. A. Pizer, Ramanujan graphs, in *Computational Perspectives on Number Theory* (American Mathematical Society, 1998), pp. 159–178
30. A. Rostovtsev, A. Stolbunov, Public-key cryptosystem based on isogenies. IACR Cryptol. ePrint Arch. **2006**, 145 (2006), http://eprint.iacr.org/2006/145
31. J. Silverman, *The Arithmetic of Elliptic Curves*, GTM, vol. 106, 2nd edn. (Springer, Berlin, 2009)
32. B. Smith, *Explicit endomorphisms and correspondences*. Ph.D. thesis, The University of Sydney (2005)
33. B. Smith, Computing low-degree isogenies in genus 2 with the Dolgachev-Lehavi method. Arith. Geom. Coding Theory Contemp. Math. **574**, 159–170 (2012)
34. A. Sutherland, Identifying supersingular elliptic curves. LMS J. Comput. Math. **15**, 317–325 (2012)
35. A.V. Sutherland, Isogeny volcanoes, in *Algorithmic Number Theory 10th International Symposium (ANTS X), Open Book Series*, vol. 1 (MSP, 2013), pp. 507–530
36. H. Tachibana, K. Takashima, T. Takagi, Constructing an efficient hash function from 3-isogenies. To appear in JSIAM Letters (2016)
37. K. Takashima, R. Yoshida, An algorithm for computing a sequence of Richelot isogenies. Bull. Korean Math. Soc. **46**(4), 789–802 (2009)
38. J. Vélu, Isogénies entre courbes elliptiques. C.R. Acad. Sc. Paris, Séries A. **273**, 238–241 (1971)
39. R. Yoshida, K. Takashima, Computing a sequence of 2-isogenies on supersingular elliptic curves. IEICE Trans. Fundam. **96-A**(1), 158–165 (2013)

Part II
Mathematics Towards Cryptography

Spectral Degeneracies in the Asymmetric Quantum Rabi Model

Cid Reyes-Bustos and Masato Wakayama

Abstract The aim of this article is to investigate certain family of (so-called constraint) polynomials which determine the quasi-exact spectrum of the asymmetric quantum Rabi model. The quantum Rabi model appears ubiquitously in various quantum systems and its potential applications include quantum computing and quantum cryptography. In (Wakayama, Symmetry of Asymmetric Quantum Rabi Models) [30], using the representation theory of the Lie algebra $\mathfrak{sl}_2$, we presented a picture of the asymmetric quantum Rabi model equivalent to the one drawn by confluent Heun ordinary differential equations. Using this description, we proved the existence of spectral degeneracies (level crossings in the spectral graph) of the asymmetric quantum Rabi model when the symmetry-breaking parameter ε equals $\frac{1}{2}$ by studying the constraint polynomials, and conjectured a formula that ensures the presence of level crossings for general $\varepsilon \in \frac{1}{2}\mathbb{Z}$. These results on level crossings generalize a result on the degenerate spectrum, given first by Kuś in 1985 for the (symmetric) quantum Rabi model. It was demonstrated numerically by Li and Batchelor in 2015, investigating an earlier empirical observation by (Braak, Phys. Rev. Lett. 107, 100401–100404, 2011) [3]. In this paper, although the proof of the conjecture has not been obtained, we deepen this conjecture and give insights together with new formulas for the target constraint polynomials.

Keywords Quantum rabi models · Degenerate eigenvalues · Level crossings · Exceptional spectrum · Juddian solutions · quasi-exact solutions · Constraint polynomials · confluent Heun differential equations · Orthogonal polynomials · discrete series · Stirling numbers of the first kind · Eulerian numbers

C. Reyes-Bustos (✉)
Graduate School of Mathematics, Kyushu University, 744 Motooka,
Nishi-ku, Fukuoka 819-0395, Japan
e-mail: c-reyes@math.kyushu-u.ac.jp

M. Wakayama
Institute of Mathematics for Industry, Kyushu University, 744 Motooka,
Nishi-ku, Fukuoka 819-0395, Japan
e-mail: wakayama@imi.kyushu-u.ac.jp

T. Takagi et al. (eds.), *Mathematical Modelling for Next-Generation Cryptography*,
Mathematics for Industry 29, DOI 10.1007/978-981-10-5065-7_7

2010 Mathematics Subject Classification: *Primary* 34L40 · *Secondary* 81Q10 · 34M05 · 81S05.

1 Introduction

The interaction of matter and light is one of the important subjects in theoretical/experimental physics and applications. Among them, the *quantum Rabi model* (QRM), the fully quantized version [15] of the original Rabi model (see also [8]), is known to be the simplest model used in quantum optics to describe such interaction, i.e., the interaction between a two-level atom and a single bosonic mode (photon). It actually appears ubiquitously in various quantum systems including cavity and circuit quantum electrodynamics, quantum dots and artificial atoms, with potential applications in quantum information technologies including quantum cryptography, quantum computing, etc. (see e.g., [11]). Although the QRM has no continuous symmetry, it has a $\mathbb{Z}_2$-symmetry (parity). Using this $\mathbb{Z}_2$-symmetry, Braak [3] has shown the integrability of the QRM in 2011 (see [25]).

The aim of this article is to study a particular structure of the spectrum of the Hamiltonian $H^{\varepsilon}_{\text{Rabi}}$ of the *asymmetric quantum Rabi model* (AQRM) [32] where the $\mathbb{Z}_2$-symmetry is broken by the presence of an additional spin-flip operator $\varepsilon\sigma_x$. The Hamiltonian of the symmetric QRM is given by H^{0}_{Rabi}. This asymmetric model actually provides a more realistic description of circuit QED experiments employing flux qubits (two-level systems) than the QRM itself [23]. The Hamiltonian $H^{\varepsilon}_{\text{Rabi}}$ ($\hbar = 1$) reads

$$H^{\varepsilon}_{\text{Rabi}} = \omega a^{\dagger}a + \Delta\sigma_z + g\sigma_x(a^{\dagger} + a) + \varepsilon\sigma_x, \tag{1}$$

where $a^{\dagger}$ and a are the creation and annihilation operators of the bosonic mode, i.e., $[a, a^{\dagger}] = 1$ and $\sigma_x = \begin{bmatrix} 0 & 1 \\ 1 & 0 \end{bmatrix}$, $\sigma_z = \begin{bmatrix} 1 & 0 \\ 0 & -1 \end{bmatrix}$ are the Pauli matrices, 2Δ is the energy difference between the two levels, and g denotes the coupling strength between the two-level system and the bosonic mode with frequency ω (subsequently, we set $\omega = 1$). There are several studies [2, 5, 7, 19, 20, 30] on the spectrum of this AQRM. It is worth noting that most of these papers are more or less emphasizing appearance of level crossings or degeneracy of the spectrum of the models.

We now recall the spectral structure of the "symmetric" QRM. Denote the spectrum (set of all eigenvalues) of the quantum Rabi model Hamiltonian H^{0}_{Rabi} by Σ_{Rabi}. It is known that Σ_{Rabi} consists of the regular spectrum $\sigma_{\text{reg.}}$, the degenerate exceptional spectrum $\sigma_{\text{deg.excep.}}$ and the non-degenerate exceptional spectrum $\sigma_{\text{non-deg.excep.}}$. Notice that the non-degenerate spectrum is the union of the sets $\sigma_{\text{reg.}}$ and $\sigma_{\text{non-deg.excep}}$. Moreover, the degenerate exceptional eigenstates, described by Kuś [18] in 1985, are called *Juddian solutions* of the QRM and the multiplicity of the eigenvalue is always equal to 2 [7, 22, 31]. These solutions are not present for arbitrary model parameters g and Δ. If one degenerate solution appears in the spectrum, the parameters satisfy a certain condition, given by a so-called *constraint polynomial* (of order, say N).

The two wave functions associated to the degenerate eigenvalue have then an especially simple form and the eigenvalue itself is simply $N - g^2$. These special solutions are therefore called *quasi-exact* and have been formally investigated by Turbiner [27].

The existence of degenerate eigenvalues for special values of the parameters leads naturally to *level crossings* in the spectral graph. For the symmetric QRM these crossings are expected to occur on general grounds (not related to their quasi-exact features) because the $\mathbb{Z}_2$-symmetry creates two dynamically invariant subspaces whose respective spectra may intersect upon changing the parameters. In the AQRM, where this symmetry is lost, no level crossings should be expected to appear. Indeed, for all values $\varepsilon \notin \frac{1}{2}\mathbb{Z}$, the spectral graph is free from any level crossings. The AQRM has also a quasi-exact spectrum with simple wave functions given by constraint polynomials but the quasi-exact eigenvalues are no longer doubly degenerate [19, 20]. It comes therefore as a surprise that for special (half-integer) values of the symmetry-breaking parameter ε, level crossings do seem to appear in the AQRM. This paper is devoted to their investigation from a mathematical point of view. (Generally, level crossings have been studied in mathematics for various interacting quantum models, e.g., [12–14, 29]). Clearly, the level crossings with nonzero ε may be considered to be an AQRM-counterpart of the Juddian solutions of the QRM [19, 30]. In [30], using the representation theory of $\mathfrak{sl}_2$ via the Bargmann picture [1] of the model, we illustrated the picture of the AQRM equivalent to the one drawn by confluent Heun ordinary differential equations (see, e.g., [24]). (This may be derived from the method developed in [27].) Furthermore, if we consider the Hurwitz-type spectral zeta function $\zeta_{H^{\varepsilon}_{\text{Rabi}}}(s, \tau)$ for the AQRM as studied in the recent paper [26] for the QRM, the "gamma factor" of the associate zeta regularized product [28] should be given by a product over the spectrum of degenerate (level crossings) and non-degenerate Juddian solutions together with the non-degenerate exceptional spectrum for the AQRM if any.

We extensively discuss a family of constraint polynomials, which determines the quasi-exact spectrum of the AQRM. A representation theoretic derivation of such constraint polynomials for the AQRM was obtained in [30]. Actually, the constraint polynomials are defined as the continuant of certain tridiagonal matrices obtained by the eigenvalue problem in the space of finite dimensional irreducible representation of $\mathfrak{sl}_2$ (see, e.g., [21]). By this description, we proved in [30] the existence of level crossings in the spectral graph of the AQRM for the case where the symmetry-breaking parameter ε equals $\frac{1}{2}$ and proposed a conjecture concerning two (different sets of ($\varepsilon \neq 0$)) constraint polynomials (for the relation between ε and $-\varepsilon$) for general $\varepsilon \in \frac{1}{2}\mathbb{Z}$. This result was demonstrated numerically by Li and Batchelor in 2015, investigating an earlier empirical observation in [3]. The main purpose of the paper is an extensive study of the constraint polynomials for the AQRM, leading to a set of conjectures. In the last section, we investigate the constraint polynomials for some special cases by studying continued fractions and discuss also certain coefficients of the polynomials by combinatorial quantities like the Stirling numbers of the first kind and the Eulerian numbers.

2 Constraint Polynomials and Conjectures

Kuś [18] obtained all Juddian solutions [16] of the QRM and proved a theorem on their number as function of g and Δ for each $N \in \mathbb{N}$. In this case, all Juddian solutions are degenerate quasi-exact and the multiplicity (the dimension of the corresponding eigenspace) is always 2 (doubly degenerate). In the case of AQRM, Li and Batchelor [19] have demonstrated the existence of (in general non-degenerate) quasi-exact solutions for any ε by proving the corresponding claim (concerning the constraint polynomials) for their number as function of g, Δ and ε for each N. More precisely, they obtained the following estimate of the quasi-exact solutions in the Appendix B [19].

Lemma 2.1 *For each k $(0 \leq k < N)$, there are $N-k$ quasi-exact solutions (as values of the parameter g) for Δ in the range*

$$\sqrt{k^2 + 2k\varepsilon} < \Delta < \sqrt{(k+1)^2 + 2(k+1)\varepsilon}.$$

For the QRM case, it follows immediately from this lemma that there are actually $N-k$ crossings when $k < \Delta < k+1$.

The constraint polynomials for the quasi-exactness for the AQRM are defined as follows. Let $N \in \mathbb{N}$. Define the two variable polynomials $P_k^{(N,\varepsilon)}(x,y)$ by the recurrence relations

$$\begin{aligned}
&P_0^{(N,\varepsilon)}(x,y) = 1,\\
&P_1^{(N,\varepsilon)}(x,y) = x + y - 1 - 2\varepsilon,\\
&P_k^{(N,\varepsilon)}(x,y) = (kx + y - k^2 - 2k\varepsilon)P_{k-1}^{(N,\varepsilon)}(x,y) - k(k-1)(N-k+1)xP_{k-2}^{(N,\varepsilon)}(x,y).
\end{aligned}$$

We sometimes omit one or both of the variables x, y and write $P_k^{(N,\varepsilon)}(x)$, $P_k^{(N,\varepsilon)}$, etc. in place of $P_k^{(N,\varepsilon)}(x,y)$, respectively, in accordance with the situation. By definition, $\deg P_k^{(N,\varepsilon)} = k$ as a polynomial with respect to any of its variables. It is also obvious that $P_k^{(N,\varepsilon)}(x,y) \in \mathbb{Z}[x,y]$ when $\varepsilon \in \frac{1}{2}\mathbb{Z}$. We also define polynomials $\tilde{P}_k^{(N,\varepsilon)}(x,y)$ by

$$\tilde{P}_k^{(N,\varepsilon)}(x,y) = P_k^{(N,-\varepsilon)}(x,y). \tag{2}$$

We call $P_k^{(N,\varepsilon)}(x,y)$ and $\tilde{P}_k^{(N,\varepsilon)}(x,y)$ the constraint polynomials for the Hamiltonian $H_{\text{Rabi}}^{\varepsilon}$ of the AQRM. For a fixed value $y = \Delta^2$ (for the given energy difference 2Δ between the two levels), it is known [19, 30] that if $x = (2g)^2$ is a root of the polynomial $P_N^{(N,\varepsilon)}(x)$ then $\lambda = N - g^2 + \varepsilon$ is an eigenvalue of $H_{\text{Rabi}}^{\varepsilon}$. Note that, as mentioned above (the constraint condition given by the polynomial) $P_N^{(N,\varepsilon)}(x) = 0$ possesses a positive root for any ε, which was shown in [19] in a way analogous to [18]. Precisely, Lemma 2.1 implies that for each k $(0 \leq k < N)$, $P_N^{(N,\varepsilon)}(x,y) = 0$ has $N-k$ different positive roots (as a polynomial in x) when $y(=\Delta^2)$ belongs to the open interval $(k^2 + 2k\varepsilon,\ (k+1)^2 + 2(k+1)\varepsilon)$. Moreover, by definition, if $x = (2g)^2$ is a root of $\tilde{P}_N^{(N,\varepsilon)}(x)$ then $\lambda = N - g^2 - \varepsilon$ is also an eigenvalue of $H_{\text{Rabi}}^{\varepsilon}$. Therefore, in particular, it follows that if the polynomials $\tilde{P}_{N+1}^{(N+1,\frac{1}{2})}(x)$ and $P_N^{(N,\frac{1}{2})}(x)$ possess a common positive root $x (= (2g)^2)$ then the multiplicity of the eigenvalue $\lambda = N - g^2 + \frac{1}{2}$ of $H_{\text{Rabi}}^{\varepsilon}$ becomes two, i.e., a level crossing occurs at the energy curves or spectral graph with respect to the parameter g. Indeed, we have proved [30] that the corresponding two eigenvectors are linearly independent by representation theory of $\mathfrak{sl}_2$. (Also, for a systematic derivation of the recurrence relations of $P_k^{(N,\varepsilon)}(x, y)$ and $\tilde{P}_k^{(N,\varepsilon)}(x, y)$ from the eigenvalue problem of the Hamiltonian $H_{\text{Rabi}}^{\varepsilon}$ using finite dimensional representation theory of the Lie algebra $\mathfrak{sl}_2$, see [30]).

The existence of such level crossing was empirically observed in [3] when $\varepsilon = \frac{1}{2}$. Recently, Li and Batchelor [19] gave explicit evidence based on numerical computation on roots of polynomials $\tilde{P}_{N+1}^{(N+1,\frac{1}{2})}(x)$ and $P_N^{(N,\frac{1}{2})}(x)$, asserting the existence of level crossings for $\varepsilon = \frac{1}{2}$. Moreover, they demonstrated that level crossings occur in general for $\varepsilon \in \frac{1}{2}\mathbb{Z}$. Their claim for the case $\varepsilon = \frac{1}{2}$ was proved in [30]. Actually, the claim follows immediately from Lemma 2.5 (Theorem 6.2 [30]) below. In [30] (Conjecture 6.4), we also presented the following conjecture. In the sequel, we assume that $x = (2g)^2$ and $y = \Delta^2$ when we discuss in the physical context.

Conjecture 2.2 *For $\ell, N \in \mathbb{N}$, there exists a polynomial $A_N^{\ell}(x, y) \in \mathbb{Z}[x, y]$ such that*

$$\tilde{P}_{N+\ell}^{(N+\ell,\ell/2)}(x, y) = A_N^{\ell}(x, y) P_N^{(N,\ell/2)}(x, y). \tag{3}$$

Moreover, the polynomial $A_N^{\ell}(x, y)$ is positive for any $x, y > 0$.

Example 1 The following are the proposed polynomials for small values of ℓ.

$$\begin{aligned}
A_N^1(x, y) &= (N+1)x + y,\\
A_N^2(x, y) &= (N+1)_2 x^2 + (3+2N)xy + y(1+y),\\
A_N^3(x, y) &= (N+1)_3 x^3 + (11 + 3N(N+4))x^2 y + (N+2)x(3y+4)y + y(2+y)^2,\\
A_N^4(x, y) &= (N+1)_4 x^4 + (50 + 4\sum_{l=1}^{N}(11 + 3l(l+4))))x^3 y\\
&\quad + ((58 + 10N(N+5)) + (35 + 6N(N+5))y)x^2 y\\
&\quad + 2(5+2N)xy(y+2)(y+3)\\
&\quad + y(3+y)^2(4+y).
\end{aligned}$$

where $(a)_n := a(a+1)\cdots(a+n-1) = \Gamma(a+n)/\Gamma(a)$.

Remark 2.3 If Conjecture 2.2 is true, we note that the polynomial $A_N^{\ell}(x, y)$ has a determinant expression from the results in [30] derived from representation theory of $\mathfrak{sl}_2$.

Lemma 2.4 *1. The constant terms of the polynomials $\tilde{P}_{N+\ell}^{(N+\ell,\ell/2)}(x,y)$ and $P_N^{(N,\ell/2)}(x,y)$ with respect to the variable x are, respectively, given by*

$$P_N^{(N,\ell/2)}(0,y)=\prod_{k=1}^{N}\big(y-k(k+\ell)\big) \quad and \quad \tilde{P}_{N+\ell}^{(N+\ell,\ell/2)}(0,y)=\prod_{k=1}^{N+\ell}\big(y-k(k-\ell)\big).$$

In particular, we have $A_N^\ell(0,y)=\prod_{k=1}^{\ell}\big(y-k(k-\ell)\big)$.

2. *The coefficients of the leading term of $\tilde{P}_{N+\ell}^{(N+\ell,\ell/2)}(x,y)$ and $P_N^{(N,\ell/2)}(x,y)$ are respectively given by $(N+\ell)!$ and $N!$. Moreover, if we assume that $A_N^\ell(x,y)$ is a polynomial in x, then the coefficient of the leading term of $A_N^\ell(x,y)$ with respect to x, i.e., the coefficient of x^ℓ, is given by $(N+1)_\ell$.*

Proof 1. The claim for the constant terms follows from the definition of the respective recurrence relations. The latter follows also immediate from $A_N^\ell(0,y)=\tilde{P}_{N+\ell}^{(N+\ell,\ell/2)}(0,y)/P_N^{(N,\ell/2)}(0,y)$.
2. If we look at the highest degree with respect to x at the recurrence relations of $\tilde{P}_{N+\ell}^{(N+\ell,\ell/2)}(x,y)$ and $P_N^{(N,\ell/2)}(x,y)$, the former claim follows immediately. Hence the latter is also obvious. □

If Conjecture 2.2 is true, since the positive roots $x=(2g)^2$ of the polynomials $P_N^{(N,\ell/2)}(x)$ and $\tilde{P}_{N+\ell}^{(N+\ell,\ell/2)}(x)$ (for a fixed $y=\Delta^2$) coincide, we find that the multiplicity of the eigenvalue $\lambda=N-g^2+\frac{\ell}{2}$ of $H_{\text{Rabi}}^{\varepsilon}$ is equal to two.

Moreover, when $\varepsilon=\frac{1}{2}$ (i.e., $\ell=1$), we have the following identity in [30] (Theorem 6.2).

Lemma 2.5 *The following formulas hold for any k $(0\le k\le N)$.*

$$\tilde{P}_{k+1}^{(N+1,\frac{1}{2})}(x)=[(k+1)x+\Delta^2]P_k^{(N,\frac{1}{2})}(x)-k(k+1)(N-k)xP_{k-1}^{(N,\frac{1}{2})}(x). \quad (4)$$

Therefore, we now generalize the conjecture above more explicit manner, namely, for a general $k=0,1,\ldots,N$, we have the following.

Conjecture 2.6 *For $\ell, N\in\mathbb{N}$, there exist polynomials $A_k^{(N,\ell/2)}(x,y)\in\mathbb{Z}[x,y]$ and $B_k^{(N,\ell/2)}(x,y)\in\mathbb{Z}[x,y]$ $(k=0,1,\ldots,N)$ such that*

$$\tilde{P}_{k+\ell}^{(N+\ell,\ell/2)}(x,y)=A_k^{(N,\ell/2)}(x,y)P_k^{(N,\ell/2)}(x,y)+B_k^{(N,\ell/2)}(x,y), \quad (5)$$

with $B_N^{(N,\ell/2)}(x,y)=B_0^{(N,\ell/2)}(x,y)=0$. Moreover, the polynomials $A_k^{(N,\ell/2)}(x,y)$ are positive for any $x,y>0$.

Remark 2.7 Note that $A_N^{(N,\ell/2)}(x,y)$ in Conjecture 2.6 is nothing but $A_N^\ell(x,y)$ in Conjecture 2.2. Obviously, Conjecture 2.2 follows immediately from Conjecture 2.6.

Remark 2.8 From Lemma 2.5, we observe

$$A_k^{(N,\frac{1}{2})}(x,y) = (k+1)x + y, \quad B_k^{(N,\frac{1}{2})}(x,y) = -k(k+1)(N-k)xP_{k-1}^{(N,\frac{1}{2})}(x,y)$$

for $k = 0, 1, \ldots, N$ upon setting $P_{-1}^{(N,\frac{1}{2})}(x,y) = 0$. Also, we notice that

$$A_0^{(N,\ell/2)}(x,y) = \tilde{P}_\ell^{(N+\ell,\ell/2)}(x,y).$$

Example 2 For $\ell = 2$, we observe

$$A_k^{(N,1)}(x,y) = (k+1)_2 x^2 + (3+2k)xy + y(1+y),$$
$$B_0^{(N,1)}(x,y) = -2Nx$$

and the case $k \geq 1$ for $B_k^{(N,1)}(x,y)$ is given by

$$B_k^{(N,1)}(x,y)$$
$$= -(k+1)(N-k)x\left(2\left(k(k+2)x + (k+1)y\right)P_{k-1}^{(N,1)}(x,y) - (k+2)Q_k^{(N,1)}(x,y)\right).$$

Here

$$Q_k^{(N,1)}(x,y) = \sum_{i=1}^{k-1} \frac{k!\,(N-i)!}{i!\,(N-k)!}\left(P_i^{(N,1)}(x,y) - (y - i(i+2))P_{i-1}^{(N,1)}(x,y)\right).$$

Remark 2.9 Suppose $x < 0$. Let us consider $P_k^{(N,\varepsilon)}(x,y)$ as polynomials in y and the recurrence relations

$$P_k^{(N,\varepsilon)}(y) = \{y - (-kx + k^2 + 2k\varepsilon)\}P_{k-1}^{(N,\varepsilon)}(y) - k(k-1)(N-k+1)xP_{k-2}^{(N,\varepsilon)}(y). \tag{6}$$

starting from $k = N+1$ with initial values $P_{N+1}^{(N,\varepsilon)}(y) = 0,\ P_{N+2}^{(N,\varepsilon)}(y) = 1$. Then, by the Favard theorem (see, e.g., [9]), there is a unique moment functional $\mathscr{M}$ such that

$$\begin{cases} \mathscr{M}(1) = (N+1)(N+2)x, \\ \mathscr{M}(P_i^{(N,\varepsilon)}(y)P_j^{(N,\varepsilon)}(y)) = c_{i,j}\delta_{ij} \quad \text{for} \quad i,j = N+2, N+3, \ldots. \end{cases}$$

More precisely, since $k(k-1)(N-k+1)x > 0$ for $k > N+1$, the moment functional $\mathscr{M}$ is positive definite, whence the set $\{P_k^{(N,\varepsilon)}(y)\}_{k=N+2,N+3,\ldots}$ defines an orthogonal family of polynomials. Mathematically, it is important to study i) the generating function $\sum_{k=N+2}^{\infty} P_k^{(N,\varepsilon)}(y)t^k$, ii) the differential equation satisfied by $P_k^{(N,\varepsilon)}(y)$, and iii) the measure $d\mu$, i.e., the function $\mu : \mathbb{R}^+ \to \mathbb{R}^+$ of bounded variation on nonnegative real half line $\mathbb{R}^+$ such that $\int_0^\infty y^n d\mu$ is finite for all

$n = 0, 1, 2, \ldots$ and $\int_0^\infty P_i^{(N,\varepsilon)}(y)P_j^{(N,\varepsilon)}(y)d\mu = c_{i,j}\delta_{ij}$ $(i, j = N+2, N+3, \ldots)$ for some nonzero real constants $c_{i,j}$. Furthermore, it is interesting if we may find any representation theoretic interpretation of these orthogonal polynomials in terms of, e.g., the discrete series representations [21] of $\mathfrak{sl}_2(\mathbb{R})$ (see [30]).

Remark 2.10 From the three terms recurrence equation (6), we have a determinant expression of the polynomials $P_k^{(N,\varepsilon)}(y)$ (and $\tilde{P}_k^{(N,\varepsilon)}(y)$) in y by a tridiagonal matrix with all off-diagonal elements equal to 1. (Note that the coefficients of the matrix are at most linear in x). Then the resulting determinant expression is obviously different from the one obtained in [30].

3 Explicit Formulas for the Constraint Polynomials

In this section, we give an explicit expression of the constraint polynomials and discuss on Conjecture 2.2. The approach we use to find the coefficients of the constrain polynomials is to compute the derivatives at all orders by differentiating the defining recurrence formula. In the sequel, the notation ∂_x^m stands for the m−th derivative with respect to the variable x.

Proposition 3.1 *For $m \in \mathbb{N}$, we have*

$$\begin{aligned}\partial_x^m P_k^{(N,\ell/2)}(x) &= mk(\partial_x^{m-1} P_{k-1}^{(N,\ell/2)}(x) - (k-1)(N-k+1)\partial_x^{m-1} P_{k-2}^{(N,\ell/2)}(x))\\ &\quad + (kx + y - k(k+\ell))\partial_x^m P_{k-1}^{(N,\ell/2)}(x) - k(k-1)(N-k+1)x\partial_x^m P_{k-2}^{(N,\ell/2)}(x).\end{aligned}$$

and

$$\begin{aligned}\partial_x^m \tilde{P}_k^{(N+\ell,\ell/2)}(x) &= mk(\partial_x^{m-1} \tilde{P}_{k-1}^{(N+\ell,\ell/2)}(x) - (k-1)(N+\ell-k+1)\partial_x^{m-1} \tilde{P}_{k-2}^{(N+\ell,\ell/2)}(x))\\ &\quad + (kx + y - k(k-\ell))\partial_x^m \tilde{P}_{k-1}^{(N+\ell,\ell/2)}(x) - k(k-1)(N+\ell-k+1)x\partial_x^m \tilde{P}_{k-2}^{(N+\ell,\ell/2)}(x).\end{aligned}$$

Proof Differentiating the defining recurrence relation for $P_k^{(N,\ell/2)}$, we obtain

$$\begin{aligned}\partial_x P_k^{(N,\ell/2)}(x) &= k(P_{k-1}^{(N,\ell/2)}(x) - (k-1)(N-k+1)P_{k-2}^{(N,\ell/2)}(x))\\ &\quad + (kx + y - k(k+\ell))\partial_x P_{k-1}^{(N,\ell/2)}(x) - k(k-1)(N-k+1)x\partial_x P_{k-2}^{(N,\ell/2)}(x).\end{aligned}$$

Repeating this process m times gives the result. The proof for the polynomials $\tilde{P}_k^{(N+\ell,\ell/2)}$ is analogous. □

To find the coefficients of the constraint polynomials, it is enough to consider the constant term of the polynomials given by the partial derivatives, in other words, the case $x = 0$. In our study of the constraint polynomials, quantities of the form

$j(j+\ell)$ and $j(j-\ell)$ for integer j have appeared repeatedly (for instance part 1. of Lemma 2.4 or Example 2). Therefore, we introduce the notations

$$c_j^{(\ell)} = j(j+\ell) \quad \text{and} \quad \tilde{c}_j^{(\ell)} = j(j-\ell). \tag{7}$$

Obviously, the relation $c_j^{(\ell)} = \tilde{c}_{j+\ell}^{(\ell)}$ holds.

Corollary 3.2 *For $m \in \mathbb{N}$, we have*

$$\begin{aligned}
&\partial_x^m P_k^{(N,\ell/2)}(0) \\
&= m \prod_{i=1}^{k} (y - c_i^{(\ell)}) \sum_{j=m}^{k} \frac{j(\partial_x^{m-1} P_{j-1}^{(N,\ell/2))}(0) - (j-1)(N-j+1)\partial_x^{m-1} P_{j-2}^{(N,\ell/2))}(0))}{\prod_{i=1}^{j}(y - c_j^{(\ell)})}.
\end{aligned}$$

In addition,

$$\begin{aligned}
&\partial_x^m \tilde{P}_k^{(N+\ell,\ell/2)}(0) \\
&= m \prod_{i=1}^{k} (y - \tilde{c}_i^{(\ell)}) \sum_{j=m}^{k} \frac{j(\partial_x^{m-1} \tilde{P}_{j-1}^{(N,\ell/2))}(0) - (j-1)(N+\ell-j+1)\partial_x^{m-1} \tilde{P}_{j-2}^{(N,\ell/2))}(0))}{\prod_{i=1}^{j}(y - \tilde{c}_j^{(\ell)})}.
\end{aligned}$$

Here we use the convention that $P_{-1}^{(N,\varepsilon)} = \tilde{P}_{-1}^{(N,\varepsilon)} = 0$.

Proof The result is obtained by expanding the recurrence relation for $\partial_x^m P_k^{(N,\ell/2)}(x)$ in Proposition 3.1 and setting $x = 0$. The proof for the case $\partial_x^m \tilde{P}_k^{(N+\ell,\ell/2)}(0)$ follows in the same manner. □

These formulas can be used to recursively compute the partial derivatives of the constraint polynomials starting from $m = 1$ by computing $\partial_x^{m-1} P_{j-1}^{(N,\ell/2))}(0) - (j-1)(N-j+1)\partial_x^{m-1} P_{j-2}^{(N,\ell/2))}(0)$ for each m. For instance, it follows from Corollary 3.2 and Lemma 2.4 that

$$\partial_x P_k^{(N,\ell/2)}(0) = \prod_{i=1}^{k} (y - c_i^{(\ell)}) \sum_{j=1}^{k} \frac{j(y-(j-1)(N+\ell))}{(y - c_j^{(\ell)})(y - c_{j-1}^{(\ell)})}$$

and

$$\partial_x \tilde{P}_k^{(N+\ell,\ell/2)}(0) = \prod_{i=1}^{k} (y - \tilde{c}_i^{(\ell)}) \sum_{j=1}^{k} \frac{j(y-(j-1)N)}{(y - \tilde{c}_j^{(\ell)})(y - \tilde{c}_{j-1}^{(\ell)})}.$$

To generalize this expression, we introduce the expressions $\psi_i(j)$ and $\tilde{\psi}_i(j)$, with $i, j \in \mathbb{N}$, given by

$$\psi_1(j) = \frac{j(y-(j-1)(N+\ell))}{(y-\tilde{c}^{(\ell)}_{\ell+j})(y-\tilde{c}^{(\ell)}_{\ell+j-1})}, \qquad \tilde{\psi}_1(j) = \frac{j(y-(j-1)(N))}{(y-\tilde{c}^{(\ell)}_{j})(y-\tilde{c}^{(\ell)}_{j-1})},$$

$$\psi_i(j) = \frac{j}{(y-\tilde{c}^{(\ell)}_{j+\ell})}\psi_{i-1}(j-1), \qquad \tilde{\psi}_i(j) = \frac{j}{(y-\tilde{c}^{(\ell)}_{j})}\tilde{\psi}_{i-1}(j-1).$$

Note that for fixed i and j, the expression is a rational function in the variable y. We extend the definition to $i, j \in \mathbb{Z}$ by setting $\psi_i(j) = 0$ whenever $i \le 0$ or $j \le 0$. With this notation, by using Corollary 3.2, we directly compute

$$\partial_x^2 P_k^{(N,l/2)}(0) = \prod_{i=1}^{k}(y-c_i^{(\ell)})\left(\sum_{j=2}^{k}\psi_2(j) + \sum_{i_1=2}^{k}\psi_1(i_1)\sum_{i_2=1}^{i_1-2}\psi_1(i_2)\right).$$

To express the general form of the derivatives of the constraint polynomials, we need a general form for sums of functions ψ_i and $\tilde{\psi}_i$ of the kind appearing in the computation of $\partial_x^2 P_k^{(N,l/2)}(0)$ above. Fix $m \le k$, a positive integer $\alpha \le m$ and a vector $\nu = (\nu_1, \nu_2, \dots, \nu_\alpha) \in \mathbb{N}^\alpha$, with $|\nu|_1 := \nu_1 + \nu_2 + \cdots + \nu_\alpha = m$. Define the rational function $\Psi_\nu^{(m)}(k) = \Psi_\nu^{(m)}(k)(y)$ in y as

$$\Psi_\nu^{(m)}(k) = \sum_{i_1=m}^{k}\psi_{\nu_1}(i_1)\sum_{i_2=m-\nu_1}^{i_1-\nu_1-1}\psi_{\nu_2}(i_2)\cdots\sum_{i_n=m-\sum_{i=1}^{n-1}\nu_j}^{i_{n-1}-\nu_{n-1}-1}\psi_{\nu_n}(i_n)\cdots\sum_{i_\alpha=m-\sum_{i=1}^{\alpha-1}\nu_j}^{i_{\alpha-1}-\nu_{\alpha-1}-1}\psi_{\nu_\alpha}(i_\alpha).$$

The rational functions $\tilde{\Psi}_\nu^{(m)}(k)$ are analogously defined. By splitting the first sum in $\Psi_\nu^{(m)}(k)$, we derive an elementary identity, which is needed later in the proof of the main result of this section. Namely, for $\nu \in \mathbb{N}^\alpha$, it holds that

$$\Psi_\nu^{(m)}(k) = \psi_{\nu_1}(k)\Psi_{\nu'}^{(m-\nu_1)}(k-1-\nu_1) + \Psi_\nu^{(m)}(k-1), \tag{8}$$

where $\nu' \in \mathbb{N}^{\alpha-1}$ is obtained by dropping the first component ν_1 of ν. If $\nu \in \mathbb{N}^1$, we set $\Psi_{\nu'}^{(i)}(j) = 1$ for any $i, j \in \mathbb{N}$.

Summarizing the discussion above, we have the following explicit expressions of the constraint polynomials.

Theorem 3.3 *We have*

$$P_k^{(N,\ell/2)}(x)(= P_k^{(N,\ell/2)}(x,y)) = \sum_{m=0}^{k}\prod_{i=1}^{k}(y-c_i^{(\ell)})\left[\sum_{\alpha=1}^{m}\sum_{\substack{\nu\in\mathbb{N}^\alpha \\ |\nu|_1=m}}\Psi_\nu^{(m)}(k)(y)\right]x^m$$

and

$$\tilde{P}_k^{(N+\ell,\ell/2)}(x)(=\tilde{P}_k^{(N+\ell,\ell/2)}(x,y)) = \sum_{m=0}^{k}\prod_{i=1}^{k}(y-\tilde{c}_i^{(\ell)})\left[\sum_{\alpha=1}^{m}\sum_{\substack{\nu\in\mathbb{N}^\alpha\\|\nu|_1=m}}\tilde{\Psi}_\nu^{(m)}(k)(y)\right]x^m.$$

Proof We prove the result only for $P_k^{(N,\ell/2)}(x)$, since the case for is $\tilde{P}_k^{(N+\ell,\ell/2)}(x)$ is analogous. To establish the result it is enough to show, for $m\le k$, that

$$\partial_x^m P_k^{(N,\ell/2)}(0) = m!\prod_{i=1}^{k}(y-c_i^{(\ell)})\left[\sum_{\alpha=1}^{m}\sum_{\substack{\nu\in\mathbb{N}^\alpha\\|\nu|_1=m}}\Psi_\nu^{(m)}(k)\right].$$

The proof is by induction, the cases $m=0,1$ were already established above. Assume the identity holds for $m\in\mathbb{N}$. Computing directly from Corollary 3.2 and using the induction hypothesis we obtain

$$\begin{aligned}
&\partial_x^{m+1}P_k^{(N,\ell/2)}(0)\\
&=(m+1)\prod_{i=1}^{k}(y-c_i^{(\ell)})\sum_{j=m+1}^{k}\frac{j(\partial_x^m P_{j-1}^{(N,\ell/2))}(0)-(j-1)(N-j+1)\partial_x^m P_{j-2}^{(N,\ell/2))}(0))}{\prod_{i=1}^{j}(y-c_j^{(\ell)})}\\
&=(m+1)!\prod_{i=1}^{k}(y-c_i^{(\ell)})\sum_{j=m+1}^{k}\frac{1}{(y-c_j)(y-c_{j-1})}\Bigg(j(y-c_{j-1})\sum_{\alpha=1}^{m}\sum_{\substack{\nu\in\mathbb{N}^\alpha\\|\nu|_1=m}}\Psi_\nu^{(m)}(j-1)\\
&\qquad -j(j-1)(N-j+1)\sum_{\alpha=1}^{m}\sum_{\substack{\nu\in\mathbb{N}^\alpha\\|\nu|_1=m}}\Psi_\nu^{(m)}(j-2)\Bigg).
\end{aligned}$$

Then applying the identity (8) and factoring, we get

$$\begin{aligned}
&(m+1)!\prod_{i=1}^{k}(y-c_i^{(\ell)})\Bigg(\sum_{\alpha=1}^{m}\sum_{\substack{\nu\in\mathbb{N}^\alpha\\|\nu|_1=m}}\sum_{j=m+1}^{k}\frac{j}{(y-c_j)}\psi_{\nu_1}(j-1)\Psi_{\nu'}^{(m-\nu_1)}(j-2-\nu_1)\\
&\qquad+\sum_{j=m+1}^{k}\frac{j(y-(j-1)(N+\ell)}{(y-c_j)(y-c_{j-1})}\sum_{\alpha=1}^{m}\sum_{\substack{\nu\in\mathbb{N}^\alpha\\|\nu|_1=m}}\Psi_\nu^{(m)}(j-2)\Bigg)\\
&=(m+1)!\prod_{i=1}^{k}(y-c_i^{(\ell)})\Bigg(\sum_{\alpha=1}^{m}\sum_{\substack{\nu\in\mathbb{N}^\alpha\\|\nu|_1=m}}\sum_{j=m+1}^{k}\psi_{\nu_1+1}(j)\Psi_{\nu'}^{(m-\nu_1)}(j-2-\nu_1)
\end{aligned}$$

$$+\sum_{\alpha=1}^{m}\sum_{\substack{\nu\in\mathbb{N}^\alpha\\|\nu|_1=m}}\sum_{j=m+1}^{k}\psi_1(j)\Psi_\nu^{(m)}(j-2)\Bigg)$$

$$=(m+1)!\prod_{i=1}^{k}(y-c_i^{(\ell)})\Bigg(\sum_{\alpha=1}^{m}\sum_{\substack{\nu\in\mathbb{N}^\alpha\\|\nu|_1=m+1\\\nu_1\neq 1}}\Psi_\nu^{(m+1)}(k)+\sum_{\alpha=2}^{m+1}\sum_{\substack{\nu\in\mathbb{N}^\alpha\\|\nu|_1=m+1\\\nu_1=1}}\Psi_\nu^{(m+1)}(k)\Bigg)$$

$$=(m+1)!\prod_{i=1}^{k}(y-c_i^{(\ell)})\sum_{\alpha=1}^{m+1}\sum_{\substack{\nu\in\mathbb{N}^\alpha\\|\nu|_1=m+1}}\Psi_\nu^{(m+1)}(k).$$

Hence the desired result follows. □

Remark 3.4 The vectors $\nu\in\mathbb{N}^\alpha$ with $|\nu|_1=m$ in the sum $\Psi_\nu^{(m)}(k)$ represent the *compositions* (ordered partitions) of the integer m that consist of α elements. For instance, for the case $m=2$ above, there are only two compositions, $\{2,1+1\}$. These are precisely the indices of the functions ψ_i involved in the sums in the expression for $\partial_x^2 P_k^{(N,l/2)}(0)$. In fact, the use of identity (8) in the proof of Theorem 3.3 resembles the way of constructing compositions of a number n starting with the compositions of $n-1$. Suppose $a_1+a_2+\ldots+a_k$ is a composition of $n-1$, then we construct two compositions $(a_1+1)+a_2+\ldots+a_k$ and $1+a_1+a_2+\ldots+a_k$ of n. It is easy to verify that this algorithm produces all compositions of n. For $n=1,2,3$ we illustrate the algorithm as a tree in Fig. 1.

We now make a small experimental discussion on the positivity of β_k for $k=0,1$. Assume that Conjecture 2.2 is true. Then it is possible to compute the polynomial $A_N^\ell(x,y)$ using simple algebraic computations. Set $A_N^\ell(x,y)=\sum_{i=0}^{\ell}\beta_i(y)x^i$. Then Lemma 2.4 gives

$$\beta_0(y)=\prod_{i=1}^{\ell}(y-\tilde{c}_i^{(\ell)}).$$

Clearly $\beta_0\in\mathbb{Z}[y]$ and it has positive integer coefficients, when $\beta_0>0$.

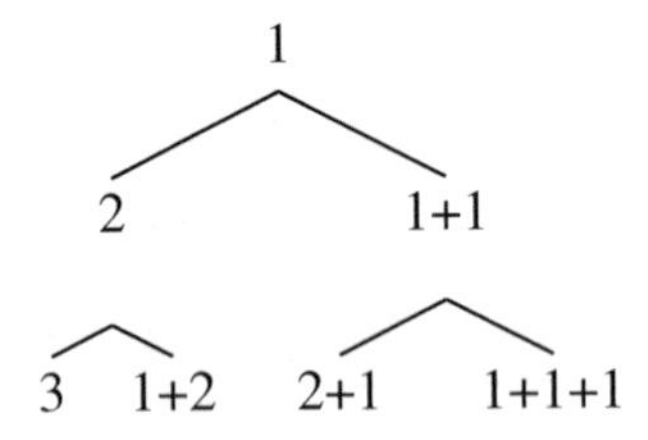

Fig. 1 Compositions for $n=1,2,3$

Similarly, we compute

$$\beta_1(y) = \prod_{i=1}^{\ell}(y - \tilde{c}_i^{(\ell)}) \left(\sum_{j=1}^{\ell} \frac{j(y-(j-1)N)}{(y-\tilde{c}_j^{(\ell)})(y-\tilde{c}_{j-1}^{(\ell)})} + \frac{\ell N}{y} \right).$$

Likewise, it is clear that $\beta_1(y) \in \mathbb{Z}[y]$ and we can show that $\beta_1(y) > 0$ for positive y. For example, we present the case $\ell = 2m$, $m \in \mathbb{N}$. Using the partial fraction decomposition we obtain

$$\beta_1(y) = (2N+1+\ell) \prod_{i=1}^{\ell}(y - \tilde{c}_i^{(\ell)}) \left(\sum_{j=1}^{\ell} \frac{j(j-\ell)}{((\ell-2j)^2-1)(y-\tilde{c}_j^{(\ell)}))} \right).$$

Denote the inner sum by $g(y)$ and rewrite it as

$$g(y) = \frac{m^2}{y+m^2} - 2\sum_{j=1}^{m-1} \frac{j(\ell-j)}{(4(m-j)^2-1)(y+j(\ell-j))}.$$

The value of $g(0)$ is given by

$$\begin{aligned} g(0) &= 1 - 2\sum_{j=1}^{m-1} \frac{1}{4(m-j)^2-1} = 1 - 2\sum_{j=1}^{m-1} \left(\frac{1}{2(2j-1)} - \frac{1}{2(2j+1)} \right) \\ &= \frac{1}{2m-1} > 0. \end{aligned}$$

Moreover, we see that the derivative of each of the summands of $(y+m^2)g(y)$ is equal to

$$\frac{j(j-\ell)(j-m)^2}{\left(4(j-m)^2-1\right)(y+j(\ell-j)))^2} > 0.$$

Consequently, it holds that $\left((y+m^2)g(y)\right)' > 0$, when $\beta_1(y) > 0$ for any positive y.

The foregoing argument, in addition to the observation that the coefficients of the proposed polynomials $A_N^{\ell}(x, y)$ ($\ell = 1, 2, 3, 4$) of Example 1 have positive coefficients, suggests that one strategy for proving Conjecture 2.2 is to prove that the expressions for the coefficients $\beta_i(y)$ of $A_N^{\ell}(x, y)$ obtained algebraically are elements of the polynomial ring $\mathbb{Z}[y]$ with positive coefficients.

4 Combinatorial Study of $A_0^\ell(x, y)$

In the previous section, we obtained the formulas for the constraint polynomials directly from the defining recurrence relations. Since the constraint polynomials appear from the study of the continuant of a tridiagonal matrix (see [30] for details), it is natural to study certain associated continued fractions. Concretely, for a fixed $\ell \in \mathbb{N}$, we have

$$\frac{\tilde{P}_k^{(\ell,\ell/2)}(x,y)}{\tilde{P}_{k-1}^{(\ell,\ell/2)}(x,y)} = b_k + \cfrac{a_k}{b_{k-1} + \cfrac{a_{k-1}}{b_{k-2} + \cfrac{a_{k-2}}{\dots + \cfrac{a_2}{b_1}}}}, \tag{9}$$

where

$$a_k = -k(k-1)(\ell - k + 1)x \quad \text{and} \quad b_k = kx + y + k(\ell - k).$$

Notice that when $N = 0$ in Conjecture 2.2, the polynomial $A_0^\ell(x, y)$ is just $\tilde{P}_\ell^{(\ell,\ell/2)}$ (x, y). The left-hand side of equation (9) is the k-th convergent of the continued fraction. The Euler–Minding formulas (see, for example, [17]) give the expression of the numerator and denominator of of the $k-$th convergent of the continued fraction as polynomials in $\{a_i, b_i\}$, $i \in \{0, 1, \dots, k\}$. In particular, for the numerator of the ℓ-th convergent we get

$$\begin{aligned} &A_0^\ell(x,y) \left(= \tilde{P}_\ell^{(\ell,\ell/2)}(x,y)\right) \\ &= b_1 b_2 \cdots b_l \left(1 + \sum_{i_1=1}^{\ell-1} \left(\frac{a_{i_1+1}}{b_{i_1} b_{i_1+1}} \right) + \sum_{i_1<i_2}^{\ell-2} \left(\frac{a_{i_1+1}}{b_{i_1} b_{i_1+1}} \right) \left(\frac{a_{i_2+2}}{b_{i_2+1} b_{i_2+2}} \right) + \cdots \right. \\ &\qquad \left. + \sum_{i_1<i_2<\dots<i_n}^{\ell-n} \prod_{j=1}^{n} \left(\frac{a_{i_j+j}}{b_{i_j+j-1} b_{i_j+j}} \right) + \cdots \right). \end{aligned} \tag{10}$$

Note that the sum inside the parenthesis consists of at most $\lfloor \ell/2 \rfloor + 1$ summands.

Next, we consider $A_0^\ell(x, y)$ as a polynomial in x with coefficients $\alpha_k(y)$. For $i \le \lfloor \ell/2 \rfloor$ the inner sum with i fractional terms $\frac{a_j}{a_{j-1}a_j}$ in (10) is given by

$$(-1)^i \ell! \sum_{j_1<j_2<\dots<j_i} x^i \prod_{n=1}^{i} (\ell - (j_n + n) + 1) \prod_{\substack{\beta=1 \\ \beta \notin \{j_\gamma+\gamma-1, j_\gamma+\gamma\} \\ \gamma=1,2,\dots,i}}^{\ell} \left(x + \frac{y}{\beta} + (\beta - \ell) \right),$$

which is a degree $\ell - i$ polynomial in x. In order to obtain the polynomial $\alpha_k(y)$, we find the coefficients of x^k in each of these expressions for $i = 0, \ldots, \lfloor \ell/2 \rfloor$. For $k \leq \lfloor \ell/2 \rfloor$, these are given by

$$\alpha_k(y) = \ell! \sum_{i=0}^{k} (-1)^i \sum_{j_1<j_2<\ldots<j_i} \prod_{n=1}^{i} (\ell - (j_n + n) + 1) \sum_{\substack{\beta_1<\beta_2<\ldots<\beta_{\ell-k-i} \\ \beta_* \notin \{j_\gamma+\gamma-1, j_\gamma+\gamma\} \\ \gamma=1,2,\ldots,i}}^{\ell} \prod_j \left(\frac{y}{\beta_j} + (\beta_j - \ell) \right)$$

and

$$\alpha_{\ell-k}(y) = \ell! \sum_{i=0}^{k} (-1)^i \sum_{j_1<j_2<\ldots<j_i} \prod_{n=1}^{i} (\ell - (j_n + n) + 1) \sum_{\substack{\beta_1<\beta_2<\ldots<\beta_{k-i} \\ \beta_* \notin \{j_\gamma+\gamma-1, j_\gamma+\gamma\} \\ \gamma=1,2,\ldots,i}}^{\ell} \prod_j \left(\frac{y}{\beta_j} + (\beta_j - \ell) \right),$$

from where we see that the degree of $\alpha_i(y)$ is $\ell - i$ for any $0 \leq i \leq \ell$.

Proposition 4.1 *As a polynomial in the variable x, the leading coefficient $\alpha_\ell(y)$ of $A_0^\ell(x, y)$ is $\ell!$ and the remaining coefficients $\alpha_k(y)$ are divisible by y.*

Proof The first claim follows directly from (10) since the coefficient of x^ℓ in $A_0^\ell(x, y)$ is equal to the coefficient of x^ℓ in $b_1 b_2 \ldots b_l$, and this is just $\ell!$. We prove the second claim by verifying that $\alpha_k(0)$ vanishes for any $0 \leq k < \ell$. For $1 \leq k \leq \lfloor \ell/2 \rfloor$, we have

$$\alpha_k(0) = \ell! \sum_{i=0}^{k} (-1)^i \sum_{j_1<j_2<\ldots<j_i} \prod_{n=1}^{i} (\ell - (j_n + n) + 1) \sum_{\substack{\beta_1<\beta_2<\ldots<\beta_{\ell-k-i} \\ \beta_* \neq j_\gamma+\gamma \\ \gamma=1,2,\ldots,i}}^{\ell} \prod_j (\ell - \beta_j).$$

Note that each summand consists of a product of exactly $\ell - k$ factors. For a fixed i and choice of $\{j_n\}_{n \in \{1,2,\ldots,i\}}$, none of the factors $(\ell - (j_n + n))$ appears in the innermost product. Therefore, for a fixed i, we rewrite the two inner sums as a number of sums with indices $n_1 < n_2 < \ldots < n_{\ell-k}$ with a subset $\Omega \subset \{1, 2, \ldots, \ell - k - 1\}$ of i elements such that $n_\eta + 1 < n_{\eta+1}$ for $\eta \in \Omega$. The elements of the subset Ω are determined by the relative position of the $\{j_n\}_{n=1,2,\ldots,i}$ with respect to the indices of the innermost sum. Concretely, we write

$$\sum_{n_1=1} \sum_{n_2=n_1+1+\delta_1} \cdots \sum_{n_{\ell-k-1}=n_{\ell-k-1}+1+\delta_{\ell-k}} \prod_{j=1}^{\ell-k} (\ell - n_j),$$

where $\delta_i \in \{0, 1\}$ for $i = 0, 1, \ldots, \ell - k - 1$, and $\delta_i = 1$ in accordance of the relative positions of $\{j_\eta\}$, as described above. It is clear that for fixed i, the vector $(\delta_0, \delta_1, \ldots, \delta_{\ell-k-1})$ contains i or $i - 1$ ones. The latter case being when one of the j_η is on the last relative position. Furthermore, all the possible combinations are accounted. We now prove that for a given vector $\nu = (\delta_0, \delta_1, \ldots, \delta_{\ell-k-1})$ corresponding to a given arrangement of $\{j_\eta\}_{\eta=1,\ldots,i}$ there is exactly one arrangement $\{j_\gamma\}_{\gamma=1,\ldots,i+1}$ corresponding to the same vector ν, leading to cancelation due to the alternation of signs. In fact, the specific arrangement is the one where the first i elements j_γ have the same relative positions as $\{j_\eta\}_{\eta=1,\ldots,i}$ and j_{i+1} is in the last relative position. The uniqueness follows from the fact that $\{j_\gamma\}_{\gamma=1,\ldots,i+1}$ in the sums are given in a strictly increasing order. To finish the proof, we show that in the case $i = k$, any sum corresponding to a vector $(\delta_0, \delta_1, \ldots, \delta_{\ell-k-1})$ with exactly k ones should vanish. Since the starting point of the last sum is $i_{\ell-k} = \ell - k + \sum_{i=0}^{\ell-k-1} \delta_i = \ell$, the product vanishes. Consequently, all the elements of the sum vanish making $\alpha_k(0) = 0$. The case of $\alpha_{\ell-k}(0)$ is proved in a similar way. □

Remark 4.2 The first claim of Proposition 4.1 together with the second claim of Lemma 2.4 supports in parts the assumption (i.e., on $A_N^\ell(x, y)$ is a polynomial in x) in Lemma 2.4.

Now, we relate some of the coefficients $\alpha_k(y)$ of the polynomial $A_0^\ell(x, y) = \sum_{k=0}^{\ell} \alpha_k(y) x^k$ with certain combinatorial quantities. We recall the definitions for convenience and refer the reader to [10] for more information and identities on the Stirling numbers and other combinatorial quantities.

Definition 1 The (unsigned) Stirling number of the first kind $s(n, k)$ is defined as the number of permutations of n elements with exactly k cycles.

We need the following elementary identity.

Lemma 4.3 *We have*

$$s(n+1, k+1) = n! \sum_{i_1 < i_2 < \ldots < i_k}^{n} \frac{1}{i_1 i_2 \cdots i_k}$$

for $k = 0, 1, \ldots, n$.

Proof We first note that the cases $k = 0, 1$ are well-known. In fact, $s(n + 1, 1) = n!$ and $s(n + 1, 2) = n! H_n$, where H_n is the harmonic number. Assume the formula holds for k. From the definition of $s(n, k)$, by fixing the number 1 in the first cycle of each permutation we get

$$\begin{aligned}
&s(n+1,k+1)\\
&= s(n,k) + \sum_{j=k}^{n-k} \frac{n!}{(n-j+1)!} s(n+1-j,k)\\
&= (n-1)! \sum_{i_1<\ldots<i_{k-1}} \frac{1}{i_1 i_2 \cdots i_{k-1}} + \sum_{j=k}^{n-k} \frac{n!}{(n-j+1)!}(n-j)! \sum_{i_1<\ldots<i_{k-1}}^{n+1-j} \frac{1}{i_1 i_2 \cdots i_{k-1}}\\
&= \frac{n!}{n} \sum_{i_1<\ldots<i_{k-1}} \frac{1}{i_1 i_2 \cdots i_{k-1}} + \sum_{j=k}^{n-k} \frac{n!}{(n-j+1)!} \sum_{i_1<\ldots<i_{k-1}}^{n+1-j} \frac{1}{i_1 i_2 \cdots i_{k-1}}\\
&= n! \sum_{j=k+1}^{n} \frac{1}{j} \sum_{i_1<\ldots<i_{k-1}}^{j-1} \frac{1}{i_1 i_2 \cdots i_{k-1}}\\
&= n! \sum_{i_1<i_2<\ldots<i_k}^{n} \frac{1}{i_1 i_2 \cdots i_k}.
\end{aligned}$$

This proves the lemma. □

Proposition 4.4 *For $k = 0, 1, \ldots, \ell$, the coefficient of $y^{\ell-k}$ in $\alpha_k(y)$ is the Stirling number of the first kind $s(\ell+1, \ell+1-k)$.*

Proof From the expression given above, the coefficient corresponding to $y^{\ell-k}$ for $\alpha_k(y)$ is equal to the one of

$$\ell! \sum_{\beta_1<\beta_2<\ldots<\beta_{\ell-k}}^{\ell} \prod_{j=1}^{\ell-k} \left(\frac{y}{\beta_j} + (\beta_j - \ell) \right),$$

that is

$$\ell! \sum_{\beta_1<\beta_2<\ldots<\beta_{\ell-k}}^{\ell} \frac{1}{\beta_1 \beta_2 \cdots \beta_{\ell-k}}.$$

Hence the claim follows from the lemma above. □

Definition 2 We say that a permutation $\sigma \in \mathfrak{S}_n$ has an *ascent* if there is an $i \in \{1, 2, \ldots, n-1\}$ such that $\sigma(i) < \sigma(i+1)$.

Definition 3 1. The Eulerian number $A_1(n,k)$ is defined as the number of permutations $\sigma \in \mathfrak{S}_n$ with exactly k ascents.
2. The Eulerian number of second order $A_2(n,k)$ is defined as the number of permutations σ of the multiset $\{1, 1, 2, 2, \ldots, n, n\}$ such that for any $m \in \{1, 2, \ldots, n\}$ if $i \leq j$ are the preimages of m then $m < \sigma(k)$ for any $i < k < j$, and such that σ contains exactly k ascents.

Remark 4.5 The Eulerian numbers of second order satisfy the following recurrence relation

$$A_2(n,k) = (k+1)A_2(n-1,k) + (2n-1-k)A_2(n-1,k-1),$$

which is sometimes used as the definition. In addition, we have $A_2(n,n-1) = n!$. We also note that Eulerian numbers should not be confused the Euler numbers E_k.

Proposition 4.6 *The coefficient of* y *in* $\alpha_{\ell-2}(y)$ *is the Eulerian number of second order* $A_2(\ell,\ell-2)$.

Proof By using the formula for $\alpha_{\ell-2}(y)$, we find that the desired coefficient is given by

$$\ell!\sum_{i=1}^{\ell-1}\frac{\ell-i}{i+1} = \ell!\sum_{i=1}^{\ell-1}\frac{i}{\ell-i+1}.$$

It is easy to verify the claim for the cases $\ell = 2,3$ and use induction to prove the result. Assume it holds for all integers up to $\ell-1$ and compute from the recurrence of Eulerian numbers of second order

$$\begin{aligned}
A_2(\ell,\ell-2) &= (\ell-1)A_2(\ell-1,\ell-2) + (\ell+1)A_2(\ell-1,\ell-3)\\
&= (\ell-1)(\ell-1)! + (\ell+1)\sum_{i=1}^{\ell-2}\frac{\ell-1-i}{i+1}\\
&= (\ell-1)!\left((\ell-1) + \sum_{i=1}^{\ell-2}\frac{\ell-1-i}{i+1}\right) + \ell!\sum_{i=1}^{\ell-2}\frac{\ell-1-i}{i+1},
\end{aligned}$$

using a change of variable, we get

$$\begin{aligned}
A_2(\ell,\ell-2) &= (\ell-1)!\left((\ell-1) + \sum_{i=1}^{\ell-2}\frac{i}{\ell-i}\right) + \ell!\sum_{i=1}^{\ell-2}\frac{i}{\ell-i}\\
&= \ell!\sum_{i=0}^{\ell-1}\frac{i-1}{\ell-i+1} + (\ell-1)!\left((\ell-1) + \sum_{i=1}^{\ell-2}\frac{i}{\ell-i}\right)\\
&= \ell!\sum_{i=1}^{\ell-1}\frac{i}{\ell-i+1} - \ell!\sum_{i=1}^{\ell-1}\frac{1}{\ell-i+1} + (\ell-1)!\left((\ell-1) + \sum_{i=1}^{\ell-2}\frac{i}{\ell-i}\right).
\end{aligned}$$

To complete the proof we show that the last two summands vanish. In fact, we have

$$
\begin{aligned}
-\ell!\sum_{i=1}^{\ell-1}\frac{1}{\ell-i+1}+(\ell-1)!\left((\ell-1)+\sum_{i=1}^{\ell-2}\frac{i}{\ell-i}\right)&\\
=(\ell-1)!\left(\sum_{i=1}^{\ell-2}\frac{i}{\ell-i}-\sum_{i=0}^{\ell-2}\frac{\ell}{\ell-i}\right)+(\ell-1)(\ell-1)!&\\
=(\ell-1)!\left(\sum_{i=1}^{\ell-2}\frac{i-\ell}{\ell-i}-1\right)+(\ell-1)(\ell-1)!&\\
=-(\ell-1)(\ell-1)!+(\ell-1)(\ell-1)!=0.&
\end{aligned}
$$

This proves the claim. □

Remark 4.7 It is expected to have similar expressions in terms of combinatorial quantities for the coefficients of the polynomial $A_0^\ell(x, y)$, and also for the cases $A_N^\ell(x, y)$ with $N \neq 0$.

Remark 4.8 In the physics literature, the spectral graph and the constraint polynomials are usually considered as functions of the coupling g for a fixed qubit energy 2Δ together with additional parameters (e.g., ε for the cases of AQRM). It would be, however, interesting to exchange these two parameters. Actually, it is interesting if we could get some efficient information on the "constraint" function $P_k^{(N,\varepsilon)}(x, y)$ from the orthogonality of $\{P_k^{(N,\varepsilon)}(x, y)\}$ as functions of y for a fixed value $x(< 0)$ (see Remark 2.9). Moreover, the comparison of (a function obtained by taking a certain limit of) $P_k^{(N,\varepsilon)}(x)$ and the G-functions $G_\pm(g, \Delta)$ for the non-degenerate exceptional eigenvalues of the QRM when $\varepsilon = 0$ [7] may hold some potential to further elucidate the possible connections among the different types of exceptional spectra. It would be further interesting to consider the "general" orthogonal polynomials (i.e., with a non-positive definite moment functional $\mathscr{M}$) in y defined by the recurrence equation for the parameter N when $N \notin \mathbb{N}$.

5 Conclusion

The unexpected appearance of level crossings (spectral degeneracies) in the asymmetric quantum model could be considered as a hint to the presence of a *hidden* $\mathbb{Z}_2$-symmetry, which does not appear on the level of definition of the Hamiltonian $H_{\text{Rabi}}^\varepsilon$: It is not easily possible to construct an operator X with $X^2 =$ identity and $[H_{\text{Rabi}}^\varepsilon, X] = 0$. Nevertheless, the level crossings strongly hint at the existence of such an operator for the AQRM and $\varepsilon \in \frac{1}{2}\mathbb{Z}$. Actually, we have observed a certain reciprocity (spherical and non-spherical representations) in [30] for any $\varepsilon \in \frac{1}{2}\mathbb{Z}$ under the truth of Conjecture 2.2. We may infer the hidden symmetry group is $\mathbb{Z}_2$ because there are only two *ladders* of intersecting eigenvalues (i.e., two invariant subspaces) just as in the QRM.

Acknowledgements The authors wish to thank Daniel Braak for many valuable comments and suggestions particularly from the physics side. This work is partially supported by Grand-in-Aid for Scientific Research (C) No. 16K05063 of JSPS, Japan. The first author was supported during the duration of the research by the Japanese Government (MONBUKAGAKUSHO: MEXT) scholarship.

References

1. V. Bargmann, On a Hilbert space of analytic functions and an associated integral transform part I. Commun. Pure Appl. Math. **14**, 187–214 (1961)
2. M.T. Batchelor, Z.-M. Li, H.-Q. Zhou, Energy landscape and conical intersection points of the driven Rabi model. J. Phys. A Math. Theor. **49**, 01LT01 (6pp) (2015)
3. D. Braak, Integrability of the Rabi model. Phys. Rev. Lett. **107**, 100401–100404 (2011)
4. D. Braak, Continued fractions and the Rabi model. J. Phys. A Math. Theor. **46**, 175301 (10pp) (2013)
5. D. Braak, A generalized G-function for the quantum Rabi model. Ann. Phys. **525**(3), L23–L28 (2013)
6. D. Braak, Solution of the Dicke model for $N = 3$. J. Phys. B At. Mol. Opt. Phys. **46**, 224007 (2013)
7. D. Braak, Analytical solutions of basic models in quantum optics, *in Applications + Practical Conceptualization + Mathematics = fruitful Innovation, Proceedings of the Forum of Mathematics for Industry 2014*, ed. by R. Anderssen, et al., vol. 11 (Mathematics for Industry Springer, Heidelberg, 2016), pp. 75–92
8. D. Braak, Q.H. Chen, M.T. Batchelor, E. Solano, Semi-classical and quantum Rabi models: in celebration of 80 years. J. Phys. A Math. Theor. **49**, 300301 (4pp) (2016)
9. T.S. Chihara, *An Introduction to Orthogonal Polynomials* (Gordon and Breach, London, 1978)
10. R.L. Graham, D.E. Knuth, O. Patashhnik, *Concrete Mathematics: A Foundation for Computer Science*, 2nd edn. (Addison-Wesley, Longman, 1994)
11. S. Haroche, J.M. Raimond, *Exploring the Quantum. Atoms, Cavities and Photons* (Oxford University Press, Oxford, 2008)
12. M. Hirokawa, The Dicke-type crossing among eigenvalues of differential operators in a class of non-commutative oscillators. Indiana Univ. Math. J. **58**, 1493–1536 (2009)
13. M. Hirokawa, F. Hiroshima, Absence of energy level crossing for the ground state energy of the Rabi model. Commun. Stoch. Anal. **8**, 551–560 (2014)
14. F. Hiroshima, I. Sasaki, Spectral analysis of non-commutative harmonic oscillators: the lowest eigenvalue and no crossing. J. Math. Anal. Appl. **105**, 595–609 (2014)
15. E.T. Jaynes, F.W. Cummings, Comparison of quantum and semiclassical radiation theories with application to the beam maser. Proc. IEEE **51**, 89–109 (1963)
16. B.R. Judd, Exact solutions to a class of Jahn-Teller systems. J. Phys. C Solid State Phys. **12**, 1685 (1979)
17. S. Khrushchev, *Orthogonal Polynomials and Continued Fractions, From Euler's Point of View* (Cambridge University Press, Cambridge, 2008)
18. M. Kuś, On the spectrum of a two-level system. J. Math. Phys. **26**, 2792–2795 (1985)
19. Z.-M. Li, M.T. Batchelor, Algebraic equations for the exceptional eigenspectrum of the generalized Rabi model. J. Phys. A: Math. Theor. **48**, 454005 (13pp) (2015)
20. Z.-M. Li, M.T. Batchelor, Addendum to Algebraic equations for the exceptional eigenspectrum of the generalized Rabi model. J. Phys. A Math. Theor. **49**, 369401 (5pp) (2016)
21. S. Lang, $SL_2(\mathbb{R})$ (Addison-Wesley, Reading, 1975)
22. A.J. Maciejewski, M. Przybylska, T. Stachowiak, Full spectrum of the Rabi model. Phys. Lett. A **378**, 16–20 (2014)

23. T. Niemczyk et al., Beyond the Jaynes-Cummings model: circuit QED in the ultrastrong coupling regime. Nat. Phys. **6**, 772–776 (2010)
24. A. Ronveaux (eds.), *Heun's Differential Equations* (Oxford University Press, Oxford, 1995)
25. E. Solano, Viewpoint: the dialogue between quantum light and matter. Physics **4**, 68–72 (2011)
26. S. Sugiyama, Spectral zeta functions for the quantum Rabi models. Nagoya Math. J. (2016). doi:10.1017/nmj.2016.62,1-47
27. A.V. Turbiner, Quasi-exactly-solvable problems and sl(2) algebra. Commun. Math. Phys. **118**, 467–474 (1988)
28. M. Wakayama, Remarks on quantum interaction models by Lie theory and modular forms via non-commutative harmonic oscillators, in *Mathematical Approach to Research Problems of Science and Technology – Theoretical Basis and Developments in Mathematical Modelling* ed. by R. Nishii, et al., Mathematics for Industry, vol. 5 (Springer, Berlin, 2014), pp. 17–34
29. M. Wakayama, Equivalence between the eigenvalue problem of non-commutative harmonic oscillators and existence of holomorphic solutions of Heun differential equations, eigenstates degeneration and the Rabi model. Int. Math. Res. Notices [rnv145 (2015)], 759–794 (2016)
30. M. Wakayama, Symmetry of Asymmetric Quantum Rabi Models. J. Phys. A: Math. Theor. **50**, 174001 (22pp) (2017)
31. M. Wakayama, T. Yamasaki, The quantum Rabi model and Lie algebra representations of sl_2. J. Phys. A Math. Theor. **47**, 335203 (17pp) (2014)
32. Q.-T. Xie, H.-H. Zhong, M.T. Batchelor, C.-H. Lee, *The Quantum Rabi Model: Solution and Dynamics*, arXiv:1609.00434

Spectra of Group-Subgroup Pair Graphs

Kazufumi Kimoto

Abstract Graphs with large isoperimetric constants play an important role in cryptography because one can utilize such graphs to construct cryptographic hash functions. Ramanujan graphs are important optimal examples of such graphs, and known explicit construction of infinite families of Ramanujan graphs are given by Cayley graphs. A group–subgroup pair graph, which is a generalization of a Cayley graph, is defined for a given triplet consisting of finite group, its subgroup, and a suitable subset of the group. We study the spectra, that is the eigenvalues of the adjacency operators, of such graphs. In fact, we give an explicit formula of the eigenvalues of such graphs when the corresponding subgroups are abelian in terms of the characters of the subgroups as well as give a lower bound estimation for the second largest eigenvalues.

Keywords Cayley graphs · Spectra of graphs · Alon–Boppana theorem · Ramanujan graphs · Group–subgroup pair graphs · Second largest eigenvalue · Biregular bigraphs · Characters

1 Introduction

Graphs with large isoperimetric constants play an important role in cryptography. Indeed, it is known that random walk on such a graph rapidly converges to the uniform distribution as the number of walk steps tends to infinity. This means that one can make use of such graphs to construct cryptographic hash functions (see [3], in which hash functions are constructed from LPS graphs [7] and Pizer graphs [9]). Hence it is desirable to obtain an infinite family of such graphs, so-called an *expander family*. Among such expander graphs, *Ramanujan graphs* are important because they provide optimal expander families as well as they are very interesting in view of number theory; a regular graph is Ramanujan if and only if its zeta function

K. Kimoto (✉)
Department of Mathematical Sciences, Faculty of Science, University of the Ryukyus,
1 Senbaru Nishihara-cho, Okinawa 903-0213, Japan
e-mail: kimoto@math.u-ryukyu.ac.jp

T. Takagi et al. (eds.), *Mathematical Modelling for Next-Generation Cryptography*,
Mathematics for Industry 29, DOI 10.1007/978-981-10-5065-7_8

satisfies an analog of the *Riemann Hypothesis*, and all known explicit constructions of infinite families of Ramanujan graphs are based on deep results in number theory (for instance, the construction of the LPS graphs due to Lubotzky, Philips and Sarnak [7] is based on the Ramanujan–Peterson conjecture on automorphic forms). They are *Cayley graphs* for certain finite groups together with carefully chosen symmetric generating sets.

It is natural to extend the notion of Ramanujan graphs to non-regular cases. In fact, there is an analogous notion of Ramanujan graphs, called *Ramanujan bigraphs*, defined for *biregular bigraphs*, which is a bipartite graph with bipartition $V = V_1 \sqcup V_2$ of vertices such that the degrees are constant on each subset V_i. Ballantine et al. [2] gave an explicit construction of an infinite family of Ramanujan bigraphs by making use of the Bruhat-Tits buildings of an inner form of $SU(\mathbb{Q}_p)$.

In this article, we deal with a special kind of graphs denoted by $\mathcal{G}(G, H, S)$ made from a given triplet (G, H, S) of a finite group, a subgroup of it and a suitable subset of it, which is introduced by Reyes–Bustos [10]; this is a generalization of the Cayley graphs and is a quasi-multipartite graph. The main purpose of the article is to determine the spectrum of $\mathcal{G}(G, H, S)$ completely in terms of the characters of H when the subgroup H is abelian (Theorem 3.5). We also try to estimate the second largest eigenvalues of such graphs (Theorem 3.8).

2 Basic Notions on Graphs

We briefly recall several basic notions on graphs. In what follows, we always assume that a graph is *finite, simple and undirected otherwise stated*.

Let $X = (V, E)$ be a graph. We write $x \sim y$ for $x, y \in V$ to imply that x and y are adjacent, that is, $\{x, y\} \in E$. The adjacency operator $A = A_X$ is a linear transformation on $L^2(V) = \{f : V \to \mathbb{C}\}$ defined by

$$(Af)(x) = \sum_{\substack{y \in V \\ y \sim x}} f(y).$$

We always regard the space $L^2(V)$ as a Hilbert space with respect to the standard inner product

$$\langle f, g \rangle_{L^2(V)} = \langle f, g \rangle = \sum_{x \in V} f(x)\overline{g(x)} \quad (f, g \in L^2(V)).$$

Notice that the eigenvalues of A are real since A is self-adjoint with respect to the standard inner product, and one can choose real-valued eigenfunctions corresponding to them. We sometimes use the symbol A to denote the adjacency matrix, the representation matrix of the adjacency operator with respect to the standard basis $\{\delta_v \,\big|\, v \in V\}$, where

$$\delta_v(x) = \begin{cases} 1 & x = v, \\ 0 & x \neq v. \end{cases}$$

The *isoperimetric constant* $h(X)$ of X is defined by

$$h(X) = \min \left\{ \frac{|\partial F|}{|F|} \,\middle|\, F \subset V,\ 0 < |F| \leq \frac{|V|}{2} \right\},$$

where $\partial F = \{e \in E \mid e = \{x, y\}, \exists x \in F, \exists y \in V \setminus F\}$ is the boundary of a given subset $F \subset V$. A graph with large isoperimetric constant is regarded as an efficient network. Let $\{X_n = (V_n, E_n)\}_{n \geq 1}$ be an infinite family of k-regular connected graphs such that $|V_n| \to \infty$ as $n \to \infty$. Such a family $\{X_n\}$ is called an *expander family* if there exists a positive constant $\varepsilon > 0$ such that $h(X_n) \geq \varepsilon$ for any $n \geq 1$. If we denote by $\lambda_i = \lambda_i(X)$ $(i = 0, 1, \ldots, |V| - 1)$ the eigenvalues of (the adjacency operator of) X in decreasing order $(\lambda_0 \geq \lambda_1 \geq \cdots \geq \lambda_{|V|-1})$, then we see that

$$\frac{k - \lambda_1}{2} \leq h(X) \leq \sqrt{2k(k - \lambda_1)}.$$

Hence, $\{X_n\}$ is an expander family if there exists a positive constant $\varepsilon > 0$ such that $k - \lambda_1(X_n) \geq \varepsilon$. Note that $\lambda_0(X) = k$ if X is k-regular and connected, so that $k - \lambda_1$ is sometimes called the *spectral gap*.

Thus, a graph X with large spectral gap (or small $\lambda_1(X)$) is desirable. However, there is a lower bound for $\lambda_1(X)$.

Theorem 2.1 (Alon–Boppana) *Let $\{X_n = (V_n, E_n)\}_{n \geq 1}$ be an infinite family of connected k-regular graphs such that $|V_n| \to \infty$ as $n \to \infty$. Then*

$$\liminf_{n \to \infty} \lambda_1(X_n) \geq \mathrm{RB}(k),$$

where $\mathrm{RB}(k) := 2\sqrt{k - 1}$.

A k-regular connected graph X is called a *Ramanujan graph* if $\lambda(X) \leq \mathrm{RB}(k)$, where $\lambda(X) = \max \{|\lambda_j(X)| \mid \lambda_j \neq \pm k\}$. In view of the Alon–Boppana theorem, an expander family $\{X_n\}$ is considered to be optimal if it is a family of Ramanujan graphs.

There is an analog of the Alon–Boppana theorem for biregular graphs. A bipartite graph $X = (V, E)$ with a bipartition $V = V_0 \sqcup V_1$ is called a *(p, q)-biregular bigraph* (or *semiregular bipartite graph of valency* (p, q)) if

$$x \in V_0 \implies |\{y \in V_1 \mid x \sim y\}| = p,$$
$$y \in V_1 \implies |\{x \in V_0 \mid x \sim y\}| = q.$$

For example, the complete bipartite graph (or biclique) $K_{p,q}$ is a (p, q)-biregular bigraph. The largest eigenvalue $\lambda_0(X)$ for such a graph is equal to $\sqrt{pq}$ (Fig. 1).

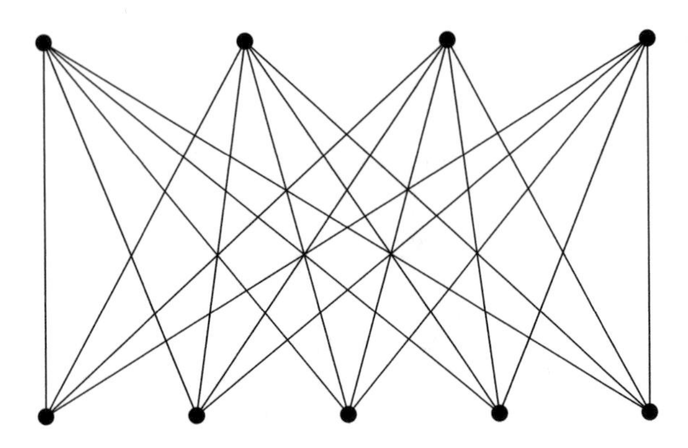

Fig. 1 Biclique $K_{5,4}$ is a $(5, 4)$-biregular bigraph

Theorem 2.2 (Feng-Li [4]) *Let $\{X_n = (V_n, E_n)\}_{n \geq 1}$ be an infinite family of connected (p, q)-biregular bigraphs such that $|V_n| \to \infty$ as $n \to \infty$. Then*

$$\liminf_{n \to \infty} \lambda_1(X_n) \geq \mathrm{RB}(p, q),$$

where $\mathrm{RB}(p, q) := \sqrt{p-1} + \sqrt{q-1}$.

Remark 2.3 Theorem 2.2 itself is not given explicitly in [4]; one can find a proof of it in [12] for instance. We give an equivalent proof in the appendix for reader's convenience.

A (p, q)-biregular connected bigraph X is called a *Ramanujan bigraph* [6, 12] if

$$\left|\sqrt{p-1} - \sqrt{q-1}\right| \leq \lambda(X) \leq \sqrt{p-1} + \sqrt{q-1},$$

which is also equivalent to

$$\left|\lambda(X)^2 - p' - q'\right| \leq 2\sqrt{p'q'} \quad (p = p' + 1, \; q = q' + 1).$$

We note that a biregular bigraph X is Ramanujan if and only if its zeta function satisfies the Riemann Hypothesis [1]. We refer to [2] for an example of an explicit construction of an infinite family of Ramanujan bigraphs. We also refer to [11] for an explicit construction of *Ramanujan hypergraphs*.

3 Cayley–Type Graphs for Group-Subgroup Pairs and Their Spectra

Let G be a finite group and S be a symmetric generating subset of G (i.e., $S^{-1} = S$ and $G = \langle S \rangle$) not containing the identity element e. Then we can define a graph, denoted by $\mathrm{Cay}(G, S)$, as follows; the set of vertices is G, and two vertices $x, y \in G$

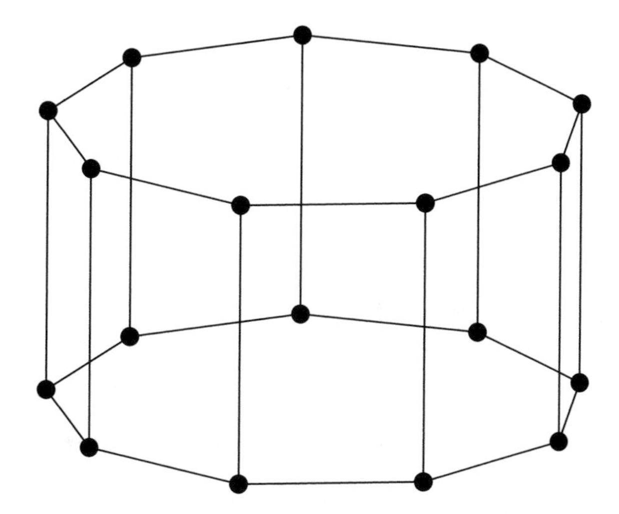

Fig. 2 $\mathrm{Cay}(D_7, \{\sigma, \sigma^{-1}, \tau\})$

are adjacent if and only if $x^{-1}y \in S$. This graph $\mathrm{Cay}(G, S)$ is called a *Cayley graph* of G.

Example 3 Let $G = D_7 = \langle \sigma, \tau \rangle$ be the dihedral group of degree 14 ($\sigma^7 = \tau^2 = e$, $\tau\sigma\tau = \sigma^{-1}$). If we take $S = \{\sigma, \sigma^{-1}, \tau\}$, then $e \notin S = S^{-1}$ and $G = \langle S \rangle$. The Cayley graph $\mathrm{Cay}(G, S)$ is visualized as in Fig. 2.

In this section, we recall the definition and basic properties of the *group-subgroup pair graphs*, which is a generalized notion of Cayley graphs, and calculate their spectra in certain special cases.

3.1 Quasi-bigraphs

Definition 4 Let $X = (V, E)$ be a graph with a nontrivial bipartition $V = V_0 \sqcup V_1$, and $D = \begin{pmatrix} d_{00} & d_{01} \\ d_{10} & d_{11} \end{pmatrix}$ be a 2×2 matrix whose entries are nonnegative integers. We say that X is a *D-regular quasi-bigraph*, or simply a *D-bigraph*, if

$$x \in V_i \implies \left|\left\{y \in V_j \,\middle|\, x \sim y\right\}\right| = d_{ij} \quad (i, j = 0, 1).$$

Note that

$$x \in V_i \implies \deg(x) = d_{i0} + d_{i1} \quad (i = 0, 1)$$

when $X = (V, E)$ is a D-bigraph (Fig. 3).

Remark 3.1 Here we use the term "quasi-bigraph" to distinguish with the ordinary bigraph (or biregular bipartite graph). In general, for a given matrix $D =$

Fig. 3 Conceptual diagram of a D-bigraph

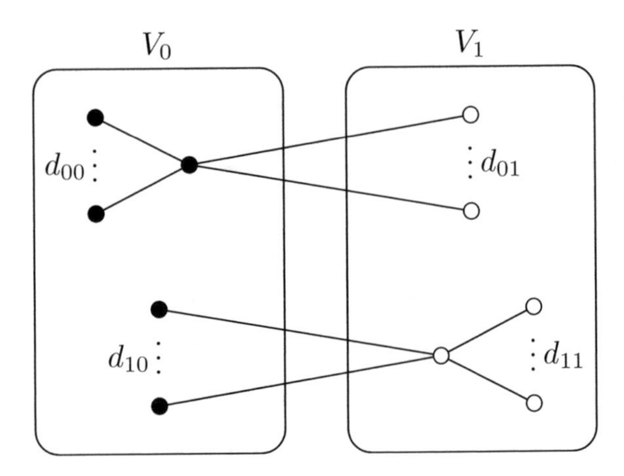

$(d_{ij})_{0\le i,j\le m} \in M_{m+1}(\mathbb{Z}_{\ge 0})$, it is natural to define that a graph $X = (V, E)$ is a *D-regular quasi-multipartite graph* if V has a multipartition $V = V_0 \sqcup V_1 \sqcup \cdots \sqcup V_m$ such that

$$x \in V_i \implies \left|\{y \in V_j \mid x \sim y\}\right| = d_{ij} \quad (i, j = 0, 1, \ldots, m).$$

The largest eigenvalue of a quasi-bigraph is easily seen as in the case of regular graphs.

Proposition 3.2 *Let X be a connected D-bigraph. Then the eigenvalues of D are also eigenvalues of X. The corresponding eigenfunctions are constant on each subset V_i. In particular, $\lambda_0(X)$ is the largest eigenvalue of D.*

Proof Let α_0, α_1 be eigenvalues of D and $\boldsymbol{v}_i = \begin{pmatrix} v_{i0} \\ v_{i1} \end{pmatrix}$ $(i = 0, 1)$ be the corresponding eigenvectors. Define the functions $f_i \in L^2(V)$ by

$$f_i(x) = \begin{cases} v_{i0} & x \in V_0, \\ v_{i1} & x \in V_1 \end{cases} \quad (i = 0, 1).$$

Then we have

$$(Af_i)(x) = \sum_{\substack{y\in V_0 \\ y\sim x}} f_i(y) + \sum_{\substack{y\in V_1 \\ y\sim x}} f_i(y) = d_{k0}v_{i1} + d_{k1}v_{i1} = \alpha_i v_{ik} = \alpha_i f_i(x) \quad (x \in V_k).$$

When X is connected, if we assume that α_0 is the largest positive eigenvalue, then we can choose an eigenvector $\boldsymbol{v}_0$ with positive entries by Peron-Frobenius theorem, and α_0 is also the Peron–Frobenius root of A since f_0 is positive-valued. $\square$

Remark 3.3 Similar result holds for general D-regular quasi-multipartite graphs.

Example 4 A $\begin{pmatrix} 0 & p \\ q & 0 \end{pmatrix}$-bigraph X is a (p, q)-biregular bigraph, and the largest eigenvalue of X is $\sqrt{pq}$.

3.2 Group–Subgroup Pair Graphs

Let G be a finite group, H a subgroup of G with index $k + 1$ and order n, and S be a nonempty subset of G. Put $N = |G| = (k + 1)n$ for short. Fix a set $\{x_0 = e, x_1, x_2, \ldots, x_k\}$ of representatives of the right cosets in G modulo H;

$$G = \bigsqcup_{i=0}^{k} V_i, \qquad V_i = Hx_i,$$

and put $S_i = Hx_i \cap S$. We also put $d_i = |S_i|$ and $d = |S|$ for later use. Assume that S_0 is symmetric, that is, $S_0^{-1} = S_0$.

Define a graph $\mathcal{G}(G, H, S)$ as follows: the vertex set is G, and two vertices $x, y \in G$ are connected by an edge if and only if $\{x, y\} = \{h, hs\}$ for some $h \in H$ and $s \in S$. This graph $\mathcal{G}(G, H, S)$ is called a *group–subgroup pair graph* (or simply *pair graph*) for a triplet (G, H, S) [10]. Notice that the subgraph of $\mathcal{G}(G, H, S)$ induced by H (the subgraph of $\mathcal{G}(G, H, S)$ such that the vertices are H and two vertices are adjacent if they are adjacent in $\mathcal{G}(G, H, S)$) is a Cayley graph $\mathrm{Cay}(H, S_0)$ of H with symmetric subset S_0 if $H = \langle S_0 \rangle$. Especially, if $G = H = \langle S \rangle$, then $\mathcal{G}(G, G, S)$ is nothing but the Cayley graph $\mathrm{Cay}(G, S)$. Hence the pair graphs are a generalization of the Cayley graphs.

Here we summarize several elementary facts on pair graphs (see [10] for the proof). We denote by $A = A_{\mathcal{G}(G,H,S)}$ the adjacency operator for $\mathcal{G}(G, H, S)$ on $L^2(G)$, and by $\lambda_i = \lambda_i(\mathcal{G}(G, H, S))$ $(i = 0, 1, \ldots, N - 1)$ be the eigenvalues of $\mathcal{G}(G, H, S)$ which are ordered in decreasing order: $\lambda_0 \geq \lambda_1 \geq \cdots \geq \lambda_{N-1}$.

- We have

$$\deg(v) = \begin{cases} d = |S| & v \in V_0 = H, \\ d_i = |S_i| & v \in V_i \quad (i = 1, \ldots, k). \end{cases}$$

In particular, $\mathcal{G}(G, H, S)$ is regular if and only if $k = 0$ or $k = 1$ and $S_0 = \emptyset$ (i.e., $S = S_1$). We also have

$$x \in Hx_i \implies \left|\{y \in Hx_j \,|\, x \sim y\}\right| = \begin{cases} d_j & i = 0, \\ d_i & j = 0, \\ 0 & \text{otherwise} \end{cases}$$

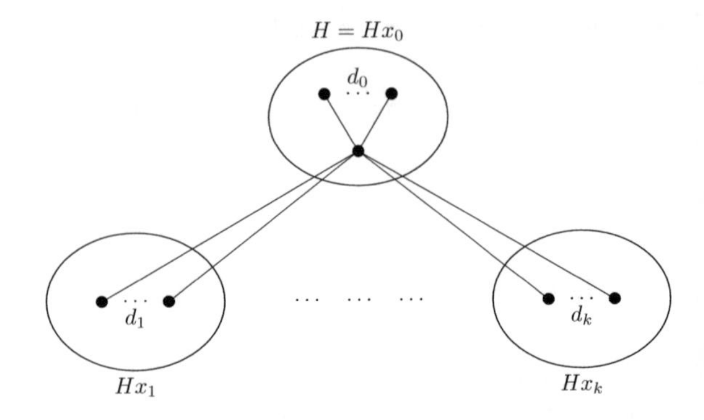

Fig. 4 Conceptual diagram of a group-subgroup pair graph

(see Fig. 4). In other words, $\mathcal{G}(G, H, S)$ is a D-regular quasi-multipartite graph for

$$D = \begin{pmatrix} d_0 & d_1 & \dots & d_k \\ d_1 & & & \\ \vdots & & O & \\ d_k & & & \end{pmatrix}.$$

- If $x \sim y$, then either x or y are in $H = V_0$. Hence, if $x \in V_i$, $y \in V_j$ for $1 \leq i < j \leq n$, then $x \nsim y$.
- $\mathcal{G}(G, H, S)$ is bipartite if and only if $S_0 = \emptyset$. The bipartition of G is then given by V_0 and $\bigcup_{i=1}^k V_i$.
- $\mathcal{G}(G, H, S)$ is connected if and only if $|S_i| \geq 1$ for all $i \geq 1$ and $S_0 \cup \bigcup_{i=1}^k S_i S_i^{-1}$ generates H (Theorem 3.3 in [10]).
- $$\mu_{\pm} = \frac{1}{2}\Big(d_0 \pm \big(d_0^2 + 4\sum_{i=1}^k d_i^2\big)^{1/2}\Big)$$

  are eigenvalues of $\mathcal{G}(G, H, S)$ called trivial eigenvalues (Theorem 5.1 in [10]). The largest eigenvalue λ_0 is equal to μ_+, which is simple if $\mathcal{G}(G, H, S)$ is connected. Notice that $-\lambda_0$ is also an eigenvalue if and only if $\mathcal{G}(G, H, S)$ is bipartite. For any eigenvalue λ of $\mathcal{G}(G, H, S)$ other than $\pm\lambda_0$, we have $|\lambda| < \lambda_0$.

When $k = 1$ (or $[G : H] = 2$), $\mathcal{G}(G, H, S)$ is a $\begin{pmatrix} d_0 & d_1 \\ d_1 & 0 \end{pmatrix}$-bigraph with a bipartition $G = Hx_0 \sqcup Hx_1$, and the largest eigenvalue $\lambda_0 = \mu_+ = \frac{1}{2}(d_0 + \sqrt{d_0^2 + 4d_1^2})$ of $\mathcal{G}(G, H, S)$ is a positive root of the quadratic equation $x^2 - d_0 x - d_1^2 = 0$. When $k > 1$, $\mathcal{G}(G, H, S)$ is not D-bigraph for a certain $D \in M_2(\mathbb{Z}_{\geq 0})$ in general (for instance, see the last graph in Fig. 5, whose vertices have degrees 2, 3 and 5).

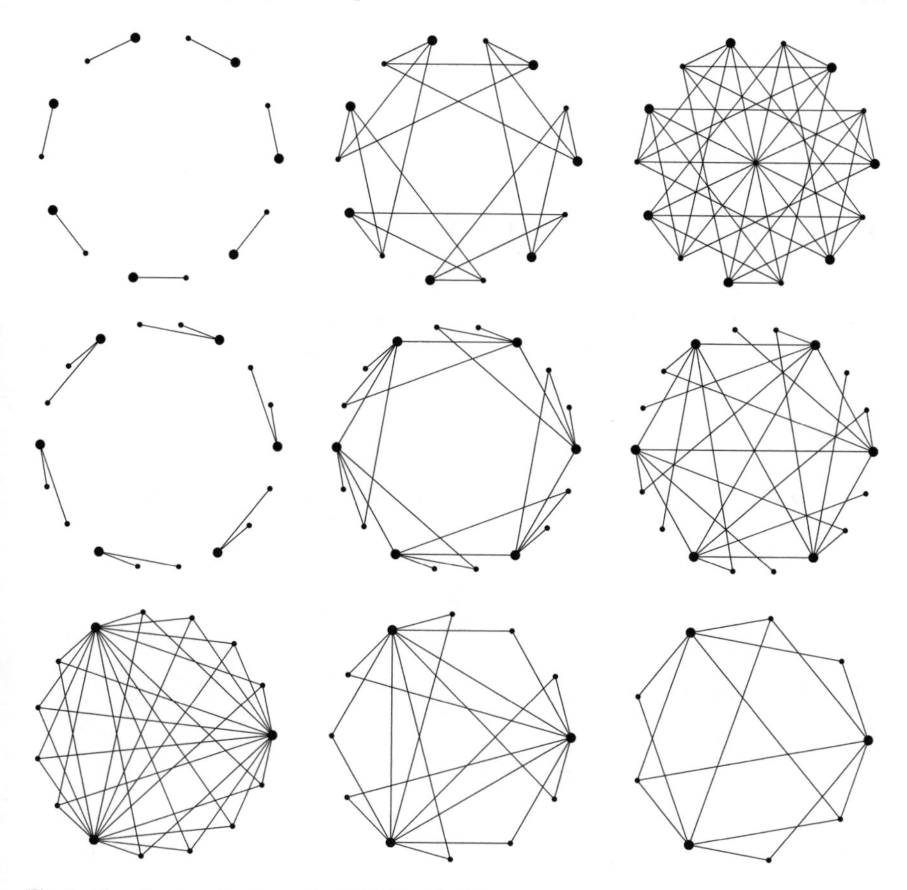

Fig. 5 Visualization of pair graphs $\mathcal{G}(\mathbb{Z}/N\mathbb{Z}, H, S)$ for several $N \in \mathbb{N}$, $H < \mathbb{Z}/N\mathbb{Z}$ and $S \subset \mathbb{Z}/N\mathbb{Z}$ (larger dots means the vertices in H)

3.3 Spectrum of a Pair Graph with Abelian Subgroup

Let $R = R_H$ be the standard matrix realization of the right regular representation of H, that is, for $x, y \in H$, the (x, y)-entry of the matrix R is given by $R(h)_{xy} = \delta_{x,yh^{-1}}$ for $h \in H$. We define the subsets $H_j \subset H$ $(j = 0, 1 \ldots, k)$ by the condition $S_j = H_j x_j$.

Let A be the adjacency matrix of the pair graph $\mathcal{G}(G, H, S)$. We can assume that A is of the form

$$A = \begin{pmatrix} A_{00} & A_{01} & \ldots & A_{0k} \\ A_{10} & & & \\ \vdots & & O & \\ A_{k0} & & & \end{pmatrix},$$

where each A_{0j} is an $n \times n$ matrix indexed by the elements of H given by

$$(A_{0j})_{xy} = \begin{cases} 1 & x \sim yx_j, \\ 0 & x \nsim yx_j \end{cases} \quad (x, y \in H),$$

and $A_{j0} = A_{0j}^*$ (the adjoint matrix of A_{j0}).

Remark 3.4 A_{00} is the adjacency matrix of $\mathrm{Cay}(H, S_0)$, and $\sum_{j=1}^{k} A_{0j} A_{0j}^*$ is the adjacency matrix of a *multigraph* such that the vertices are H and the number of edges between two edges $x, y \in H$ is equal to the number of paths from x to y via $G \setminus H$ of length 2.

Since

$$x \sim yx_j \iff \exists h \in H_j;\ yx_j = xhx_j \iff \exists h \in H_j;\ x = yh^{-1}$$

for $x, y \in H$, it follows that

$$A_{0j} = \sum_{h \in H_j} R(h).$$

Using the identity

$$\det \begin{pmatrix} A & B \\ C & D \end{pmatrix} = \det D \det(A - BD^{-1}C),$$

we have

$$\begin{aligned} \det(xI_{(k+1)n} - A) &= \det \begin{pmatrix} xI_n - A_{00} & -A_{01} \\ -A_{10} & xI_{kn} \end{pmatrix} \\ &= x^{(k-1)n} \det\Big(x^2 I_n - xA_{00} - \sum_{j=1}^{k} A_{0j} A_{0j}^*\Big) \\ &= x^{(k-1)n} \det\Big(x^2 I_n - x \sum_{h \in H_0} R(h) - \sum_{j=1}^{k} \sum_{h \in H_j} R(h) \sum_{h \in H_j} R(h)^*\Big). \end{aligned}$$

We can decompose the regular representation R into irreducible ones as

$$R(h) = U\Big(\bigoplus_{\pi \in \widehat{H}} \pi(H)^{\oplus \deg \pi}\Big) U^*,$$

where U is a certain unitary matrix (which represents an intertwining operator), $\widehat{H}$ is the unitary dual of H, that is, the set of all equivalent classes of the irreducible unitary representations of H, and $\deg \pi$ is the degree of the representation π. Thus, we have

$$\det(xI_{(k+1)n} - A)$$
$$= x^{(k-1)n} \prod_{\pi \in \widehat{H}} \det\Big(x^2 I_{\deg \pi} - x \sum_{h \in H_0} \pi(h) - \sum_{j=1}^{k} \sum_{h \in H_j} \pi(h) \sum_{h \in H_j} \pi(h)^*\Big)^{\deg \pi}.$$

In particular, if H *is abelian*, then every irreducible representation of H is one-dimensional so that all the $R(h)$'s are simultaneously *diagonalized* as

$$UR(h)U^* = \mathrm{diag}(\varphi(h))_{\varphi \in H^*}$$

for all $h \in H$, where $H^* = \mathrm{Hom}(H, \mathbb{C}^\times)$ is the dual group of H, and hence it follows that

$$\det(xI_{(k+1)n} - A) = x^{(k-1)n} \prod_{\varphi \in H^*} \Big(x^2 - x \sum_{h \in H_0} \varphi(h) - \sum_{j=1}^{k} \Big|\sum_{h \in H_j} \varphi(h)\Big|^2\Big).$$

Hence we obtain the following.

Theorem 3.5 *If H is abelian, then the eigenvalues of $\mathcal{G}(G, H, S)$ are given by*

$$\lambda_{\varphi,\pm} = \frac{1}{2}\Bigg(\sum_{h \in H_0} \varphi(h) \pm \Bigg(\Big(\sum_{h \in H_0} \varphi(h)\Big)^2 + 4\sum_{j=1}^{k}\Big|\sum_{h \in H_i} \varphi(h)\Big|^2\Bigg)^{1/2}\Bigg) \quad (\varphi \in H^*)$$

and zeros whose multiplicity is at least $(k-1)n$.

Remark 3.6 The largest eigenvalue λ_0 is equal to $\lambda_{1,+}$ which corresponds to the trivial character $1 \in H^*$.

Corollary 3.7 *If $\mathcal{G}(G, H, S)$ is bipartite (i.e., $H_0 = \emptyset$) and H is abelian, then its eigenvalues are given by*

$$\pm\sqrt{\sum_{j=1}^{k}\Big|\sum_{h \in H_i} \varphi(h)\Big|^2} \quad (\varphi \in H^*)$$

and zeros.

Example 5 Let $G = \langle \sigma \rangle$ be the cyclic group of order $2n$ ($n \geq 3$) generated by σ. We take $H = \langle \sigma^2 \rangle$, and $S = \{\sigma, \sigma^3, \sigma^5, \sigma^2, \sigma^{-2}\}$, and set $x_0 = e$ and $x_1 = \sigma$. We have $S_0 = \{\sigma^2, \sigma^{-2}\} (= H_0)$ and $S_1 = \{\sigma, \sigma^3, \sigma^5\}$ ($H_1 = \{e, \sigma^2, \sigma^4\}$). The dual group H^* is given by $H^* = \langle \varphi \rangle$, where $\varphi(\sigma^{2r}) = \exp \frac{2\pi i r}{n}$. Thus, the eigenvalues of $\mathcal{G}(G, H, S)$ (see Fig. 6) are

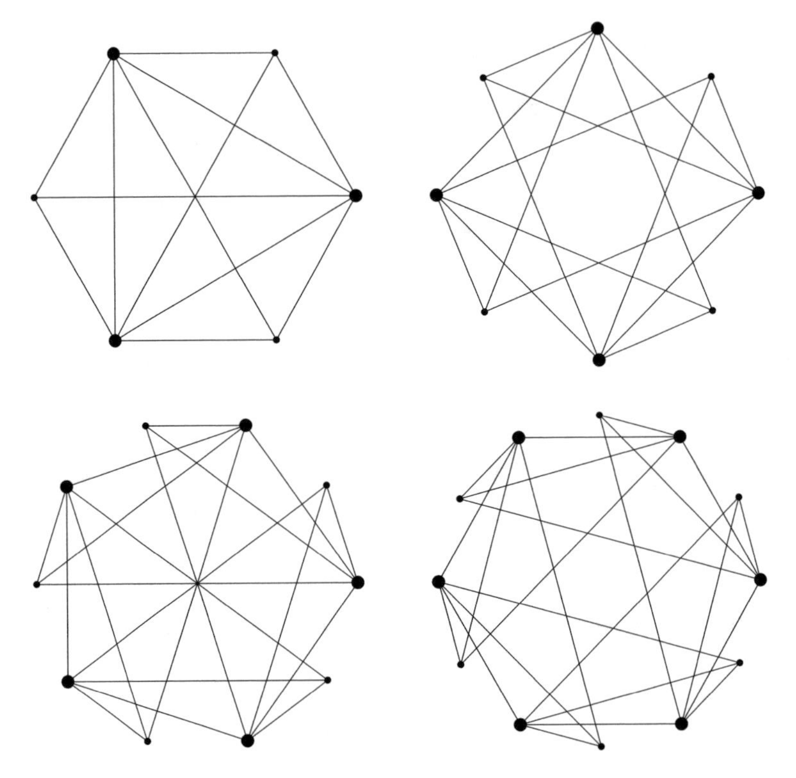

Fig. 6 $\mathcal{G}(G, H, S)$ in Example 5 for $n = 3, 4, 5, 6$

$$\lambda_{j,\pm} = \cos\frac{j\pi}{n} \pm \left(\cos^2\frac{j\pi}{n} + \left(2\cos\frac{2\pi j}{n} + 1\right)^2\right)^{1/2} \quad (j = 0, 1, \ldots, n-1).$$

The largest eigenvalue is $\lambda_0 = \lambda_{0,+} = 1 + \sqrt{10}$, and the second largest eigenvalue is $\lambda_1 = \lambda_{1,+} = \cos\frac{\pi}{n} + \left(\cos^2\frac{\pi}{n} + \left(2\cos\frac{2\pi}{n} + 1\right)^2\right)^{1/2}$. We note that the spectral gap $\lambda_0 - \lambda_1$ tends to 0 as $n \to \infty$.

3.4 An Estimation of λ_1 of Pair Graphs for Abelian Groups

Let us consider the quasi-bipartite case, that is, the case where $k = 1$ and G is *abelian*; we have $G = H \sqcup Hx_1$, $S = S_0 \sqcup S_1$ and $\mathcal{G}(G, H, S)$ is $\begin{pmatrix} d_0 & d_1 \\ d_1 & 0 \end{pmatrix}$-regular. The neighborhood of $h \in H$ is $\{hs \mid s \in S\}$, and that of $x \in Hx_1$ is $\{xs^{-1} \mid s \in S_1\}$.

Thus, for $f \in L^2(G)$, we have

$$(Af)(x) = \sum_{\substack{y \in G \\ y \sim x}} f(y) = \begin{cases} \sum_{s \in S} f(xs) & x \in H, \\ \sum_{s \in S_1} f(xs^{-1}) & x \in Hx_1. \end{cases}$$

Fix a subset $\Xi = \{\xi_0 = 1, \xi_1, \ldots, \xi_{n-1}\} \subset G^*$ such that $G^* = \Xi \sqcup \Xi\eta$, where $\eta \in G^*$ is given by

$$\eta(x) = \begin{cases} 1 & x \in H, \\ -1 & x \in G \setminus H. \end{cases}$$

We put $\xi_{n+j} = \xi_j \eta$ for convenience. For $\xi \in G^*$, we have

$$(A\xi)(x) = \sum_{\substack{y \in G \\ y \sim x}} \xi(y) = \begin{cases} (\sum_{s \in S} \xi(s))\xi(x) & x \in H, \\ (\sum_{s \in S_1} \xi(s^{-1}))\xi(x) & x \in Hx_1. \end{cases}$$

Let $f = \sum_{j=0}^{2n-1} z_j \xi_j \in L^2(G)$. We have

$$\langle f, f \rangle = 2n \sum_{i=0}^{2n-1} |z_i|^2$$

and

$$\begin{aligned} \langle Af, f \rangle &= \sum_{x \in G} (Af)(x)\overline{f(x)} \\ &= \sum_{h \in H} (Af)(h)\overline{f(h)} + \sum_{x \in Hx_1} (Af)(x)\overline{f(x)} \\ &= \sum_{i,j=0}^{2n-1} z_i \overline{z_j} \sum_{h \in H} (A\xi_i)(h)\overline{\xi_j(h)} + \sum_{i,j=0}^{2n-1} z_i \overline{z_j} \sum_{x \in H} (A\xi_i)(hx_1)\overline{\xi_j(hx_1)} \\ &= \sum_{i,j=0}^{2n-1} z_i \overline{z_j} \left\langle \xi_i|_H, \xi_j|_H \right\rangle_{L^2(H)} \left(\sum_{s \in S} \xi_i(s) + \xi_i(x_1)\overline{\xi_j(x_1)} \sum_{s \in S_1} \xi_i(s^{-1}) \right). \end{aligned}$$

Since

$$\left\langle \xi_i|_H, \xi_j|_H \right\rangle_{L^2(H)} = \begin{cases} n & \xi_j = \xi_i \text{ or } \xi_i\eta, \\ 0 & \text{otherwise}, \end{cases}$$

$$\xi_i(x_1)\overline{(\xi_i\eta^j)(x_1)} = (-1)^j \quad (j = 0, 1),$$

we have

$$\begin{aligned}
\langle Af, f\rangle &= n\sum_{i=0}^{2n-1}\sum_{j=0}^{1} z_i\overline{z_{i+jn}}\Big(\sum_{s\in S_0}\xi_i(s)+\sum_{s\in S_1}\xi_i(s)+(-1)^j\sum_{s\in S_1}\xi_i(s^{-1})\Big)\\
&= n\sum_{i=0}^{2n-1}|z_i|^2\Big(\sum_{s\in S_0}\xi_i(s)+\sum_{s\in S_1}(\xi_i(s)+\xi_i(s^{-1}))\Big)\\
&\quad + n\sum_{i=0}^{2n-1} z_{i+n}\overline{z_i}\Big(\sum_{s\in S_0}\xi_i(s)-\sum_{s\in S_1}(\xi_i(s)-\xi_i(s^{-1}))\Big).
\end{aligned}$$

Here the indices of $z_0, z_1, \ldots, z_{2n-1}$ are understood modulo $2n$. Recall that $\xi_{n+j} = \xi_j\eta$, and notice that

$$\xi_{n+j}(s) = \begin{cases} \xi_j(s) & s \in S_0, \\ -\xi_j(s) & s \in S_1. \end{cases}$$

Hence

$$\begin{aligned}
\langle Af, f\rangle &= n\sum_{i=0}^{n-1}|z_i|^2\Big(\sum_{s\in S_0}\xi_i(s)+\sum_{s\in S_1}(\xi_i(s)+\xi_i(s^{-1}))\Big)\\
&\quad + n\sum_{i=0}^{n-1}|z_{n+i}|^2\Big(\sum_{s\in S_0}\xi_i(s)-\sum_{s\in S_1}(\xi_i(s)+\xi_i(s^{-1}))\Big)\\
&\quad + n\sum_{i=0}^{n-1}z_i\overline{z_{i+n}}\Big(\sum_{s\in S_0}\xi_i(s)+\sum_{s\in S_1}(\xi_i(s)-\xi_i(s^{-1}))\Big)\\
&\quad + n\sum_{i=0}^{n-1}z_i\overline{z_{i+n}}\Big(\sum_{s\in S_0}\xi_i(s)-\sum_{s\in S_1}(\xi_i(s)-\xi_i(s^{-1}))\Big)\\
&= n\sum_{i=0}^{n-1}(|z_i|^2+z_i\overline{z_{n+i}}+z_{n+i}\overline{z_i}+|z_{n+i}|^2)\sum_{s\in S_0}\xi_i(s)\\
&\quad + n\sum_{i=0}^{n-1}(|z_i|^2-|z_{n+i}|^2)\sum_{s\in S_1}(\xi_i(s)+\xi_i(s^{-1}))\\
&\quad + n\sum_{i=0}^{n-1}(z_i\overline{z_{n+i}}-z_{n+i}\overline{z_i})\sum_{s\in S_1}(\xi_i(s)-\xi_i(s^{-1})).
\end{aligned}$$

If we assume that $z_{n+i} = \alpha z_i$ with $|\alpha| = 1$, then

$$
\begin{aligned}
\langle Af, f\rangle &= n\sum_{i=0}^{n-1}|z_i|^2\Big(|1+\alpha|^2\sum_{s\in S_0}\xi_i(s)+(\overline{\alpha}-\alpha)\sum_{s\in S_1}(\xi_i(s)-\xi_i(s^{-1}))\Big)\\
&= n\sum_{i=0}^{n-1}|z_i|^2\Big(|1+\alpha|^2\sum_{s\in S_0}\Re\xi_i(s)+4\Im\alpha\sum_{s\in S_1}\Im\xi_i(s)\Big),
\end{aligned}
$$

where $\Re z$ and $\Im z$ is the real and imaginary part of $z \in \mathbb{C}$ respectively, and

$$
\langle f, f\rangle = 4n\sum_{i=0}^{n-1}|z_i|^2 .
$$

Notice that $A\,|1+\alpha|^2 + 4B\,\Im\alpha$, where $A, B \in \mathbb{R}$ are constants, has the maximum value $2(A+\sqrt{A^2+4B^2})$ as a function in α on the unit circle $|\alpha| = 1$. Since the eigenfunctions for the trivial eigenvalues are linear combinations of $\xi_0 = 1$ and $\xi_n = \eta$, we have the

Theorem 3.8 *Put*

$$
a_j = |1+\alpha|^2\sum_{s\in S_0}\Re\xi_j(s)+4\Im\alpha\sum_{s\in S_1}\Im\xi_j(s) \in \mathbb{R}
$$

for $j = 1, 2, \ldots, n-1$. *Then*

$$
\lambda_1(x) \geq \frac{\sum_{i=1}^{n-1} a_i t_i^2}{4\sum_{i=1}^{n-1} t_i^2}
$$

for any $(t_1, \ldots, t_{n-1})(\neq (0, \ldots, 0)) \in \mathbb{R}^{n-1}$. *In particular, the inequality*

$$
\lambda_1(X) \geq \max\{a_1', \ldots, a_{n-1}'\}
$$

holds for

$$
a_j' = \frac{1}{2}\Big(\sum_{s\in S_0}\Re\xi_j(s)+\Big(\Big(\sum_{s\in S_0}\Re\xi_j(s)\Big)^2+4\Big(\sum_{s\in S_1}\Im\xi_j(s)\Big)^2\Big)^{1/2}\Big).
$$

4 Concluding Remarks

4.1 A Generalization of Pair Graphs

Let G be a finite group, H its subgroup of index 2. Fix a collection of representatives $\{x_0 = e, x_1\}$ of $H\backslash G$. Take subsets S_{ij} $(i, j = 0, 1)$ of G such that $e \notin S_{ij} = S_{ji}^{-1} \subset$

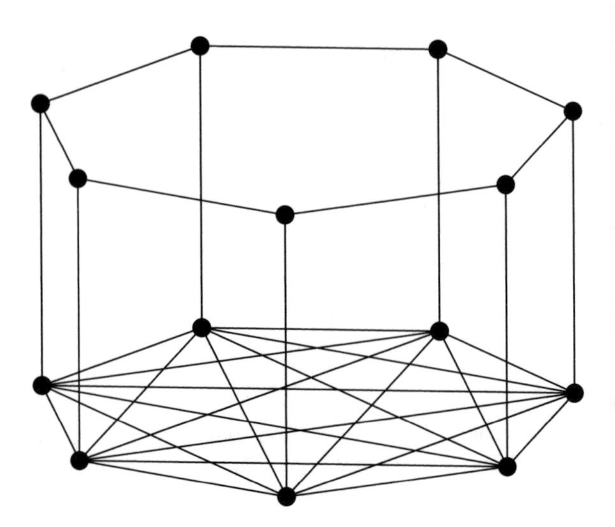

Fig. 7 $\mathcal{G}(G, H, \mathcal{S})$ in Example 6

$Hx_{|i-j|}$. For two vertices $x, y \in G$, we combine these two by an edge if and only if $y = xs$ for some $s \in S_{ij}$ when $x \in Hx_i$ and $y \in Hx_j$ $(i, j = 0, 1)$. We denote this graph by $\mathcal{G}(G, H, \mathcal{S})$ $(\mathcal{S} = \{S_{ij}\})$. This is a $\begin{pmatrix} d_{00} & d_{01} \\ d_{10} & d_{11} \end{pmatrix}$-bigraph, where $d_{ij} = |S_{ij}|$. When $S_{11} = \emptyset$, $\mathcal{G}(G, H, \mathcal{S})$ is reduced to the original group-subgroup pair graph.

If H is a *normal* subgroup of G, then we can similarly define a graph $\mathcal{G}(G, H, \mathcal{S})$ for a suitable collection of subsets $\mathcal{S} = \{S_{ij}\}_{i,j=0}^{k}$, which is a (d_{ij})-regular quasi-multipartite graph for $d_{ij} = |S_{ij}|$.

Example 6 Let $G = D_7 = \langle \sigma, \tau \rangle$ be the dihedral group of degree 14 ($\sigma^7 = \tau^2 = e$, $\tau\sigma\tau = \sigma^{-1}$). We take $H = \langle \sigma \rangle$ and $x_0 = e, x_1 = \tau$. Put $S_{00} = \{\sigma, \sigma^{-1}\}$, $S_{01} = S_{10} = \{\tau\}$, $S_{11} = H \setminus \{e\}$. Then $\mathcal{G}(G, H, \mathcal{S})$ is a $\begin{pmatrix} 2 & 1 \\ 1 & 6 \end{pmatrix}$-bigraph shown in Fig. 7.

When H is abelian, we can determine the spectrum of $\mathcal{G}(G, H, S)$ by the same way as in Theorem 3.5.

4.2 *The Alon–Boppana Theorem for Quasi-bigraphs*

In the article, we give an explicit determination of the eigenvalues of $\mathcal{G}(G, H, S)$ when H is abelian as well as an lower bound estimation of λ_1 (when G is also abelian) in terms of the characters of H. To obtain an analog of the Alon–Boppana theorem for quasi-bigraphs in general (which is important to seek an analogous formulation of Ramanujan graphs), an estimation for the second largest eigenvalue $\lambda_1(X)$ of the form

$$\lambda_1(X) > \mathrm{RB}(D) - \frac{c}{\mathrm{diam}(X)}$$

is desirable for D-bigraphs X, where $\mathrm{RB}(D)$ and c are constants depending only on D. See [5, 8] for trials to obtain an Alon–Boppana type estimation for the second largest eigenvalues.

5 Appendix

We give a proof of Theorem 2.2, which is not given explicitly in [4], by using the following theorem.

Theorem 5.1 ([4, Theorem 2]) *Let $X = (V, E)$ be a k-regular multigraph (X has multi-edges but no loops). Suppose that there exists a constant $g \in \mathbb{Z}_{\geq 0}$ such that*

$$x \sim y \implies \left|\{z \in V \mid z \sim x,\ z \sim y\}\right| \geq g$$

for $x, y \in V$. If $\mathrm{diam}(X) \geq 4$, then

$$\lambda_1(X) > g + 2\sqrt{k-g-1} - \frac{2\sqrt{k-g-1}-1}{\left\lfloor \frac{1}{2}\mathrm{diam}(X) - 1 \right\rfloor}.$$

Theorem 2.2 follows from the

Proposition 5.2 *Let $X = (V, E)$ be a connected (p, q)-biregular bigraph with bipartition $V = V_0 \sqcup V_1$. Suppose that $\mathrm{diam}(X)$ is sufficiently large. Then*

$$\lambda_1(X) > \sqrt{p-1} + \sqrt{q-1} - \frac{c}{\mathrm{diam}(X)}$$

for a certain positive constant c which depends only on p and q.

Proof Let $m = |V_0|$ and $n = |V_1|$. For convenience, we set $V_0 = \{v_1, \ldots, v_m\}$ and $V_1 = \{v_{m+1}, \ldots, v_{m+n}\}$. Put a_{ij} to be the number of edges between v_i and v_j, then $A = (a_{ij})$ is the adjacency matrix of X. Notice that the (i, j)-entry of A^2 is the number of paths of length 2 from v_i to v_j. Let us write

$$A = \begin{pmatrix} O & B \\ {}^tB & O \end{pmatrix},$$

where B is $m \times n$. We see that

$$A^2 = \begin{pmatrix} B{}^tB & O \\ O & {}^tBB \end{pmatrix},$$

and hence $\lambda_1(X)^2$ is equal to the second largest eigenvalue of B^tB.

Let X' be a multigraph such that the vertices are $V' = V_0$ and the number of edges between two vertices $x, y \in V'$ is equal to the number of non-backtracking path from x to y on X with length 2. Notice that $2\,\mathrm{diam}(X') \leq \mathrm{diam}(X) \leq 2\,\mathrm{diam}(X') + 2$. We see that X' is a $p(q-1)$-regular graph, and $A' = B^tB - pI_m$ is the adjacency matrix of X'. Moreover, if $x, y \in V'$ are adjacent in X', then there are at least $q-2$ vertices adjacent to both x and y. Hence, by Theorem 5.1, we have

$$\begin{aligned}\lambda_1(X') &> (q-2) + 2\sqrt{p(q-1) - (q-2) - 1} - \frac{c}{\mathrm{diam}(X')}\\ &\geq (\sqrt{p-1} + \sqrt{q-1})^2 - p - \frac{2c}{\mathrm{diam}(X)},\end{aligned}$$

where $c = 2\sqrt{(p-1)(q-1)} - 1$. Since $\lambda_1(X') = \lambda_1(X)^2 - p$, we have

$$\lambda_1(X)^2 > (\sqrt{p-1} + \sqrt{q-1})^2 - \frac{2c}{\mathrm{diam}(X)},$$

from which we have

$$\lambda_1(X) > \sqrt{p-1} + \sqrt{q-1} - \frac{c'}{\mathrm{diam}(X)},$$

where $c' = 2c/(\sqrt{p-1} + \sqrt{q-1})$, as we desired. □

Remark 5.3 Similar approach to the Alon–Boppana type lower bound estimation does not seem to work in the case of general D-bigraphs. Let $X = (V, E)$ be a D-bigraph with bipartition $V = V_0 \sqcup V_1$, and set $V_0 = \{v_1, \ldots, v_m\}$ and $V_1 = \{v_{m+1}, \ldots, v_{m+n}\}$ as above. The adjacency matrix A of X is of the form

$$A = \begin{pmatrix} A_{00} & A_{01} \\ A_{10} & A_{11} \end{pmatrix},$$

where A_{00} is $m \times m$ and A_{11} is $n \times n$. Then

$$A^2 = \begin{pmatrix} A_{00}^2 + A_{01}A_{10} & A_{00}A_{01} + A_{01}A_{11} \\ A_{10}A_{00} + A_{11}A_{10} & A_{10}A_{01} + A_{11}^2 \end{pmatrix}.$$

Let X_i be a graph whose vertices are V_i and edges are non-backtracking paths of length 2. Then the adjacency matrix of X_0 and X_1 is given by $A_{X_0} = A_{00}^2 + A_{01}A_{10} - d_{01}I_m$ and $A_{X_1} = A_{10}A_{01} + A_{11}^2 - d_{10}I_n$ respectively. Though we can estimate the second largest eigenvalues of A_{X_i} by Theorem 5.1, the estimation is not used to obtain a lower bound of $\lambda_1(X)$ as in the case of bigraphs.

References

1. C. Ballantine, D. Ciubotaru, Ramanujan bigraphs associated with $SU(3)$ over a p-adic field. Proc. Amer. Math. Soc. **139**, 1939–1953 (2011)
2. C. Ballantine, B. Feigon, R. Ganapathy, J. Kool, K. Maurischat, A. Wooding, Explicit construction of Ramanujan bigraphs, in *Women in Numbers Europe: Research Directions in Number Theory, vol. 2, Association for Women in Mathematics Series*, ed. by M.J. Bertin, et al. (Springer, Berlin, 2015), pp. 1–16
3. D. Charles, K. Lauter, E. Goren, Cryptographic hash functions from expander graphs. J. Cryptol. **22**, 93–113 (2009)
4. K. Feng, W.-C.W. Li, Spectra of hypergraphs and applications. J. Number Theory **60**(1), 1–22 (1996)
5. J. Friedman, J.-P. Tillich, Generalized alon-boppana theorems and error-correcting codes. SIAM J. Discret. Math. **19**, 700–718 (2005)
6. K. Hashimoto, Zeta functions of finite graphs and representations of p-adic groups. Automorphic forms and geometry of arithmetic varieties, 211–280, Adv. Stud. Pure Math. **15**. Academic Press, Boston (1989)
7. A. Lubotzky, R. Phillips, P. Sarnak, Ramanujan graphs. Combinatorica **8**, 261–277 (1988)
8. B. Mohar, A strengthening and a multipartite generalization of the alon-boppana-Serre theorem. Proc. Amer. Math. Soc. **138**, 3899–3909 (2010)
9. A.K. Pizer, Ramanujan graphs and Hecke operators. Bull. Am. Math. Soc. (N.S.) **23**, 127–137 (1990)
10. C. Reyes-Bustos, Cayley-type graphs for group-subgroup pairs. Linear Algebr. Appl. **488**, 320–349 (2016)
11. A. Sarveniazi, Explicit construction of a ramanujan $(n_1, n_2, \ldots, n_{d-1})$-regular hypergraph. Duke Math. J. **139**, 141–171 (2007)
12. P. Solé, Ramanujan hypergraphs and Ramanujan geometries. Emerging applications of number theory (Minneapolis, MN, 1996), 583–590, IMA Vol. Math. Appl. **109** (Springer, New York, 1999)

Ramanujan Cayley Graphs of the Generalized Quaternion Groups and the Hardy–Littlewood Conjecture

Yoshinori Yamasaki

Abstract Ramanujan graphs are graphs which have many connections with various mathematical fields including an application to the cryptography. In this article, as a continuous work of our research on Ramanujan graphs, we investigate the bound of the valency of the Cayley graphs of the generalized quaternion groups which guarantees to be Ramanujan. As is the cases of the cyclic and dihedral groups, we show that the determination of the bound in a special setting is related to the classical Hardy–Littlewood conjecture for primes represented by a quadratic polynomial.

Keywords Ramanujan graphs · Generalized quaternion groups · Hardy–Littlewood conjecture

1 Introduction

Expander graph is a sparse graph having strong connectivity properties. Because of its rich theory with many applications, it is widely studied in various fields of mathematics such as combinatorics, group theory, differential geometry and number theory (see [7, 9] for survey of the expander graphs). In particular, Ramanujan graph, which is an optimal expander graph in the sense of Alon-Boppana's theorem and was first defined in [10], plays an important role in not only pure mathematics but also applied mathematics. Actually, because a graph is Ramanujan if and only if the associated Ihara zeta function satisfies the "Riemann hypothesis", it has a special interest for number theorists, especially who study zeta functions (see, e.g., [11]). Moreover, from the fact that a random walk on a Ramanujan graph quickly converges to the uniform distribution, it is used to construct a cryptographic hash function [3]. From these reasons, it is worth finding or constructing Ramanujan graphs as many as possible, however, it is in general difficult. In fact, as it is explained in detail in [8], we can not obtain any families of expander graphs of a fixed degree, which are

Y. Yamasaki (✉)
Graduate School of Science and Engineering, Ehime University,
2-5 Bunkyo-cho, Matsuyama, Ehime 790-8577, Japan
e-mail: yamasaki@math.sci.ehime-u.ac.jp

T. Takagi et al. (eds.), *Mathematical Modelling for Next-Generation Cryptography*,
Mathematics for Industry 29, DOI 10.1007/978-981-10-5065-7_9

needed for applications in computer science, as families of Cayley graphs of some elementary groups.

In this paper, we consider the following problem on Ramanujan graphs. Naively, one easily imagines that, because a Ramanujan graph has a strong connectivity property, if we have a Ramanujan graph, then there expects to be another Ramanujan graph around it (cf. [1]). This means that, even if we get rid of some edges from the given Ramanujan graph anyhow, it may remain to be Ramanujan. Now our problem is to clarify how many edges we can freely remove from the given Ramanujan graph with remaining to be Ramanujan in a given family of graphs. In particular, as a first stage, we consider this problem starting from the trivial Ramanujan graph, that is, the complete graph, in a family of Cayley graphs of a fixed group. Notice that, in this setting, removing edges corresponds to reducing elements of a Cayley subset of the group. See the end of Sect. 2 for more precise mathematical formulation of our problem.

In [5], we first investigated this problem for the cyclic groups. Moreover, in [6], we studied it for the dihedral groups, which are non-abelian (simplest) extensions of the previous case (we actually consider this problem for groups in the class of the Frobenius groups in [6], which contains for example the semi-direct product of the cyclic groups and hence, especially, the dihedral groups). In both cases, we showed that the determination of the above maximal number of removable edges (it corresponds to $\tilde{l}$ in our formulation) is related to the classical Hardy–Littlewood conjecture on analytic number theory, which asserts that every quadratic polynomial expresses infinitely many primes under some standard conditions, if the order of the group is odd prime (resp. twice odd prime) in the case of the cyclic group (resp. the dihedral group). In succession to these cases, in the present paper, we work the same problem for the generalized quaternion group Q_{4m} and actually obtain the similar result (Theorem 4.9) if we choose the set of Cayley graphs suitably. Notice that we indeed consider a wider class of groups in the sense that Q_{4m} can not be expressed as any semi-directed product of the cyclic groups. We also remark that our discussion may be applied to groups whose maximal degree of the irreducible representations is at most two.

We use the following notations in this paper. The set of all real numbers, integers and odd primes are denoted by $\mathbb{R}$, $\mathbb{Z}$ and $\mathbb{P}$, respectively. For $x \in \mathbb{R}$, $\lfloor x \rfloor$ (resp. $\lceil x \rceil$) denote the largest (resp. smallest) integer less (resp. greater) than or equal to x. We also note that the most of our numerical computations are performed by using Mathematica.

2 Preliminary

In this section, we prepare some definitions and notations of graph theory, which are necessary for our discussion (see, more precisely, [8]). Throughout this paper, all graphs are assumed to be finite, undirected, connected, simple and regular.

Let X be a k-regular graph with m-vertices. The adjacency matrix A_X of X is the symmetric matrix of size m whose entry is 1 if the corresponding pair of vertices are connected by an edge and 0 otherwise. We call the eigenvalues of A_X the eigenvalues of X. Let $\Lambda(X)$ be the set of all eigenvalues of X. We know that it can be written as $\Lambda(X) = \{k = \lambda_0 > \lambda_1 \geq \cdots \geq \lambda_{m-1}\} \subset [-k, k]$. Let $\lambda(X)$ be the largest non-trivial eigenvalue of X in the sense of absolute value; $\lambda(X) = \max\{|\lambda| \mid \lambda \in \Lambda(X),\ |\lambda| \neq k\}$. Then, X is called Ramanujan if the inequality $\lambda(X) \leq 2\sqrt{k-1}$ holds. Here the constant $2\sqrt{k-1}$ is called the Ramanujan bound for X and is denoted by $\mathrm{RB}(X)$.

Let G be a finite group with the identity element 1. Let S be a Cayley subset of G, that is, S is a symmetric generating subset of G without 1. We denote by $X(S)$ the Cayley graph of G with respect to the Cayley subset S. This is $|S|$-regular graph whose vertex set is G and edge set $\{(x, y) \in G^2 \mid x^{-1}y \in S\}$. Let $\mathscr{S}_G$ be the set of all Cayley subsets of G. In what follows, for $S \in \mathscr{S}_G$, we write $\Lambda(S) = \Lambda(X(S))$, $\lambda(S) = \lambda(X(S))$, $\mathrm{RB}(S) = \mathrm{RB}(X(S))$, and so on. It is well known that the eigenvalues of $X(S)$ can be described in terms of the irreducible representations of G as follows.

Lemma 2.1 (cf. [2]) *Let G be a finite group and $\mathrm{Irr}(G)$ the set of all equivalence classes of the irreducible representations of G. Then, for $S \in \mathscr{S}_G$, we have*

$$\Lambda(S) = \bigcup_{\pi \in \mathrm{Irr}(G)} d_\pi \cdot \mathrm{Spec}(M_\pi(S)),$$

where, for $\pi \in \mathrm{Irr}(G)$, d_π is the degree of π, $M_\pi(S) = \sum_{s \in S} \pi(s)$ and $\mathrm{Spec}(M_\pi(S))$ is the set of all eigenvalues of $M_\pi(S)$. Here, we understand that an element in $\mathrm{Spec}(M_\pi(S))$ is counted d_π times in $d_\pi \cdot \mathrm{Spec}(M_\pi(S))$.

We here explain our problem on Ramanujan graphs. For a set $\mathscr{S} \subset \mathscr{S}_G$ of Cayley subsets of G, let $\mathscr{L} = \mathscr{L}_{G,\mathscr{S}} = \{l(S) \mid S \in \mathscr{S}\}$ where $l(S) = |G \setminus S| = |G| - |S|$ is the covalency of $S \in \mathscr{S}$. Then, we have the decomposition $\mathscr{S} = \sqcup_{l \in \mathscr{L}} \mathscr{S}_l$ with $\mathscr{S}_l = \{S \in \mathscr{S} \mid l(S) = l\}$. Now our aim is to determine the bound

$$\tilde{l} = \tilde{l}_{G,\mathscr{S}} = \max\{l \in \mathscr{L} \mid X(S) \text{ is Ramanujan for all } S \in \mathscr{S}_k (1 \leq k \leq l)\}.$$

Remark that $\tilde{l} \geq 1$ if $G \setminus \{1\} \in \mathscr{S}_1$ because $X(G \setminus \{1\})$ is the complete graph $K_{|G|}$ with $|G|$ vertices, which is a (trivial) Ramanujan graph. Hence, in this case, roughly speaking, $\tilde{l}$ represents the maximal number of removable edges from the complete graph $K_{|G|}$ keeping to be Ramanujan.

In this paper, we investigate $\tilde{l}$ when G is the generalized quaternion group.

3 Cayley Graphs of the Generalized Quaternion Groups

For a positive integer m, the generalized quaternion group Q_{4m} is defined by

$$Q_{4m} = \langle x, y \mid x^{2m} = 1,\ x^m = y^2,\ y^{-1}xy = x^{-1} \rangle.$$

This is non-commutative unless $m = 1$ and can not be expressed as a semi-direct product of any pair of subgroups of Q_{4m}. One easily sees that the order of Q_{4m} is $4m$ because it has the expression

$$Q_{4m} = \{x^k y^l \mid 0 \le k \le 2m-1,\ l = 0, 1\} = \langle x \rangle \sqcup \langle x \rangle y,$$

where $\langle x \rangle = \{x^k \mid 0 \le k \le 2m-1\}$ and $\langle x \rangle y = \{x^k y \mid 0 \le k \le 2m-1\}$. Notice that $(x^k)^{-1} = x^{2m-k}$ and $(x^k y)^{-1} = x^{m+k} y$.

To calculate the eigenvalues of the Cayley graph of Q_{4m}, we need the information about the conjugacy classes and the irreducible representations of Q_{4m}. For $z \in Q_{4m}$, let $C(z)$ be the conjugacy class of Q_{4m} containing z. Then, the following exhausts all conjugacy classes of Q_{4m}; $C(1) = \{1\}$, $C(x^k) = \{x^k, x^{2m-k}\}$ $(1 \le k \le m-1)$, $C(x^m) = \{x^m\}$, $C(y) = \{x^{2k}y \mid 0 \le k \le m-1\}$ and $C(xy) = \{x^{2k+1}y \mid 0 \le k \le m-1\}$. Moreover, the irreducible representations of Q_{4m} are given as follows; $\chi_1 = \mathbf{1}$ (the trivial character), χ_2, χ_3 and χ_4 which are of degree 1 and φ_j $(1 \le j \le m-1)$ of degree 2. We give the values of these representations in Table 1. Here, $\omega = e^{\frac{2\pi i}{2m}}$.

From now on, we let $\mathscr{S} = \mathscr{S}_{Q_{4m}}$ be the set of all Cayley subsets of Q_{4m}. Let us calculate the eigenvalues of the Cayley graph $X(S)$ for $S \in \mathscr{S}$. Put $S_1 = S \cap \langle x \rangle$ and $S_2 = S \cap \langle x \rangle y$ so that we can write

$$S = S_1 \sqcup S_2.$$

Moreover, put $l_1(S) = 2m - |S_1|$ and $l_2(S) = 2m - |S_2|$ so that $l(S) = l_1(S) + l_2(S)$. Notice that $S_1 \ne \langle x \rangle$ since $1 \notin S$ and hence $l_1(S) > 0$ and $S_2 \ne \emptyset$ because

Table 1 The tables of the values of the irreducible representations of Q_{4m}: the *left* one is the case of odd m and the *right* one is of even m

	x^k	$x^k y$
χ_1	1	1
χ_2	1	-1
χ_3	$(-1)^k$	$i(-1)^k$
χ_4	$(-1)^k$	$i(-1)^{k+1}$
φ_j	$\begin{bmatrix} \omega^{jk} & 0 \\ 0 & \omega^{-jk} \end{bmatrix}$	$\begin{bmatrix} 0 & \omega^{jk} \\ (-1)^j \omega^{-jk} & 0 \end{bmatrix}$

	x^k	$x^k y$
χ_1	1	1
χ_2	1	-1
χ_3	$(-1)^k$	$(-1)^k$
χ_4	$(-1)^k$	$(-1)^{k+1}$
φ_j	$\begin{bmatrix} \omega^{jk} & 0 \\ 0 & \omega^{-jk} \end{bmatrix}$	$\begin{bmatrix} 0 & \omega^{jk} \\ (-1)^j \omega^{-jk} & 0 \end{bmatrix}$

S generates Q_{4m} and therefore $l_2(S) < 2m$. One sees that, because S is symmetric, both S_1 and S_2 are also symmetric. This implies that they are respectively expressed as

$$\begin{aligned} S_1 &= \bigsqcup_{\substack{x^{k_1} \in S \\ 1 \le k_1 \le m-1}} \left\{x^{k_1}, x^{2m-k_1}\right\} \sqcup \left\{x^m\right\}^{\delta}, \\ S_2 &= \bigsqcup_{\substack{x^{k_2} y \in S \\ 0 \le k_2 \le m-1}} \left\{x^{k_2} y, x^{m+k_2} y\right\}, \end{aligned} \tag{1}$$

where $\delta = \delta(S) = 1$ if $x^m \in S$ and 0 otherwise. Here, we understand that $A^0 = \emptyset$ and $A^1 = A$ for any set A. From these expressions, we have $l_2(S) \equiv 0 \pmod 2$ and $l_1(S) \equiv l(S) \equiv \delta \pmod 2$. Based on the above expression, we obtain the following decomposition of $\mathscr{S}$;

$$\mathscr{S} = \bigsqcup_{l \in \mathscr{L}} \mathscr{S}_l = \bigsqcup_{l \in \mathscr{L}} \bigsqcup_{(l_1, l_2) \in \mathscr{L}_l} \mathscr{S}_{l_1, l_2}, \tag{2}$$

where

$$\mathscr{L}_l = \left\{ (l_1, l_2) \in \mathbb{Z}^2 \;\middle|\; \begin{array}{l} 0 < l_1 \le 2m, \ l_1 \equiv l \pmod 2, \\ 0 \le l_2 < 2m, \ l_2 \equiv 0 \pmod 2, \end{array} \ l_1 + l_2 = l \right\}$$

and $\mathscr{S}_{l_1, l_2} = \{S \in \mathscr{S} \,|\, l_1(S) = l_1, \ l_2(S) = l_2\}$ for $l \in \mathscr{L}$ and $(l_1, l_2) \in \mathscr{L}_l$.

Put

$$\begin{aligned} \sigma_1^{\mathrm{e}} &= \sigma_1^{\mathrm{e}}(S) = \#\{k_1 \in \mathbb{Z} \,|\, 1 \le k_1 \le m-1, x^{k_1} \in S, k_1 \equiv 0 \pmod 2\}, \\ \sigma_1^{\mathrm{o}} &= \sigma_1^{\mathrm{o}}(S) = \#\{k_1 \in \mathbb{Z} \,|\, 1 \le k_1 \le m-1, x^{k_1} \in S, k_1 \equiv 1 \pmod 2\}, \\ \sigma_2^{\mathrm{e}} &= \sigma_2^{\mathrm{e}}(S) = \#\{k_2 \in \mathbb{Z} \,|\, 0 \le k_2 \le m-1, x^{k_2} y \in S, k_2 \equiv 0 \pmod 2\}, \\ \sigma_2^{\mathrm{o}} &= \sigma_2^{\mathrm{o}}(S) = \#\{k_2 \in \mathbb{Z} \,|\, 0 \le k_2 \le m-1, x^{k_2} y \in S, k_2 \equiv 1 \pmod 2\}. \end{aligned}$$

Using these notations, it can be written as $|S_1| = 2(\sigma_1^{\mathrm{e}} + \sigma_1^{\mathrm{o}}) + \delta$ and $|S_2| = 2(\sigma_2^{\mathrm{e}} + \sigma_2^{\mathrm{o}})$. Note that

$$0 \le \sigma_1^{\mathrm{e}} \le \frac{m-1}{2}, \ 0 \le \sigma_1^{\mathrm{o}} \le \frac{m-1}{2}, \ 0 \le \sigma_2^{\mathrm{e}} \le \frac{m+1}{2}, \ 0 \le \sigma_2^{\mathrm{o}} \le \frac{m-1}{2}$$

if m is odd and

$$0 \le \sigma_1^{\mathrm{e}} \le \frac{m}{2} - 1, \ 0 \le \sigma_1^{\mathrm{o}} \le \frac{m}{2}, \ 0 \le \sigma_2^{\mathrm{e}} \le \frac{m}{2}, \ 0 \le \sigma_2^{\mathrm{o}} \le \frac{m}{2}$$

otherwise.

From Lemma 2.1 together with the expression (1) of the Cayley subset, one can explicitly calculate the eigenvalues of $X(S)$ as follows.

Lemma 3.1 *For $S \in \mathscr{S}$, we have*

$$\Lambda(S) = 1 \cdot \{\lambda_1, \lambda_2, \lambda_3, \lambda_4\} \cup \bigcup_{j=1}^{m-1} 2 \cdot \left\{\mu_j^+, \mu_j^-\right\}.$$

Here, for $1 \leq j \leq m-1$, $\mu_j^\pm = \mu_j^\pm(S)$ is given by $\mu_j^\pm = z_j \pm |w_j|$ with

$$z_j = z_j(S) = \sum_{\substack{x^{k_1} \in S \\ 0 \leq k_1 \leq 2m-1}} \omega^{jk_1} = \sum_{\substack{x^{k_1} \in S \\ 1 \leq k_1 \leq m-1}} (\omega^{jk_1} + \omega^{-jk_1}) + \delta(-1)^j,$$

$$w_j = w_j(S) = \sum_{\substack{x^{k_2} y \in S \\ 0 \leq k_2 \leq 2m-1}} \omega^{jk_2} = \left(1 + (-1)^j\right) \sum_{\substack{x^{k_2} y \in S \\ 0 \leq k_2 \leq m-1}} \omega^{jk_2},$$

and, for $1 \leq i \leq 4$, $\lambda_i = \lambda_i(S)$ are respectively given as follows.

(1) When m is odd,

$$\begin{aligned}
\lambda_1 &= 2(\sigma_1^{\mathrm{e}} + \sigma_1^{\mathrm{o}}) + \delta + 2(\sigma_2^{\mathrm{e}} + \sigma_2^{\mathrm{o}}), \\
\lambda_2 &= 2(\sigma_1^{\mathrm{e}} + \sigma_1^{\mathrm{o}}) + \delta - 2(\sigma_2^{\mathrm{e}} + \sigma_2^{\mathrm{o}}), \\
\lambda_3 &= 2(\sigma_1^{\mathrm{e}} - \sigma_1^{\mathrm{o}}) - \delta, \\
\lambda_4 &= 2(\sigma_1^{\mathrm{e}} - \sigma_1^{\mathrm{o}}) - \delta.
\end{aligned}$$

(2) When m is even,

$$\begin{aligned}
\lambda_1 &= 2(\sigma_1^{\mathrm{e}} + \sigma_1^{\mathrm{o}}) + \delta + 2(\sigma_2^{\mathrm{e}} + \sigma_2^{\mathrm{o}}), \\
\lambda_2 &= 2(\sigma_1^{\mathrm{e}} + \sigma_1^{\mathrm{o}}) + \delta - 2(\sigma_2^{\mathrm{e}} + \sigma_2^{\mathrm{o}}), \\
\lambda_3 &= 2(\sigma_1^{\mathrm{e}} - \sigma_1^{\mathrm{o}}) + \delta + 2(\sigma_2^{\mathrm{e}} - \sigma_2^{\mathrm{o}}), \\
\lambda_4 &= 2(\sigma_1^{\mathrm{e}} - \sigma_1^{\mathrm{o}}) + \delta - 2(\sigma_2^{\mathrm{e}} - \sigma_2^{\mathrm{o}}).
\end{aligned}$$

Proof These are directly obtained from the above tables of the values of the irreducible representations of Q_{4m}. We notice that $\lambda_i = M_{\chi_i}(S)$ and $\{\mu_j^+, \mu_j^-\} = \mathrm{Spec}(M_{\varphi_j}(S)) = \mathrm{Spec}\left(\begin{bmatrix} z_j & w_j \\ \overline{w_j} & z_j \end{bmatrix}\right)$. □

Remark that $\lambda_1 = |S|$, which corresponds to the trivial character, is the largest eigenvalue of $X(S)$. We also notice that $z_j \in \mathbb{R}$ and $w_j = 0$ if j is odd.

The following lemma is useful in the case of estimating the eigenvalues of $X(S)$ corresponding to λ_i.

Lemma 3.2 *Fix $l \in \mathscr{L}$ and $(l_1, l_2) \in \mathscr{L}_l$. Let $S \in \mathscr{S}_{l_1,l_2}$. Then, we have $\lambda_1 = 4m - l$ and $\lambda_2 = -l + 2l_2$. Moreover,*

(1) when m is odd and

(i) l is odd, we have $-(l-l_2) \le \lambda_3 = \lambda_4 \le l - l_2 - 2$. The absolute values $|\lambda_3| = |\lambda_4|$ of λ_3 and λ_4 take the maximum value $l - l_2$ if and only if $(\sigma_1^{\mathrm{e}}, \sigma_1^{\mathrm{o}}) = (\frac{m+1}{2} - \frac{l-l_2+1}{2}, \frac{m-1}{2})$.

(ii) l is even, we have $-(l-l_2-2) \le \lambda_3 = \lambda_4 \le l - l_2 - 2$. The absolute values $|\lambda_3| = |\lambda_4|$ of λ_3 and λ_4 take the maximum value $l - l_2 - 2$ if and only if $(\sigma_1^{\mathrm{e}}, \sigma_1^{\mathrm{o}}) = (\frac{m+1}{2} - \frac{l-l_2}{2}, \frac{m-1}{2})$ or $(\frac{m-1}{2}, \frac{m+1}{2} - \frac{l-l_2}{2})$.

(2) when m is even, we have $-l \le \lambda_3 \le l + 2\delta - 4$ and $-l \le \lambda_4 \le l + 2\delta - 4$. The absolute value $|\lambda_3|$ of λ_3 takes the maximum value l if and only if $(\sigma_1^{\mathrm{e}}, \sigma_1^{\mathrm{o}}) = (\frac{m}{2} - \frac{l-l_2+\delta}{2}, \frac{m}{2})$ and $(\sigma_2^{\mathrm{e}}, \sigma_2^{\mathrm{o}}) = (\frac{m}{2} - \frac{l_2}{2}, \frac{m}{2})$. Similarly, the absolute value $|\lambda_4|$ of λ_4 takes the maximum value l if and only if $(\sigma_1^{\mathrm{e}}, \sigma_1^{\mathrm{o}}) = (\frac{m}{2} - \frac{l-l_2+\delta}{2}, \frac{m}{2})$ and $(\sigma_2^{\mathrm{e}}, \sigma_2^{\mathrm{o}}) = (\frac{m}{2}, \frac{m}{2} - \frac{l_2}{2})$.

Proof The claims on λ_1 and λ_2 are clear from Lemma 3.1 with the expressions $l_1 = 2m - (2(\sigma_1^{\mathrm{e}} + \sigma_1^{\mathrm{o}}) + \delta)$, $l_2 = 2m - 2(\sigma_2^{\mathrm{o}} + \sigma_2^{\mathrm{o}})$ and $l_1 = l - l_2$. Now, let us consider the other cases.

When m is odd, since $0 \le \sigma_1^{\mathrm{e}} \le \frac{m-1}{2}$ and $0 \le \sigma_1^{\mathrm{o}} \le \frac{m-1}{2}$, we see that $\sigma_1^{\mathrm{e}} + \sigma_1^{\mathrm{o}} = m - \frac{l-l_2+\delta}{2}$ implies that $-\frac{l-l_2+\delta-2}{2} \le \sigma_1^{\mathrm{e}} - \sigma_1^{\mathrm{o}} \le \frac{l-l_2+\delta-2}{2}$. This shows that $-(l - l_2 - 2) - 2\delta \le \lambda_3 = \lambda_4 \le l - l_2 - 2$. Now $|\lambda_3| = |\lambda_4|$ takes maximum value $l - l_2$ if l is odd, which is indeed realized when $-\frac{l-l_2+\delta-2}{2} = \sigma_1^{\mathrm{e}} - \sigma_1^{\mathrm{o}}$ (with $\delta = 1$), and $l - l_2 - 2$ otherwise, which is realized when $\sigma_1^{\mathrm{e}} - \sigma_1^{\mathrm{o}} = \pm\frac{l-l_2+\delta-2}{2}$ (with $\delta = 0$).

When m is even, since $0 \le \sigma_1^{\mathrm{e}} \le \frac{m}{2} - 1, 0 \le \sigma_1^{\mathrm{o}} \le \frac{m}{2}, 0 \le \sigma_2^{\mathrm{e}} \le \frac{m}{2}$ and $0 \le \sigma_2^{\mathrm{o}} \le \frac{m}{2}$, we see that $\sigma_1^{\mathrm{e}} + \sigma_1^{\mathrm{o}} = m - \frac{l-l_2+\delta}{2}$ and $\sigma_2^{\mathrm{e}} + \sigma_2^{\mathrm{o}} = m - \frac{l_2}{2}$ imply that $-\frac{l-l_2+\delta}{2} \le \sigma_1^{\mathrm{e}} - \sigma_1^{\mathrm{o}} \le \frac{l-l_2+\delta-4}{2}$ and $-\frac{l_2}{2} \le \sigma_2^{\mathrm{e}} - \sigma_2^{\mathrm{o}} \le \frac{l_2}{2}$, respectively. This shows that $-l \le \lambda_3 \le l + 2\delta - 4$ and $-l \le \lambda_4 \le l + 2\delta - 4$. Similarly as the above, $|\lambda_3|$ takes maximum value l if $-\frac{l-l_2+\delta}{2} = \sigma_1^{\mathrm{e}} - \sigma_1^{\mathrm{o}}$ and $-\frac{l_2}{2} = \sigma_2^{\mathrm{e}} - \sigma_2^{\mathrm{o}}$ and $|\lambda_4|$ takes l if $-\frac{l-l_2+\delta}{2} = \sigma_1^{\mathrm{e}} - \sigma_1^{\mathrm{o}}$ and $\sigma_2^{\mathrm{e}} - \sigma_2^{\mathrm{o}} = \frac{l_2}{2}$. □

4 Main Results

4.1 Trivial Lower Bound of $\tilde{l}$

We first show that a lower bound of $\tilde{l}$ is obtained by using the trivial estimate of the eigenvalues of Cayley graphs.

Lemma 4.1 *Assume $|S| \ge 2m$. Then, for all $\lambda \in \Lambda(S)$ with $|\lambda| \ne |S|$, we have $|\lambda| \le l(S)$.*

Proof The claim is clear for the cases $\lambda = \lambda_i$ for $2 \le i \le 4$. Actually, since $\lambda_i = \sum_{s\in S} \chi_i(s) = -\sum_{s\notin S} \chi_i(s)$, by the orthogonality of characters, it holds that $|\lambda_i| \le$

$\min\{|S|, l(S)\} = l(S)$. We next consider the cases $\lambda = \mu_j^{\pm}$ for $1 \le j \le m-1$. Let $|\mu_j| = \max\{|\mu_j^+|, |\mu_j^-|\}$. As is the case of the dihedral groups [6], we see that

$$|\mu_j| = |z_j| + |w_j|. \tag{3}$$

Hence, since

$$z_j = \sum_{\substack{x^{k_1} \in S \\ 0 \le k_1 \le 2m-1}} \omega^{jk_1} = - \sum_{\substack{x^{k_1} \notin S \\ 0 \le k_1 \le 2m-1}} \omega^{jk_1},$$
$$w_j = \sum_{\substack{x^{k_2} y \in S \\ 0 \le k_2 \le 2m-1}} \omega^{jk_2} = - \sum_{\substack{x^{k_2} y \notin S \\ 0 \le k_2 \le 2m-1}} \omega^{jk_2},$$

we have $|\mu_j| \le \min\{|S_1|, l_1(S)\} + \min\{|S_2|, l_2(S)\}$. Now, it is easy to see that the right-hand side of the inequality is bounded above by $l(S)$. □

Proposition 4.2 *Let $l_0 = \lfloor 4\sqrt{m} \rfloor - 2$. Then, we have $\tilde{l} \ge l_0$.*

Proof We see from Lemma 4.1 that $X(S)$ is Ramanujan if $l(S) \le \mathrm{RB}(S) = 2\sqrt{|S|-1} = 2\sqrt{(4m - l(S)) - 1}$. Now one sees that this is equivalent to $l(S) \le 4\sqrt{m} - 2$ and hence obtain the desired result. Remark that $l(S) \le 4\sqrt{m} - 2$ implies that $l(S) \le 2m$, that is, $|S| \ge 2m$ for all $m \ge 1$. □

We call l_0 a trivial bound of $\tilde{l}$. Using Lemma 3.2, we can easily determine the bound $\tilde{l}$ in the case of $\mathscr{S} = \mathscr{S}_{Q_{4m}}$.

Theorem 4.3 *We have $\tilde{l} = l_0$.*

Proof Take any $S \in \mathscr{S}_{l_0+1}$ with $l_2(S) = 0$, that is, $S \in \mathscr{S}_{l_0+1,0}$. Then, from Lemma 3.2, we have $|\lambda_2| = l_0 + 1$ and hence, by the definition of l_0, $|\lambda_2| > \mathrm{RB}(S)$. This means that $X(S)$ is not Ramanujan. □

4.2 A Modification

From Theorem 4.3, in the case of $\mathscr{S} = \mathscr{S}_{Q_{4m}}$, we may not expect a connection between our problem on Ramanujan graphs and a problem on analytic number theory, as our previous studies in the cases of the cyclic and dihedral groups [5, 6]. So, we next take another set of Cayley subsets of Q_{4m}, that is,

$$\mathscr{S}' = \{S \in \mathscr{S}_{Q_{4m}} \mid l_2(S) \ne 0\}.$$

Notice that $l_2(S) \ne 0$ is equivalent to $S_2 \ne \langle x \rangle y$. This means that the setting on $\mathscr{S}'$ is reasonable in the sense that we do not consider the extreme case $S_2 = \langle x \rangle y$.

Furthermore, put $\mathscr{L}' = \{l(S) \mid S \in \mathscr{S}'\}$ and $\mathscr{S}'_l = \mathscr{S}_l \cap \mathscr{S}'$. Now our new purpose is to determine the bound

$$\tilde{l}' = \max\{l \in \mathscr{L}' \mid X(S) \text{ is Ramanujan for all } S \in \mathscr{S}'_k (1 \le k \le l)\}.$$

It is clear that

$$\tilde{l}' \ge l_0. \tag{4}$$

Moreover, it holds that

Theorem 4.4 *When m is even, we have $\tilde{l}' = l_0$.*

Proof From Lemma 3.2 (2), we can find $S \in \mathscr{S}'_{l_0+1}$ with $l_2(S) \neq 0$ satisfying $|\lambda_3| = l_0 + 1 > \mathrm{RB}(S)$ (or $|\lambda_4| = l_0 + 1 > \mathrm{RB}(S)$). This immediately shows that $X(S)$ is not Ramanujan. □

From this theorem, we may assume in what follows that m is odd. Remark that, in this case, from Lemma 3.2 again, we have $|\lambda_i| < l$ for $2 \le i \le 4$ for any $l \in \mathscr{L}'$ and $S \in \mathscr{S}'_l$.

4.3 An Upper Bound of $\tilde{l}'$

As is the case of $\mathscr{S}$, it is convenient to decompose $\mathscr{S}'$ as follows;

$$\mathscr{S}' = \bigsqcup_{l \in \mathscr{L}'} \mathscr{S}'_l = \bigsqcup_{l \in \mathscr{L}'} \bigsqcup_{(l_1,l_2) \in \mathscr{L}'_l} \mathscr{S}'_{l_1,l_2}, \tag{5}$$

where

$$\mathscr{L}'_l = \left\{ (l_1, l_2) \in \mathbb{Z}^2 \;\middle|\; \begin{matrix} 0 < l_1 \le 2m, \; l_1 \equiv l \pmod 2, \\ 0 < l_2 < 2m, \; l_2 \equiv 0 \pmod 2, \end{matrix} \; l_1 + l_2 = l \right\}$$

and $\mathscr{S}'_{l_1,l_2} = \mathscr{S}_{l_1,l_2} \cap \mathscr{S}'$ for $l \in \mathscr{L}'$ and $(l_1, l_2) \in \mathscr{L}'_l$.

The aim of this subsection is to show the following result.

Proposition 4.5 *For $m \ge 65$, we have $\tilde{l}' = l_0$ or $\tilde{l}' = l_0 + 1$.*

Let $l \in \mathscr{L}'$. To prove Proposition 4.5, we first construct $S^{(l_1,l_2)} \in \mathscr{S}'_{l_1,l_2}$ for each $(l_1, l_2) \in \mathscr{L}'_l$ such that $X(S^{(l_1,l_2)})$ may have the maximal eigenvalue (in the sense of absolute value) among $X(S)$ with $S \in \mathscr{S}'_{l_1,l_2}$. Let $(l_1, l_2) \in \mathscr{L}'_l$. We define $S^{(l_1,l_2)} = S_1^{(l_1)} \sqcup S_2^{(l_2)} \in \mathscr{S}'_{l_1,l_2}$ by

$$S_1^{(l_1)} = \langle x \rangle \setminus \{1, x^{\pm 1}, \ldots, x^{\pm \frac{l_1-2+\delta}{2}}\} \cup \{x^m\}^{1-\delta},$$
$$S_2^{(l_2)} = \langle x \rangle y \setminus \{y, xy, \ldots, x^{\frac{l_2}{2}-1}y, x^m y, x^{m+1}y, \ldots, x^{m+\frac{l_2}{2}-1}y\},$$

where $\delta = 1$ if l is odd and 0 otherwise. We respectively write z_j, w_j and $|\mu_j|$ as $z_j^{(l_1,l_2)}$, $w_j^{(l_1,l_2)}$ and $|\mu_j^{(l_1,l_2)}|$ when $S = S^{(l_1,l_2)}$. Recall that $w_j^{(l_1,l_2)} = 0$ when j is odd. On the other hand when j is even, we have

$$w_j^{(l_1,l_2)} = -2 \sum_{k_2=0}^{\frac{l_2}{2}-1} e^{\frac{2\pi i j k_2}{2m}} = -2e^{\frac{\pi i j (l_2-2)}{4m}} \frac{\sin \frac{\pi j l_2}{4m}}{\sin \frac{\pi j}{2m}}.$$

Moreover, $z_j^{(l_1,l_2)}$ is calculated as

$$z_j^{(l_1,l_2)} = -\left(\sum_{k_1=-\frac{l_1-2+\delta}{2}}^{\frac{l_1-2+\delta}{2}} e^{\frac{2\pi i j k_1}{2m}} + (1-\delta)(-1)^j \right)$$
$$= -\left(\frac{\sin \frac{\pi j (l_1-1+\delta)}{2m}}{\sin \frac{\pi j}{2m}} + (1-\delta)(-1)^j \right).$$

Hence we have

$$|\mu_j^{(l_1,l_2)}| = \left(\frac{\sin \frac{\pi j (l_1-1+\delta)}{2m}}{\sin \frac{\pi j}{2m}} + (1-\delta)(-1)^j \right) + \delta_j \left(2 \frac{\sin \frac{\pi j l_2}{4m}}{\sin \frac{\pi j}{2m}} \right), \tag{6}$$

where $\delta_j = 1$ if j is even and 0 otherwise. We now focus on the case of $j = 2$.

Lemma 4.6 *Let $l \in \mathscr{L}'$. When $l \equiv r \pmod 6$ for $0 \le r \le 5$, we have*

$$\max \left\{ |\mu_2^{(l_1,l_2)}| \;\middle|\; (l_1, l_2) \in \mathscr{L}_l' \right\} = |\mu_2^{(\check{l}_1,\check{l}_2)}|, \tag{7}$$

where $(\check{l}_1, \check{l}_2) = (\frac{l+a_r}{3}, \frac{2l-a_r}{3}) \in \mathscr{L}_l'$ with

$$a_1 = 2, \;\; a_3 = 0, \;\; a_5 = -2, \;\; a_0 = 0, \;\; a_2 = 4, \;\; a_4 = 2.$$

Proof It holds that

$$\frac{\partial}{\partial l_2} |\mu_2^{(l_1,l_2)}| = \frac{\frac{2\pi}{m}}{\sin \frac{\pi}{m}} \sin \frac{\pi(2(l-1+\delta)-l_2)}{4m} \sin \frac{\pi(2(l-1+\delta)-3l_2)}{4m}.$$

Hence, noting that l, l_1 and l_2 are small enough rather than m, we see that, as a continuous function of l_2, $\frac{\partial}{\partial l_2}|\mu_2^{(l_1,l_2)}| = 0$ on $[1, l]$ if and only if $l_2 = \frac{2(l-1+\delta)}{3}$, which means that $|\mu_2^{(l_1,l_2)}|$ is monotone increasing on $[1, \frac{2(l-1+\delta)}{3}]$ and decreasing on $[\frac{2(l-1+\delta)}{3}, l]$. Let us find $(l_1^{\pm}, l_2^{\pm}) \in \mathscr{L}_l'$ such that l_2^- is the maximum and l_2^+ the minimum integer satisfying $l_2^- \le \frac{2(l-1+\delta)}{3} \le l_2^+$ (notice that $l_1^{\pm}$ are automatically determined from $l_2^{\pm}$ by $l_1^{\pm} + l_2^{\pm} = l$). If we write $l = 6k + r$, then one sees that these are respectively given as follows:

r	1	3	5
(l_1^-, l_2^-)	$(2k+1, 4k)$	$(2k+1, 4k+2)$	$(2k+3, 4k+2)$
(l_1^+, l_2^+)	$(2k-1, 4k+2)$	$(2k+1, 4k+2)$	$(2k+1, 4k+4)$

r	0	2	4
(l_1^-, l_2^-)	$(2k+2, 4k-2)$	$(2k+2, 4k)$	$(2k+2, 4k+2)$
(l_1^+, l_2^+)	$(2k, 4k)$	$(2k, 4k+2)$	$(2k+2, 4k+2)$

Now the result follows from the facts $|\mu_2^{(l_1^-,l_2^-)}| > |\mu_2^{(l_1^+,l_2^+)}|$ for $r = 1, 2$, $|\mu_2^{(l_1^-,l_2^-)}| = |\mu_2^{(l_1^+,l_2^+)}|$ for $r = 3, 4$ and $|\mu_2^{(l_1^-,l_2^-)}| < |\mu_2^{(l_1^+,l_2^+)}|$ for $r = 5, 0$. Namely, $(\check{l}_1, \check{l}_2) = (l_1^-, l_2^-)$ for $r = 1, 2$, $(l_1^-, l_2^-) = (l_1^+, l_2^+)$ for $r = 3, 4$ and (l_1^+, l_2^+) for $r = 5, 0$. □

Using Lemma 4.6, we give a proof of Proposition 4.5.

Proof of Proposition 4.5. It is sufficient to show that there exists $S \in \mathscr{L}_{l_0+2}'$ such that $X(S)$ is not Ramanujan. Actually, let $l_0 = \lfloor 4\sqrt{m} \rfloor - 2 \equiv r \pmod 6$ for $0 \le r \le 5$. Take $S^{(\check{l}_1,\check{l}_2)} \in \mathscr{S}_{l_0+2}'$ with $(\check{l}_1, \check{l}_2) = (\frac{l_0+2+a_{r+2}}{3}, \frac{2(l_0+2)-a_{r+2}}{3}) \in \mathscr{L}_{l_0+2}'$. Here the index of a_r is considered modulo 6. Then, noticing that $4\sqrt{m} - 1 < l_0 + 2 \le 4\sqrt{m}$, we have

$$
\begin{aligned}
&|\mu_2^{(\check{l}_1,\check{l}_2)}| - \mathrm{RB}(S^{(\check{l}_1,\check{l}_2)}) \\
&= \frac{\sin \frac{\pi(\check{l}_1-1+\delta)}{m}}{\sin \frac{\pi}{m}} + (1-\delta) + 2\frac{\sin \frac{\pi \check{l}_2}{2m}}{\sin \frac{\pi}{m}} - 2\sqrt{4m - (\check{l}_1 + \check{l}_2) - 1} \\
&= \frac{\sin\left(\frac{\pi}{m}\left(\frac{l_0+2+a_{r+2}}{3} - 1 + \delta\right)\right)}{\sin \frac{\pi}{m}} + (1-\delta) + 2\frac{\sin\left(\frac{\pi}{2m}\frac{2(l_0+2)-a_{r+2}}{3}\right)}{\sin \frac{\pi}{m}} \\
&\quad - 2\sqrt{4m - (l_0+2) - 1} \\
&> \frac{\sin \frac{\pi}{m}\left(\frac{4\sqrt{m}-1+a_{r+2}}{3} - 1 + \delta\right)}{\sin \frac{\pi}{m}} + (1-\delta) + 2\frac{\sin \frac{\pi}{2m}\frac{2(4\sqrt{m}-1)-a_{r+2}}{3}}{\sin \frac{\pi}{m}} \\
&\quad - 2\sqrt{4m - (4\sqrt{m} - 1) - 1} \\
&= 1 - \frac{64\pi^2 - 27}{54} m^{-\frac{1}{2}} + O(m^{-1})
\end{aligned}
$$

as $m \to \infty$. This shows that $|\mu_2^{(\check{l}_1,\check{l}_2)}| > \mathrm{RB}(S^{(\check{l}_1,\check{l}_2)})$ for sufficiently large m and hence concludes that the corresponding Cayley graph $X(S^{(\check{l}_1,\check{l}_2)})$ is not Ramanujan. Actually, one can check that the inequality holds for $m \ge 105$. Moreover, we can numerically see that $|\mu_2^{(\check{l}_1,\check{l}_2)}| - \mathrm{RB}(S^{(\check{l}_1,\check{l}_2)}) > 0$ for $65 \le m \le 103$ (however it does not hold when $m = 63$).

4.4 A Characterization of Exceptional Primes

From now on, we concentrate on the case where $m = p$ is odd prime (we can perform the similar discussion for general m as in [5], though it may be complicated). We know from Proposition 4.5 that it can be written as $\tilde{l}' = l_0 + \varepsilon$ for some $\varepsilon = \varepsilon_p \in \{0, 1\}$. As is the case of the cyclic and dihedral graphs [5, 6], we call p exceptional if $\varepsilon = 1$ and ordinary otherwise. Now our task is to clarify which $p \in \mathbb{P}$ is exceptional.

For $l \in \mathcal{L}'$, let $\lambda(l) = \max\{\lambda(S) \mid S \in \mathcal{S}'_l\}$ and $\mathrm{RB}(l) = 2\sqrt{4p - l - 1}$, which is nothing but the Ramanujan bound of $X(S)$ for $S \in \mathcal{S}'_l$. From the definition, p is exceptional if and only if $\lambda(l_0 + 1) \le \mathrm{RB}(l_0 + 1)$.

Lemma 4.7 *Let $l \in \mathcal{L}'$. For $(l_1, l_2) \in \mathcal{L}'_l$, let $\lambda(l_1, l_2) = \max\{\lambda(S) \mid S \in \mathcal{S}'_{l_1,l_2}\}$. Then, we have $\lambda(l_1, l_2) = |\mu_2^{(l_1,l_2)}|$ for sufficiently large p.*

Proof Take any $S \in \mathcal{S}'_{l_1,l_2}$. When j is odd, since $w_j = 0$, we have $|\mu_j| = |z_j| \le |z_1^{(l_1,l_2)}| = |\mu_1^{(l_1,l_2)}|$ because p is prime. On the other hand when j is even, since jk is always even modulo $2p$ for any k, it holds that $|\mu_j| \le |\mu_2^{(l_1,l_2)}|$ by the same reason as above. Moreover, since

$$|\mu_1^{(l_1,l_2)}| = \frac{\sin \frac{\pi(l_1-1+\delta)}{2p}}{\sin \frac{\pi}{2p}} - (1-\delta) = (-2 + 2\delta + l_1) + O(p^{-2}),$$

$$|\mu_2^{(l_1,l_2)}| = \frac{\sin \frac{\pi(l_1-1+\delta)}{p}}{\sin \frac{\pi}{p}} + (1-\delta) + 2\frac{\sin \frac{\pi l_2}{2p}}{\sin \frac{\pi}{p}} = (l_1 + l_2) + O(p^{-2}),$$

we see that $|\mu_2^{(l_1,l_2)}| - |\mu_1^{(l_1,l_2)}| = l_2 + 2 - 2\delta + O(p^{-2})$ as $p \to \infty$. Hence, under the condition $l_2 > 0$, we have $|\mu_2^{(l_1,l_2)}| > |\mu_1^{(l_1,l_2)}|$ for sufficiently large p. Combining this together with the fact $\max\{|\lambda_i| \mid 2 \le i \le 4,\ S \in \mathcal{S}'_{l_1,l_2}\} = l_1 + l_2 - 2 < |\mu_2^{(l_1,l_2)}|$ for sufficiently large p, one obtains the claim. □

Proposition 4.8 *Let $p \ge 67$. When $l_0 \equiv r \pmod 6$ for $0 \le r \le 5$, we have*

$$\lambda(l_0 + 1) = |\mu_2^{(\check{l}_1,\check{l}_2)}|,$$

where $(\check{l}_1, \check{l}_2) = (\frac{l_0+1+a_{r+1}}{3}, \frac{2(l_0+1)-a_{r+1}}{3}) \in \mathcal{L}'_{l_0+1}$.

Proof This follows immediately from Lemmas 4.6 and 4.7. Remark that the inequality $|\mu_2^{(\check{l}_1,\check{l}_2)}| - |\mu_1^{(\check{l}_1,\check{l}_2)}| > 0$ in fact holds for $p \ge 67$. □

Write $l_0 = \lfloor 4\sqrt{p} \rfloor - 2$ as

$$l_0 = 24k + r$$

for $k \geq 0$ and $0 \leq r \leq 23$. In this case, we see that $p \in I_{r,k} \cap \mathbb{P}$ where

$$\begin{aligned} I_{r,k} &= \left\{ t \in \mathbb{R} \,\middle|\, \lfloor 4\sqrt{t} \rfloor - 2 = 24k + r \right\} \\ &= \left[36k^2 + 3(r+2)k + \frac{(r+2)^2}{16}, 36k^2 + 3(r+3)k + \frac{(r+3)^2}{16} \right). \end{aligned}$$

In other words, p can be written as $p = f_{r,c}(k)$ for some integers $k \geq 0$ and $c \in \mathbb{Z}$ with $f_{r,c}(x)$ being a quadratic polynomial defined by

$$f_{r,c}(x) = 36x^2 + 3(r+3)x + c$$

and $-3k + \lceil \frac{(r+2)^2}{16} \rceil \leq c \leq \lfloor \frac{(r+3)^2}{16} \rfloor$.

For $0 \leq r \leq 23$, let $I_r = \bigsqcup_{k \geq 0} I_{r,k} \cap \mathbb{P}$ and $C_r = \{ \lfloor \frac{(r+3)^2}{16} \rfloor + s \mid -5 \leq s \leq 0 \}$. Moreover, let $C'_r = \{ c \in C_r \mid f_{r,c}(x) \text{ is irreducible over } \mathbb{Z} \}$. Furthermore, for $c \in C'_r$, define $k_{r,c} \in \mathbb{Z}$ as in Table 2. The following is our main result, which gives a characterization for the exceptional primes.

Theorem 4.9 *A prime $p \in I_r$ with $p \geq 67$ is exceptional if and only if it is of the form of $p = f_{r,c}(k)$ for some $c \in C'_r$ and $k \geq k_{r,c}$.*

Proof We first notice that, from the previous discussion with Proposition 4.8, p is exceptional if and only if $|\mu_2(\check{l}_1, \check{l}_2)| \leq \mathrm{RB}(l_0 + 1)$. To clarify when this inequality holds, we introduce an interpolation function $F_r(t)$ of the difference between $|\mu_2(\check{l}_1, \check{l}_2)|$ and $\mathrm{RB}(l_0 + 1)$ on $I_{r,k}$, that is,

$$\begin{aligned} F_r(t) = &\frac{\sin \frac{\pi(8k + \frac{r+1+a_{r+1}}{3} - 1 + \delta)}{t}}{\sin \frac{\pi}{t}} + (1 - \delta) + 2 \frac{\sin \frac{\pi(16k + \frac{2(r+1) - a_{r+1}}{3})}{2t}}{\sin \frac{\pi}{t}} \\ &- 2\sqrt{4t - (24k + r + 1)} - 1. \end{aligned}$$

Notice that $(\check{l}_1, \check{l}_2) = \left(8k + \frac{r+1+a_{r+1}}{3}, 16k + \frac{2(r+1) - a_{r+1}}{3} \right)$ when $l_0 = 24k + r$. One can see that $F_r(t)$ is monotone decreasing on $I_{r,k}$ for sufficiently large k. Moreover, at $t = p = f_{r,c}(k) \in I_{r,k} \cap \mathbb{P}$, one has

$$F_r(p) = \frac{27(r+3)^2 - 432c - 256\pi^2}{1296} k^{-1} + O(k^{-2})$$

as $k \to \infty$ because

$$
\begin{aligned}
|\mu_2(\check{l}_1, \check{l}_2)| &= \frac{\sin \frac{\pi(8k+\frac{r+1+a_{r+1}}{3}-1+\delta)}{36k^2+3(r+3)k+c}}{\sin \frac{\pi}{36k^2+3(r+3)k+c}} + (1-\delta) + 2\frac{\sin \frac{\pi(16k+\frac{2(r+1)-a_{r+1}}{3})}{2(36k^2+3(r+3)k+c)}}{\sin \frac{\pi}{36k^2+3(r+3)k+c}} \\
&= 24k + (1+r) - \frac{16\pi^2}{81}k^{-1} + O(k^{-2}), \\
\mathrm{RB}(l_0+1) &= 2\sqrt{4(36k^2+3(r+3)k+c)} - (24k+r+1) - 1 \\
&= 24k + (1+r) - \frac{(r+3)^2-16c}{48}k^{-1} + O(k^{-2}).
\end{aligned}
$$

This shows that $F_r(p) < 0$ for sufficiently large k if and only if $27(r+3)^2 - 432c - 256\pi^2 < 0$, in other words, $\lceil \frac{27(r+3)^2-256\pi^2}{432} \rceil \leq c$. Here, we see that $\lceil \frac{27(r+3)^2-256\pi^2}{432} \rceil = \lfloor \frac{(r+3)^2}{16} \rfloor - 5$ for all $0 \leq r \leq 23$, which means that $c \in C_r$. Moreover, since $f_{r,c}(k)$ does not express any prime if $f_{r,c}(x)$ is not irreducible over $\mathbb{Z}$, c must be in C'_r. Furthermore, it is checked that, for each $0 \leq r \leq 23$ and $c \in C'_r$, the inequalities $f_{r,c}(k) \geq 67$ and $F_r(p) < 0$ for $p = f_{r,c}(k)$ hold if and only if $k \geq k_{r,c}$. This completes the proof of the theorem. $\square$

For $0 \leq r \leq 23$ and $c \in C'_r$, let

$$J_{r,c} = \{p \mid p = f_{r,c}(k) \in I_r \text{for some } k \geq k_{r,c}\}.$$

Namely, $J_{r,c}$ is the set of exceptional primes p of the form of $p = f_{r,c}(k)$. We show the first five such primes in Table 2 for each r and c.

The classical Hardy–Littlewood conjecture [4, Conjecture F] asserts that if a quadratic polynomial $f(x) = ax^2 + bx + c$ with $a, b, c \in \mathbb{Z}$ satisfies the conditions that $a > 0$, a, b and c are relatively prime, $a + b$ and c are not both even and the discriminant $D(f) = b^2 - 4ac$ of f is not a square, then there are infinitely many primes represented by $f(x)$ and, moreover, that

$$\pi(f; x) = \#\{p \leq x \mid p = f(k) \in \mathbb{P} \text{ for some } k \geq 0\}$$

obeys the asymptotic behavior

$$\pi(f; x) \sim \frac{\varepsilon(f)C(f)}{\sqrt{a}} \prod_{\substack{p|a,\ p|b \\ p\geq 3}} \frac{p}{p-1} \cdot \frac{\sqrt{x}}{\log x}$$

as $x \to \infty$ where $\varepsilon(f)$ is 1 if $a + b$ is odd and 2 otherwise and

$$C(f) = \prod_{\substack{p \nmid a \\ p\geq 3}} \left(1 - \frac{\left(\frac{D(f)}{p}\right)}{p-1}\right)$$

Table 2 The fifty-four quadratic polynomials $f_{r,c}(x)$ for $0 \leq r \leq 23$ and $c \in C'_r$

r	$c \in C'_r$	$f_{r,c}(x)$	$k_{r,c}$	$J_{r,c}$	$N_{r,c}$	$\frac{C(f_{r,c})}{2\delta_r}$
0	−5	$36x^2 + 9x - 5$	9	7177, 11821, 20947, 52321, 121621	9597	0.24501
0	−4	$36x^2 + 9x - 4$	2	347, 941, 1823, 4451, 6197	17722	0.45086
0	−2	$36x^2 + 9x - 2$	2	349, 6199, 8233, 16063, 19249	11061	0.28123
1	−1	$36x^2 + 12x - 1$	2	167, 359, 1367, 1847, 2399	24414	0.61666
2	−4	$36x^2 + 15x - 4$	9	4517, 16187, 22871, 30707, 44621	9685	0.24501
2	−2	$36x^2 + 15x - 2$	1	367, 1867, 3049, 4519, 6277	13501	0.34106
2	−1	$36x^2 + 15x - 1$	1	173, 2423, 11933, 14699, 28643	11181	0.28123
3	−1	$36x^2 + 18x - 1$	2	179, 647, 1889, 2447, 3779	31692	0.80725
3	1	$36x^2 + 18x + 1$	1	181, 379, 991, 3079, 7309	23288	0.59109
4	−1	$36x^2 + 21x - 1$	1	659, 7349, 9551, 12041, 33029	10633	0.26894
4	1	$36x^2 + 21x + 1$	1	661, 1423, 2473, 5437, 7351	15712	0.40086
4	2	$36x^2 + 21x + 2$	2	389, 1913, 6359, 13397, 16319	15405	0.39341
5	−1	$36x^2 + 24x - 1$	2	191, 1019, 1439, 1931, 5471	23332	0.59109
5	1	$36x^2 + 24x + 1$	2	193, 397, 673, 1021, 1933	27255	0.69166
6	1	$36x^2 + 27x + 1$	1	199, 1459, 2521, 9649, 33211	10609	0.26894
6	4	$36x^2 + 27x + 4$	1	67, 409, 1039, 3163, 4657	15494	0.39341
7	1	$36x^2 + 30x + 1$	2	1051, 3187, 7477, 9697, 13567	18210	0.46393
7	5	$36x^2 + 30x + 5$	1	71, 419, 701, 1481, 1979	23192	0.59109
8	2	$36x^2 + 33x + 2$	9	4721, 8597, 23327, 61871, 81077	9591	0.24501
8	4	$36x^2 + 33x + 4$	1	73, 1069, 1999, 3217, 4723	13526	0.34106
8	5	$36x^2 + 33x + 5$	1	1499, 7523, 9749, 12263, 29153	10933	0.28123
9	7	$36x^2 + 36x + 7$	1	79, 223, 439, 727, 1087	24281	0.61666
10	5	$36x^2 + 39x + 5$	9	5657, 7607, 18287, 65147, 99377	9537	0.24501
10	7	$36x^2 + 39x + 7$	1	229, 739, 5659, 12373, 15187	13322	0.34106
10	8	$36x^2 + 39x + 8$	1	83, 449, 1103, 4793, 6599	11175	0.28123
11	7	$36x^2 + 42x + 7$	2	457, 751, 1117, 2647, 3301	18110	0.46393
11	11	$36x^2 + 42x + 11$	1	89, 239, 461, 1559, 2069	23297	0.59109
12	10	$36x^2 + 45x + 10$	1	2089, 3331, 4861, 6679, 16831	10588	0.26894
12	13	$36x^2 + 45x + 13$	1	769, 1579, 2677, 5737, 7699	15505	0.39341
13	11	$36x^2 + 48x + 11$	1	251, 479, 1151, 2111, 2699	23137	0.59109
13	13	$36x^2 + 48x + 13$	1	97, 1153, 1597, 2113, 3361	27257	0.69166
14	14	$36x^2 + 51x + 14$	1	101, 491, 3389, 4931, 6761	10559	0.26894
14	16	$36x^2 + 51x + 16$	1	103, 1171, 2137, 3391, 4933	15790	0.40086
14	17	$36x^2 + 51x + 17$	1	263, 797, 1619, 2729, 4127	15393	0.39341
15	17	$36x^2 + 54x + 17$	1	107, 269, 503, 809, 1187	31685	0.80725
15	19	$36x^2 + 54x + 19$	1	109, 271, 811, 2161, 4159	23208	0.59109

(continued)

Table 2 (continued)

r	$c \in C'_r$	$f_{r,c}(x)$	$k_{r,c}$	$J_{r,c}$	$N_{r,c}$	$\frac{C(f_{r,c})}{2\delta_r}$
16	17	$36x^2 + 57x + 17$	8	2777, 29837, 34127, 54167, 72221	9606	0.24501
16	19	$36x^2 + 57x + 19$	1	277, 823, 1657, 7873, 15559	13448	0.34106
16	20	$36x^2 + 57x + 20$	1	113, 3449, 5003, 11393, 17093	11096	0.28123
17	23	$36x^2 + 60x + 23$	1	839, 1223, 2207, 5039, 5927	24229	0.61666
18	22	$36x^2 + 63x + 22$	8	9067, 11497, 24097, 27967, 36571	9662	0.24501
18	23	$36x^2 + 63x + 23$	1	293, 1697, 4253, 10247, 12821	17614	0.45086
18	25	$36x^2 + 63x + 25$	1	853, 1699, 2833, 7963, 12823	10918	0.28123
19	25	$36x^2 + 66x + 25$	1	127, 547, 2251, 2857, 5107	18271	0.46393
19	29	$36x^2 + 66x + 29$	1	131, 1259, 1721, 2861, 3539	23270	0.59109
20	29	$36x^2 + 69x + 29$	1	311, 881, 15809, 34499, 43991	10567	0.26894
20	31	$36x^2 + 69x + 31$	1	313, 883, 1741, 2887, 6043	15875	0.40086
20	32	$36x^2 + 69x + 32$	1	137, 563, 1277, 5147, 7013	15649	0.39341
21	31	$36x^2 + 72x + 31$	1	139, 571, 1291, 1759, 5179	23262	0.59109
22	35	$36x^2 + 75x + 35$	1	911, 2939, 13049, 22571, 26321	10591	0.26894
22	37	$36x^2 + 75x + 37$	1	331, 1783, 6121, 10453, 15937	15764	0.40086
22	38	$36x^2 + 75x + 38$	1	149, 587, 11717, 17489, 20807	15460	0.39341
23	37	$36x^2 + 78x + 37$	2	337, 1327, 1801, 2347, 10501	18177	0.46393
23	41	$36x^2 + 78x + 41$	1	599, 929, 2351, 2969, 3659	23223	0.59109

with $\left(\frac{D}{p}\right)$ being the Legendre symbol. The constant $C(f)$ is called the Hardy–Littlewood constant of f. Because the polynomial $f_{r,c}(x)$ satisfies the above conditions, we can expect that it indeed represents infinitely many primes. In our case, it may hold that

$$\pi(f_{r,c}; x) \sim \frac{C(f_{r,c})}{2\delta_r} \frac{\sqrt{x}}{\log x}, \quad C(f_{r,c}) = \prod_{p \geq 5} \left(1 - \frac{\left(\frac{(r+3)^2 - 16c}{p}\right)}{p-1} \right).$$

We also show both the numerical value of $\frac{C(f_{r,c})}{2\delta_r}$ and the exact number $N_{r,c} = \#\{p \leq 10^{12} \mid p = f_{r,c}(k) \in I_r \text{ for some } k \geq k_{r,c}\}$ in Table 2. Notice that $\frac{\sqrt{x}}{\log x} = 36191.20\ldots$ when $x = 10^{12}$.

The following is immediate from Theorem 4.9.

Corollary 4.10 *There exists infinitely many exceptional primes if the Hardy–Littlewood conjecture is true for at least one of $f_{r,c}(x)$ for $0 \leq r \leq 23$ and $c \in C'_r$.*

We notice that, if we can show that there exists infinitely many exceptional primes (in the frame work of the graph theory) on the other hand, then it implies that at least one of $f_{r,c}(t)$ represents infinitely many primes.

We also remark that though we omit to show here but the existence of infinitely many ordinary primes is similarly verified by using Dirichlet's theorem on arithmetic progressions as [5, 6].

5 Open Problems

We finally give some problems related to our study on Ramanujan graphs.

(1) Study the same problem for general (more complicated) groups. More precisely, determine the bound $\tilde{l} = \tilde{l}_{G,\mathscr{S}}$ explicitly for a given finite group G and a set $\mathscr{S}$ of Cayley subsets of G. Can you describe it in terms of data of G such that the degrees of irreducible representations and the number or cardinality of the conjugacy classes of G? Moreover, is the determination of it related to some arithmetic problems?
(2) More generally, study the distribution of Ramanujan graphs around a given Ramanujan graph. For example, can you obtain a Ramanujan graph by removing some edges from the so-called LPS-Ramanujan graph $X^{p,q}$ constructed in [10]?
(3) Study an application of our problem in general setting to computer science, e.g., to the cryptographic hash function.

Acknowledgements The author thanks the referee for helpful comments on the manuscript. Moreover, the author also would like to thank Miki Hirano and Kohei Katata for variable discussion.

References

1. N. Alon, Y. Roichman, Random Cayley graphs and expanders. Random Struct. Algorithms **5**, 271–284 (1994)
2. L. Babai, Spectra of Cayley graphs. J. Comb. Theory Ser. B **27**, 180–189 (1979)
3. D. Charles, K. Lauter, E. Goren, Cryptographic hash functions from expander graphs. J. Cryptol. **22**, 93–113 (2009)
4. G.H. Hardy, J.E. Littlewood, Some problems of 'Partitio numerorum'; III: on the expression of a number as a sum of primes. Acta Math. **44**, 1–70 (1923)
5. M. Hirano, K. Katata, Y. Yamasaki, Ramanujan circulant graphs and the conjecture of Hardy-Littlewood and Bateman-Horn (submitted) (2016)
6. M. Hirano, K. Katata, Y. Yamasaki, Ramanujan Cayley graphs of Frobenius groups. Bull. Aust. Math. Soc. **94**, 373–383 (2016)
7. S. Hoory, N. Linial, A. Wigderson, Expander graphs and their applications. Bull. Am. Math. Soc. **43**, 439–561 (2006)
8. M. Krebs, A. Shaheen, *Expander Families and Cayley Graphs. A Beginner's Guide* (Oxford University Press, Oxford, 2011)
9. A. Lubotzky, Expander graphs in pure and applied mathematics. Bull. Am. Math. Soc. **49**, 113–162 (2012)
10. A. Lubotzky, R. Phillips, P. Sarnak, Ramanujan graphs. Combinatorica **8**, 261–277 (1988)
11. A. Terras, *Zeta Functions of Graphs, A Stroll Through the Garden*. Cambridge Studies in Advanced Mathematics, vol. 128 (Cambridge University Press, Cambridge, 2011)

Uniform Random Number Generation and Secret Key Agreement for General Sources by Using Sparse Matrices

Jun Muramatsu and Shigeki Miyake

Abstract In this paper, we investigate the problems of uniform random number generation, independent uniform random number generation, and secret key agreement, which provide the information theoretic security. We consider the strong uniformity and strong independence, where it has been unclear whether or not sparse matrices can be applied to these problems for general (correlated) sources with respect to these criteria. To prove the theorems, we first introduce the notion of the balanced-coloring property and the collision-resistance property. We next apply these properties to the problems. Since an ensemble of sparse matrices (with logarithmic column weight) over a finite field satisfies these properties, we can construct a code achieving the fundamental limits by using sparse matrices.

Keywords Information theory · Information theoretic security · Uniform random number generation · Independent uniform random number gneneration · Secret key agreement · Ensemble of sparse matrices over a finite field

---

This paper is based on *Proc. 2012 IEEE Inform. Theory Workshop*, Lausanne, Switzerland, Sep. 3–7, 2012, pp. 612–616.

J. Muramatsu (✉)
NTT Communication Science Laboratories, NTT Corporation, Hikaridai 2-4, Seika-cho, Soraku-gun, Kyoto 619-0237, Japan
e-mail: muramatsu.jun@lab.ntt.co.jp

S. Miyake
NTT Network Innovation Laboratories, NTT Corporation, Hikarinooka 1-1, Yokosuka-shi, Kanagawa 239-0847, Japan
e-mail: miyake.shigeki@lab.ntt.co.jp

T. Takagi et al. (eds.), *Mathematical Modelling for Next-Generation Cryptography*, Mathematics for Industry 29, DOI 10.1007/978-981-10-5065-7_10

1 Introduction, Formal Description of Problems, and Previous Results

This paper investigates the problems of uniform random number generation, independent uniform random number generation, and secret key agreement, which have been studied in [1, 2, 6, 11, 18, 19, 22, 26, 27, 29, 30]. We prove that the fundamental limits of the generation rate for general (correlated) sources are achievable by using sparse matrices over a finite field, where we consider the case of strong uniformity and strong independence.

We use the following definitions throughout the paper. Let μ_{UV} be the joint probability distribution of random variables U and V. Let μ_U and μ_V be the respective marginal distributions and $\mu_{U|V}$ be the conditional probability distribution. Then the entropy $H(U)$, the conditional entropy $H(U|V)$, and the mutual information $I(U;V)$ of random variables are defined as $H(U) \equiv \sum_u \mu_U(u)\log(1/\mu_U(u))$, $H(U|V) \equiv \sum_{u,v} \mu_{UV}(u,v)\log(1/\mu_{U|V}(u|v))$, and $I(U;V) \equiv H(U) - H(U|V)$, respectively, where we assume that the base of the logarithm is 2 throughout the paper.

In this paper, we consider the general source introduced in [14, 15, 30]. Let $P(\cdot)$ denote the probability of an event. Let $\{\mathcal{U}_n\}_{n=1}^{n}$ be a sequence of sets. Let $\mathbf{U} \equiv \{U_n\}_{n=1}^{\infty}$ be a sequence of random variables $U_n \in \mathcal{U}_n$. We call $\mathbf{U}$ a *general source*. When $\mathcal{U}_n$ is equal to a product set $\mathcal{U}^n$, we denote $\mathbf{U} \equiv \{U^n\}_{n=1}^{\infty}$, and we do not assume the consistency condition of U^n and $U^{n'}$, namely that the distribution of U^n is the same as that of the first n components of $U^{n'}$ for any $n < n'$. For a general source $\mathbf{U}$, let $\underline{H}(\mathbf{U})$ be the spectrum inf-entropy rate of $\mathbf{U}$ defined as

$$\underline{H}(\mathbf{U}) \equiv \sup\left\{\theta : \lim_{n\to\infty} P\left(\frac{1}{n}\log\frac{1}{\mu_{U_n}(U_n)} < \theta\right) = 0\right\}.$$

Let $(\mathbf{U},\mathbf{V}) \equiv \{(U_n,V_n)\}_{n=1}^{\infty}$ be a pair of general sources. Let $\overline{H}(\mathbf{U}|\mathbf{V})$ and $\underline{H}(\mathbf{U}|\mathbf{V})$ be the spectrum conditional sup-entropy rate and the spectrum conditional inf-entropy rate of $\mathbf{U}$ given $\mathbf{V}$ defined as

$$\overline{H}(\mathbf{U}|\mathbf{V}) \equiv \inf\left\{\theta : \lim_{n\to\infty} P\left(\frac{1}{n}\log\frac{1}{\mu_{U_n|V_n}(U_n|V_n)} > \theta\right) = 0\right\}$$

$$\underline{H}(\mathbf{U}|\mathbf{V}) \equiv \sup\left\{\theta : \lim_{n\to\infty} P\left(\frac{1}{n}\log\frac{1}{\mu_{U_n|V_n}(U_n|V_n)} < \theta\right) = 0\right\},$$

respectively.

It is known that $\underline{H}(\mathbf{U})$ is equal to the entropy rate of $\mathbf{U}$ when $\mathbf{U}$ is stationary ergodic. In particular, $\underline{H}(\mathbf{U})$ is equal to the entropy when $\mathbf{U}$ is stationary memoryless. When $(\mathbf{U},\mathbf{V})$ is stationary ergodic, both $\overline{H}(\mathbf{U}|\mathbf{V})$ and $\underline{H}(\mathbf{U}|\mathbf{V})$ are equal to the conditional entropy rate of **U** given **V**. When (**U**, **V**) is stationary memoryless, both $\overline{H}(\mathbf{U}|\mathbf{V})$ and $\underline{H}(\mathbf{U}|\mathbf{V})$ are equal to the conditional entropy. Thus, results for general sources can be directly applied to stationary ergodic/memoryless sources by replacing the entropy functions.

1.1 Measures for Uniformity and Independence

In this section, we introduce two types of measures of uniformity and independence. The uniformity of a random variable is measured in terms of the difference between its distribution and the uniform distribution on the same range. The independence of two random variables is measured in terms of the difference between its joint distribution and the product of marginal distributions. In the discussion of information theoretic security, secrecy is defined in terms of one of these measures.

First, we introduce two types of measure of the difference between two probability distributions p and q on $\mathcal{M}$. They are the variational distance $d(p,q) \equiv \sum_{\mathbf{m}\in\mathcal{M}} |p(\mathbf{m}) - q(\mathbf{m})|/2$ and the divergence $D(p\|q) \equiv \sum_{\mathbf{m}\in\mathcal{M}} p(\mathbf{m}) \log(p(\mathbf{m})/q(\mathbf{m}))$.

Next, we introduce two types of measure of uniformity. Let $p_{\overline{M}}$ be the uniform distribution on $\mathcal{M}$. Then the uniformity can be measured with $d(\mu_M, p_{\overline{M}})$ or $D(\mu_M\|p_{\overline{M}}) = \log|\mathcal{M}| - H(M)$. From the inequality $d(p,q) \leq \sqrt{2\ln(2)D(p\|q)}$ (see [9, Lemma 11.6.1]), we have the fact that if μ_M is close to the uniform distribution with respect to the divergence then it is also close to the uniform distribution with respect to the variational distance.

For a general source **M**, we introduce three types of measure of asymptotical uniformity. We say **M** is *strongly uniform with respect to the variational distance* if $\lim_{n\to\infty} d(\mu_{M_n}, p_{\overline{M}_n}) = 0$. We say **M** is *strongly uniform with respect to the divergence* if $\lim_{n\to\infty} D(\mu_{M_n}\|p_{\overline{M}_n}) = 0$, which is equivalent to $\lim_{n\to\infty}\left[\log|\mathcal{M}_n| - H(M_n)\right] = 0$ because $D(\mu_{M_n}\|p_{\overline{M}_n}) = \log|\mathcal{M}_n| - H(M_n)$. We say **M** is *weakly uniform (with respect to the normalized divergence)* if $\lim_{n\to\infty} D(\mu_{M_n}\|p_{\overline{M}_n})/n = \lim_{n\to\infty}[\log|\mathcal{M}_n| - H(M_n)]/n = 0$. The following lemma asserts that strong uniformity with respect to the divergence can be obtained when the convergence speed of $d(\mu_{M_n}, p_{\overline{M}_n}) \to 0$ is sufficiently rapid.

Lemma 1.1 *If* $\log|\mathcal{M}_n| = O(n)$ *and* $d(\mu_{M_n}, p_{\overline{M}_n}) = o(1/n)$*, then* $D(\mu_{M_n}\|p_{\overline{M}_n}) \leq o(1)$.

The proof is given in Appendix 1.

Next, we introduce three types of measure of independence. For a joint probability distribution $\mu_{M_nZ^n}$ on $\mathcal{M}_n \times \mathcal{Z}^n$ and its marginal distributions μ_{M_n} and μ_{Z^n}, let $\mu_{M_n} \times \mu_{Z^n}$ be defined as $\mu_{M_n} \times \mu_{Z^n}(\mathbf{m}, \mathbf{z}) \equiv \mu_{M_n}(\mathbf{m})\mu_{Z^n}(\mathbf{z})$. We call **M** *strongly independent of* **Z** *with respect to the variational distance* if $\lim_{n\to\infty} d(\mu_{M_nZ^n}, \mu_{M_n} \times \mu_{Z^n}) = 0$. We call **M** *strongly independent of* **Z** *with respect to the divergence (mutual information)* if $\lim_{n\to\infty} D(\mu_{M_nZ^n}\|\mu_{M_n} \times \mu_{Z^n}) = 0$, which is equivalent to $\lim_{n\to\infty} I(M_n; Z^n) = 0$ because $D(\mu_{M_nZ^n}\|\mu_{M_n} \times \mu_{Z^n}) = I(M_n; Z^n)$. We call **M** *weakly independent of* $\mathbf{Z}$ *with respect to the normalized divergence (mutual information)* if $\lim_{n\to\infty} D(\mu_{M_n Z^n} \| \mu_{M_n} \times \mu_{Z^n})/n = \lim_{n\to\infty} I(M_n; Z^n)/n = 0$.

To measure uniformity and independence simultaneously, we introduce three types of measures $d(\mu_{M_n Z^n}, p_{\overline{M}_n} \times \mu_{Z^n})$, $D(\mu_{M_n Z^n} \| p_{\overline{M}_n} \times \mu_{Z^n})$, and $D(\mu_{M_n Z^n} \| p_{\overline{M}_n} \times \mu_{Z^n})/n$. Since

$$\begin{aligned}
d(\mu_{M_n}, p_{\overline{M}_n}) &= \frac{1}{2} \sum_{\mathbf{m}} \left| \mu_{M_n}(\mathbf{m}) - p_{\overline{M}_n}(\mathbf{m}) \right| \\
&= \frac{1}{2} \sum_{\mathbf{m}} \left| \sum_{\mathbf{z}} \left[\mu_{M_n Z^n}(\mathbf{m}, \mathbf{z}) - p_{\overline{M}_n}(\mathbf{m}) \mu_{Z^n}(\mathbf{z}) \right] \right| \\
&\leq \frac{1}{2} \sum_{\mathbf{m}} \sum_{\mathbf{z}} \left| \mu_{M_n Z^n}(\mathbf{m}, \mathbf{z}) - p_{\overline{M}_n}(\mathbf{m}) \mu_{Z^n}(\mathbf{z}) \right| \\
&= d(\mu_{M_n Z^n}, p_{\overline{M}_n} \times \mu_{Z^n}), \qquad (1) \\
d(\mu_{M_n Z^n}, \mu_{M_n} \times \mu_{Z^n}) &\leq d(\mu_{M_n Z^n}, p_{\overline{M}_n} \times \mu_{Z^n}) + d(p_{\overline{M}_n} \times \mu_{Z^n}, \mu_{M_n} \times \mu_{Z^n}) \\
&= d(\mu_{M_n Z^n}, p_{\overline{M}_n} \times \mu_{Z^n}) + d(p_{\overline{M}_n}, \mu_{M_n}) \\
&\leq 2d(\mu_{M_n Z^n}, p_{\overline{M}_n} \times \mu_{Z^n}), \qquad (2) \\
d(\mu_{M_n Z^n}, p_{\overline{M}_n} \times \mu_{Z^n}) &\leq d(\mu_{M_n Z^n}, \mu_{M_n} \times \mu_{Z^n}) + d(\mu_{M_n} \times \mu_{Z^n}, p_{\overline{M}_n} \times \mu_{Z^n}) \\
&= d(\mu_{M_n Z^n}, \mu_{M_n} \times \mu_{Z^n}) + d(p_{\overline{M}_n}, \mu_{M_n}), \qquad (3)
\end{aligned}$$

we have the fact that $\lim_{n\to\infty} d(\mu_{M_n Z^n}, p_{\overline{M}_n} \times \mu_{Z^n}) = 0$ is equivalent to the condition that $\mathbf{M}$ is strongly uniform and strongly independent of $\mathbf{Z}$ with respect to the variational distance. Since

$$\begin{aligned}
&\max\{D(\mu_{M_n} \| p_{\overline{M}_n}), D(\mu_{M_n Z^n} \| \mu_{M_n} \times \mu_{Z^n})\} \\
&\leq D(\mu_{M_n} \| p_{\overline{M}_n}) + D(\mu_{M_n Z^n} \| \mu_{M_n} \times \mu_{Z^n}) = D(\mu_{M_n Z^n} \| p_{\overline{M}_n} \times \mu_{Z^n}) \qquad (4)
\end{aligned}$$

we have the fact that $\lim_{n\to\infty} D(\mu_{M_n Z^n} \| p_{\overline{M}_n} \times \mu_{Z^n}) = 0$ is equivalent to the condition that $\mathbf{M}$ is strongly uniform and strongly independent of $\mathbf{Z}$ with respect to the divergence. The following lemma asserts that strong uniformity and strong independence with respect to divergence can be obtained when the convergence speed of $d(\mu_{M_n Z^n}, p_{\overline{M}_n} \times \mu_{Z^n}) \to 0$ is sufficiently rapid.

Lemma 1.2 *Let* $\mu_{M_n Z^n}$ *be a joint distribution on* $\mathcal{M}_n \times \mathcal{Z}^n$ *and* μ_{Z^n} *be its marginal distribution on* $\mathcal{Z}^n$. *Let* $p_{\overline{M}_n}$ *be the uniform distribution on* $\mathcal{M}_n$. *If* $\log|\mathcal{M}_n| = O(n)$ *and* $d(\mu_{M_n Z^n}, p_{\overline{M}_n} \times \mu_{Z^n}) = o(1/n)$, *then*

$$\log|\mathcal{M}_n| - H(M_n) \leq o(1) \qquad (5)$$

$$I(M_n; Z^n) \leq o(1). \qquad (6)$$

The proof is given in Appendix 1. Historically, conditions (5) and (6) are referred as *strong secrecy* and (6) can be replaced by an equivalent condition $H(M_n) - H(M_n|Z^n) \leq o(1)$.

1.2 Uniform Random Number Generation

In this section, we consider the problem of uniform random number generation from a nonuniform random source.

Let μ_{X^n} be a (nonuniform) probability distribution of random variable $X^n \in \mathscr{X}^n$. A *fixed-rate random number generator* is a mapping $\varphi_n : \mathscr{X}^n \to \mathscr{M}_n$. The aim of this problem is to obtain uniform random numbers. We call rate $R \equiv \log(|\mathscr{M}_n|)/n$ *achievable* if for all $\delta > 0$ and all sufficiently large n, there is a generator φ_n such that $\rho(\mu_{M_n}, p_{\overline{M}_n}) \leq \delta$, where μ_{M_n} is the probability distribution of a random variable $M_n \equiv \varphi_n(X^n) \in \mathscr{M}_n$, $p_{\overline{M}_n}$ is the uniform distribution on the set $\mathscr{M}_n$, and the function ρ denotes one of the difference measures (variational distance, divergence, or normalized divergence) between two distributions. For a general source $\mathbf{X}$, the supremum $\mathrm{IR}(\mathbf{X})$ of achievable rate R is called the *intrinsic randomness* of $\mathbf{X}$.

When the difference measure ρ is the variational distance or the normalized divergence, the information theoretic formula of intrinsic randomness for a general source is proved in [30] for a case where $\mathscr{M}_n$ is a set of fixed length binary sequences. It was extended in [15] to the general set $\mathscr{M}_n$. When ρ is the divergence, the information theoretic formula of intrinsic randomness for a stationary memoryless source is obtained implicitly in [11] for a stationary memoryless source in the context of generating mutually independent random numbers.

Proposition 1.3 ([11][15, Theorem 2.2.2][30]) *Let* $\mathbf{X}$ *be a general source* $\mathbf{X}$. *Then* $\mathrm{IR}(\mathbf{X}) = \underline{H}(\mathbf{X})$, *where the difference measure* ρ *is the variational distance or the normalized divergence.*

In this paper, we prove that intrinsic randomness is achievable by using a sparse matrix, where the difference measure ρ is the variational distance.

1.3 Independent Random Number Generation

In this section, we consider the problem of generating a uniform random number from one of two correlated sources, where the generated random number is independent of the other source.

In the following, we assume that a source $X^n \in \mathscr{X}^n$, which is correlated with $Z^n \in \mathscr{Z}^n$, is available for a random number generator $\varphi_n : \mathscr{X}^n \to \mathscr{M}_n$. The aim of this problem is to obtain uniform random numbers that are asymptotically independent of Z^n. We call rate $R \equiv \log(|\mathscr{M}_n|)/n$ *achievable* if there is a random number generator φ_n satisfying

$$\rho(\mu_{M_n Z^n}, p_{\overline{M}_n} \times \mu_{Z^n}) \leq \delta \tag{7}$$

for any $\delta > 0$ and all sufficiently large n, where μ_{M_n} is the probability distribution of random variable $M_n \equiv \varphi_n(X^n)$, $p_{\overline{M}_n}$ is the uniform distribution on $\mathcal{M}_n$, and the function ρ denotes one of the differences (variational distance, divergence, or normalized divergence) between two distributions. For a pair of general sources $(\mathbf{X}, \mathbf{Z}) \equiv \{(X^n, Z^n)\}_{n=1}^{\infty}$, we call the supremum $\mathrm{IIR}(\mathbf{X}|\mathbf{Z})$ of achievable rate R the *independent intrinsic randommness* of $\mathbf{X}$ of $\mathbf{Z}$.

The information theoretic expression of independent intrinsic randomness has been obtained implicitly in the context of the secret key agreement and the privacy amplification studied in [2, 7, 18], where (X, Z) is assumed to be stationary memoryless and the difference measure ρ is the normalized divergence. The relationship between this problem, secret key agreement, and privacy amplification will be discussed in Sect. 5. The same expression is obtained in [11] where (X, Z) is assumed to be stationary memoryless, and the difference measure ρ is the variational distance or the divergence. When ρ is the normalized divergence, the result is extended in [22] to a case where $H(X^n)/n$, $H(Z^n)/n$, $H(X^n, Z^n)/n$ have limits. When ρ is the variational distance, the result is extended in [5] to general correlated sources are as follows:

Proposition 1.4 ([5, Prop. 2]) *Let* $(\mathbf{X}, \mathbf{Z})$ *be a general source. Then* $\mathrm{IIR}(\mathbf{X}|\mathbf{Z}) = \underline{H}(\mathbf{X}|\mathbf{Z})$*, where the difference measure* ρ *is the variational distance.*

In this paper, we prove that independent intrinsic randomness is acheivable by using a sparse matrix, where the difference measure ρ is the variational distance.

1.4 Secret Key Agreement

In this section, we review the problem of the secret key agreement from the correlated source outputs introduced in [2, 18].

Assume that a sender, a legitimate receiver, and an eavesdropper have access to correlated sources X^n, Y^n, and Z^n, respectively. The aim of the sender and the legitimate receiver is to agree on a secret key that is completely secret from the eavesdropper. In this paper, we consider the scenario illustrated in Fig. 1 where only the sender sends information over a public channel, which may be monitored by the eavesdropper, and the sender and the legitimate receiver generate a key M_n that is secret from the eavesdropper. It should be noted that this type of protocol is studied in [7]. Bidirectional communication is allowed in the general case discussed in [2, 18].

Now we formally define the (one-way) protocol and the secret key capacity. Let $\mathcal{M}_n$ be the range of keys M_n and M'_n generated by the sender and the receiver, respectively. We call a triplet of random variables (C_n, M_n, M'_n) a *protocol* if it satisfies the following Markov conditions

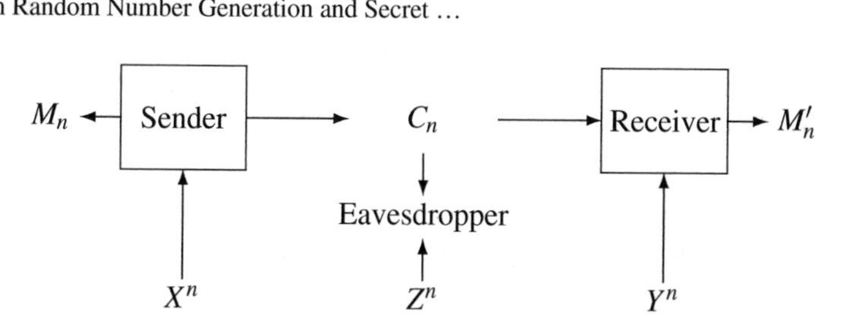

Fig. 1 Secret Key Agreement from Correlated Source Outputs (One-way Protocol)

- $Y^n Z^n M'_n \leftrightarrow X^n \leftrightarrow C_n M_n$, that is, the sender calculates (C_n, M_n) by using a channel (nondeterministic map) with inputs X^n,
- $X^n Z^n M_n \leftrightarrow Y^n C_n \leftrightarrow M'_n$, that is, the receiver calculates M'_n by using a channel (nondeterministic map) with inputs Y^n and C_n.

We call rate $R \equiv \log(|\mathscr{M}_n|)/n$ *secretly achievable* if for any $\delta > 0$ there are a number n and a protocol (C_n, M_n, M'_n) such that

$$P(M_n \neq M'_n) \leq \delta \tag{8}$$

$$\rho(p_{M_n Z^n C_n}, p_{\overline{M}_n} \times p_{Z^n C_n}) \leq \delta, \tag{9}$$

where ρ denotes one of the difference measures (variational distance, divergence, or normalized divergence) and $p_{\overline{M}_n}$ is the uniform distribution on $\mathscr{M}_n$. The first inequality indicates that the decoding error for the receiver is arbitrarily small. The second inequality indicates that random variables M_n and (Z^n, C_n) are almost mutually independent, which implies that the amount of information obtained by the eavesdropper is arbitrarily small. The *forward secret key capacity* $S_{\mathrm{f}}(\mathbf{X}; \mathbf{Y} \| \mathbf{Z})$ is defined by the supremum of the achievable rate R.

In the original definition introduced in [2, 18], condition (9) is replaced by two conditions

$$\frac{H(M_n)}{n} \geq R - \delta \quad \text{and} \quad \frac{I(M_n; Z^n, C_n)}{n} \leq \delta. \tag{10}$$

In the definition introduced in [11], condition (9) is replaced by two conditions

$$H(M_n) \geq nR - \delta \quad \text{and} \quad I(M_n; Z^n, C_n) \leq \delta. \tag{11}$$

In Sect. 1.1, we discussed the fact that condition (10) corresponds to a case of weak security, that is, the difference measure ρ is the normalized divergence, and condition (11) corresponds to a case of strong security, that is, the difference measure ρ is the divergence. It should be noted that condition (9) implies that M_n is required to be strongly uniform and strongly secure from (Z^n, C_n) when $\delta = o(1/n)$ and ρ is the variational distance. In fact, from Lemma 1.2, condition (9) implies that $nR - H(M_n) \leq o(1)$ and $I(M_n; Z^n, C_n) \leq o(1)$. It should be noted that the case where the difference measure ρ is the divergence has been considered implicitly in [2]. We prove the following theorem, where ρ is variational distance.

Theorem 1.5 *Let* $(\mathbf{X}, \mathbf{Y}, \mathbf{Z})$ *be a triplet of general sources. Then*

$$S_{\mathrm{f}}(\mathbf{X}; \mathbf{Y}\|\mathbf{Z}) = \sup_{\mathbf{W}, \widehat{\mathbf{X}}} \left[\underline{H}(\widehat{\mathbf{X}}|\mathbf{Z}, \mathbf{W}) - \overline{H}(\widehat{\mathbf{X}}|\mathbf{Y}, \mathbf{W}) \right], \tag{12}$$

where the supremum is taken over all general sources $\widehat{\mathbf{X}} \equiv \{\widehat{X}_n\}_{n=1}^{\infty}$ *and* $\mathbf{W} \equiv \{W_n\}_{n=1}^{\infty}$, *which satisfy the Markov condition*

$$W_n \widehat{X}_n \leftrightarrow X^n \leftrightarrow Y^n Z^n \quad \textit{for all } n. \tag{13}$$

It should be noted that $S_{\mathrm{f}}(\mathbf{X}, \mathbf{Y}|\mathbf{Z})$ is derived in [2, 11][12, Theorem 17.21] when $(\mathbf{X}, \mathbf{Y}, \mathbf{Z})$ is stationary memoryless, where the supremum is taken over all stationary memoryless sources $\widehat{\mathbf{X}}$ and $\mathbf{W}$ satisfying the Markov condition. For a general source, the upper and lower bounds are derived in [6]. A similar expression to (12) is derived in [29] by using the smooth Rényi entropy.

In the following, we consider a scenario where correlated sources $(\mathbf{X}, \mathbf{Y}, \mathbf{Z})$ satisfy $\mathbf{X} = \mathbf{Y}$. It should be noted that this condition corresponds to the assumption of privacy amplification studied in [4]. Since $\mathbf{X} = \mathbf{Y}$, condition (8) is satisfied by assuming that the sender and the receiver use the same function for key generation. This implies that we can consider this scenario to be the independent random number generation discussed in Sect. 1.3, where M_n is the output of a uniform random number generator independent of Z^n and it is unnecessary for the sender to send a message C_n before the key generation. In fact, condition (9) is equivalent to (7) by letting C_n be a constant. It should be noted that, for stationary memoryless correlated sources (X, Y, Z), the secret key capacity $S(X; Y\|Z)$, which is the optimal key generation rate, of this scenario is given as $S(X; Y\|Z) = I(X; Y|Z) = H(X|Z)$ (see [2, 7, 18, 26]).

2 Balanced-Coloring Property, Collision-Resistance Property, and Hash Property

In this section, we review the definitions of the balanced-coloring property, the collision-resistance property, and the hash property for an ensemble of functions introduced in [20, 21, 24, 28].[1]

Throughout this paper, we use the following definitions and notations. The cardinality of a set $\mathcal{U}$ is denoted by $|\mathcal{U}|$, $\mathcal{U}^c$ denotes the complement of $\mathcal{U}$, and $\mathcal{U} \setminus \mathcal{V} \equiv \mathcal{U} \cap \mathcal{V}^c$ denotes the set difference. Column vectors and sequences are denoted in boldface. Let $F\mathbf{u}$ denote a value taken by a function F at $\mathbf{u} \in \mathcal{U}^n$, where $\mathcal{U}^n$ is the domain of the function. It should be noted that F may be nonlinear. When

[1] In [24], it is called 'strong hash property'.

F is a linear function expressed by an $l \times n$ matrix, we assume that $\mathcal{U} \equiv \mathrm{GF}(q)$ is a finite field and the range of functions is $\mathcal{U}^l$. For a set $\mathcal{F}_n$ of functions on $\mathcal{U}^n$, let $\mathrm{Im}\mathcal{F}_n$ be defined as $\mathrm{Im}\mathcal{F}_n \equiv \bigcup_{F \in \mathcal{F}_n} \{F\mathbf{u} : \mathbf{u} \in \mathcal{U}^n\}$. We define a set $\mathcal{C}_F(\mathbf{c})$ as $\mathcal{C}_F(\mathbf{c}) \equiv \{\mathbf{u} : F\mathbf{u} = \mathbf{c}\}$. The random variables of a function F and a vector $\mathbf{c}$ are denoted by sans serif letters F and c, respectively. On the other hand, the random variable of a vector $\mathbf{u}$ is denoted by U^n as used previously when it is not used to represent a matrix. The expected value of a function of a random variables F is denoted by E_{F}.

Here, we introduce the hash property for an ensemble of functions. It requires stronger conditions than those introduced in [23].

Definition 2.1 ([20, 21, 24, 28]) Let $\{\mathcal{F}_n\}_{n=1}^{\infty}$ be a sequence of sets such that $\mathcal{F}_n$ is a set of functions on $\mathcal{U}^n$. For a probability distribution $p_{\mathsf{F},n}$ on $\mathcal{F}_n$, we call a sequence $\{(\mathcal{F}_n, p_{\mathsf{F},n})\}_{n=1}^{\infty}$ an *ensemble*. Consider the following conditions for two sequences $\alpha_{\mathsf{F}} \equiv \{\alpha_{\mathsf{F}}(n)\}_{n=1}^{\infty}$ and $\beta_{\mathsf{F}} \equiv \{\beta_{\mathsf{F}}(n)\}_{n=1}^{\infty}$

$$\sum_{\mathbf{u}' \in \mathcal{U}^n \setminus \{\mathbf{u}\}:\ p_{\mathsf{F},n}(\{F : F\mathbf{u} = F\mathbf{u}'\}) > \frac{\alpha_{\mathsf{F}}(n)}{|\mathrm{Im}\mathcal{F}_n|}} p_{\mathsf{F},n}\left(\{F : F\mathbf{u} = F\mathbf{u}'\}\right) \leq \beta_{\mathsf{F}}(n) \tag{14}$$

for any $\mathbf{u} \in \mathcal{U}^n$ and

$$\lim_{n\to\infty} \alpha_{\mathsf{F}}(n) = 1 \tag{15}$$

$$\lim_{n\to\infty} \frac{1}{n} \log(1 + \beta_{\mathsf{F}}(n)) = 0 \tag{16}$$

$$\lim_{n\to\infty} \frac{1}{n} \log \alpha_{\mathsf{F}}(n) = 0 \tag{17}$$

$$\lim_{n\to\infty} \beta_{\mathsf{F}}(n) = 0. \tag{18}$$

Then, we say that $\{(\mathcal{F}_n, p_{\mathsf{F},n})\}_{n=1}^{\infty}$ has an $(\alpha_{\mathsf{F}}, \beta_{\mathsf{F}})$*-balanced-coloring property* if there are α_{F} and β_{F} depending on $\{p_{\mathsf{F},n}\}_{n=1}^{\infty}$ satisfying (14), (15), and (16). We say that $\{(\mathcal{F}_n, p_{\mathsf{F},n})\}_{n=1}^{\infty}$ has an $(\alpha_{\mathsf{F}}, \beta_{\mathsf{F}})$*-collision-resistance property* if there are α_{F} and β_{F} depending on $\{p_{\mathsf{F},n}\}_{n=1}^{\infty}$ satisfying (14), (17), and (18). We say that $\{(\mathcal{F}_n, p_{\mathsf{F},n})\}_{n=1}^{\infty}$ has an $(\alpha_{\mathsf{F}}, \beta_{\mathsf{F}})$*-hash property* if there are α_{F} and β_{F} depending on $\{p_{\mathsf{F},n}\}_{n=1}^{\infty}$ satisfying (14), (15), and (18). Throughout this paper, we omit the dependence on n of α_{F} and β_{F}.

It should be noted that the hash property is equivalent to the condition that both the balanced-coloring property and the collision-resistance property are satisfied. Before introducing examples of ensemble, we introduce the following lemmas, where it is unnecessary to assume the linearity of functions.

The following lemma is related to the *balanced-coloring property*, which is an extension of the leftover hash lemma [16] and the balanced-coloring lemma [3, Lemma 3.1][12, Lemma 17.3]. This lemma implies that there is a function F such that $\mathcal{T}$ is equally partitioned by F with respect to a measure Q.

Lemma 2.2 ([20, Lemma 5][25, Lemma 4]) *If* $(\mathscr{F}_n, p_{\mathsf{F},n})$ *satisfies (14), then*

$$E_{\mathsf{F}}\left[\sum_{\mathbf{c}}\left|\frac{Q\left(\mathscr{T}\cap\mathscr{C}_{\mathsf{F}}(\mathbf{c})\right)}{Q(\mathscr{T})}-\frac{1}{|\mathrm{Im}\mathscr{F}_n|}\right|\right]\leq\sqrt{\alpha_{\mathsf{F}}-1+\frac{[\beta_{\mathsf{F}}+1]|\mathrm{Im}\mathscr{F}_n|\max_{\mathbf{u}\in\mathscr{T}}Q(\mathbf{u})}{Q(\mathscr{T})}} \tag{19}$$

for any function $Q : \mathscr{U}^n \to [0, \infty)$ *and* $\mathscr{T} \subset \mathscr{U}^n$, *where* $Q(\mathscr{T}) \equiv \sum_{\mathbf{u}\in\mathscr{T}} Q(\mathbf{u})$.

The following lemma is related to the *collision-resistance property*, that is, if the number of bins is greater than the number of items then there is an assignment such that every bin contains at most one item.

Lemma 2.3 ([20, Lemma 4][23, Lemma 1])[2] *If* $(\mathscr{F}_n, p_{\mathsf{F},n})$ *satisfies (14), then*

$$p_{\mathsf{F},n}\left(\{F : [\mathscr{G}\setminus\{\mathbf{u}\}]\cap\mathscr{C}_F(F\mathbf{u})\neq\emptyset\}\right)\leq\frac{|\mathscr{G}|\alpha_{\mathsf{F}}}{|\mathrm{Im}\mathscr{F}_n|}+\beta_{\mathsf{F}}$$

for any $\mathscr{G} \subset \mathscr{U}^n$ *and* $\mathbf{u} \in \mathscr{U}^n$.

Here, we introduce examples of ensemble satisfying the hash property. When $\mathscr{F}_n$ is a two-universal class of hash functions [13] and $p_{\mathsf{F},n}$ is the uniform distribution on $\mathscr{F}_n$, then $\{(\mathscr{F}_n, p_{\mathsf{F},n})\}_{n=1}^{\infty}$ has a strong $(\mathbf{1}, \mathbf{0})$-hash property, where random binning [8] and the set of all linear functions [10] are examples of a two-universal class of hash functions.

In the following, we introduce an ensemble of sparse matrices satisfying the hash property. A q-ary $l \times n$ matrix is generated by using the following procedure starting from the all-zero matrix:

- For each $j \in \{1, \ldots, n\}$, repeat the following procedure τ_n times:
  1. Choose $(i, a) \in \{1, \ldots, l\} \times [\mathrm{GF}(q) \setminus \{0\}]$ uniformly at random.
  2. Add a to the (i, j) component of the matrix.

We have the following lemma, where a generated matrix is sparse in the sense that the maximum row weight τ_n satisfies $\tau_n/n \to 0$.

Lemma 2.4 ([20, Sect. 3-B][23, Sects. 4 and 6-C][24, Sect. 3-B]) *For a given* $R > 0$, *let*[3] $l \equiv nR$ *and* $\tau_n \equiv 2\lceil\ln(n^2 R)\rceil$, *where* $\lceil\theta\rceil$ *denotes the minimum integer not less than* θ. *Then, there are* $\alpha_{\mathsf{F}} \equiv \{\alpha_{\mathsf{F}}(n)\}_{n=1}^{\infty}$ *and* $\beta_{\mathsf{F}} \equiv \{\beta_{\mathsf{F}}(n)\}_{n=1}^{\infty}$ *such that the above ensemble of matrices satisfies* $(\alpha_{\mathsf{F}}, \beta_{\mathsf{F}})$*-hash property.*

It should be noted that the above examples also satisfy both the balanced-coloring property and the collision-resistance property. It is proved in [21, Sect. 4-B] that an ensemble of systematic[4] sparse matrices have a balanced-coloring property.

---

[2] In [23, Lemma 1][20, Lemma 4], this lemma is shown under a weaker condition [23, Eq. (H3)] [20, Eq. (H3')].

[3] See [23, Eq. (52)].

[4] A square part of the matrix is identity.

3 Uniform Random Number Generation

In this section, we construct a uniform random number generator from a nonuniform random source.

For a given function $A : \mathcal{X}^n \to \mathrm{Im}\mathcal{A}_n$, a generator φ_n is defined as $\varphi_n(\mathbf{x}) \equiv A\mathbf{x}$ for each $\mathbf{x} \in \mathcal{X}^n$. By letting $\mathcal{M}_n \equiv \mathrm{Im}\mathcal{A}_n$, the generation rate R is given as

$$R = \frac{\log |\mathrm{Im}\mathcal{A}_n|}{n}. \tag{20}$$

Let us assume that a (nonuniform) probability distribution μ_{X^n} of random variable $X^n \in \mathcal{X}^n$ is known prior to code construction. We have the following theorem which asserts that the fundamental limit $\underline{H}(\mathbf{X})$ is achievable with the generator using sparse matrix.

Theorem 3.1 *Let* $\mathbf{X}$ *be a general source. For a given* R *defined by (20) such that* $R < \underline{H}(\mathbf{X})$, *assume that* $\{(\mathcal{A}_n, p_{\mathsf{A},n})\}_{n=1}^{\infty}$ *has a balanced-coloring property. Then for all* $\delta > 0$ *and all sufficiently large* n *there is a function (sparse matrix)* $A \in \mathcal{A}_n$ *satisfying* $d(\mu_{M_n}, p_{\overline{M}_n}) \le \delta$.

4 Independent Random Number Generation

In this section, we construct a uniform random number generator from one of two correlated sources, where the generated random number is independent of the other source.

For a given function $A : \mathcal{X}^n \to \mathrm{Im}\mathcal{A}_n$, a generator φ_n is defined as $\varphi_n(\mathbf{x}) \equiv A\mathbf{x}$ for each $\mathbf{x} \in \mathcal{X}^n$. By letting $\mathcal{M}_n \equiv \mathrm{Im}\mathcal{A}_n$, the generation rate R is given as

$$R = \frac{\log |\mathrm{Im}\mathcal{A}_n|}{n}. \tag{21}$$

Let us assume that the probability distribution of (X^n, Z^n) is known before the construction of a code. We have the following theorem which asserts that the fundamental limit $\underline{H}(\mathbf{X}|\mathbf{Z})$ is achievable with the generator using sparse matrix.

Theorem 4.1 *Let* $(\mathbf{X}, \mathbf{Z})$ *be a pair of correlated general sources. For a given* R *satisfying (21) and* $R < \underline{H}(\mathbf{X}|\mathbf{Z})$, *assume that* $\{(\mathcal{A}_n, p_{\mathsf{A},n})\}_{n=1}^{\infty}$ *has a balanced-coloring property. Then for all* $\delta > 0$ *and all sufficiently large* n *there is a function (sparse matrix)* $A \in \mathcal{A}_n$ *such that* $d(\mu_{M_n Z^n}, p_{\overline{M}_n} \times \mu_{Z^n}) \le \delta$.

5 Secret Key Agreement

In this section, we construct a protocol for secret key agreement from the correlated source outputs introduced in [2, 18]. The following construction is based on [26]. It should be noted that the result presented in this paper is proof of strong security while weak security was proved in [26].

For a given joint distribution of (X^n, Y^n, Z^n), we fix functions[5] $A : \mathcal{X}^n \to \mathrm{Im}\mathcal{A}_n$ and $B : \mathcal{X}^n \to \mathrm{Im}\mathcal{B}_n$, which are available for an encoder, a decoder, and an eavesdropper. Let (C_n, M_n, M'_n) be a protocol defined as

$$\begin{aligned} C_n &\equiv BX^n \\ M_n &\equiv AX^n \\ M'_n &\equiv Ag_B(C_n|Y^n), \end{aligned}$$

where g_A is defined as

$$g_A(\mathbf{c}|\mathbf{y}) \equiv \arg\max_{\mathbf{x}' \in \mathcal{C}_A(\mathbf{c})} \mu_{X|Y}(\mathbf{x}'|\mathbf{y}). \tag{22}$$

By letting $\mathcal{M}_n \equiv \mathrm{Im}\mathcal{A}_n$, the rate R is given as

$$R \equiv \frac{\log |\mathrm{Im}\mathcal{A}_n|}{n}. \tag{23}$$

We define the encoding rate r of C_n as

$$r \equiv \frac{\log |\mathrm{Im}\mathcal{B}_n|}{n}. \tag{24}$$

Let us assume that the probability distribution of (X^n, Y^n, Z^n) is known prior to code construction. We have the following theorem under weaker conditions than those in [27], which assumes the hash property of both ensembles $\{(\mathcal{A}_n, p_{\mathsf{A},n})\}_{n=1}^{\infty}$ and $\{(\mathcal{B}_n, p_{\mathsf{B},n})\}_{n=1}^{\infty}$.

Theorem 5.1 *Let* $(\mathbf{X}, \mathbf{Y}, \mathbf{Z})$ *be a triplet of correlated general sources satisfying* $\underline{H}(\mathbf{X}|\mathbf{Z}) - \overline{H}(\mathbf{X}|\mathbf{Y}) > 0$. *For given r and R satisfying (23), (24), and*

$$\begin{aligned} r &> \overline{H}(\mathbf{X}|\mathbf{Y}) && (25) \\ r + R &< \underline{H}(\mathbf{X}|\mathbf{Z}), && (26) \end{aligned}$$

we assume that ensembles $\{(\mathcal{A}_n, p_{\mathsf{A},n})\}_{n=1}^{\infty}$ *and* $\{(\mathcal{B}_n, p_{\mathsf{B},n})\}_{n=1}^{\infty}$ *have a balanced-coloring property and a collision-resistance property, respectively. Then for all* $\delta > 0$

[5] It should be noted that the roles of A and B are the reverse of those in [26].

and all sufficiently large n, there are functions (sparse matrices) $A \in \mathscr{A}_n$ *and* $B \in \mathscr{B}_n$ *such that the proposed protocol* (C_n, M_n, M'_n) *satisfies (8) and*

$$d(p_{MZ^nC}, p_{\overline{M}} \times p_{Z^nC}) \le \delta. \tag{27}$$

This implies that the rate $\underline{H}(\mathbf{X}|\mathbf{Z}) - \overline{H}(\mathbf{X}|\mathbf{Y})$ *is secretly achievable with the proposed protocol by letting* $r \to \overline{\overline{H}}(\mathbf{Y}|\mathbf{X})$ *and* $R \to \underline{H}(\mathbf{X}|\mathbf{Z}) - \overline{H}(\mathbf{X}|\mathbf{Y})$.

Next, we combine the proposed protocol with the advantage distillation protocol to achieve $S_f(\mathbf{X}; \mathbf{Y}\|\mathbf{Z})$. For given correlated sources (X^n, Y^n, Z^n), we assume that random variables $(\widehat{X}_n, W_n)$ satisfy (13) and attain the maximum on the right hand side of (12), where $\widehat{X}_n$ is the output of a channel (nondeterministic map) with input X^n, and W_n is the output of a channel (nondeterministic map) with input $\widehat{X}_n$. Then by applying Theorem 5.1 to the correlated general source $(\widehat{\mathbf{X}}, (\mathbf{Y}, \mathbf{W}), (\mathbf{Z}, \mathbf{W}))$, we have the fact that $\underline{H}(\widehat{\mathbf{X}}|\mathbf{Z}, \mathbf{W}) - \overline{H}(\widehat{\mathbf{X}}|\mathbf{Y}, \mathbf{W})$ is achievable with functions A, B, and g_B. Since $\widehat{\mathbf{X}}$ and $\mathbf{W}$ can be obtained by the sender, we define a protocol for the correlated sources $(\mathbf{X}, \mathbf{Y}, \mathbf{Z})$ as

$$\begin{aligned} C_n &\equiv (B\widehat{X}^n, W^n) \\ M_n &\equiv A\widehat{X}^n \\ M'_n &\equiv Ag_B(B\widehat{X}^n|Y^n, W^n), \end{aligned}$$

where we define $\widehat{X}^n \equiv f_n(\widehat{X}_n) \in \widehat{\mathscr{X}^n}$ by using a bijection $f_n : \widehat{\mathscr{X}_n} \to \widehat{\mathscr{X}^n}$. From the fact that

$$S_f(\mathbf{X}; \mathbf{Y}\|\mathbf{Z}) \le \sup_{\mathbf{W}, \widehat{\mathbf{X}}} \left[\underline{H}(\widehat{\mathbf{X}}|\mathbf{Z}, \mathbf{W}) - \overline{H}(\widehat{\mathbf{X}}|\mathbf{Y}, \mathbf{W})\right], \tag{28}$$

we have Theorem 1.5 and the following corollary.

Corollary 5.2 *The forward secret key capacity* $S_f(\mathbf{X}; \mathbf{Y}\|\mathbf{Z})$ *is achievable with the proposed protocol.*

It should be noted that this corollary extends the results reported in [19, 26] for stationary memoryless sources. The proofs of Theorem 5.1 and (28) are given in the next section.

6 Proof of Theorems

In this section, we prove the theorems. We omit the dependence on n of X, Y, Z, and C when they appear in the subscript of $\overline{\mathscr{T}}$ and $\underline{\mathscr{T}}$.

6.1 Proof of Theorems 3.1 and 4.1

In the following, we prove Theorem 4.1. Theorem 3.1 can be proved by assuming that Z^n is a constant random variable, namely that there is a $\mathbf{z}' \in \mathcal{Z}^n$ such that $\mu_{X^nZ^n}(\mathbf{x}, \mathbf{z}') = \mu_{X^n}(\mathbf{x})$ and $\mu_{X^nZ^n}(\mathbf{x}, \mathbf{z}) = 0$ for any $\mathbf{x} \in \mathcal{X}^n$ and $\mathbf{z} \neq \mathbf{z}'$. Under this assumption, we have $\underline{H}(\mathbf{X}|\mathbf{Z}) = \underline{H}(\mathbf{X})$.

For a given $\zeta > 0$, let $\underline{\mathcal{T}}_{X|Z} \subset \mathcal{X}^n \times \mathcal{Z}^n$ and $\underline{\mathcal{T}}_{X|Z}(\mathbf{z}) \subset \mathcal{X}^n$ be defined as

$$\underline{\mathcal{T}}_{X|Z} \equiv \left\{ (\mathbf{x}, \mathbf{z}) : \frac{1}{n} \log \frac{1}{\mu_{X^n|Z^n}(\mathbf{x}|\mathbf{z})} \geq \underline{H}(\mathbf{X}|\mathbf{Z}) - \zeta \right\}$$

$$\underline{\mathcal{T}}_{X|Z}(\mathbf{z}) \equiv \left\{ \mathbf{x} : (\mathbf{x}, \mathbf{z}) \in \underline{\mathcal{T}}_{X|Z} \right\}.$$

First, we show the following lemma, which is proved in Appendix 2.

Lemma 6.1 ([20, Eq. (70)]) *Assume that $(\mathcal{A}_n, p_{\mathsf{A},n})$ satisfy (14). Then*

$$E_{\mathsf{A}} \left[\sum_{\mathbf{m},\mathbf{z}} \left| \mu_{X^nZ^n}(\mathcal{C}_{\mathsf{A}}(\mathbf{m}), \mathbf{z}) - \frac{\mu_{Z^n}(\mathbf{z})}{|\mathrm{Im}\mathcal{A}_n|} \right| \right]$$
$$\leq \sqrt{\alpha_{\mathsf{A}} - 1 + [\beta_{\mathsf{A}} + 1]|\mathrm{Im}\mathcal{A}_n| 2^{-n[\underline{H}(\mathbf{X}|\mathbf{Z}) - \zeta]}} + 2\mu_{X^nZ^n}([\underline{\mathcal{T}}_{X|Z}]^c).$$

Next, we prove Theorem 4.1. From the assumption $R < \underline{H}(\mathbf{X}|\mathbf{Z})$, we have the fact that there is $\zeta > 0$ such that

$$R + \zeta < \underline{H}(\mathbf{X}|\mathbf{Z}). \tag{29}$$

From Lemma 6.1 and (21), we have the fact that there is a function (sparse matrix) A such that

$$\sum_{\mathbf{m},\mathbf{z}} \left| \sum_{\mathbf{x} \in \mathcal{C}_A(\mathbf{m})} \mu_{X^nZ^n}(\mathbf{x}, \mathbf{z}) - \sum_{\mathbf{x}} \frac{\mu_{X^nZ^n}(\mathbf{x}, \mathbf{z})}{|\mathrm{Im}\mathcal{A}_n|} \right|$$
$$\leq \sqrt{\alpha_{\mathsf{A}} - 1 + [\beta_{\mathsf{A}} + 1] 2^{-n[\underline{H}(\mathbf{X}|\mathbf{Z}) - R - \zeta]}} + 2\mu_{X^nZ^n}([\underline{\mathcal{T}}_{X|Z}]^c).$$

Then the variational distance is upper bounded as

$$d(\mu_{M_nZ^n}, p_{\overline{M}_n} \times \mu_{Z^n})$$
$$= \frac{1}{2} \sum_{\mathbf{m},\mathbf{z}} \left| \sum_{\mathbf{x} \in \mathcal{C}_A(\mathbf{m})} \mu_{X^nZ^n}(\mathbf{x}, \mathbf{z}) - \frac{\mu_{Z^n}(\mathbf{z})}{|\mathrm{Im}\mathcal{A}_n|} \right|$$
$$\leq \frac{1}{2} \sqrt{\alpha_{\mathsf{A}} - 1 + [\beta_{\mathsf{A}} + 1] 2^{-n[\underline{H}(\mathbf{X}|\mathbf{Z}) - R - \zeta]}} + \mu_{X^nZ^n}([\underline{\mathcal{T}}_{X|Z}]^c). \tag{30}$$

From (29) and the fact that $\alpha_{\mathsf{A}} \to 1$, $\log(\beta_{\mathsf{A}} + 1)/n \to 0$, $\mu_{X^nZ^n}([\underline{\mathcal{T}}_{X|Z}]^c) \to 0$ as $n \to \infty$, we have $d(\mu_{M_nZ^n}, p_{\overline{M}_n} \times \mu_{Z^n}) \leq \delta$ for all $\delta > 0$ and all sufficiently large n. □

6.2 Proof of Theorems 5.1

For a given $\zeta > 0$, let $\overline{\mathcal{T}}_{X|Y} \subset \mathcal{X}^n \times \mathcal{Y}^n$ and $\overline{\mathcal{T}}_{X|Y}(\mathbf{y}) \subset \mathcal{X}^n$ be defined as

$$\overline{\mathcal{T}}_{X|Y} \equiv \left\{(\mathbf{x}, \mathbf{y}) : \frac{1}{n} \log \frac{1}{\mu_{X^n|Y^n}(\mathbf{x}|\mathbf{y})} \leq \overline{H}(\mathbf{X}|\mathbf{Y}) + \zeta\right\}$$
$$\overline{\mathcal{T}}_{X|Y}(\mathbf{y}) \equiv \left\{\mathbf{x} : (\mathbf{x}, \mathbf{y}) \in \overline{\mathcal{T}}_{X|Y}\right\}.$$

We use the following lemma, which is shown in Appendix 3.

Lemma 6.2 ([20, Eq. (58)]) *Assume that $(\mathcal{B}_n, p_{\mathsf{B},n})$ satisfy (14). Then*

$$E_{\mathsf{B}}\left[\sum_{\mathbf{x},\mathbf{y}} \mu_{X^nY^n}(\mathbf{x}, \mathbf{y})\chi(g_{\mathsf{B}}(\mathsf{B}\mathbf{x}|\mathbf{y}) \neq \mathbf{x})\right] \leq \frac{2^{n[\overline{H}(\mathbf{X}|\mathbf{Y})+\zeta]}\alpha_{\mathsf{B}}}{|\mathrm{Im}\mathcal{B}_n|} + \beta_{\mathsf{B}} + \mu_{X^nY^n}([\overline{\mathcal{T}}_{X|Y}]^c),$$

where g_B is defined by (22).

Here, we prove the theorem. From (25) and (26), we have the fact that there is $\gamma > 0$ such that

$$r + \zeta > \overline{H}(\mathbf{X}|\mathbf{Y}) \tag{31}$$
$$r + R + \zeta < \underline{H}(\mathbf{X}|\mathbf{Z}). \tag{32}$$

First, we evaluate the average of error probability $\mathrm{Error}_{X^nY^n}(B) \equiv P(M_n \neq M'_n)$ over a random selection of a function $B \in \mathcal{B}_n$. Since the key agreement is successful when $g_B(B\mathbf{x}|\mathbf{y}) = \mathbf{x}$, we have

$$\begin{aligned} E_{\mathsf{B}}\left[\mathrm{Error}_{X^nY^n}(\mathsf{B})\right] &\leq E_{\mathsf{B}}\left[\sum_{\mathbf{x},\mathbf{y}} \mu_{X^nY^n}(\mathbf{x}, \mathbf{y})\chi(g_{\mathsf{B}}(\mathsf{B}\mathbf{x}|\mathbf{y}) \neq \mathbf{x})\right] \\ &\leq \frac{2^{-n[\overline{H}(\mathbf{X}|\mathbf{Y})-\zeta]}\alpha_{\mathsf{B}}}{|\mathrm{Im}\mathcal{B}_n|} + \beta_{\mathsf{B}} + \mu_{X^nY^n}([\overline{\mathcal{T}}_{X|Y}]^c) \\ &= 2^{-n[r-\overline{H}(\mathbf{X}|\mathbf{Y})-\zeta]}\alpha_{\mathsf{B}} + \beta_{\mathsf{B}} + \mu_{X^nY^n}([\overline{\mathcal{T}}_{X|Y}]^c), \end{aligned} \tag{33}$$

where the second inequality comes from Lemma 6.2, and the equality comes from (24). Then we have the fact that there is a function (sparse matrix) $B \in \mathcal{B}$ such that

$$\mathrm{Error}_{X^nY^n}(B) \leq 2^{-n[r-\overline{H}(\mathbf{X}|\mathbf{Y})-\zeta]}\alpha_{\mathsf{B}} + \beta_{\mathsf{B}} + \mu_{X^nY^n}([\overline{\mathcal{T}}_{X|Y}]^c). \tag{34}$$

In the following, we assume that A satisfies (34) and let $\mathbf{C} \equiv \{C_n\}_{n=1}^{\infty}$ be defined as $C_n \equiv BX^n$. Then, similarly to the proof of Theorem 4.1 (by applying Lemma 6.1 and (23) to general correlated sources $(\mathbf{X}, (\mathbf{Z}, \mathbf{C}))$), we have the fact that there is a function (sparse matrix) $A \in \mathscr{A}_n$, which depends on B, such that

$$d(\mu_{M_n Z^n C_n}, p_{\overline{M}_n} \times \mu_{Z^n C_n}) \\ \leq \frac{1}{2}\sqrt{\alpha_{\mathsf{A}} - 1 + [\beta_{\mathsf{A}} + 1]2^{-n[\underline{H}(\mathbf{X}|\mathbf{Z},\mathbf{C}) - R - \zeta]}} + \mu_{X^n Z^n C_n}([\underline{\mathcal{T}}_{X|ZC}]^c). \tag{35}$$

From (24) and the fact that $C_n \in \mathrm{Im}\mathscr{B}_n$, we have $\overline{H}(\mathbf{C}|\mathbf{Z}) \leq r$. Then, we have

$$\begin{aligned} R + \zeta &< \underline{H}(\mathbf{X}|\mathbf{Z}) - r \\ &\leq \underline{H}(\mathbf{X}, \mathbf{C}|\mathbf{Z}) - \overline{H}(\mathbf{C}|\mathbf{Z}) \\ &\leq \underline{H}(\mathbf{X}|\mathbf{Z}, \mathbf{C}) \end{aligned} \tag{36}$$

where the first inequality comes from (32), the second inequality comes from the fact that $\underline{H}(\mathbf{X}|\mathbf{Z}) = \underline{H}(\mathbf{X}, \mathbf{C}|\mathbf{Z})$, and $\overline{H}(\mathbf{C}|\mathbf{Z}) \leq r$. Thus, from (31), (34)–(36), and the fact that $\alpha_{\mathsf{A}} \to 1$, $\log(\beta_{\mathsf{A}} + 1)/n \to 0$, $\log(\alpha_{\mathsf{B}})/n \to 0$, $\beta_{\mathsf{B}} \to 0$, $\mu_{X^n Y^n}([\overline{\mathcal{T}}_{X|Y}]^c) \to 0$, $\mu_{X^n Z^n C_n}([\underline{\mathcal{T}}_{X|ZC}]^c) \to 0$ as $n \to \infty$, we have the fact that A and B satisfy (8) and (27) for all $\delta > 0$ and all sufficiently large n. □

6.3 *Proof of (28)*

We use the following lemmas, where Lemma 6.4 is shown in Appendix 4.

Lemma 6.3 ([17, Lemma 4][20, Lemma 7]) *Let* $(\mathbf{U}, \mathbf{V}) \equiv \{(U_n, V_n)\}_{n=1}^{\infty}$ *be a pair of general sources. If there is* $\{\psi_n\}_{n=1}^{\infty}$ *such that* $\lim_{n\to\infty} P(\psi_n(V_n) \neq U_n) = 0$, *then* $\overline{H}(\mathbf{U}|\mathbf{V}) = 0$.

Lemma 6.4 ([5, Prop. 2][27, Eq. (15)]) *Let* $(\mathbf{U}, \mathbf{V}) \equiv \{(U_n, V_n)\}_{n=1}^{\infty}$ *be a pair of general sources. If there is a function* $\varphi_n : \mathscr{U}^n \to \mathscr{M}_n$ *such that* $\lim_{n\to\infty} d(\mu_{M_n V^n}, p_{\overline{M}_n} \times \mu_{V^n}) = 0$, *then* $R \leq \underline{H}(\mathbf{U}|\mathbf{V})$.

Here, we show (28). Since M'_n is generated from Y^n and C_n, the joint distribution $\mu_{M_n M'_n Y^n C_n}$ is given as

$$\mu_{M_n M'_n Y^n C_n}(\mathbf{m}, \mathbf{m}', \mathbf{y}, \mathbf{c}) = \mu_{M_n|Y^n C_n}(\mathbf{m}|\mathbf{y}, \mathbf{c})\mu_{M'_n|Y^n C_n}(\mathbf{m}'|\mathbf{y}, \mathbf{c})\mu_{Y^n C_n}(\mathbf{y}, \mathbf{c}).$$

Let $\psi_n(\mathbf{y}, \mathbf{c}) \equiv \arg\max_{\mathbf{m}\in\mathrm{Im}\mathscr{A}_n} \mu_{M_n|Y^n C_n}(\mathbf{m}|\mathbf{y}, \mathbf{c})$. Then we have

$$\begin{aligned}P(M_n = M'_n) &= \sum_{\mathbf{m},\mathbf{y},\mathbf{c}} \mu_{M_n|Y^nC_n}(\mathbf{m}|\mathbf{y},\mathbf{c})\mu_{M'_n|Y^nC_n}(\mathbf{m}|\mathbf{y},\mathbf{c})\mu_{Y^nC_n}(\mathbf{y},\mathbf{c})\\ &\le \sum_{\mathbf{m},\mathbf{y},\mathbf{c}} \mu_{M_n|Y^nC_n}(\psi_n(\mathbf{y},\mathbf{c})|\mathbf{y},\mathbf{c})\mu_{M'_n|Y^nC_n}(\mathbf{m}|\mathbf{y},\mathbf{c})\mu_{Y^nC_n}(\mathbf{y},\mathbf{c})\\ &= \sum_{\mathbf{y},\mathbf{c}} \mu_{M_n|Y^nC_n}(\psi_n(\mathbf{y},\mathbf{c})|\mathbf{y},\mathbf{c})\mu_{Y^nC_n}(\mathbf{y},\mathbf{c})\\ &= P(\psi_n(Y^n, C_n) = M_n). \end{aligned} \tag{37}$$

This implies that $0 \le P(\psi_n(Y^n, C_n) \ne M_n) \le P(M_n \ne M'_n)$. We have $\lim_{n\to\infty} P(\psi_n(Y^n, C_n) \ne M_n) = 0$ from (8) by letting $n \to \infty$ and $\delta \to 0$. Then we have $\overline{H}(\mathbf{M}|\mathbf{Y},\mathbf{C}) = 0$ from Lemma 6.3 and

$$\begin{aligned}R &\le \underline{H}(\mathbf{M}|\mathbf{Z},\mathbf{C})\\ &= \underline{H}(\mathbf{M}|\mathbf{Z},\mathbf{C}) - \overline{H}(\mathbf{M}|\mathbf{Y},\mathbf{C}),\end{aligned} \tag{38}$$

where the first inequality comes from Lemma 6.4 and (9). Thus, (28) follows immediately from this inequality by letting $\widehat{\mathbf{X}} \equiv \mathbf{M}$ and $\mathbf{W} \equiv \mathbf{C}$. □

Appendix 1: Proof of Lemmas 1.1 and 1.2

In the following proofs, we use the following lemma.

Lemma 6.5 ([31]) *Let p and q be probability distributions on the same set $\mathcal{U}$. Then*

$$|H(p) - H(q)| \le d(p,q)\log(|\mathcal{U}| - 1) + h(d(p,q)),$$

where $H(\mu) \equiv \sum_u \mu(u)\log(1/\mu(u))$ and $h(\theta) \equiv -\theta\log\theta - [1-\theta]\log(1-\theta)$.

First, we prove Lemma 1.1. From the fact that $H(p_{\overline{M}_n}) = \log|\mathcal{M}_n|$, we have

$$\begin{aligned}D(\mu_{M_n}\|p_{\overline{M}_n}) &= \sum_{\mathbf{m}\in\mathcal{M}_n} \mu_{M_n}(\mathbf{m})\log\frac{\mu_{M_n}(\mathbf{m})}{p_{\overline{M}_n}(\mathbf{m})}\\ &= \log|\mathcal{M}_n| - H(\mu_{M_n})\\ &= \left|H(p_{\overline{M}_n}) - H(\mu_{M_n})\right|\\ &\le d(\mu_{M_n}, p_{\overline{M}_n})\log|\mathcal{M}_n| + h(d(\mu_{M_n}, p_{\overline{M}_n}))\\ &= o(1),\end{aligned} \tag{39}$$

where the inequality comes from Lemma 6.5. □

Next, we prove Lemma 1.2. The inequality (5) is shown immediately from (1) and Lemma 1.1. From Lemma 6.5, we have

$$\begin{aligned}
&\left|H(p_{\overline{M}_n}) - H(\mu_{M_n})\right| \\
&\leq d(\mu_{M_n}, p_{\overline{M}_n}) \log|\mathcal{M}_n| + h(d(\mu_{M_n}, p_{\overline{M}_n})) \\
&\leq d(\mu_{M_nZ^n}, p_{\overline{M}_n} \times \mu_{Z^n}) \log|\mathcal{M}_n| + h(d(\mu_{M_nZ^n}, p_{\overline{M}_n} \times \mu_{Z^n})), \qquad (40)
\end{aligned}$$

where the second inequality comes from (1). Furthermore, from Lemma 6.5, we have

$$\begin{aligned}
&\sum_{\mathbf{z}} \mu_{Z^n}(\mathbf{z}) \left|H(p_{\overline{M}_n}) - H(\mu_{M_n|Z^n}(\cdot|\mathbf{z}))\right| \\
&\leq \sum_{\mathbf{z}} \mu_{Z^n}(\mathbf{z}) d(p_{\overline{M}_n}, \mu_{M_n|Z^n}(\cdot|\mathbf{z})) \log|\mathcal{M}_n| + \sum_{\mathbf{z}} \mu_{Z^n}(\mathbf{z}) h(d(p_{\overline{M}_n}, \mu_{M_n|Z^n}(\cdot, \mathbf{z}))) \\
&\leq d(\mu_{M_nZ^n}, p_{\overline{M}_n} \times \mu_{Z^n}) \log|\mathcal{M}_n| + h(d(\mu_{M_nZ^n}, p_{\overline{M}_n} \times \mu_{Z^n})), \qquad (41)
\end{aligned}$$

where the second inequality comes from the convexity of h. Then we have (6) as

$$\begin{aligned}
I(M_n; Z^n) &= \sum_{\mathbf{z}} \mu_{Z^n}(\mathbf{z}) \left[H(\mu_{M_n}) - H(\mu_{M_n|Z^n}(\cdot|\mathbf{z}))\right] \\
&\leq \sum_{\mathbf{z}} \mu_{Z^n}(\mathbf{z}) \left|H(\mu_{M_n}) - H(p_{\overline{M}_n})\right| + \sum_{\mathbf{z}} \mu_{Z^n}(\mathbf{z}) \left|H(p_{\overline{M}_n}) - H(\mu_{M_n|Z^n}(\cdot|\mathbf{z}))\right| \\
&\leq 2d(\mu_{M_nZ^n}, p_{\overline{M}_n} \times \mu_{Z^n}) \log|\mathcal{M}_n| + 2h(d(\mu_{M_nZ^n}, p_{\overline{M}_n} \times \mu_{Z^n})) \\
&\leq o(1), \qquad (42)
\end{aligned}$$

where the second inequality comes from (40) and (41). □

Appendix 2: Proof of Lemma 6.1

When $\mu_{X^n|Z^n}(\underline{\mathcal{T}}_{X|Z}(\mathbf{z})|\mathbf{z}) \neq 0$, we have

$$\begin{aligned}
&E_{\mathsf{A}}\left[\sum_{\mathbf{m}} \left|\frac{\mu_{X^n|Z^n}(\underline{\mathcal{T}}_{X|Z}(\mathbf{z}) \cap \mathcal{C}_{\mathsf{A}}(\mathbf{m})|\mathbf{z})}{\mu_{X^n|Z^n}(\underline{\mathcal{T}}_{X|Z}(\mathbf{z})|\mathbf{z})} - \frac{1}{|\mathrm{Im}\mathcal{A}_n|}\right|\right] \\
&\leq \sqrt{\alpha_{\mathsf{A}} - 1 + \frac{[\beta_{\mathsf{A}} + 1]|\mathrm{Im}\mathcal{A}_n| \max_{\mathbf{x} \in \underline{\mathcal{T}}_{X|Z}(\mathbf{z})} \mu_{X^n|Z^n}(\mathbf{x}|\mathbf{z})}{\mu_{X^n|Z^n}(\underline{\mathcal{T}}_{X|Z}(\mathbf{z})|\mathbf{z})}} \\
&\leq \frac{\sqrt{\alpha_{\mathsf{A}} - 1 + [\beta_{\mathsf{A}} + 1]|\mathrm{Im}\mathcal{A}_n| 2^{-n[\underline{H}(\mathbf{X}|\mathbf{Z}) - \zeta]}}}{\mu_{X^n|Z^n}(\underline{\mathcal{T}}_{X|Z}(\mathbf{z})|\mathbf{z})}, \qquad (43)
\end{aligned}$$

where the first inequality comes from Lemma 2.2 and the second inequality comes from the definition of $\underline{\mathcal{T}}_{X|Z}(\mathbf{z})$ and the fact that $\mu_{X^n|Z^n}(\underline{\mathcal{T}}_{X|Z}(\mathbf{z})|\mathbf{z}) \leq 1$. Then we have

$$
\begin{aligned}
&E_{\mathsf{A}}\left[\sum_{\mathbf{m},\mathbf{z}}\left|\mu_{X^nZ^n}(\mathscr{C}_A(\mathbf{m}),\mathbf{z})-\frac{\mu_{Z^n}(\mathbf{z})}{|\mathrm{Im}\mathscr{A}_n|}\right|\right]\\
&\leq E_{\mathsf{A}}\left[\sum_{\mathbf{m},\mathbf{z}}\left|\mu_{X^nZ^n}(\mathscr{C}_{\mathsf{A}}(\mathbf{m})\cap\underline{\mathscr{T}}_{X|Z}(\mathbf{z}),\mathbf{z})-\frac{\mu_{X^nZ^n}(\underline{\mathscr{T}}_{X|Z}(\mathbf{z}),\mathbf{z})}{|\mathrm{Im}\mathscr{A}_n|}\right|\right]\\
&\quad+E_{\mathsf{A}}\left[\sum_{\mathbf{m},\mathbf{z}}\mu_{X^nZ^n}(\mathscr{C}_{\mathsf{A}}(\mathbf{m})\cap[\underline{\mathscr{T}}_{X|Z}(\mathbf{z})]^c,\mathbf{z})\right]+E_{\mathsf{A}}\left[\sum_{\mathbf{m},\mathbf{z}}\frac{\mu_{X^nZ^n}([\underline{\mathscr{T}}_{X|Z}(\mathbf{z})]^c,\mathbf{z})}{|\mathrm{Im}\mathscr{A}_n|}\right]\\
&=\sum_{\substack{\mathbf{z}:\\ \mu_Z(\mathbf{z})\neq 0\\ \mu_{X^n|Z^n}(\underline{\mathscr{T}}_{X|Z}(\mathbf{z})|\mathbf{z})\neq 0}}\mu_{X^nZ^n}(\underline{\mathscr{T}}_{X|Z}(\mathbf{z}),\mathbf{z})E_{\mathsf{A}}\left[\sum_{\mathbf{m}}\left|\frac{\mu_{X^n|Z^n}(\underline{\mathscr{T}}_{X|Z}(\mathbf{z})\cap\mathscr{C}_{\mathsf{A}}(\mathbf{m})|\mathbf{z})}{\mu_{X^n|Z^n}(\underline{\mathscr{T}}_{X|Z}(\mathbf{z})|\mathbf{z})}-\frac{1}{|\mathrm{Im}\mathscr{A}_n|}\right|\right]\\
&\quad+2\sum_{\mathbf{z}}\mu_{X^nZ^n}([\underline{\mathscr{T}}_{X|Z}(\mathbf{z})]^c,\mathbf{z})\\
&\leq\sqrt{\alpha_{\mathsf{A}}-1+[\beta_{\mathsf{A}}+1]|\mathrm{Im}\mathscr{A}_n|2^{-n[\underline{H}(\mathbf{X}|\mathbf{Z})-\zeta]}}+2\mu_{X^nZ^n}([\underline{\mathscr{T}}_{X|Z}]^c),
\end{aligned}
\tag{44}
$$

which concludes the lemma. □

Appendix 3: Proof of Lemma 6.2

If $g_B(B\mathbf{x}|\mathbf{y})\neq\mathbf{x}$, there is $\mathbf{x}'\in\mathscr{C}_B(B\mathbf{x})$ such that $\mathbf{x}'\neq\mathbf{x}$ and

$$
\begin{aligned}
\mu_{X^n|Y^n}(\mathbf{x}'|\mathbf{y})&\geq\mu_{X^n|Y^n}(\mathbf{x}|\mathbf{y})\\
&\geq 2^{-n[\overline{H}(\mathbf{X}|\mathbf{Y})+\zeta]},
\end{aligned}
\tag{45}
$$

where the second inequality comes from the definition of $\overline{\mathscr{T}}_{X|Y}$. This implies that $[\overline{\mathscr{T}}_{X|Y}(\mathbf{y})\setminus\{\mathbf{x}\}]\cap\mathscr{C}_B(B\mathbf{x})\neq\emptyset$. Then we have

$$
\begin{aligned}
p_{\mathsf{B},n}\left(\{B:g_B(B\mathbf{x}|\mathbf{y})\neq\mathbf{x}\}\right)&\leq p_{\mathsf{B},n}\left(\left\{B:\left[\overline{\mathscr{T}}_{X|Y}(\mathbf{y})\setminus\{\mathbf{x}\}\right]\cap\mathscr{C}_B(B\mathbf{x})\neq\emptyset\right\}\right)\\
&\leq\frac{2^{n[\overline{H}(\mathbf{X}|\mathbf{Y})+\zeta]}\alpha_{\mathsf{B}}}{|\mathrm{Im}\mathscr{B}_n|}+\beta_{\mathsf{B}},
\end{aligned}
\tag{46}
$$

where the second inequality comes from Lemma 2.3 and the fact that $|\overline{\mathscr{T}}_{X|Y}(\mathbf{y})|\leq 2^{n[\overline{H}(\mathbf{X}|\mathbf{Y})+\zeta]}$. We have

$$
\begin{aligned}
&E_{\mathsf{B}}\left[\sum_{\mathbf{x},\mathbf{y}}\mu_{X^nY^n}(\mathbf{x},\mathbf{y})\chi(g_{\mathsf{B}}(\mathsf{B}\mathbf{x}|\mathbf{y})\neq\mathbf{x})\right] \\
&\leq \sum_{(\mathbf{x},\mathbf{y})\in\overline{\mathscr{T}}_{X|Y}}\mu_{X^nY^n}(\mathbf{x},\mathbf{y})p_{\mathsf{B}}(\{A:g_B(B\mathbf{x}|\mathbf{y})\neq\mathbf{x}\})+\sum_{(\mathbf{x},\mathbf{y})\in[\overline{\mathscr{T}}_{X|Y}]^c}\mu_{X^nY^n}(\mathbf{x},\mathbf{y}) \\
&\leq \frac{2^{n[\overline{H}(\mathbf{X}|\mathbf{Y})+\zeta]}\alpha_{\mathsf{B}}}{|\mathrm{Im}\mathscr{B}_n|}+\beta_{\mathsf{B}}+\mu_{X^nY^n}([\overline{\mathscr{T}}_{X|Y}]^c), \qquad (47)
\end{aligned}
$$

which concludes the lemma. □

Appendix 4: Proof of Lemma 6.4

Let $\underline{\mathscr{T}}'$ and $\underline{\mathscr{T}}'(\mathbf{z})$ be defined as:

$$
\begin{aligned}
\underline{\mathscr{T}}' &\equiv \left\{(\mathbf{x},\mathbf{z}) : \frac{1}{n}\log\frac{1}{\mu_{X^n|Z^n}(\mathbf{x}|\mathbf{z})}\geq R-\zeta\right\} \\
\underline{\mathscr{T}}'(\mathbf{z}) &\equiv \left\{\mathbf{x} : (\mathbf{x},\mathbf{z})\in\underline{\mathscr{T}}'\right\}.
\end{aligned}
$$

Since $\varphi_n(\mathbf{x})\in\varphi_n([\underline{\mathscr{T}}'(\mathbf{z})]^c)$ and $\mathbf{x}\in\varphi_n^{-1}(\varphi_n(\mathbf{x}))$ for all $\mathbf{x}\in[\underline{\mathscr{T}}'(\mathbf{z})]^c$, we have

$$
\begin{aligned}
\sum_{\mathbf{z}}\sum_{\mathbf{m}\in\varphi_n([\underline{\mathscr{T}}'(\mathbf{z})]^c)}\sum_{\mathbf{x}\in\varphi_n^{-1}(\mathbf{m})}\mu_{X^nZ^n}(\mathbf{x},\mathbf{z}) &\geq \sum_{\mathbf{z}}\sum_{\mathbf{x}\in[\underline{\mathscr{T}}'(\mathbf{z})]^c}\mu_{X^nZ^n}(\mathbf{x},\mathbf{z}) \\
&= \mu_{X^nZ^n}(\underline{\mathscr{T}}'^c). \qquad (48)
\end{aligned}
$$

On the other hand, we have

$$
\begin{aligned}
\sum_{\mathbf{z}}\sum_{\mathbf{m}\in\varphi_n([\underline{\mathscr{T}}'(\mathbf{z})]^c)}\frac{\mu_{Z^n}(\mathbf{z})}{|\mathscr{M}_n|} &= \sum_{\mathbf{z}}\frac{\left|\varphi_n([\underline{\mathscr{T}}'(\mathbf{z})]^c)\right|\mu_{Z^n}(\mathbf{z})}{|\mathscr{M}_n|} \\
&\leq \sum_{\mathbf{z}}\frac{|[\underline{\mathscr{T}}'(\mathbf{z})]^c|\mu_{Z^n}(\mathbf{z})}{|\mathscr{M}_n|} \\
&\leq 2^{-n\zeta}, \qquad (49)
\end{aligned}
$$

where the last inequality comes from the fact that $|\mathscr{M}_n|=2^{nR}$ and $|[\underline{\mathscr{T}}'(\mathbf{z})]^c|\leq 2^{n[R-\zeta]}$. From (48) and (49), we have

$$d(\mu_{M_n Z^n}, p_{\overline{M}_n} \times \mu_{Z^n}) = \frac{1}{2} \sum_{\mathbf{m},\mathbf{z}} \left| \sum_{\mathbf{x}\in\varphi^{-1}(\mathbf{m})} \mu_{X^n Z^n}(\mathbf{x}, \mathbf{z}) - \frac{\mu_{Z^n}(\mathbf{z})}{|\mathcal{M}_n|} \right|$$
$$\geq \frac{1}{2} \sum_{\mathbf{z}} \sum_{\mathbf{m}\in\varphi_n([\underline{\mathcal{T}}'(\mathbf{z})]^c)} \left[\sum_{\mathbf{x}\in\varphi_n^{-1}(\mathbf{m})} \mu_{X^n Z^n}(\mathbf{x}, \mathbf{z}) - \frac{\mu_{Z^n}(\mathbf{z})}{|\mathcal{M}_n|} \right]$$
$$\geq \frac{1}{2} \left[\mu_{X^n Z^n}(\underline{\mathcal{T}}'^c) - 2^{-n\zeta} \right]. \tag{50}$$

Then we have the fact that

$$P\left(\frac{1}{n} \log \frac{1}{\mu_{X^n|Z^n}(X^n|Z^n)} < R - \zeta \right) = \mu_{X^n Z^n}(\underline{\mathcal{T}}'^c)$$
$$\leq 2d(\mu_{M_n Z^n}, p_{\overline{M}_n} \times \mu_{Z^n}) + 2^{-n\zeta} \tag{51}$$

and

$$\lim_{n\to\infty} P\left(\frac{1}{n} \log \frac{1}{\mu_{X^n|Z^n}(X^n|Z^n)} < R - \zeta \right) = 0$$

by letting $n \to \infty$. From the definition of $\underline{H}(\mathbf{X}|\mathbf{Z})$, we have

$$R - \zeta \leq \underline{H}(\mathbf{X}|\mathbf{Z}),$$

which concludes the lemma by letting $\zeta \to 0$. □

References

1. E. Abbe, Polarization and randomness extraction, in *Proceedings of 2011 IEEE International Symposium Information Theory* (St. Petersburg, Russia, 31 July–5 Aug 2011), pp. 184–188
2. R. Ahlswede, I. Csiszár, Common randomness in information theory and cryptography — Part I: secret sharing. IEEE Trans. Inform. Theory **IT–39**(4), 1121–1132 (1993)
3. R. Ahlswede, I. Csiszár, Common randomness in information theory and cryptography — Part II: CR capacity. IEEE Trans. Inform. Theory **IT–44**(1), 225–240 (1998)
4. C.H. Bennett, G. Brassard, C. Crepeau, U. Maurer, Generalized privacy amplification. IEEE Trans. Inform. Theory **IT–41**(6), 1915–1923 (1995)
5. M. Bloch, Channel intrinsic randomness, in *Proceedings of 2010 IEEE International Symposium Information Theory* (Austin, USA, 13–18 June 2010), pp. 2607–2611
6. M. Bloch, J.N. Laneman, Strong secrecy from channel resolvability. IEEE Trans. Inform Theory **IT–59**(12), 8077–8098 (2013)
7. C. Cachin, U.M. Maurer, Linking information reconciliation and privacy amplification. J. Cryptol. **10**, 97–110 (1997)
8. T.M. Cover, A proof of the data compression theorem of Slepian and Wolf for ergodic sources. IEEE Trans. Inform. Theory **IT–21**(2), 226–228 (1975)
9. T.M. Cover, J.A. Thomas, *Elements of Information Theory*, 2nd edn. (Wiley, Hoboken, 2006)
10. I. Csiszár, Linear codes for sources and source networks: error exponents, universal coding. IEEE Trans. Inform. Theory **IT–28**(4), 585–592 (1982)

11. I. Csiszár, Almost independence and secrecy capacity. Probl. Inform. Transm. **32**(1), 40–47 (1996)
12. I. Csiszár, J. Körner, *Information Theory: Coding Theorems for Discrete Memoryless Systems*, 2nd edn. (Cambridge University Press, Cambridge, 2011)
13. J.L. Carter, M.N. Wegman, Universal classes of hash functions. J. Comput. Syst. Sci. **18**, 143–154 (1979)
14. T.S. Han, S. Verdú, Apploximation theory of output statistics. IEEE Trans. Inform. Theory **IT–39**(3), 752–772 (1993)
15. T.S. Han, *Information-Spectrum Methods in Information Theory* (Springer, Berlin, 2003)
16. R. Impagliazzo, D. Zuckerman, How to recycle random bits, in *30th IEEE Symposium of Fundations of Computer Science* (30 Oct–1 Nov 1989), pp. 248–253
17. H. Koga, Coding theorem on the threshold scheme for a general source. IEEE Trans. Inform. Theory **IT–52**(4), 2658–2677 (2008)
18. U.M. Maurer, Secret key agreement by public discussion from common information. IEEE Trans. Inform. Theory **IT–39**(3), 733–742 (1993)
19. J. Muramatsu, Secret key agreement from correlated source outputs using low density parity check matrices. IEICE Trans. Fundam. **E89–A**(7), 2036–2046 (2006)
20. J. Muramatsu, Channel coding and lossy source coding using a generator of constrained random numbers. IEEE Trans. Inform. Theory **IT–60**(5), 2667–2686 (2014)
21. J. Muramatsu, Variable-length lossy source cod using a constrained-random-number generator. IEEE Trans. Inform. Theory **IT–61**(6), 3574–3592 (2014)
22. J. Muramatsu, T. Koga, T. Mukouchi, On the problem of generating mutually independent random sequences. IEICE Trans. Fundam. **E86–A**(5), 1275–1284 (2003)
23. J. Muramatsu, S. Miyake, Hash property and coding theorems for sparse matrices and maximal-likelihood coding. IEEE Trans. Inform. Theory **IT–56**(5), 2143–2167 (2010). Corrections: **IT-56**(9), 4762 (2010); **IT-59**(10), 6952–6953 (2013)
24. J. Muramatsu, S. Miyake, Construction of Slepian-Wolf source code and broadcast channel code based on hash property (2010), arXiv:1006.5271 [CS.IT]
25. J. Muramatsu, S. Miyake, Construction of strongly secure wiretap channel code based on hash property, in *Proceedings 2011 IEEE International Symposium on Information Theory* (St. Petersburg, Russia, 31 July–5 Aug 2011), pp. 612–616
26. J. Muramatsu, S. Miyake, Construction of codes for wiretap channel and secret key agreement from correlated source outputs based on hash property. IEEE Trans. Inform. Theory **IT–58**(2), 671–692 (2012)
27. J. Muramatsu, S. Miyake, Uniform random number generation by using sparse matrix, *Proceedings of IEEE Information Theory Workshop* (Lausanne, Switzerland, 3–7 Sept 2012), pp. 612–616
28. J. Muramatsu, S. Miyake, Construction of a channel code from an arbitrary source code with decoder side information, in *Proceedings of International Symposium Information Theory and Its Applications* (Monterey, CA, USA, 30 Oct–2 Nov, 2016), pp. 176–180
29. R. Renner, S. Wolf, Simple and tight bounds for information reconciliation and privacy amplification. Lect. Notes Comput. Sci. **3788**, 199–216 (2005)
30. S. Vembu, S. Verdú, Generating random bits from an arbitrary source. IEEE Trans. Inform. Theory **IT–41**(5), 1322–1332 (1995)
31. Z. Zhang, Estimating mutual information via Kolmogorov distance. IEEE Trans. Inform. Theory **IT–53**(9), 3280–3283 (2007)

Mathematical Approach for Recovering Secret Key from Its Noisy Version

Noboru Kunihiro

Abstract In this paper, we discuss how to recover the RSA secret key from a noisy version of the secret key obtained through physical attacks such as cold boot and side channel attacks. For example, consider a cold boot attack to extract the RSA secret key stored in the memory. The attacker can obtain a degraded version of the secret key so that some bits are erased. In principle, if many erasures occur, the key recovery for the secret key becomes rather difficult. To date, many noise models other than the erasure model have been introduced. For the discrete noise case, the binary erasure model, binary error model, and binary erasure and error model have been introduced. Effective algorithms have been proposed for each noise model, and the conditions for noise which the original secret key can be recovered in polynomial time have been derived. Research has also been conducted on models that can obtain continuous leakage. In this case, several algorithms have been proposed according to the degree of knowledge of the leakage model. Many studies have been conducted on by taking heuristic approaches. In this paper, we provide a survey of existing research and then attempt to explain it within a unified framework.

Keywords RSA key recovery · Noisy secret key · Noise/leakage model

1 Introduction

1.1 Background and Motivation

Side channel attacks are a significant concern for security analysis in the both of public key cryptography and symmetric cryptography. In the typical scenario of the side channel attacks, attackers attempt to recover full secret keys when they can measure some leaked information from cryptographic devices. Following the

---

N. Kunihiro (✉)
School of Frontier Sciences, University of Tokyo, 5-1-5 Kashiwanoha, Kashiwa-shi, Chiba 277-8561, Japan
e-mail: kunihiro@k.u-tokyo.ac.jp

T. Takagi et al. (eds.), *Mathematical Modelling for Next-Generation Cryptography*, Mathematics for Industry 29, DOI 10.1007/978-981-10-5065-7_11

proposal of *differential power analysis (DPA)* by Kocher et al. [10], there have been many extensive studies of side channel attacks.

We mainly focus on side channel attacks on the RSA cryptosystem. In the RSA cryptosystem [19], a public modulus N is chosen as the product of two distinct primes p and q. The key-pair $(e, d) \in \mathbb{Z}^2$ satisfies $ed \equiv 1 \pmod{(p-1)(q-1)}$. The encryption keys are (N, e) and the decryption keys are (N, d). The PKCS#1 standard [18] specifies that the RSA secret key includes $(p, q, d, d_p, d_q, q^{-1} \bmod p)$ in addition to d, which allows for fast decryption using the Chinese Remainder Theorem.

1.2 Review of Existing Works

We briefly review previous works for recovering the RSA secret key from its noisy version. The cold boot attack was proposed by Halderman et al. [7] at USENIX Security 2008. They demonstrated that DRAM remanence effects make possible practical, nondestructive attacks that recover a noisy version of secret keys stored in a computer's memory. They showed how to reconstruct the full secret key from the noisy version for some encryption schemes: DES, AES, tweakable encryption modes, and RSA.

Inspired by cold boot attacks [7], much research has been conducted on recovering an RSA secret key from a noisy version of the secret key. At Crypto 2009, Heninger and Shacham [9] proposed an algorithm that efficiently recovers secret keys (p, q, d, d_p, d_q) given a random fraction of their bits. Specifically, they showed that if more than 27% of the secret key bits are leaked at random, the full secret key can be recovered. Conversely, this means that even if 73% of the correct secret bits are erased, the key can be recovered. In contrast with the Heninger–Shacham algorithm for correcting erasures, Henecka et al. [8] proposed an algorithm for correcting symmetric error bits of secret keys at Crypto 2010. They showed that the secret key (p, q, d, d_p, d_q) can be fully recovered if the probability of bit-flip is less than 0.237. Paterson et al. proposed an algorithm to correct error bits that occur asymmetrically at Asiacrypt 2012 [16]. They adopted a coding theoretical approach for designing a new algorithm and analyzing its performance. Kunihiro et al. [14] proposed an algorithm that generalized the work of [8, 9], and considered a combined erasure and error setting. Kunihiro [11] improved the bound obtained in [14, 16] by using tighter inequalities than Hoeffding bounds.

The works explained thus far have considered the discrete leakage. Kunihiro and Honda introduced analog leakage of RSA secret key bits [12]. They proposed two algorithms: the ML-based algorithm and DPA-like algorithm for key recovery with analog leakage and obtained their success conditions. Note that ML stands for maximum likelihood and DPA stands for differential power analysis. Kunihiro and Takahashi [15] improved the results of [12]. They proposed another two algorithms: one is effective when the estimation of true leakage distribution is known and the other works when the variances of leakage are known.

In this paper, we explain these results using the same framework. In our framework, we maintain the bit information of candidates for the secret key using a binary tree and search for a correct key to perform pruning using a score function based on likelihood. Note that the score function is constructed from the noise model and leakage model. We derive the success conditions for recovering the correct secret key.

2 Preliminaries

2.1 Notation

This section presents an overview of the methods [8, 9, 16] using binary trees to recover the secret key of the RSA cryptosystem. We use similar notation to that in [8]. For an n-bit sequence $\mathbf{x} = (x_{n-1}, \ldots, x_0) \in \{0, 1\}^n$, we denote the i-th bit of $\mathbf{x}$ by $x[i] = x_i$, where $x[0]$ is the least significant bit of $\mathbf{x}$. We denote by $\ln Z$ the natural logarithm of positive number Z to the base e and by $\log Z$ the logarithm of Z to the base 2. We denote the expectation of random variable X by $\mathrm{E}[X]$ and the variance of X by $\mathrm{Var}[X]$.

2.2 Noise/Leakage Model

In this section, we review the noise/leakage model discussed in this paper. Suppose that attackers attempt to physically reveal secret keys stored in some storage. If the attackers succeed in obtaining the correct values of the stored secret keys, they already have the secret keys and need to do no more. However, in a practical attack situation, the attackers cannot obtain the correct secret keys, but will obtain their noisy variants. We introduce several noise models. In this paper, we identify the noise/leakage model for the side channel attack with a noisy communication channel. The noisy channel can be roughly categorized into two types: the discrete noise channel and the analog noise channel. All the existing works consider memoryless channels in which the output probability distribution only depends on the current channel input.

The discrete noise channel can be further divided into four classes:

1. binary erasure channel (BEC)
2. binary symmetric channel (BSC)
3. binary asymmetric channel (BASC)
4. erasure and error channel (BEEC)

We will explain them in more detail.

In the BEC model, the original key bit is erased with some probability, and the attacker cannot know whether it was originally 0 or 1. Denoting the erasure symbol by ?, the received symbol set is given by {0, 1, ?}. The property of this channel is characterized by

$$\Pr(0 \to ?) = \delta_0,\ \Pr(0 \to 0) = 1 - \delta_0,\ \Pr(0 \to 1) = 0$$

and

$$\Pr(1 \to ?) = \delta_1,\ \Pr(1 \to 1) = 1 - \delta_1,\ \Pr(1 \to 0) = 0$$

for nonnegative δ_0 and δ_1.

We next explain the channel with bit-flip, which is used for the BSC and BASC models. The received symbol set is given by {0, 1}. The property of this channel is characterized by

$$\Pr(0 \to 1) = \varepsilon_0,\ \Pr(0 \to 0) = 1 - \varepsilon_0$$

and

$$\Pr(1 \to 0) = \varepsilon_1,\ \Pr(1 \to 1) = 1 - \varepsilon_1.$$

This channel is categorized into two subclasses. The binary channel with the same crossover probability ($\varepsilon_0 = \varepsilon_1$) is called a BSC. The other class of binary channel ($\varepsilon_0 \neq \varepsilon_1$) is called a BASC. Consider the BASC with $\varepsilon_0 = 0$ as a special case. For this channel, a crossover $1 \to 0$ occurs with positive probability, whereas the crossover $0 \to 1$ never occurs. This particular channel is especially called a Z-channel.

We next discuss the combined channel of erasure and bit-flip, called the BEEC. The received symbol set is given by {0, 1, ?} and this channel is characterized by

$$\Pr(0 \to ?) = \delta_0,\ \Pr(0 \to 1) = \varepsilon_0,\ \Pr(0 \to 0) = 1 - \delta_0 - \varepsilon_0$$

and

$$\Pr(1 \to ?) = \delta_1,\ \Pr(1 \to 0) = \varepsilon_1,\ \Pr(1 \to 1) = 1 - \delta_1 - \varepsilon_1.$$

We then consider on the analog leakage model. Let F_0 and F_1 be probability functions of an observed value when the correct secret key bits are 0 and 1, respectively. This means that we assume that each observed value x follows the fixed probability function F_b when the correct secret key bit is $b \in \{0, 1\}$. In what follows, we assume that F_0 and F_1 have probability density functions f_0 and f_1, respectively. Without loss of generality, we assume that the mean of F_b is $(-1)^b$. We say that the probability density functions f_0 and f_1 are imbalanced when $f_0(x)$ and $f_1(-x)$ are very different. Suppose that $f_0 = \mathcal{N}(+1, \sigma_0^2)$, and $f_1 = \mathcal{N}(-1, \sigma_1^2)$. We say that f_0 and f_1 are imbalanced when $\sigma_0 \ll \sigma_1$. We say that the distributions are symmetric when $f_1(y) = f_0(-y)$. A typical example of the symmetric distribution is *symmetric additive noise*: the sample can be written as the sum of a deterministic part and symmetric random noise part.

If the noise is too large, it is impossible to recover the original secret key from that of the noisy variant. For example, consider the case when $\delta_0 = \delta_1 = 1$ in the BEC model and the case when $\varepsilon_0 = \varepsilon_1 = 1/2$ in the BSC model. In this case, because all information of the original secret key is lost, it is impossible to recover the original key bits. Conversely, when if there is no noise, or very little noise in the observed secret bit sequence, we can recover the secret key.

2.3 Useful Functions from Information Theory

We first introduce the binary entropy and Kullback–Leibler divergence for a binary information source.

Definition 2.1 (*Binary Entropy* [4]) The binary entropy function $H(x)$ is defined by $H(x) := -x \log x - (1-x)\log(1-x)$.

Definition 2.2 (*Kullback–Leibler Divergence* [4]) Consider two distributions P and Q on $\{0, 1\}$. Let $P(0) = 1-p$, $P(1) = p$, $Q(0) = 1-q$, $Q(1) = q$ for some probabilities p and q. The Kullback–Leibler divergence between P and Q is defined by

$$D(P||Q) := p\log\frac{p}{q} + (1-p)\log\frac{1-p}{1-q}.$$

For simplicity, we use notation $D(p, q)$ instead of $D(P||Q)$.

It follows by Definition 2.2 that the Kullback–Leibler divergence is always non-negative and $D(p, q) = 0$ if and only if $p = q$.

We denote a uniform distribution on $\{0, 1\}$ by U, which implies $U(0) = U(1) = 1/2$. It directly follows from Definition 2.2 that we have

$$D(P||U) = D(p, 1/2) = p\log(2p) + (1-p)\log 2(1-p) = 1 - H(p).$$

We then introduce the entropy and Kullback–Leibler divergence of a continuous probability density function.

Definition 2.3 The differential entropy $h(f)$ of a probability density function f is defined as

$$h(f) = -\int_{-\infty}^{\infty} f(y)\log f(y)\mathrm{d}y.$$

We then introduce the Kullback–Leibler divergence [4].

Definition 2.4 The Kullback–Leibler divergence $D(p||q)$ of probability density functions p and q is defined as

$$D(p||q) = \int_{-\infty}^{\infty} p(y)\log\frac{p(y)}{q(y)}\mathrm{d}y.$$

It is well known that the Kullback–Leibler divergence $D(p||q)$ is nonnegative, and it is zero if and only if $p = q$. It is considered as some type of the distance between p and q.

3 Construction of a Binary Tree from the RSA Public Key

In this section, we review how to recover the RSA secret key using a binary tree. Our explanation is almost the same as previous works (e.g., [8, 9]). We review the key setting of the RSA cryptosystem [19], particular for the PKCS #1 standard [18]. Let (N, e) be the RSA public key, where N is an n-bit RSA modulus, and $\mathbf{sk} = (p, q, d, d_p, d_q, q^{-1} \bmod p)$ be the RSA secret key. We assume that p and q are of the same bit length.

As in all existing works, we ignore the last component $q^{-1} \bmod p$ in the secret key. The public and secret keys have the following four equations:

$$N = pq,\ ed \equiv 1 \pmod{(p-1)(q-1)},\ ed_p \equiv 1 \pmod{p-1},\ ed_q \equiv 1 \pmod{q-1}.$$

There exist integers k, k_p and k_q such that

$$N = pq,\ ed = 1 + k(p-1)(q-1),\ ed_p = 1 + k_p(p-1),\ ed_q = 1 + k_q(q-1). \quad (1)$$

There are five unknowns (p, q, d, d_p, d_q) in the four equations in Eq. (1) because we can compute the exact values of k, k_p and k_q, as will be discussed later.

A small public exponent e is usually used in practical applications [23]. Thus, we suppose that e is sufficiently small such that $e = 2^{16} + 1$, as is the case in [8, 9]. Note that if e is sufficiently large, none of the algorithms presented in this paper work in polynomial time. We first show how to compute k. As is easily verified, we have $0 < k < e$ because $0 < k < e\frac{d}{\phi(n)} < e$, where $\phi(n) = (p-1)(q-1)$. Hence, there exist at most $e - 1$ possible candidates for k. Hence, when e is sufficiently small (say, $e = 2^{16} + 1$), we can perform an exhaustive search over all candidate values for k. For each candidate k', we set

$$\tilde{d}(k') = \lfloor \frac{1 + k'(N+1)}{e} \rfloor.$$

As Boneh, Durfee, and Frankle [3] showed,

$$0 \le \tilde{d}(k') - d \le \frac{k(p+q)}{e} < p + q < 3\sqrt{N}$$

holds if $k' = k$. Then, for the correct choice of k, the values of d and $d(k)$ are almost the same for half their most significant bits. Hence, we can determine the correct k by checking whether $\tilde{d}(k') - d < 3\sqrt{N}$ holds. Furthermore, once the correct value of k is determined, we can compute k_p and k_q. For details of how to compute k_p, k_q, see [8, 9].

In all the methods presented in this paper, a secret key **sk** is recovered using a binary tree-based technique. We explain how to recover secret keys, considering $\mathbf{sk} = (p, q, d, d_p, d_q)$ as an example.

First, we discuss the generation of the tree. Because p and q are $n/2$-bit prime numbers, there exist at most $2^{n/2}$ candidates of secret key (p, q, d, d_p, d_q). Heninger and Shacham [9] introduced the concept of the *slice*. Let $\tau(M)$ denote the largest exponent such that $2^{\tau(M)} | M$ for some positive integer M. We define the i-th bit slice for each bit index i as

$$\mathbf{slice}(i) := (p[i], q[i], d[i + \tau(k)], d_p[i + \tau(k_p)], d_q[i + \tau(k_q)]).$$

Assume that we have computed a partial solution $\mathbf{sk}' = (p', q', d', d'_p, d'_q)$ up to $\mathbf{slice}(i - 1)$. Heninger and Shacham [9] applied Hensel's lemma to Eq. (1) and obtained the following equations:

$$\begin{aligned}
p[i] + q[i] &= (N - p'q')[i] \bmod 2,\\
d[i + \tau(k)] + p[i] + q[i] &= (k(N + 1) + 1 - k(p' + q') - ed')[i + \tau(k)] \bmod 2,\\
d_p[i + \tau(k_p)] + p[i] &= (k_p(p' - 1) + 1 - ed'_p)[i + \tau(k_p)] \bmod 2,\\
d_q[i + \tau(k_q)] + q[i] &= (k_q(q' - 1) + 1 - ed'_q)[i + \tau(k_q)] \bmod 2.
\end{aligned}$$

We can easily see that $p[i], q[i], d[i + \tau(k)], d_p[i + \tau(k_p)]$, and $d_q[i + \tau(k_q)]$ are not independent. Each Hensel lift, therefore, yields exactly two candidate solutions. Thus, the total number of candidates is given by at most $2^{n/2}$.

4 General Framework for Key Recovery and Intuitive Understanding

We present a general framework for a key recovery algorithm from its noisy variant and provide an intuitive understanding of how it works. By following the previous section, we can construct a set of the secret key candidates only from the public key (N, e). We introduce the distance between each candidate and noisy variant of the secret key, and take a strategy to consider a candidate with a minimum distance to noisy secret key as the correct secret key. We briefly digress by presenting a general problem from the RSA key recovery. Let $k \in \{0, 1\}^n$ be the secret key to be recovered. The following is our framework for key recovery:

(1) Key Expansion The secret key k is expanded by some injective function $f : \{0, 1\}^n \to \{0, 1\}^{n'}$, where $n < n'$. Suppose that f is invertible from the element in $f(\{0, 1\}^n)$. The key expansion function f depends on the target scheme and

its public key. Denote the expanded key by $K \in f(\{0,1\}^n) \subset \{0,1\}^{n'}$. Hence, the key expansion function f can be written as

$$f : k \in \{0,1\}^n \mapsto K \in f(\{0,1\}^n) \subset \{0,1\}^{n'}.$$

(2) Observation The observed sequence is obtained through the noise model. Let the output alphabet through some physical observation be $\mathscr{A}$. The observed data $\hat{K} \in \mathscr{A}^{n'}$ is obtained through the noise model from K. The attack goal is to recover K, and finally k, from $\hat{K}$.

(3) Score Function A distance function D between $\hat{K}$ and K is designed using a priori knowledge of the noise model and its estimation. The distance function D is written as

$$D : \{0,1\}^{n'} \times \mathscr{A}^{n'} \to \mathbb{R}$$

In what follows, we call a score function instead of distance.

(4) Key Recovery The attack is executed by determining $K \in f(\{0,1\}^n)$ that minimizes $D(K, \hat{K})$ given $\hat{K}$. Equivalently, we can say that it is executed by determining $k \in \{0,1\}^n$ that minimizes $D(f(k), \hat{K})$ given $\hat{K}$. To summarize, in this framework, the attack goal is to determine the argument

$$\arg\min_{k \in \{0,1\}^n} D(f(k), \hat{K}).$$

The expansion function for the RSA scheme was provided in Sect. 3. For example, once p is given, (p, q, d, d_p, d_q) is determined. Note that p corresponds to secret key k in the framework and (p, q, d, d_p, d_q) corresponds to expanded key K. This type of attack succeeds because of the redundancy of the expanded key.

The design of the score function is important in this study. It should be designed based on the noise/leakage model. We will discuss how to design the score function in Sect. 6. In the definition of the score function, its relative scale is more important than its absolute value. We can use a simpler description of a more easily computable score function provided the order of returned values is maintained.

Consider the complexity of the key recovery step. The most straightforward approach for finding k that minimizes the score function is an exhaustive search for all candidates $k \in \{0,1\}^n$. Its complexity is evaluated as 2^n, which is not the polynomial of n. If f has a special structure, it is possible to reduce the complexity of the key recovery step. The expansion function for the RSA scheme has such a special property; it can be represented by a binary tree, as explained in Sect. 3.

In this problem setting, we can verify whether the candidate of the secret key is correct by checking whether $N = \tilde{p}\tilde{q}$ holds, where $(\tilde{p}, \tilde{q})$ is a secret key candidate. Hence, the key recovery step does not necessarily output a single candidate but multiple candidates.

5 Practical Algorithms: Branch-and-Bound Approach

5.1 Expansion and Pruning

In this section, we restrict the discussion to key recovery of the RSA scheme. The candidates of the secret key are maintained using the binary tree, which is enabled to do so as discussed in Sect. 3.

As discussed in Sect. 4, it is impossible in the computational sense to manipulate all the candidates simultaneously. Hence, we use a so-called branch-and-bound approach. First, some leaves are generated from each surviving leaf. Moreover, some leaves are pruned according to a prefixed rule. To do this, we introduce a *truncated* score function, which is a truncated variant of the score function and corresponds to the score from the root to the current leaf.

This procedure is iterated using prefixed numbers. There exist two rules for pruning leaves in existing works.

- A leaf with the score less than a threshold C is pruned.
- A leaf with the rank by the score function larger than L is pruned.

It does not make a large difference which of pruning rule is adopted. For the first rule, the judgment of whether a left survives or is pruned can be executed spontaneously. By contrast, it is hard to estimate the maximal size of memory used in the key recovery process. For the second rule, the storage for keeping the node information is guaranteed as L, and then it is easy to manage memory. By contrast, we must wait until all the score values of the leaf are calculated. For the both rules, thresholds C and L need to be set adequately.

Let us focus on the first rule. Suppose that there exist M leaves at some step. By expanding a tree by t times, the number of leaves is $M2^t$ since it doubles for one expansion. A leaf whose score is less than the prefixed threshold C is pruned. The value C needs to be determined so that the average of the surviving leaves is M or that of the final step falls in a polynomial of input size.

For the second rule, by expanding a tree by t bits, the number of leaves is $L2^t$. In the rule, a leaf whose rank of the score is larger than the prefixed threshold L is pruned. If we use a small L, we can conduct the key recovery with a small amount of memory. However, if L is too small, the probability that the correct candidate is pruned becomes higher. Hence, in that case, we use the available large L to reduce that probability.

6 Implication

In this section, we present the individual analysis of each noise model. We focus on the discussion of how to set the score function, and the success condition when the score function is used. We first review the syntax of the score functions.

Definition 6.1 (*Syntax of the Score Function*) The score function receives a candidate sequence $\boldsymbol{b} = (b_1, \dots, b_n) \in \{0, 1\}^n$ and corresponding observed sequence $\boldsymbol{x} = (x_1, \dots, x_n) \in \mathbb{R}^n$ and outputs a real number. We use the following notation: $\mathbf{Score}(\boldsymbol{b}; \boldsymbol{x})$.

We denote by m the number of associated secret keys. For example, $m = 5$ if $\mathbf{sk} = (p, q, d, d_p, d_q)$, $m = 3$ if $\mathbf{sk} = (p, q, d)$, and $m = 2$ if $\mathbf{sk} = (p, q)$. This m corresponds to the expansion rate of secret key.

6.1 Analyses of Discrete Noise

We first discuss a common strategy for designing the score function. All the score functions are based on the *conditional probability*. We denote a candidate sequence by $\boldsymbol{b}$ and an observed sequence by $\boldsymbol{x}$, which corresponds to the noisy variant of the RSA secret key. We further denote the probability that an original sequence is $\boldsymbol{b}$ under the condition that $\boldsymbol{x}$ is observed by $\Pr(\boldsymbol{b}|\boldsymbol{x})$. Denoting the set of candidates $\{\boldsymbol{b}^{(1)}, \dots, \boldsymbol{b}^{(L)}\}$, we compute the argument

$$\arg\max_{1 \le i \le L} \Pr(\boldsymbol{b}^{(i)}|\boldsymbol{x}) \tag{2}$$

and consider that it is the correct secret key.

Using Bayes' theorem, this conditional probability can be rewritten as

$$\Pr(\boldsymbol{b}|\boldsymbol{x}) = \frac{\Pr(\boldsymbol{x}|\boldsymbol{b})\Pr(\boldsymbol{b})}{\Pr(\boldsymbol{x})}.$$

$\Pr(\boldsymbol{x})$ does not depend on the choice of $\boldsymbol{b}$ and can be ignored. Furthermore, $\Pr(\boldsymbol{b})$ can be also ignored by assuming that $\boldsymbol{b}$ is constant in the valid candidate set. We finally set $\Pr(\boldsymbol{x}|\boldsymbol{b})$ as a score function.

We denote the number of positions where $\boldsymbol{b}$ has an i and $\boldsymbol{x}$ has a j by n_{ij} for $i \in \{0, 1\}$ and $j \in \{0, 1, ?\}$ in the noise/leakage model discussed in Sect. 2. The conditional probability $\Pr(\boldsymbol{x}|\boldsymbol{b})$ is explicitly given by

$$\Pr(\boldsymbol{x}|\boldsymbol{b}) = (1 - \varepsilon_0 - \delta_0)^{n_{00}} \varepsilon_0^{n_{01}} \varepsilon_1^{n_{10}} (1 - \varepsilon_1 - \delta_1)^{n_{11}} \delta_0^{n_{0?}} \delta_1^{n_{1?}}.$$

Equivalently, we may use the log of the probability,

$$\begin{aligned}\log \Pr(\boldsymbol{x}|\boldsymbol{b}) =& n_{00} \log(1 - \varepsilon_0 - \delta_0) + n_{01} \log \varepsilon_0 + n_{10} \log \varepsilon_1 + n_{11} \log(1 - \varepsilon_1 - \delta_1) \\ &+ n_{0?} \log \delta_0 + n_{1?} \log \delta_1\end{aligned} \tag{3}$$

as a score function. We simplify the score function in the accordance with the noise/leakage model. In the following discussion, an expression of the form $x \log y$ is considered to be equal to zero whenever $x = y = 0$.

6.1.1 Erasure Channel Model Case

We first examine the erasure channel model case. In the pruning step, we use the property that no error occurs in the model, that is, the candidate sequence with a contradiction to the observed sequence is discarded. Let us return to the score function. By substituting $\varepsilon_0 = \varepsilon_1 = 0$ into Eq. (3), we obtain

$$\log \Pr(\boldsymbol{x}|\boldsymbol{b}) = (n_{01} + n_{10}) \log 0 + C$$

for some real value C. When $n_{01} = n_{10} = 0$ holds, $\log \Pr(\boldsymbol{x}|\boldsymbol{b})$ is within the finite value. By contrast, when $n_{01} \neq 0$ or $n_{10} \neq 0$ hold, it is an infinitely small value. In this case, the candidate sequence with an infinitely small score is pruned.

Assume $\delta_0 = \delta_1 (:= \delta)$. The success condition for key recovery is given by

$$\delta < 2^{\frac{m-1}{m}} - 1,$$

where m is a For $m = 5, m = 3$, and $m = 2$, when $\delta < 0.74$, $\delta < 0.58$, and $\delta < 0.41$, key recovery is successful.

6.1.2 Binary Symmetric Channel Case

Next, we examine the BSC case. Substituting $\delta_0 = \delta_1 = 0$, $\varepsilon_0 = \varepsilon_1 = \varepsilon$ into Eq. (3), we obtain

$$\log \Pr(\boldsymbol{x}|\boldsymbol{b}) = n_{00} \log(1-\varepsilon) + n_{01} \log \varepsilon + n_{10} \log \varepsilon + n_{11} \log(1-\varepsilon).$$

We denote the length of the sequence by n and the Hamming distance between $\boldsymbol{x}$ and $\boldsymbol{b}$ by $\mathrm{Hw}(\boldsymbol{x}, \boldsymbol{b})$, respectively. Because $n_{10} + n_{01} = \mathrm{Hw}(\boldsymbol{x}, \boldsymbol{b})$ and $n_{00} + n_{11} = n - \mathrm{Hw}(\boldsymbol{x}, \boldsymbol{b})$, we have

$$\begin{aligned} \log \Pr(\boldsymbol{x}|\boldsymbol{b}) &= (n - \mathrm{Hw}(\boldsymbol{x}, \boldsymbol{b})) \log(1-\varepsilon) + \mathrm{Hw}(\boldsymbol{x}, \boldsymbol{b}) \log \varepsilon \\ &= n \log(1-\varepsilon) - \mathrm{Hw}(\boldsymbol{x}, \boldsymbol{b}) \log \frac{1-\varepsilon}{\varepsilon}. \end{aligned}$$

Note that $\log \frac{1-\varepsilon}{\varepsilon} > 0$ when $\varepsilon < 1/2$. The score depends on $\mathrm{Hw}(\boldsymbol{x}, \boldsymbol{b})$. Then, the score function can be replaced by the Hamming distance. The candidate sequence with a small Hamming distance survives the pruning phase. Henecka et al. [8] showed that key recovery succeeds when

$$\varepsilon < \frac{1}{2} - \sqrt{\frac{\ln 2}{2m}}, \tag{4}$$

where m is the number of associated secret keys. For $m = 5$, the condition is given by $\varepsilon < 0.237$. Paterson et al. [16] improved the bound to

$$\varepsilon < 0.243.$$

6.1.3 Binary Asymmetric Error Case

Paterson et al. provided an analysis of the binary asymmetric error case [16]. They use an error correcting technique for their analysis. Substituting $\delta_0 = \delta_1 = 0$ into Eq. (3), we obtain

$$\log \Pr(\boldsymbol{x}|\boldsymbol{b}) = n_{00} \log(1 - \varepsilon_0) + n_{01} \log \varepsilon_0 + n_{10} \log \varepsilon_1 + n_{11} \log(1 - \varepsilon_1)$$

for the score function. Paterson et al. claimed that when $\varepsilon_0 = 0$ and

$$\log(1 + (1 - \varepsilon_1)\varepsilon_1^{\varepsilon_1/(1-\varepsilon_1)}) > \frac{1}{m}, \tag{5}$$

key recovery succeeds. They obtained the condition by calculating the channel capacity for the BASC.

Kunihiro [11] indicated a small flaw in the analysis in [16] and obtained the precise condition. He claimed that a symmetric channel capacity is critical instead of the usual channel capacity. He then obtained the following condition:

$$H\left(\frac{1}{2} + \frac{\varepsilon_1 - \varepsilon_0}{2}\right) - \frac{H(\varepsilon_0)}{2} - \frac{H(\varepsilon_1)}{2} > \frac{1}{m}, \tag{6}$$

where H is the binary entropy defined in Definition 2.1. In the special case of $\varepsilon_0 = 0$, the condition can be rewritten as

$$H\left(\frac{1}{2} + \frac{\varepsilon_1}{2}\right) - \frac{H(\varepsilon_1)}{2} > \frac{1}{m}. \tag{7}$$

The conditions in Eqs. (5) and (7) are almost the same. For $m = 5$, the conditions are given by $\varepsilon_1 < 0.304$ in Eq. (5) and 0.294 in Eq. (7).

6.1.4 Error and Erasure Case

Kunihiro et al. [14] examined the error and erasure case. They showed that the success condition for key recovery is given by

$$\varepsilon + \frac{\delta}{2} \le \frac{1}{2} - \sqrt{\frac{(1 - \delta) \ln 2}{2m}}. \tag{8}$$

Kunihiro [11] improved the bound to

$$(1-\delta)\left(1-H\left(\frac{\varepsilon}{1-\delta}\right)\right) \geq \frac{1}{m} \tag{9}$$

by changing the setting of the threshold.

Consider the case of 2048-bit RSA and $\delta = 0.5$. The algorithm proposed in [14] recovers the original secret key from the noisy key bits with up to $328(= 0.064 \times 1024 \times 5)$ errors. By contrast, the algorithm proposed in [11] recovers the original secret key from the noisy key bits with up to $374(= 0.073 \times 1024 \times 5)$ errors.

Kunihiro et al. [14] and Kunihiro [11] presented the relation between Eqs. (8) and (9). The binary entropy function H can be represented by the following sum of infinite series:

$$H(x) = 1 - \frac{1}{\ln 2}\sum_{u=1}^{\infty}\frac{(1-2x)^{2u}}{2u(2u-1)}.$$

Equation (8) corresponds to the second-order truncation of Eq. (9).

6.2 Analyses of Analog Leakage

The key recovery algorithm for analog leakage was studied in [12, 15]. Their algorithms are categorized into the same framework as the discrete noise case. We refer to the probability density function f_b as leakage distribution throughout this subsection. Hereafter, we focus the discussion on the design of the score function. We consider the four cases according to knowledge of leakage distributions.

1. The leakage distributions f_0 and f_1 are exactly known.
2. The estimated distributions $f_0^{(\mathrm{E})}$ and $f_1^{(\mathrm{E})}$ of true leakage are known.
3. The variances $\mathrm{Var}[f_0]$ and $\mathrm{Var}[f_1]$ of leakage distributions are known.
4. No information about the leakage distributions is known.

Cases 1 and 4 were studied in [12] and cases 2 and 3 were studied in [15].

Let R be a score function. Note that its definition is given in Definition 6.1. Let $B \in \{0, 1\}$ be a random variable uniformly distributed over $\{0, 1\}$. We define $X \in \mathbb{R}$ as a random variable that follows probability function F_B given B.

The success condition is given by

$$\mathrm{E}[R(B; X)] > \frac{1}{m}. \tag{10}$$

This implies that once the score function is determined, we can calculate the success condition for key recovery when the score function is used. Note that $\mathrm{E}[R(B; X)]$ is called the mutual information between B and X.

6.2.1 Case 1: The Noise Distributions f_0 and f_1 are Exactly Known

In [12], the score function is given as

$$\mathrm{ML}(\boldsymbol{b}; \boldsymbol{x}) := \sum_{i=1}^{n} \log \frac{f_{b_i}(x_i)}{g(x_i)}, \tag{11}$$

where $g(x) = (f_0(x) + f_1(x))/2$.

The success condition when using this score function is given by

$$h\left(\frac{f_0 + f_1}{2}\right) - \frac{h(f_0) + h(f_1)}{2} > \frac{1}{m} \tag{12}$$

from Eq. (10).

Consider the symmetric leakage case ($f_1(x) = f_0(-x)$). Because $h(f_1) = h(f_0)$, this condition can be simply rewritten as

$$h(g) - h(f_0) > \frac{1}{m}. \tag{13}$$

6.2.2 Case 2: The Estimated Distributions $f_0^{(\mathrm{E})}$ and $f_1^{(\mathrm{E})}$ are known

In actual attack situations, the attacker does not know the exact form of f_b. Thus, we cannot apply the ML-based score from Eq. (11) directly. By contrast, if we obtain a closer estimation of the leakage distributions, we can attain the key recovery from larger noise. The second best strategy is then (i) to estimate f_b in some way and (ii) to run the key recovery algorithm with the score function designed by the estimated probability density functions. We denote the estimated distributions of f_0 and f_1 by $f_0^{(\mathrm{E})}$ and $f_1^{(\mathrm{E})}$, respectively. In [15], the score function is set as

$$\text{e-ML}(\boldsymbol{b}; \boldsymbol{x}) := \sum_{i=1}^{n} \log f_{b_i}^{(\mathrm{E})}(x_i). \tag{14}$$

The success condition is given by

$$\left(h\left(\frac{f_0 + f_1}{2}\right) - \frac{h(f_0) + h(f_1)}{2}\right) - \frac{D(f_0 || f_0^{(\mathrm{E})}) + D(f_1 || f_1^{(\mathrm{E})})}{2} > \frac{1}{m}, \tag{15}$$

where D is the Kullback–Leibler divergence defined in Definition 2.4.

On the former half of the left hand side in Eq. (15), $h((f_0 + f_1)/2) - (h(f_0) + h(f_1))/2$ is equivalent to the condition when the true distributions are known, and $(D(f_0 || f_0^{(\mathrm{E})}) + D(f_1 || f_1^{(\mathrm{E})}))/2$ corresponds to the information loss or penalty caused by misestimations. From the property of Kullback–Leibler divergence, it is always nonnegative. If the probability density function is correctly estimated, the information loss vanishes because $D(f_0||f_0^{(\mathrm{E})}) = D(f_1||f_1^{(\mathrm{E})}) = 0$. Conversely, if the accurate estimation fails, the success condition is much worse than expected because of the information loss caused by the misestimation of f_0 and f_1.

6.2.3 Case 4: No Information About the Noise Distributions Is Known

Kunihiro and Honda [12] studied the case where no information about the noise distributions is known. They introduced the following DPA-like score:

$$\mathrm{DPA}(\boldsymbol{b}; \boldsymbol{x}) := \sum_{i=1}^{n} (-1)^{b_i} x_i. \tag{16}$$

Note that this score can be calculated without knowledge of the specific form of the noise distribution. They indicated that this function is similar to the DPA selection function used in Differential Power Analysis [10].

The success condition for key recovery is given by

$$h\left(\frac{f_0 + f_1}{2}\right) - \log\sqrt{\pi \mathrm{e}(\sigma_0^2 + \sigma_1^2)} > \frac{1}{m}. \tag{17}$$

6.2.4 Case 3: The Variances Var[f_0] and Var[f_1] of the Noise Distributions Are Known

The DPA-like algorithm works with only observed data, even if the probability density functions are not known. From the nature of the DPA-like score, it cannot use any other side information of the probability density function such as variances, even if it is available. Kunihiro and Takahashi [15] introduced the weighted variant of the DPA-like score function. Letting $\sigma_0^2 = \mathrm{Var}[f_0]$ and $\sigma_1^2 = \mathrm{Var}[f_1]$, they proposed

$$\mathrm{V}(\boldsymbol{b}; \boldsymbol{x}) := \sum_{i=1}^{n} \frac{(-1)^{b_i} x_i}{\sigma_{b_i}^2} \tag{18}$$

as a score function. The success condition for key recovery is given by

$$h\left(\frac{f_0 + f_1}{2}\right) - \log\sqrt{2\pi \mathrm{e}\sigma_0\sigma_1} > \frac{1}{m}. \tag{19}$$

We provide a comparison between the DPA-like algorithm and V-based algorithm. The difference between the left hand side of the two inequalities: Eqs. (17) and (19) is given by

$$\log \sqrt{\pi e(\sigma_0^2 + \sigma_1^2)} - \log \sqrt{2\pi e \sigma_0 \sigma_1} = \frac{1}{2} \log \frac{\sigma_0^2 + \sigma_1^2}{2\sigma_0\sigma_1} \geq 0.$$

Then, the difference is always nonnegative. This shows that the V-based algorithm is superior to the DPA-like algorithm, except for the case that $\sigma_0 = \sigma_1$.

The score function $V(\boldsymbol{b}; \boldsymbol{x})$ requires the following additional inputs: variances σ_0^2 and σ_1^2. This is a significant disadvantage compared with the DPA-like algorithm. To solve this problem, in [15], the authors used the help of the expectation–maximization (EM) algorithm [2, 6] to estimate the variances from the observed data as a preprocessing for the V-based algorithm. This enables us to recover of the secret key using only the observed data, as well as the DPA-like algorithm.

The V-based score can be regarded as a weighted variant of the DPA-like score. It is natural to consider the weighted variant of the DPA-like score, which is defined by

$$\text{w-DPA}(\boldsymbol{b}; \boldsymbol{x}) := \sum_{i=1}^{n} \frac{(-1)^{b_i} x_i}{w_{b_i}} \tag{20}$$

for some type of weights w_0 and w_1. In [15], the authors showed that the V-based score ($w_0 = \mathrm{Var}[f_0]$ and $w_1 = \mathrm{Var}[f_1]$) is optimal in the weighted variant of the DPA-like score.

Both the DPA-like score and V-based score are obtained by adopting Gaussian distributions as the estimation. The DPA-like score is obtained by setting f_0, f_1 as Gaussian distributions with the same variance, whereas the V-based score is obtained by setting f_0 and f_1 as Gaussian distributions with the variance $\mathrm{Var}[f_0]$ and $\mathrm{Var}[f_1]$.

6.2.5 Extension to Discrete Noise

We have assumed that probability function F_b has probability density function f_b to enhance readability. For general cases, we can prove the theorem in the same manner by replacing the likelihood ratio $f_b(y)/g(y)$ with the Radon–Nikodym derivative $\mathrm{d}F_b/\mathrm{d}G$, and the integral $f_b(y)\mathrm{d}y$ with the Lebesgue–Stieltjes integral $\mathrm{d}F_b(y)$. For details, see [12].

7 Application to the Other Schemes

In the discussion of the key recovery of RSA, the expansion phase can be regarded as a key expansion from p to (p, q, d, d_p, d_q). This type of idea is applicable to attacks for the other encryption schemes. In this section, we present such examples.

Application to low-weight CRT-RSA At CHES2012, Sarkar and Maitra showed the cryptanalysis for CRT-RSA low-weight decryption exponents [20]. They presented some modifications that can improve the performance significantly.

Application to Discrete Log-based Schemes At CT-RSA2015, Poettering and Sibborn presented attacks for discrete log-based schemes under several implementations based on the window method [17]. OpenSSL and PolarSSL use the windowed NAF and a modified comb technique, respectively. In their attack scenario, Poettering and Sibborn assumed that an adversary had mounted a cold boot attack and obtained noisy versions of the key and its re-encoding (NAF or comb). They proposed an algorithm that used the redundancy in the original secret key and its encoding.

Application to Block Ciphers Several attacks have been demonstrated for symmetric encryptions, such as DES and AES [5]. The original secret key k has been expanded by key scheduling in several symmetric encryptions. Attacks such that the noisy variant of that expanded key is given have been demonstrated.
Halderman et al. developed a recovery algorithm for AES-128 that recovers keys from 30% erasure data [7]. At SAC2009, Tsow improved Halderman et al.'s attack [22]. His proposed algorithm recovers all secret keys from a 70% one-way bit-flip in less than 20 min and works at more than twice the bit-flip rate, with almost double the success rate in the same computational time.
At ACNS2011, Albrecht and Cid addressed attacks for AES, Serpent, and Twofish [1]. In their attack scenario, binary asymmetric errors occur. They modeled a new family of problems: solving systems of multivariate algebraic equations with noise, and proposed a method for solving problems from this family. They used the method for solving the problems to implement key recovery from the noisy version of the expanded key.
At ICISC2015, Tanigaki and Kunihiro presented attacks for the binary asymmetric error model by modifying Tsow's algorithm [21]. They showed the attack for AES-128, 192, and 256.

Table 1 summarizes the list of existing works.

8 Future Works and Open Problems

The approach presented in this paper is potentially applicable when secret keys have some type of redundancy. The cryptosystem with the key expansion process is the most typical approach. Finding new attacks for the other schemes is a challenging task. The usual implementation of a discrete log-based scheme has no redundancy in the secret key. However, when considering the actual implementation, the original secret and its re-encoding have some type of redundancy. As with this example, it is important to find an attack when considering an actual implementation.

Table 1 Summary of Existing Works

RSA	Discrete Noise	BEC	Halderman et al. '08 [7]
			Heninger, Shacham '09 [9]
		BSC	Henecka, May, Meurer '10 [8]
		BASC	Paterson, Polychroniadou, Sibborn '12 [16]
		BEEC	Kunihiro, Shinohara, Izu '13 [14]
			Kunihiro '15 [11]
	Analog Leakage	-	Kunihiro, Honda '14 [12]
			Kunihiro, Takahashi '17 [15]
RSA with low Hamming weight	Discrete Noise	BASC (Z)	Sarkar, Maitra '12 [20]
DL based	Discrete Noise	-	Poettering, Sibborn '15 [17]
AES	Discrete Noise	BASC (Z)	Halderman et al. '08 [7], Albrecht, Cid '11 [1]
		BASC	Tsow '09 [22] , Tanigaki, Kunihiro '15 [21]

In our analysis, we only considered memoryless noise, which may not fit to real noise. For example, the noise models described in this paper do not cover Markov source noise. It is important to propose an effective algorithm against a noise model that fits to real attack scenarios.

References

1. M. Albrecht, C. Cid, Cold Boot Key Recovery by Solving Polynomial Systems with Noise, in *Proceedings of ACNS2011*, vol. 6715 (LNCS, 2011) pp. 57–72
2. C.M. Bishop, *Pattern Recognition and Machine Learning* (Springer, Berlin, 2006)
3. D. Boneh, G. Durfee, Y. Frankel, An attack on RSA given a small fraction of the private key bits, in *Proceeding of Asiacrypt'98*, vol. 1514 (LNCS,1998), pp. 25–34
4. C.M. Cover, J.A. Thomas, *Elements of Information Theory*, 2nd edn. (Wiley-Interscience, Hoboken, 2006)
5. J. Daemen, V. Rijmen, *The Design of Rijndael* (Springer, Berlin, 2002)
6. A.P. Dempster, N.M. Laird, D.B. Rubin, Maximum likelihood from incomplete data via the EM algorithm. J. R. Stat. Soc. Ser. B **39**(1), 1–38 (1977)
7. J.A. Halderman, S.D. Schoen, N. Heninger, W. Clarkson, W. Paul, J.A. Calandrino, A.J. Feldman, J. Appelbaum, E.W. Felten, Lest we remember: cold boot attacks on encryption keys. Proc. USENIX Secur. Symp. **2008**, 45–60 (2008)
8. W. Henecka, A. May, A. Meurer, Correcting errors in RSA private keys, in *Proceedings of Crypto2010*, vol. 6223 (LNCS, 2010), pp. 351–369
9. N. Heninger, H. Shacham, Reconstructing RSA private keys from random key bits, in *Proceeding of Crypto2009*, vol. 5677 (LNCS,2009), pp. 1–17

10. P. Kocher, J. Jaffe, B. Jun, Differential power analysis, in *Proc. of CRYPTO'99*, vol. 1666 (LNCS, 1999), pp. 388–397
11. N. Kunihiro, An improved attack for recovering noisy RSA secret keys and its countermeasure, in *Proceeding of ProvSec2015*, vol. 9451 (LNCS, 2015), pp. 61–81
12. N. Kunihiro, J. Honda, RSA meets DPA: Recovering RSA secret keys from noisy analog data, in *Proceedings of CHES2014*, vol. 8731 (LNCS, 2014), pp. 261–278
13. N. Kunihiro, J. Honda, RSA meets DPA: recovering RSA secret keys from noisy analog data, in *IACR* (2014), arXiv:eprint:2014/513
14. N. Kunihiro, N. Shinohara, T. Izu, Recovering RSA Secret Keys from Noisy Key Bits with Erasures and Errors, in *Proceedings of PKC2013*, vol. 7778(LNCS, 2013), pp. 180–197
15. N. Kunihiro, Y. Takahashi Improved key recovery algorithms from noisy RSA secret keys with analog noise, in *Proceedings of CT-RSA2017*, vol. 10159 (LNCS, 2017), pp. 328–343
16. K.G. Paterson, A. Polychroniadou, D.L. Sibborn, A coding-theoretic approach to recovering noisy RSA keys, in *Proc. of Asiacrypt2012*, vol. 7658 (LNCS, 2012), pp. 386–403
17. B. Poettering, D.L. Sibborn, Cold Boot Attacks in the Discrete Logarithm Setting, in *Proceedings of CT-RSA2015*, vol. 9048 (LNCS, 2015), pp. 449–465
18. PKCS #1: RSA Cryptography Specifications Version 2.0, http://www.ietf.org/rfc/rfc2437.txt
19. R. Rivest, A. Shamir, L. Adleman, A method for obtaining digital signatures and public-key cryptosystems. Commun. ACM **21**(2), 120–126 (1978)
20. S. Sarkar, S. Maitra, Side channel attack to actual cryptanalysis: breaking CRT-RSA with low weight decryption exponents, in *Proceeding of CHES2012*, vol. 7428 (LNCS, 2012) pp. 476–493
21. T. Tanigaki, N. Kunihiro, Maximum likelihood-based key recovery algorithm from decayed key schedules, in *Proceedings of ICISC2015*, vol. 9558 (LNCS, 2015), pp. 314–328
22. A. Tsow, An improved recovery algorithm for decayed AES key schedule images, in *Proceedings of SAC2009*, vol. 5867 (LNCS, 2009), pp. 215–230
23. S. Yilek, E. Rescorla, H. Shacham, B. Enright, S. Savage, When private keys are public: results from the 2008 debian openssl vulnerability. IMC2009 (ACM Press, 2009), pp. 15–27

Part III
Lattices and Cryptography

Simple Analysis of Key Recovery Attack Against LWE

Masaya Yasuda

Abstract Recently, the learning with errors (LWE) problem has become a central building block to construct modern schemes in lattice-based cryptography. The security of such schemes relies on the hardness of the LWE problem. In particular, LWE-based cryptography has been paid attention as a candidate of post-quantum cryptography. In 2015, Laine and Lauter analyzed a key recovery attack against the search variant of the LWE problem. Their analysis is based on a generalization of the Boneh–Venkatesan method for the hidden number problem to the LWE problem. They adopted the LLL algorithm and Babai's nearest plane method in the attack, and they also demonstrated a successful range of the attack by experiments for hundreds of LWE instances. In this paper, we give a simple analysis of the attack. While Laine and Lauter's analysis gives explicit information about the effective approximation factor in the LLL algorithm and Babai's nearest plane method, our analysis is useful to estimate which LWE instances can be solved by the key recovery attack.

Keywords Lattices · Lattice basis reduction · Learning with errors (LWE)

1 Introduction

The LWE problem was first introduced by Regev [34] in 2005, and it is a generalization of the learning parity with noise (LPN) problem [20] into large moduli q. For a prime q, we let $\mathbb{Z}_q$ denote a set of representatives of integers modulo q (note that it is not the ring of q-adic integers in this paper), and $[a]_q \in \mathbb{Z}_q$ the reduction of an integer a by modulo q. The definition of the LWE problem is as follows:

This is a fully revised paper of [21]. In particular, we apply our estimation to LWE challenge [39] in Sect. 5.3.

M. Yasuda (✉)
Institute of Mathematics, Kyushu University, 744 Motooka Nishi-ku,
Fukuoka 819-0395, Japan
e-mail: yasuda@imi.kyushu-u.ac.jp

T. Takagi et al. (eds.), *Mathematical Modelling for Next-Generation Cryptography*,
Mathematics for Industry 29, DOI 10.1007/978-981-10-5065-7_12

Definition 1.1 (*LWE*) Let n be a security parameter, q a prime modulus parameter with $r = \lfloor \log_2(q) \rfloor$, and χ an error distribution over $\mathbb{Z}$. Let $\mathbf{s} \in \mathbb{Z}_q^n$ denote a secret vector with each entry chosen uniformly at random. Given d LWE samples

$$\left(\mathbf{a}_i, [\langle \mathbf{a}_i, \mathbf{s} \rangle + e_i]_q\right) \in \mathbb{Z}_q^n \times \mathbb{Z}_q \text{ for } 0 \leq i \leq d-1, \tag{1}$$

where $\mathbf{a}_i$'s are uniformly chosen at random from $\mathbb{Z}_q^n$ and e_i's are sampled from χ for all $0 \leq i \leq d-1$. Then two questions are asked;

- The *decision-LWE* problem is to distinguish whether given $\mathbf{b} = (b_0, \ldots, b_{d-1})$ is obtained from (1) with $b_i = [\langle \mathbf{a}_i, \mathbf{s} \rangle + e_i]_q$, or uniformly at random from $\mathbb{Z}_q^d$.
- The *search-LWE* problem is to recover $\mathbf{s}$ from LWE samples (1).

In practice, for the distribution χ, we take the discrete Gaussian distribution $D_{\mathbb{Z},\sigma}$ with standard deviation $\sigma > 0$. In this case, for example, we write the search-LWE problem as "search-LWE$_{n,r,d,\sigma}$."

Given LWE samples (1), let $\mathbf{e} = (e_0, \ldots, e_{d-1})^T \in \mathbb{Z}^d$, $\mathbf{b} = (b_0, \ldots, b_{d-1})^T \in \mathbb{Z}_q^d$ with $b_i = [\langle \mathbf{a}_i, \mathbf{s} \rangle + e_i]_q$, and $\mathbf{A}^T = [\mathbf{a}_0, \ldots, \mathbf{a}_{d-1}] \in \mathbb{Z}_q^{n \times d}$ (we always write vectors in column format). Then we can simply rewrite the LWE samples (1) as the pair

$$(\mathbf{A}, \mathbf{b}) \text{ satisfying } \mathbf{b} \equiv \mathbf{A}\mathbf{s} + \mathbf{e} \bmod q. \tag{2}$$

The LWE problem is proven to be as hard as worst-case lattice problems [10, 35]. The hardness of the LWE problem assures the security of modern lattice-based schemes such as [31, 33, 35]. Recently, the LWE problem has become a central building block to construct encryption schemes with high functionality, such as fully homomorphic encryption [9, 11, 12] and candidates of multi-linear maps [18]. On the other hand, LWE-based cryptography has been paid much attention as a candidate of post-quantum cryptography [29].

According to [5], there are three main strategies to solve concrete LWE instances; reductions to (a) *short integer solutions (SIS)* and (b) *bounded distance decoding (BDD)*, and (c) *Arora–Ge method* using Gröbner bases. Consider the LWE pair (2).

(a) In the SIS strategy, we need to find a short vector $\mathbf{v} \in \mathbb{Z}^d$ such that $\mathbf{A}^T\mathbf{v} \equiv \mathbf{0} \bmod q$. Then $\langle \mathbf{v}, \mathbf{b} \rangle \equiv \langle \mathbf{v}, \mathbf{A}\mathbf{s} \rangle + \langle \mathbf{v}, \mathbf{e} \rangle \equiv \langle \mathbf{v}, \mathbf{e} \rangle \bmod q$, and $|\langle \mathbf{v}, \mathbf{e} \rangle| \approx \|\mathbf{v}\| \cdot \sigma$ if all entries of $\mathbf{e}$ are sampled from $D_{\mathbb{Z},\sigma}$. If $\|\mathbf{v}\| \cdot \sigma \ll q$, we can solve the decision-LWE problem by distinguishing whether the value $\langle \mathbf{v}, \mathbf{b} \rangle$ is small or not in $\mathbb{Z}_q$. The distinguishing attack proposed by Micciancio–Regev [27] is efficient to solve the decision-LWE problem, and the Blum–Kalai–Wasserman (BKW) algorithm [7] might be efficient for extremely large number d of samples (see [1] for the complexity of BKW).

(b) In the BDD strategy, we regard $\mathbf{b}$ as a vector bounded in distance from a lattice point $\mathbf{A}\mathbf{s} \in \mathscr{L}(\mathbf{A})$, where let $\mathscr{L}(\mathbf{A})$ denote the lattice generated by the column vectors of $\mathbf{A}$. Hence we can view the LWE pair (2) as a BDD instance in this lattice. Since BDD is a particular case of the *closest vector problem (CVP)*, we can apply approximate-CVP methods such as Babai's nearest plane method [6] to BDD. On the other hand, there are several methods such as Kannan's embedding technique [19] to reduce BDD to the *unique shortest vector problem*, which is to find a nonzero shortest vector in a lattice L under $\lambda_2(L) > \gamma\lambda_1(L)$ for some γ (here $\lambda_i(L)$ denotes the i-th successive minimum [13, Sect. 1.3]).

(c) In Arora–Ge method [2, 4], we directly recover $\mathbf{s} = (s_0, \ldots, s_{n-1})^T$. Specifically, for $\mathbf{A} = (a_{ij}) \in \mathbb{Z}_q^{d\times n}$, we set $f_i(x_0, \ldots, x_{n-1}) = b_i - \sum_{j=0}^{n-1} a_{ij}x_j \in \mathbb{Z}[x_0, \ldots, x_{n-1}]$ for $0 \le i \le d-1$. When $|e_i| \le T$ for some bound $T > 0$, we construct

$$F_i = f_i \cdot \prod_{k=1}^{T}(f_i + k)(f_i - k) \text{ for } 0 \le i \le d-1.$$

The system of nonlinear equations $F_i \equiv 0 \bmod q$ for $0 \le i \le d-1$ has the solution $(x_0, \ldots, x_{n-1}) = (s_0, \ldots, s_{n-1})$ since $f_i(s_0, \ldots, s_{n-1}) = b_i - \langle \mathbf{a}_i, \mathbf{s}\rangle \equiv e_i \bmod q$. Hence we can recover $\mathbf{s}$ by solving the system with Gröbner bases.

We focus on the key recovery attack against the search-LWE problem, which is the most basic in the BDD strategy with CVP methods (see also [16, Example 19.7.6] for the attack). Laine and Lauter [22] in 2015 analyzed the attack, and their analysis is based on a generalization of the Boneh–Venkatesan method [8] for the hidden number problem to the LWE problem. In their analysis, they considered only a combination of the LLL algorithm [23] and Babai's nearest plane method [6]. Given an LWE instance (n, q, d, σ), they analyzed that the key recovery attack is successful with overwhelming probability when $q \ge 2^{O(n)}$. They also ran the attack for hundreds of LWE instances, and demonstrated a successful range of the attack.

In this paper, we give a simple analysis of the key recovery attack. Our analysis is heuristic but practical to estimate which LWE instances (n, q, d, σ) can be solved by the attack. In the key recovery attack, we construct so-called a q-ary lattice L from an LWE pair (2), obtain a reduced basis $\mathbf{B}$ of L by lattice basis reduction, and reduce the search-LWE problem to CVP over L. As lattice basis reduction algorithms, we take the LLL algorithm [23] and the BKZ algorithm [38] with blocksize $\beta = 20$ (in practice, $\beta = 20$ can achieve the best time/quality compromise in the BKZ algorithm). The success of the attack depends on the shape of the reduced basis $\mathbf{B}$. Let $\mathbf{B}^*$ denote the Gram–Schmidt orthogonalization of $\mathbf{B}$. Our strategy is to analyze the shape of the fundamental domain $\mathscr{P}(\mathbf{B}^*)$ of $\mathbf{B}^*$ in order to determine whether Babai's nearest plane method over L is succeeded, or not. Different from Laine and Lauter's analysis, ours is useful to estimate a successful range of LWE instances (n, q, d, σ) by the key recovery attack with practical lattice basis reduction such as the BKZ algorithm.

Notation: For an odd prime q, we denote by $\mathbb{Z}_q$ a set of representatives of integers modulo q in the interval $(-q/2, q/2)$, that is, $\mathbb{Z}_q = \mathbb{Z} \cap (-q/2, q/2)$. Let $[z]_q \in \mathbb{Z}_q$ denote the reduction of an integer z by modulo q. For two vectors $\mathbf{a} = (a_1, \ldots, a_n)^T$

and $\mathbf{b} = (b_1, \ldots, b_n)^T \in \mathbb{R}^n$, we let $\langle \mathbf{a}, \mathbf{b} \rangle = \sum_{i=1}^n a_i b_i$ denote the inner product between $\mathbf{a}$ and $\mathbf{b}$. We also let $\|\mathbf{a}\|$ denote the Euclidean norm of $\mathbf{a} \in \mathbb{R}^n$ defined by $\|\mathbf{a}\|^2 = \sum_{i=1}^n a_i^2$. Denote by $\lceil a \rfloor$ the rounding of $a \in \mathbb{Q}$ to the nearest integer.

2 Preliminaries

In this section, we briefly review lattices, lattice basis reduction, and approximate-CVP methods, which are needed in our later discussion.

2.1 *Lattices*

Given two positive integers m and n with $m \geq n$, let $\mathbf{B} \in \mathbb{R}^{m \times n}$ be a matrix and $\mathbf{b}_i \in \mathbb{R}^m$ denote its i-th column vector for $1 \leq i \leq n$. Denote by

$$\mathscr{L}(\mathbf{B}) = \left\{ \sum_{i=1}^n x_i \mathbf{b}_i : x_i \in \mathbb{Z} \text{ for } 1 \leq i \leq n \right\}$$

the set of all integral linear combinations of the $\mathbf{b}_i$'s. The set $\mathscr{L}(\mathbf{B})$ clearly gives a subgroup of $\mathbb{R}^m$. The subgroup $\mathscr{L}(\mathbf{B})$ is called a *lattice of dimension n* if all $\mathbf{b}_i$'s are linearly independent over $\mathbb{R}$. In this case, the matrix $\mathbf{B}$ is called a *basis* of the lattice. In particular, the lattice $\mathscr{L}(\mathbf{B})$ is said to be of *full-rank* when $m = n$. Note that every lattice has infinitely many lattice bases; If $\mathbf{B}_1$ and $\mathbf{B}_2$ are two bases of a lattice, then there exists a unimodular matrix $\mathbf{T} \in \mathrm{GL}_n(\mathbb{Z})$ satisfying $\mathbf{B}_1 = \mathbf{B}_2\mathbf{T}$. Since $\det(\mathbf{T}) = \pm 1$, the value $\sqrt{\det(\mathbf{B}^T \cdot \mathbf{B})}$ is invariant for any basis $\mathbf{B}$, denoted by $\det(L)$. Note that we have $\det(L) = |\det(\mathbf{B})|$ for any basis $\mathbf{B}$ when $m = n$. Given a basis $\mathbf{B}$, we let

$$\mathscr{P}(\mathbf{B}) = \left\{ \sum_{i=1}^n x_i \mathbf{b}_i : -\frac{1}{2} < x_i \leq \frac{1}{2} \text{ for } 1 \leq i \leq n \right\}$$

denote the associated half-open parallelepiped. The volume of $\mathscr{P}(\mathbf{B})$ is precisely equal to $\det(L)$.

Gram–Schmidt Orthogonalization (GSO): The GSO of a basis $\mathbf{B} = [\mathbf{b}_1, \ldots, \mathbf{b}_n]$ is the orthogonal family $[\mathbf{b}_1^*, \ldots, \mathbf{b}_n^*]$, recursively defined by

$$\mathbf{b}_i^* = \mathbf{b}_i - \sum_{j=1}^{i-1} \mu_{i,j} \mathbf{b}_j^*, \text{ where } \mu_{i,j} = \frac{\langle \mathbf{b}_i, \mathbf{b}_j^* \rangle}{\|\mathbf{b}_j^*\|^2} \text{ for } 1 \leq j < i \leq n.$$

Let $\mathbf{B}^* = [\mathbf{b}_1^*, \ldots, \mathbf{b}_n^*] \in \mathbb{R}^{m \times n}$ and $\mathbf{U} = (\mu_{i,j})_{1 \leq i,j \leq n} \in \mathbb{R}^{n \times n}$, where set $\mu_{i,i} = 1$ for all i and $\mu_{i,j} = 0$ for $j > i$. Then $\mathbf{B} = \mathbf{B}^* \mathbf{U}^T$ and $\mathrm{vol}(L) = \prod_{i=1}^{n} \|\mathbf{b}_i^*\|$. For each $1 \leq \ell \leq n$, let $\pi_\ell : \mathbb{R}^m \longrightarrow H^\perp$ denote the orthogonal projection from $\mathbb{R}^m$ over the orthogonal supplement of the $\mathbb{R}$-vector space $H := \langle \mathbf{b}_1, \ldots, \mathbf{b}_{\ell-1} \rangle_{\mathbb{R}}$. In particular, we have $\pi_1 = \mathrm{id}$ and $\mathbf{b}_\ell^* = \pi_\ell(\mathbf{b}_\ell)$ for any $1 \leq \ell \leq n$.

2.2 Lattice Basis Reduction

Given a basis of a lattice L of dimension n, *lattice basis reduction* aims to output a basis $\mathbf{B} = [\mathbf{b}_1, \ldots, \mathbf{b}_n]$ of L with short and nearly orthogonal vectors $\mathbf{b}_1, \ldots, \mathbf{b}_n$. It gives a powerful tool to break lattice-based cryptosystems. The most famous one is the LLL algorithm [23], and its blockwise generalization is known as the BKZ algorithm [38]. Here we introduce the LLL algorithm; Let us first recall the notion of LLL-reduction (e.g., see [23, Chap. 2] or [13, Sect. 4]).

Definition 2.1 (*LLL-reduced*) Let $\mathbf{B} = [\mathbf{b}_1, \ldots, \mathbf{b}_n]$ be a basis, and $\mathbf{B}^* = [\mathbf{b}_1^*, \ldots, \mathbf{b}_n^*]$ denote its GSO with coefficients $\mu_{i,j}$. Given a reduction parameter $\frac{1}{4} < \alpha < 1$, the basis $\mathbf{B}$ is called *α-LLL-reduced* if the following two conditions are satisfied:

- (Size-reduced) $|\mu_{i,j}| \leq \frac{1}{2}$ for any $1 \leq j < i \leq n$.
- (Lovász condition) $\alpha \|\mathbf{b}_{k-1}^*\|^2 \leq \|\pi_{k-1}(\mathbf{b}_k)\|^2$ for any $2 \leq k \leq n$. Since $\pi_{k-1}(\mathbf{b}_k) = \mathbf{b}_k^* + \mu_{k,k-1}\mathbf{b}_{k-1}^*$, this condition can be rewritten as

$$\|\mathbf{b}_k^*\|^2 \geq (\alpha - \mu_{k,k-1}^2)\|\mathbf{b}_{k-1}^*\|^2. \tag{3}$$

The LLL algorithm takes as input a basis $\mathbf{B} = [\mathbf{b}_1, \ldots, \mathbf{b}_n]$ of a lattice L and a reduction parameter $\frac{1}{4} < \alpha < 1$ (default $\alpha = 0.99$ in cryptanalysis), and outputs an α-LLL-reduced basis of L. The procedure of the basic LLL algorithm is as follows; Set $k \leftarrow 2$. While $k \leq n$ do:

1. Size-reduce $\mathbf{B} = [\mathbf{b}_1, \ldots, \mathbf{b}_n]$ (e.g., [28, Algorithm 3 in Chap. 2]). This makes the GSO coefficients $\mu_{i,j}$ smaller than $\frac{1}{2}$ in absolute value by modifying the basis vectors $\mathbf{b}_i$, but it does not change any GSO vectors $\mathbf{b}_i^*$.
2. Swap $\mathbf{b}_k$ with $\mathbf{b}_{k-1}$ if the Lovász condition (3) is not satisfied. In this case, set $k \leftarrow \max(2, k-1)$. Otherwise set $k \leftarrow k+1$. Then go back to step 1.

Hermite Factor: The *Hermite factor* δ of a lattice basis reduction algorithm is defined by

$$\delta = \|\mathbf{b}_1\| / \det(L)^{1/n}$$

with the output basis $[\mathbf{b}_1, \ldots, \mathbf{b}_n]$ (we assume $\|\mathbf{b}_1\| \leq \|\mathbf{b}_i\|$ for all $2 \leq i \leq n$). This factor gives an index to measure the output quality of a lattice basis reduction algorithm. The output quality becomes better as δ is smaller.

- Gama–Nguyen's experimental results [17, Fig. 4] show that the Hermite factor of LLL for random lattices is practically 1.0219^n on average in $n \geq 100$.
- BKZ uses a blockwise parameter β, and larger β improves the output quality but increases the running time. In practice, $\beta \approx 20$ can achieve the best time/quality compromise. It follows from [17, Sect. 5.2] that the Hermite factor with $\beta = 20$ for random lattices is practically 1.0128^n on average.

Geometric Series Assumption (GSA): Except when a lattice has a special structure, practical lattice basis reduction algorithms output a basis $\mathbf{B} = [\mathbf{b}_1, \ldots, \mathbf{b}_n]$ such that $\|\mathbf{b}_i^*\|/\|\mathbf{b}_{i+1}^*\| \approx c$ for any $1 \leq i \leq n-1$, that is, $\|\mathbf{b}_i^*\| \approx c^{1-i}\|\mathbf{b}_1\|$ for $1 \leq i \leq n$, where the constant c depends on algorithms. The GSA assumes that the values $\log_2(\|\mathbf{b}_1\|/\|\mathbf{b}_i^*\|)$'s are on a straight line (e.g., see [37, Fig. 1]). According to [23], we have $c \approx 1.02^2 \approx 1.04$ (resp. $c \approx 1.025$) in practice for LLL (resp. BKZ with $\beta = 20$) for random lattices.

2.3 Approximate-CVP Methods

There are two strategies to solve CVP; *exact CVP* and *approximate CVP* methods. Currently known exact CVP methods enable one to solve CVP exactly, but they require exponential operations. Here we present classical approximate-CVP methods, which solve CVP approximately but they run in polynomial-time. Let $\mathbf{B} = [\mathbf{b}_1, \ldots, \mathbf{b}_n] \in \mathbb{R}^{n \times n}$ be a basis of a full-rank lattice L, and let $\mathbf{b} \in \mathbb{R}^n$ be a target vector (assume $\mathbf{b} \notin L$). Let $\mathbf{w} \in L$ be the closest lattice vector to $\mathbf{b}$, and $\mathbf{e} = \mathbf{b} - \mathbf{w}$ the error vector.

- *Babai's rounding method* [6] is a polynomial-time algorithm, which outputs the unique lattice vector $\mathbf{v} \in L$ such that $\mathbf{b} - \mathbf{v}$ is inside the parallelepiped $\mathscr{P}(\mathbf{B})$. Hence $\mathbf{v}$ is consistent with $\mathbf{w}$ if and only if $\mathbf{e} \in \mathscr{P}(\mathbf{B})$. When we write $\mathbf{b} = \sum_{i=1}^{n} \ell_i \mathbf{b}_i$ with $\ell_i \in \mathbb{R}$, this method outputs the vector $\mathbf{v} = \sum_{i=1}^{n} \lceil \ell_i \rfloor \mathbf{b}_i \in L$.
- *Babai's nearest plane method* [6] is also a polynomial-time algorithm. It outputs the unique lattice vector $\mathbf{v} \in L$ such that $\mathbf{b} - \mathbf{v}$ is inside the parallelepiped $\mathscr{P}(\mathbf{B}^*)$. Then $\mathbf{v}$ is consistent with $\mathbf{w}$ if and only if $\mathbf{e} \in \mathscr{P}(\mathbf{B}^*)$. The procedure of this method is as follows (see [16, Algorithm 25]);

  1. Compute the GSO vectors $\mathbf{B}^* = [\mathbf{b}_1^*, \ldots, \mathbf{b}_n^*]$ of $\mathbf{B}$, and set $\mathbf{w}_n \leftarrow \mathbf{b}$
  2. For $i = n$ downto 1 do:
     a. Compute $\ell_i = \langle \mathbf{w}_i, \mathbf{b}_i^* \rangle / \|\mathbf{b}_i^*\|^2$, and set $\mathbf{y}_i \leftarrow \lceil \ell_i \rfloor \mathbf{b}_i$
     b. Set $\mathbf{w}_{i-1} \leftarrow \mathbf{w}_i - (\ell_i - \lceil \ell_i \rfloor)\, \mathbf{b}_i^* - \lceil \ell_i \rfloor \mathbf{b}_i$
  3. Return $\mathbf{v} \leftarrow \mathbf{y}_1 + \cdots + \mathbf{y}_n$

In a typical reduced basis, the output vector of Babai's nearest plane method is closer to $\mathbf{b}$ than the rounding method (but the rounding method is much faster). As for improvements of Babai's methods, Lindner–Peikert [25] proposed improvements of Babai's nearest plane method with limited exhaustive search, and their methods were further improved by Liu–Nguyen [24] with pruned enumeration.

3 Key Recovery Attack Against Search-LWE Problem

In this section, we introduce the key recovery attack against the search-LWE problem. In order to make the attack work, we assume the following [22, Definition 6]:

Assumption 3.1 In the LWE problem with samples (1), we assume

$$[\langle \mathbf{a}_i, \mathbf{s}\rangle + e_i]_q = \langle \mathbf{a}_i, \mathbf{s}\rangle_q + e_i \ for\ any\ 0 \le i \le d-1. \tag{4}$$

This assumption holds for almost all LWE samples in practice if all errors e_i are very small compared to q. (Under Assumption 3.1, Laine and Lauter [22] analyzed the key recovery attack.) On the other hand, Assumption 3.1 is required for correct decryption in LWE-based homomorphic encryption schemes such as [9, 11, 12].

3.1 Key Recovery Attack

In this subsection, we give an outline of the key recovery attack against search-$\mathrm{LWE}_{n,r,d,\chi}$ (see [22, Sect. 4] or [16, Example 19.7.6] for the attack). The procedure can be divided into the following four steps (see [22, Section 4] for details); Given d LWE samples (1) satisfying (4) (note that $d > n$ is required for the attack).

1. We first construct a $d \times (n+d)$-matrix

$$\mathbf{C} = \left(qI_{d\times d} \mid \mathbf{A}\right) \text{ with } \mathbf{A}^T = \left[\mathbf{a}_0, \mathbf{a}_1, \ldots, \mathbf{a}_{d-1}\right] \in \mathbb{Z}_q^{n\times d}, \tag{5}$$

   where let $I_{d\times d}$ denote the $d \times d$-identity matrix.
2. Let $\mathbf{A}_q$ denote the column-Hermite normal form of $\mathbf{C}$ (this operation removes the dependency of the columns of $\mathbf{C}$ so that all the columns of $\mathbf{A}_q$ are linearly independent over $\mathbb{R}$). Let L denote the lattice $\mathscr{L}(\mathbf{A}_q)$ of dimension d.
3. We then reduce $\mathbf{A}_q$ by lattice basis reduction (e.g., LLL or BKZ algorithm) to obtain a reduced basis $\mathbf{B}$ of L.
4. Set $\mathbf{b} = (b_0, b_1, \ldots, b_{d-1})^T \in \mathbb{Z}_q^d$ with $b_i = [\langle \mathbf{a}_i, \mathbf{s}\rangle + e_i]_q$ for $0 \le i \le d-1$. We try to solve CVP over L with instance $(\mathbf{B}, \mathbf{b})$ by adopting an approximate-CVP method (e.g. Babai's nearest plane or rounding method). Then the CVP method with input $(\mathbf{B}, \mathbf{b})$ may output the lattice vector

$$\mathbf{v} = \left(\langle \mathbf{a}_0, \mathbf{s}\rangle_q, \langle \mathbf{a}_1, \mathbf{s}\rangle_q, \ldots, \langle \mathbf{a}_{d-1}, \mathbf{s}\rangle_q\right)^T \in L \tag{6}$$

   if $\mathbf{B}$ is sufficiently reduced and $\mathbf{b}$ is sufficiently close to $\mathbf{v}$. Once we obtain correct $\mathbf{v} \in L$, we can recover the secret vector $\mathbf{s} \equiv \mathbf{A}^{-1}\mathbf{v} \bmod q \in \mathbb{Z}_q^n$.

Principle of Attack: Here we describe why the above procedure can solve the search-LWE problem. For each $0 \leq i \leq d-1$, there exists an integer k_i satisfying $\langle \mathbf{a}_i, \mathbf{s} \rangle_q = \langle \mathbf{a}_i, \mathbf{s} \rangle + qk_i$ in $\mathbb{Z}$. Set

$$\mathbf{w} = (k_0, \ldots, k_{d-1}, s_0, \ldots, s_{n-1})^T \in \mathbb{Z}^{n+d},$$

where we write $\mathbf{s} = (s_0, \ldots, s_{n-1})^T \in \mathbb{Z}_q^n$. Since $\mathbf{v}$ given by (6) is rewritten as $\mathbf{Cw}$, the vector $\mathbf{v}$ is contained in L. Note that $L = \mathscr{L}(\mathbf{C}) = \mathscr{L}(\mathbf{A}_q)$ as sets. (We also remark that the columns of $\mathbf{C}$ are not a basis of L, but the columns of $\mathbf{A}_q$ give a basis.) The difference between $\mathbf{b}$ and $\mathbf{v}$ is given by

$$\mathbf{e} = \mathbf{b} - \mathbf{v} = (e_0, e_1, \ldots, e_{d-1})^T \in \mathbb{Z}^d \tag{7}$$

under Assumption 3.1. Recall that each entry e_i is sampled from the error distribution χ, and it is considerably small compared to the modulus parameter q in practice. When the norm size of the noise vector $\mathbf{e} \in \mathbb{Z}^d$ is considerably small (that is, when the vector $\mathbf{v}$ is the closest lattice vector to the target vector $\mathbf{b}$), we may obtain $\mathbf{v}$ by an approximate-CVP method with input $(\mathbf{B}, \mathbf{b})$ if $\mathbf{B}$ is sufficiently reduced (see the next section for more details).

3.2 Analysis of [22] on Success Probability

Laine and Lauter [22] analyzed the success probability of the key recover attack for search-LWE$_{n,r,d,\sigma}$ when the LLL algorithm and Babai's nearest plane method are adopted. In this subsection, we briefly describe their analysis on the success probability of the attack. We begin to recall the following well-known result by Babai [6] (see also [16, Theorem 18.1.6] or [22, Theorem 2]):

Theorem 3.2 (LLL-Babai) *For an LLL-reduced basis of a lattice L of dimension N, Babai's nearest plane method on $\mathbf{z} \in \mathbb{R}^N$ (with $\mathbf{z} \notin L$) outputs $\mathbf{x} \in L$ such that*

$$\|\mathbf{z} - \mathbf{x}\| < 2^{\mu N} \cdot \|\mathbf{z} - \mathbf{y}\|$$

for all $\mathbf{y} \in L$ and some constant $\mu > 0$. In particular, when we take $\alpha = 3/4$ as the LLL-reduction parameter, a value of the effective approximate factor $\mu = 1/2$ is guaranteed in theory. However, significantly smaller μ is expected in practice for any fixed $1/4 < \alpha < 1$.

As in Theorem 3.2, let μ denote the effective approximate factor when we apply the LLL-Babai algorithm to $(\mathbf{B}, \mathbf{b})$ in the key recovery attack. The analysis in [22, Sect. 4] shows that we can expect to need

$$\log_2(1-p) + r(d-n) > d \log_2 \left[2\left(1 + 2^{\mu d}\right) \sigma \sqrt{d} + 1 \right]$$

in order to succeed to solve search-LWE$_{n,r,d,\sigma}$ with probability at least p (see the Eq. (24) in [22]). Furthermore, Laine and Lauter estimate that we expect to have a certain upper bound μ_{LLL} with $\mu \leq \mu_{\text{LLL}}$ if the attack for search-LWE$_{n,r,d,\sigma}$ is succeeded, where set

$$\mu_{\text{LLL}} \approx \frac{1}{d} \log_2 \left(\frac{1}{2\sqrt{d}} \cdot \frac{q}{\sigma} \cdot \frac{1}{q^{n/d}} \right).$$

With this estimate, Laine and Lauter measured the effective performance of LLL-Babai by their exhaustive experiments for the search-LWE problem with various LWE instances (n, r, d, σ) (see [22, Table 1] for their experimental results).

4 Our Analysis of Key Recovery Attack

In this section, we give a simple analysis of the key recovery attack. While Laine–Lauter's analysis is used to measure the effectiveness of the LLL-Babai algorithm, ours is useful to estimate which instances of search-LWE$_{n,r,d,\sigma}$ can be solved by the key recovery attack with practical lattice basis reduction and Babai's nearest plane method. In this section, we focus on the key recovery attack with the LLL algorithm and the BKZ algorithm with blocksize $\beta = 20$.

4.1 Heuristic Analysis on Successful Condition of Attack

Let $\mathbf{B} = [\mathbf{b}_1, \ldots, \mathbf{b}_d]$ denote a reduced basis of the lattice $L = \mathscr{L}(\mathbf{A}_q)$ of dimension d, obtained in Step 3 of the key recovery attack. We let $\mathbf{B}^* = [\mathbf{b}_1^*, \ldots, \mathbf{b}_d^*]$ denote its GSO. Recall that for a target vector $\mathbf{b}$, Babai's nearest plane method with input $(\mathbf{B}, \mathbf{b})$ outputs the unique lattice vector $\mathbf{z} \in L$ such that $\mathbf{b} - \mathbf{z}$ is included in the parallelepiped $\mathscr{P}(\mathbf{B}^*)$. Let $\mathbf{e} = \mathbf{b} - \mathbf{v} = (e_0, \ldots, e_{d-1})^T \in \mathbb{Z}^d$ denote the noise vector given by (7). As in [22], we assume that each entry e_i is sampled by the discrete Gaussian distribution $\chi = D_{\mathbb{Z},\sigma}$. We write

$$\mathbf{e} = \sum_{i=1}^{d} x_i \mathbf{b}_i^* \text{ with } x_i \in \mathbb{R} \text{ for } 1 \leq i \leq d.$$

Then the key recovery attack by Babai's nearest plane method on the instance $(\mathbf{B}, \mathbf{b})$ is succeeded if and only if the noise vector $\mathbf{e}$ is contained in the parallelepiped $\mathscr{P}(\mathbf{B}^*)$. In other words, the attack is succeeded if and only if $-\frac{1}{2} < x_i \leq \frac{1}{2}$ for all $1 \leq i \leq d$. Since $\langle \mathbf{e}, \mathbf{b}_i^* \rangle = x_i \cdot \|\mathbf{b}_i^*\|^2$ for $1 \leq i \leq d$, we clearly have

$$|x_i| < \frac{1}{2} \Longleftrightarrow \frac{|\langle \mathbf{e}, \mathbf{b}_i^* \rangle|}{\|\mathbf{b}_i^*\|^2} < \frac{1}{2}.$$

Heuristically, we have

$$\frac{|\langle \mathbf{e}, \mathbf{b}_i^* \rangle|}{\|\mathbf{b}_i^*\|^2} \approx \frac{\|\mathbf{e}\| \cdot \|\mathbf{b}_i^*\|}{\sqrt{d} \cdot \|\mathbf{b}_i^*\|^2} = \frac{\|\mathbf{e}\|}{\sqrt{d} \cdot \|\mathbf{b}_i^*\|}.$$

On the other hand, we can estimate $\|\mathbf{e}\| \approx \sigma\sqrt{d}$ since each entry e_i is sampled by the distribution $\chi = D_{\mathbb{Z},\sigma}$. Then we can estimate that the key recovery attack by Babai's nearest plane method is succeeded if and only if

$$2\sigma < \|\mathbf{b}_i^*\| \text{ for all } 1 \le i \le d \Longleftrightarrow 2\sigma < \min_{1 \le i \le d} \|\mathbf{b}_i^*\|. \tag{8}$$

Then it is the most important in our analysis to precisely evaluate the size of $\min_{1 \le i \le d} \|\mathbf{b}_i^*\|$, which depends on lattice basis reduction.

4.2 Experimental Estimation for min $\|\mathbf{b}_i^*\|$

If $\mathbf{B} = [\mathbf{b}_1, \ldots, \mathbf{b}_d]$ were obtained by the LLL algorithm or the BKZ algorithm with $\beta = 20$ for a basis of a *random lattice* L, we could estimate from Gama–Nguyen's experimental results [17] for random lattices that in the LLL case, we have

$$\|\mathbf{b}_d^*\| \approx (1.04)^{-(d-1)} \cdot \|\mathbf{b}_1\| \text{ and } \|\mathbf{b}_1\| \approx (1.0219)^d \cdot \mathrm{vol}(L)^{1/d}, \tag{9}$$

or in the case of BKZ with $\beta = 20$, we have

$$\|\mathbf{b}_d^*\| \approx (1.025)^{-(d-1)} \cdot \|\mathbf{b}_1\| \text{ and } \|\mathbf{b}_1\| \approx (1.0128)^d \cdot \mathrm{vol}(L)^{1/d} \tag{10}$$

on average for large $d \ge 100$ (see Sect. 2.2). Then we estimate for the random lattice L that $\min_{1 \le i \le d} \|\mathbf{b}_i^*\| \approx \|\mathbf{b}_d^*\|$. However, in the key recovery attack, we need to handle the special lattice $L = \mathscr{L}(\mathbf{A}_q)$ defined in Step 2 of the attack. More specifically, the lattice L can be rewritten as so-called the *q-ary lattice*

$$L = \left\{ \mathbf{y} \in \mathbb{Z}^d : \mathbf{y} \equiv \mathbf{A}\mathbf{z} \bmod q \text{ for some } \mathbf{z} \in \mathbb{Z}^n \right\},$$

where $\mathbf{A} \in \mathbb{Z}_q^{d \times n}$ is given by (5). Then we cannot apply Gama–Nguyen's results to the q-ary lattice L. In the following, we give experimental results of LLL and BKZ with $\beta = 20$ for q-ary lattices in order to estimate the size of $\min_{1 \le i \le d} \|\mathbf{b}_i^*\|$.

4.2.1 Case of LLL

We give experimental results for an LLL-reduced basis $\mathbf{B}$ of $L = \mathscr{L}(\mathbf{A}_q)$. Recall that the q-ary lattice L has volume $\mathrm{vol}(L) \approx q^{d-n}$ with overwhelming high probability (e.g., see [25]). Then we set

$$c_{\mathrm{LLL}} = \left(\frac{\min_{1 \le i \le d} \|\mathbf{b}_i^*\|}{q^{(d-n)/d}} \right)^{1/d}, \tag{11}$$

where $\mathbf{B}^* = [\mathbf{b}_1^*, \ldots, \mathbf{b}_d^*]$ denotes the GSO vectors of $\mathbf{B}$. For our simple analysis, we assume GSA for L, but it does not hold for any q-ary lattices in practice (see the full version of [25] for details. Specifically, different from random lattices, the last several GSO vectors $\mathbf{B}^*$ of q-ary lattices L are typically very short). From (11), we have

$$\min_{1 \le i \le d} \|\mathbf{b}_i^*\| = c_{\mathrm{LLL}}^d \cdot q^{(d-n)/d} \approx c_{\mathrm{LLL}}^d \cdot \mathrm{vol}(\Lambda)^{1/d} \tag{12}$$

(cf. by (9), we may expect $c_{\mathrm{LLL}} \approx \frac{1.0219}{1.04} \approx 0.982596$ on average for a random lattice, but this constant cannot be applied for the q-ary lattice L).

In [21, Figs. 2, 3, 4], we give experimental results on c_{LLL} for the q-ary lattice $L = \mathscr{L}(\mathbf{A}_q)$. Specifically, in each of [21, Figs. 2, 3, 4], we show histogram of c_{LLL} for L with $r = \lfloor \log_2(q) \rfloor = 15, 30$ and 50. For our experiments, we conducted the following procedure for each LWE instance:

1. Given (n, r, d, σ), we generate d LWE samples (1) satisfying (4).
2. We construct $\mathbf{C}$ as in (5), and compute its column-Hermite normal form to obtain $\mathbf{A}_q \in \mathbb{Z}^{d \times d}$ for the lattice $L = \mathscr{L}(\mathbf{A}_q)$ as in Step 2 of the attack.
3. As in Step 3, we reduce $\mathbf{A}_q^T$ by the LLL algorithm to obtain an LLL-reduced basis $\mathbf{B}$ for L, and compute its GSO vectors $\mathbf{B}^* = [\mathbf{b}_1^*, \ldots, \mathbf{b}_d^*]$. We compute the value c_{LLL} for the basis $\mathbf{B}$.

Implementation and LWE instances: We implemented the above procedure using Sage [36] (version 6.8). In particular, we used the function "LWE" in Sage for generating LWE samples, and "FP_LLL" for the LLL algorithm with reduction parameter $\alpha = 0.99$ (FP_LLL is the floating point implementation of LLL in the fpLLL library, which is also included in Sage). In our experiments, for LWE instances, we took

$$(n, r, d, \sigma) = \begin{cases} (80, 15, 255, 8/\sqrt{2\pi}), \ (100, 15, 300, 8/\sqrt{2\pi}), \\ (80, 30, 255, 8/\sqrt{2\pi}), \ (100, 30, 300, 8/\sqrt{2\pi}), \\ (80, 50, 255, 8/\sqrt{2\pi}), \ (100, 50, 300, 8/\sqrt{2\pi}). \end{cases} \tag{13}$$

The triples (n, d, σ) of the above LWE instances were used in Laine and Lauter's experiments [22] (see [22, Table 1] for details). For each LWE instance (n, r, d, σ), we calculated the value c_{LLL} 100 times by generating different LWE samples.

Constant for c_{LLL}: Here we shall fix a value of c_{LLL} for $L = \mathscr{L}(\mathbf{A}_q)$ from our experimental results. Note that c_{LLL} is independent of the parameter σ by (11), namely, c_{LLL} depends on the parameters (n, r, d). W see from [21, Figs. 2, 3, 4] that c_{LLL} does not depend on n and d, and it depends mainly on $r = \lfloor \log_2(q) \rfloor$. Since [21, Fig. 3] resembles [21, Fig. 4], we estimate that histograms of c_{LLL} are almost the same for all $r \geq 30$. Here we fix

$$c_{\text{LLL}} = 0.9775 \tag{14}$$

for any q-ary lattice L with $r \geq 15$. This constant is the minimum value in our experimental results. We expect that the minimum value would give a boundary between the success and the failure of the key recovery attack.

4.2.2 Case of BKZ with $\beta = 20$

Here we give experimental results for a basis $\mathbf{B}$ reduced by BKZ with $\beta = 20$ of $L = \mathscr{L}(\mathbf{A}_q)$. As well as the LLL case, we can define c_{BKZ20} by (11). In [21, Figs. 5, 6, 7], we show histogram of c_{BKZ20}. For our experiments, we took the same LWE instances (13). As in the LLL case, we calculated c_{BKZ20} 100 times for each LWE instance. In our implementation, we used the NTL-implementation BKZ function in Sage, which adapts *full enumeration* as a subroutine for the shortest vector problem (SVP) of dimension β (cf. BKZ 2.0 [15] uses pruned enumeration). Therefore we expect that this function could return the best output quality of the BKZ algorithm with fixed β. From our experimental results, we fix

$$c_{\text{BKZ20}} = 0.9863 \tag{15}$$

for L (cf. by (10), we may expect $c_{\text{BKZ20}} \approx \frac{1.0128}{1.025} \approx 0.9881$ on average for random lattices). The constant (15) is the minimum value in our experimental results. We expect that the constant is a limit of the BKZ algorithm with $\beta = 20$.

Summary of Experimental Results: In Table 1, we summarize values $\left(\frac{\min \|\mathbf{b}_i^*\|}{q^{(d-n)/d}}\right)^{1/d}$ for reduced bases $\mathbf{B} = [\mathbf{b}_1, \ldots, \mathbf{b}_d]$ of random lattices and q-ary lattices. The values over q-ary lattices are small compared to random lattices.

Table 1 Values $\left(\frac{\min \|\mathbf{b}_i^*\|}{q^{(d-n)/d}}\right)^{1/d}$ for reduced bases $\mathbf{B} = [\mathbf{b}_1, \ldots, \mathbf{b}_d]$ ($r = \lfloor \log_2(q) \rfloor$)

	LLL	BKZ with $\beta = 20$
Random lattices	Average: $\frac{1.0219}{1.04} \approx 0.9826$ [17]	Average: $\frac{1.0128}{1.025} \approx 0.9881$ [17]
q-ary lattices	Average: 0.9789 ($r \geq 30$) minimum: $c_{\text{LLL}} = 0.9775$	Average: 0.9868 ($r \geq 15$) minimum: $c_{\text{BKZ20}} = 0.9863$

5 Our Estimation for Key Recovery Attack

In this section, we give an estimation of the boundary between the success and the failure of the key recovery attack against search-LWE$_{n,r,d,\sigma}$. We also compare our estimation with experimental results by Laine and Lauter [22].

5.1 *Our Estimation and Comparison with [22]*

By combining (8) and (12), we can estimate that the key recovery attack against search-LWE$_{n,r,d,\sigma}$ by the LLL algorithm and Babai's nearest plane method is succeeded if and only if (recall $r = \lfloor \log_2(q) \rfloor$ and $d > n$)

$$2\sigma < c_{\mathrm{LLL}}^{d} \cdot q^{(d-n)/d} \Longleftrightarrow \log_2(2\sigma) < d \log_2 (c_{\mathrm{LLL}}) + \frac{r(d-n)}{d}$$
$$\Longleftrightarrow r > \left(\frac{d}{d-n}\right) \left\{\log_2(2\sigma) - d \log_2 (c_{\mathrm{LLL}})\right\}. \quad (16)$$

With our constant $c_{\mathrm{LLL}} = 0.9775$ by (14), the inequality (16) gives a boundary to determine which LWE instances (n, r, d, σ) can be solved by the key recovery attack with the LLL algorithm and Babai's nearest plane method. In using BKZ with $\beta = 20$ for the key recovery attack, we use $c_{\mathrm{BKZ20}} = 0.9863$ given by (15), instead of c_{LLL} (larger blocksize β can make the constant of the BKZ algorithm closer to 1).

By their exhaustive experiments, Laine and Lauter demonstrated a successful range of LWE instances (n, r, d, σ) by the key recovery attack [22, Table 1]. In their experiments, they fixed $\sigma = 8/\sqrt{2\pi} \approx 3.192$, and used the floating point variant of the LLL algorithm (with reduction parameter $\alpha = 0.99$) in PARI/GP [30].

- For $(n, d) = (200, 500)$, they showed that $r \geq 32$ can be solved (cf. the attack failed for $(n, r, d) = (200, 31, 505)$). In contrast, for such (n, d), our estimation (16) for LLL implies that $r > 31.817$ can be solved.
- For $(n, d) = (350, 805)$, they showed that $r \geq 52$ can be solved (cf. the attack failed for $(n, r, d) = (350, 51, 810)$). In contrast, for such (n, d), our estimation (16) for LLL implies that $r > 51.491$ can be solved.

As in the above comparison, we can verify that our estimation (16) for LLL coincides with all experimental results [22, Table 1]. Therefore we expect that our estimation (16) could give a practical boundary between the success and the failure of the key recovery attack for any LWE instance (n, r, d, σ).

Optimal Number of Samples for Attack: For fixed n and σ, our estimation (16) enables one to obtain the optimal number of LWE samples to solve search-LWE$_{n,r,d,\sigma}$ by the key recovery attack. In fact, the optimal number $d_{\mathrm{LLL}}^{\mathrm{opt}}$ in using the LLL algorithm is given by

$$d_{\mathrm{LLL}}^{\mathrm{opt}} = \left\lceil n + \sqrt{n^2 + n\left(\frac{c_1}{c_2}\right)} \right\rceil \in \mathbb{Z} \tag{17}$$

where $c_1 = \log_2(2\sigma)$ and $c_2 = -\log_2(c_{\mathrm{LLL}}) = -\log_2(0.9775) > 0$. Note that the optimal number $d_{\mathrm{LLL}}^{\mathrm{opt}}$ minimizes the right hand side value of our estimation (16). For example, we have the following results on the value $d_{\mathrm{LLL}}^{\mathrm{opt}}$:

- For $(n, \sigma) = (200, 8/\sqrt{2\pi})$, we have $d_{\mathrm{LLL}}^{\mathrm{opt}} = 437$ by (17). For this $d_{\mathrm{LLL}}^{\mathrm{opt}}$, our estimation (16) shows that $r > 31.386$ can be solved by the key recovery attack with the LLL algorithm and Babai's nearest plane method.
- For $(n, \sigma) = (350, 8/\sqrt{2\pi})$, we have $d_{\mathrm{LLL}}^{\mathrm{opt}} = 739$ by (17). As in the above case, for this $d_{\mathrm{LLL}}^{\mathrm{opt}}$, our estimation (16) shows that $r > 51.173$ can be solved.

In using BKZ with $\beta = 20$ for the attack, we obtain the optimal number of samples $d_{\mathrm{BKZ20}}^{\mathrm{opt}}$ as (17) by using $c_{\mathrm{BKZ20}} = 0.9863$ given by (15).

5.2 *Successful Range of Key Recovery Attack by Our Estimation*

In Fig. 1, we show our estimation of the successful range of key recovery attack against search-$\mathrm{LWE}_{n,r,d,\sigma}$. Specifically, we show successful range (n, r) of the attack

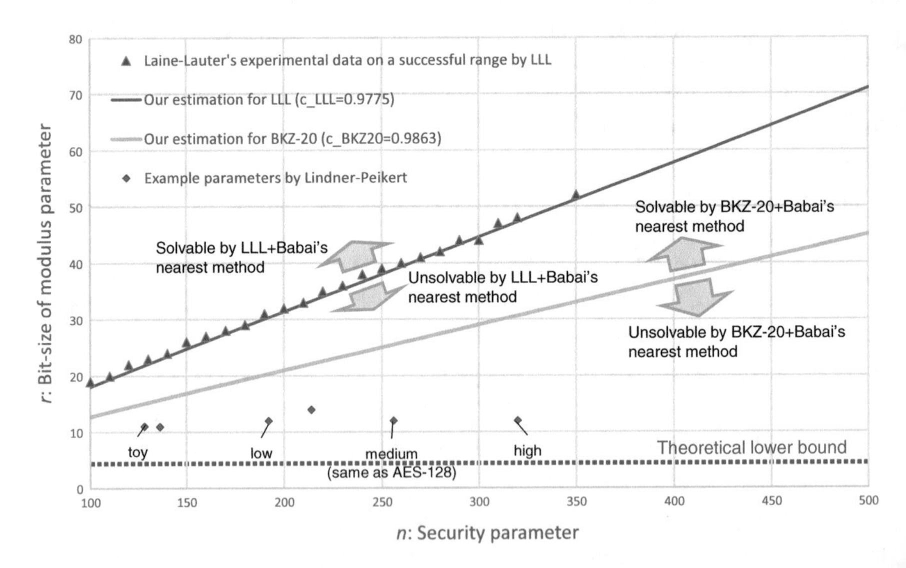

Fig. 1 Our estimation of the successful range (n, r) of the key recovery attack against search-$\mathrm{LWE}_{n,d,r,\sigma}$ by LLL and BKZ with $\beta = 20$ ($d = d_{\mathrm{LLL}}^{\mathrm{opt}}$ and $d_{\mathrm{BKZ20}}^{\mathrm{opt}}$, $\sigma = 8/\sqrt{2\pi}$). The theoretical lower bound is given by (18)

by LLL and BKZ with $\beta = 20$ with optimal sample number $d = d_{\mathrm{LLL}}^{\mathrm{opt}}$ and $d_{\mathrm{BKZ20}}^{\mathrm{opt}}$, respectively, (we set $\sigma = 8/\sqrt{2\pi}$ as in [22]). Since we take the minimum output values for c_{LLL} and c_{BKZ20}, the lines in Fig. 1 give a boundary between success and failure of the attack. For example, in case of $n = 350$, we estimate from Fig. 1 that $r = \lfloor \log_2(q) \rfloor \geq 52$ is solvable by LLL and Babai's nearest plane method, but $r \leq 51$ is unsolvable. We also estimate that $r \geq 34$ is solvable by BKZ with $\beta = 20$ and Babai's nearest plane method, but $r \leq 33$ is unsolvable.

In Fig. 1, we also show Laine and Lauter's experimental data [22, Table 1] on a successful range of the attack using LLL. We see from Fig. 1 that our estimation for LLL coincides with their experimental data. We plot example parameters (n, r) of LWE-based encryption by Lindner and Peikert [25, Fig. 3] (including two parameters by [27]), in which $\sigma \approx 8$ is used (cf. we take $\sigma = 8/\sqrt{2\pi} \approx 3.192$ for our estimation, and our estimation lines in Fig. 1 move up slightly when we take $\sigma = 8$). In particular, Lindner and Peikert [25] estimate that $(n, r) = (256, 12)$ has the almost same security level as AES-128.

Remark 5.1 For any basis $\mathbf{B} = [\mathbf{b}_1, \ldots, \mathbf{b}_d]$ of a q-ary lattice L, we have $\mathrm{vol}(L) = \prod_{i=1}^{d} \|\mathbf{b}_i^*\| \geq \left(\min_{1 \leq i \leq d} \|\mathbf{b}_i^*\| \right)^d$. Hence it follows that $\left(\frac{\min \|\mathbf{b}_i^*\|}{q^{(d-n)/d}} \right)^{1/d} \leq 1$ under the assumption $\mathrm{vol}(L) = q^{d-n}$. By inequality (16), this gives a theoretical lower bound (note $d > n$)

$$r > \left(\frac{d}{d-n} \right) \cdot \log_2(2\sigma) > \log_2(2\sigma) \tag{18}$$

for the successful range (n, r) of the key recovery attack.

5.3 Application of Our Estimation to LWE Challenge

The website [39] presents LWE challenge problems for testing algorithms that solve LWE problems. All challenge problems were created using multiparty computation executed among three universities (see [14] for details). Each challenge problem can be downloaded by selecting a pair (n, α), where n and α denote the dimension and the relative error size, respectively. Given (n, α), other LWE parameters q, m, σ are set as follows; q is the minimum prime exceeding n^2, $m = n^2$, and $\sigma = \alpha q$, where m denotes the maximal number of LWE samples.

In this subsection, we apply our estimation to determine which LWE challenge pairs (n, α) can be solved by the key recovery attack with Babai's nearest plane method. Given an LWE challenge pair (n, α), we here take $d = 2n$ as a suitable number of LWE samples to solve the LWE problem. (From the Eq. (17), such d is almost optimal in practice.) Note that we do not need to use all LWE samples, and we extract d samples from $m = n^2$ samples for the attack. We let $c(n, \alpha)$ denote the constant of lattice basis reduction defined by (11), required to solve the LWE problem by the key recovery attack. Then by our estimation (16), we now have

$$r > \left(\frac{d}{d-n}\right)\left\{\log_2(2\sigma) - d\log_2\left(c(n,\alpha)\right)\right\} \text{ with } d = 2n$$
$$\Longleftrightarrow \log_2\left(c(n,\alpha)\right) > \frac{\log_2(\alpha) + \log_2(n) + 1}{2n} \quad (19)$$

since $r = \lfloor \log_2(q) \rfloor \approx 2\log_2(n)$ and $\sigma = \alpha q \approx \alpha n^2$ in LWE challenge problems. For examples, by the inequality (19), we obtain a lower bound of $c(n,\alpha)$ as follows; (i) For $(n,\alpha) = (40, 0.005)$, we have $c(n,\alpha) > 0.98861$. (ii) For $(n,\alpha) = (40, 0.010)$, we have $c(n,\alpha) > 0.99721$. (iii) For $(n,\alpha) = (45, 0.005)$, we have $c(n,\alpha) > 0.99117$. Since $c_{\text{BKZ20}} = 0.9863$ at minimum (0.9868 on average) from Table 1, we estimate from (i) that we could not solve even the smallest LWE challenge pair $(n,\alpha) = (40, 0.005)$ by the key recovery attack with the BKZ algorithm with $\beta = 20$ and Babai's nearest plane method. Furthermore, by comparing (ii) and (iii), we estimate that larger α would be more difficult than larger n to solve LWE challenge problems by the key recovery attack.

Open Problems and Research Directions

From the above observation, we need to adopt stronger lattice basis reduction algorithms or/and improved Babai's nearest plane methods (e.g., see [24, 25]) to solve LWE challenge problems. Actually, pruned enumeration techniques or/and improved BKZ algorithms (e.g., progressive BKZ [3]) have been used for currently solved problems (see [39] for details). There are several strategies for solving LWE as introduced in Sect. 1, but one open problem is to find which strategy is the best for a given LWE instance (moreover, which algorithms should be adopted in the strategy?). Another open problem is to evaluate the computational hardness of the ring-LWE problem [32] (which is a special version of LWE), and to find a difference between LWE and ring-LWE problems.

Acknowledgements A part of this work was also supported by JSPS KAKENHI Grant Number 16H02830. The author thanks Momonari Kudo, Yang Guo, and Junpei Yamaguchi for their collecting experimental data.

References

1. M.A. Albrecht, C. Cid, J.-C. Faugère, R. Fitzpartrick, L. Perret, On the complexity of the BKW algorithm on LWE. Des. Codes Cryptogr. **74**, 325–354 (2015)
2. M.A. Albrecht, C. Cid, J.-C. Faugère, L. Perret, Algebraic algorithms for LWE, IACR ePrint 2014/1018
3. Y. Aono, Y. Wang, T. Hayashi, T. Takagi, Improved progressive BKZ algorithms and their precise cost estimation by sharp simulator, in *Advances in Cryptology-EUROCRYPT 2016*, vol. 9665 (Springer LNCS, 2016), pp. 789–819
4. S. Arora, R. Ge, New algorithms for learning in presence of errors, in *Automata, Languages and Programming* vol. 6755 (Springer LNCS, 2011), pp. 403–415

5. M.R. Albrecht, R. Player, S. Scott, On the concrete hardness of learning with errors. J. Math. Cryptol. **9**(3), 169–203 (2015)
6. L. Babai, On Lovász' lattice reduction and the nearest lattice point problem. Combinatorica **6**(1), 1–13 (1986)
7. A. Blum, A. Kalai, H. Wasserman, Noise-tolerant learning, the parity problem, and the statistical query model. J. ACM **50**(4), 506–519 (2003)
8. D. Boneh, R. Venkatesan, Hardness of computing the most significant bits of secret keys in Diffie-Hellman and related schemes, in *Advances in Cryptology-CRYPTO 1996*, vol. 1109 (Springer LNCS, 1996), pp. 129–142
9. Z. Brakerski, C. Gentry, V. Vaikuntanathan, (Leveled) fully homomorphic encryption without bootstrapping, in *Innovations in Theoretical Computer Science–ITCS 2012* (ACM, 2012), pp. 309–325
10. Z. Brakerski, A. Langlois, C. Peikert, O. Regev, D. Stehlé, Classical hardness of learning with errors, in *Theory of Computing–STOC 2013* (ACM, 2013), pp. 575–584
11. Z. Brakerski, V. Vaikuntanathan, Fully homomorphic encryption from ring-LWE and security for key dependent messages, in *Advances in Cryptology-CRYPTO 2011*, vol. 6841 (Springer LNCS, 2011), pp. 505–524
12. Z. Brakerski, V. Vaikuntanathan, Efficient fully homomorphic encryption from (standard) LWE, in *Foundations of Computer Science–FOCS 2011* (IEEE, 2011), pp. 97–106
13. M.R. Bremner, *Lattice Basis Reduction: An Introduction to the LLL Algorithm and its Applications* (CRC Press, Boca Raton, 2011)
14. J. Buchmann et al., Creating cryptographic challenges using muti-party computation: the LWE challenge, in *AsiaPKC 2016* (ACM, 2016), pp. 11–20
15. Y. Chen, P.Q. Nguyen, BKZ 2.0: Better lattice security estimates, in *Advances in Cryptology–ASIACRYPT 2011*, vol. 7073 (Springer LNCS, 2011), pp. 1–20
16. S.D. Galbraith, *Mathematics of Public Key Cryptography* (Cambridge University Press, Cambridge, 2012)
17. N. Gama, P.Q. Nguyen, Predicting lattice reduction, in *Advances in Cryptology-EUROCRYPT 2008*, vol. 4965 (Springer LNCS, 2008), pp. 31–51
18. C. Gentry, S. Gorbunov, S. Halevi, Graph-induced multilinear maps from lattices, in *Theory of Cryptography-TCC 2015*, vol. 9015 (Springer LNCS, 2015) pp. 498–527
19. R. Kannan, Minkowski's convex body theorem and integer programming. Math. Oper. Res. **12**(3), 415–440 (1987)
20. M.J. Kearns, Y. Mansour, D. Ron, R. Rubinfeld, R.E. Schapire, L. Sellie, On the learnability of discrete distributions, in *Theory of Computing–STOC 1994* (ACM, 1994) pp. 273–282
21. M. Kudo, J. Yamaguchi, Y. Guo, M. Yasuda, Practical analysis of key recovery attack against search-LWE problem, in *International Workshop on Security-IWSEC 2016*, vol. 9836 (Springer LNCS, 2016), pp. 164–181
22. K. Laine, K. Lauter, Key recovery for LWE in polynomial time, IACR ePrint 2015/176, (2015)
23. A.K. Lenstra, H.W. Lenstra, L. Lovász, Factoring polynomials with rational coefficients. Math. Ann. **261**(4), 515–534 (1982)
24. M. Liu, P.Q. Nguyen, Solving BDD by enumeration: a update, in *Topics in Cryptology-CT-RSA 2013*, vol. 7779 (Springer LNCS, 2013), pp. 293–309
25. R. Lindner, C. Peikert, Better key sizes (and attacks) for LWE-based encryption, in *Topics in Cryptology-CT-RSA 2011*, vol. 6558 (Springer LNCS, 2011), pp. 319–339
26. D. Miccincio, C. Peikert, Trapdoors for lattices: Simpler, tighter, faster, smaller, in *Advances in Cryptology-EUROCRYPT 2012*, vol.7237 (Springer LNCS, 2012), pp. 700–718 (2012)
27. D. Micciancio, O. Regev, Lattice-based cryptography, in *Post Quantum Cryptography–PQCrypto 2009* (Springer, 2009), pp. 147–191
28. P.Q. Nguyen, B. Vallée, *The LLL Algorithm, Information Security and Cryptography* (Springer, Berlin, 2010)
29. National Institute of Standards and Technology (NIST), *Report on post-quantum cryptography*, http://csrc.nist.gov/publications/drafts/nistir-8105/nistir_8105_draft.pdf
30. The PARI Group, *PARI/GP*, http://pari.math.u-bordeaux.fr/

31. C. Peikert, Public-key cryptosystems from the worst-case shortest vector problem: Extended abstract, in *Theory of Computing–STOC 2009* (ACM, 2009), pp. 333–342
32. C. Peikert, Challenges for Ring-LWE, http://web.eecs.umich.edu/cpeikert/rlwe-challenges/
33. C. Peikert, B. Waters, Lossy trapdoor functions and their applications. SIAM J. Comput. **40**(6), 1803–1844 (2011)
34. O. Regev, On lattices, learning with errors, random linear codes, and cryptography, in *Theory of Computing–STOC 2005* (ACM, 2005), pp. 84–93
35. O. Regev, On lattices, learning with errors, random linear codes, and cryptography. J. ACM **56**(6) (2009), Article No. 34
36. The Sage Group, *SageMath: Open-Source Mathematical Software System*, http://www.sagemath.org/
37. C.P. Schnorr, Lattice reduction by random sampling and birthday methods, in *STACS 2003* (Springer LNCS 2606, 2003) pp. 145–156
38. C.P. Schnorr, M. Euchner, Lattice basis reduction: improved practical algorithms and solving subset sum problems. Math. Program. **66**, 181–199 (1994)
39. TU Darmstadt, LWE Challenge, https://www.latticechallenge.org/lwe_challenge/challenge.php

A Mixed Integer Quadratic Formulation for the Shortest Vector Problem

Keiji Kimura and Hayato Waki

Abstract Lattice-based cryptography is based on the hardness of the lattice problems, e.g., the shortest vector problem and the closed vector problem. In fact, these mathematical optimization problems are known to be NP-hard. Our interest is to know how large-scale shortest vector problems can be solved. For this, we provide a mixed integer quadratic programming formulation for the shortest vector problem and propose a technique to restrict the search space of the shortest vector problem. This approach is a potential technique to improve the performance of the state-of-the-art software for mixed integer programming problems. In fact, we observe that this technique improves the numerical performance for TU Darmstadt's benchmark instances with the dimension up to 49.

Keywords Shortest vector problem · Integer program · Mixed integer program · Presolve · Optimization-based bound tightening · Second-order cone program

1 Introduction

The shortest vector problem (SVP) is the problem of finding the nonzero shortest vector constructed from a basis of a given integer lattice B in $\mathbb{R}^n$. The NP-hardness of SVP is shown by Ajtai [3]. On the other hand, LLL algorithm [9] is powerful to find a nonzero vector within $2^{(n-1)/2}$ factor of the shortest vector length. To solve SVP, Schnorr-Euchner's enumeration [12], Ajtai-Kumar-Sivakumar algorithm [4] and random sampling [7] are proposed as well as LLL algorithm.

K. Kimura (✉)
Graduate School of Mathematics, Kyushu University, 774 Motooka, Nishi-ku, Fukuoka 819-0395, Japan
e-mail: k-kimura@math.kyushu-u.ac.jp

H. Waki
Institute of Mathematics for Industry, Kyushu University, 774 Motooka, Nishi-ku, Fukuoka 819-0395, Japan
e-mail: waki@imi.kyushu-u.ac.jp

T. Takagi et al. (eds.), *Mathematical Modelling for Next-Generation Cryptography*, Mathematics for Industry 29, DOI 10.1007/978-981-10-5065-7_13

The purpose of this chapter is to provide a mixed integer quadratic programming (MIQP) formulation of SVP and propose a technique to restrict its search space of SVP. MIQP is the problem of minimizing a quadratic objective function over all points in a polyhedron whose elements are integer. MIQP problems are getting solvable by recent environment of computing and optimization technologies, for instance see [6]. One of the advantages of our MIQP formulation is to provide a guarantee that the computed vector is the shortest. We show in the numerical experiments that our MIQP formulation with the restriction of search space is a potential technique to improve the performance of CPLEX [14], which is the state-of-the-art software to solve mixed integer problems. This approach is one of the optimizations based on bound tightening and exploits second-order programming (SOCP) problems, which are the nonlinear programming problems. In particular, we prove that both the generated SOCP problems and their duals satisfy a sufficient condition to have their optimal solutions. It is guaranteed that both SOCP problems and their dual have optimal solutions and the duality gap is zero. Hence we can expect that their SOCP problems can be solved without numerical difficulties.

The organization of this chapter is as follows: Section 3 gives a brief introduction of a branch-and-bound algorithm, which is an algorithm to solve mixed integer programming problems including MIQP problems. Section 3 provides an MIQP formulation of SVP. We discuss a technique to restrict the search space of SVP in Sect. 4, and show the numerical performance of the technique proposed in Sect. 5. Appendix of this chapter provides a detail of SOCP and the proof of the strong feasibility of SOCP problems used for our proposed restriction.

We denote the sets of natural numbers, integer numbers, and real numbers by $\mathbb{N}$, $\mathbb{Z}$, and $\mathbb{R}$, respectively.

2 A General Description of a Branch-and-Bound Algorithm

This section presents briefly *a branch-and-bound* algorithm [1] that is used to solve optimization problems with integer variables. First, we consider the following optimization problem:

$$P_X : \min_x \{f(x) : x \in X,\ x \in \mathbb{R}^n,\ x_j \in \mathbb{Z}\ (j \in I)\}, \tag{1}$$

with an objective function $f : X \to \mathbb{R}$, a subset X of $\mathbb{R}^n$, and a subset $I \subseteq N = \{1, \ldots, n\}$ of the variable index set. If $x \in \mathbb{R}^n$ satisfies $x \in X$ and $x_j \in \mathbb{Z}$ for all $j \in I$, then x is said to be *feasible for* (1), *or a feasible solution of* (1). In addition, (1) said to be *feasible* if (1) has a feasible solution. A vector x^* is *an optimal solution of* (1) if x^* is feasible for (1) and $f(x^*) \leq f(x)$ for all feasible solutions x of (1).

We call the objective value $f(x^*)$ at an optimal solution x^* of (1) *the optimal value* of (1).[1]

Next, we consider the following problem without integrality restrictions:

$$\tilde{P}_X : \min_x \{f(x) : x \in X, \ x \in \mathbb{R}^n\}.$$

This problem is called *a relaxation problem* of P_X. Since $\tilde{P}_X$ has all feasible solutions of P_X, the optimal value $\tilde{\theta}$ of $\tilde{P}_X$ is less than or equal to the optimal value θ of P_X, i.e., $\tilde{\theta}$ is a lower bound of θ.

The branch-and-bound algorithm is a very general and widely used method in optimization problems. We summarize the branch-and-bound algorithm for a given optimization problem P_{X^0} on Algorithm 7. The idea is to successively divide the given problem until the all generated problems are solved. The problems generated by dividing are called *subproblems*. The best of the subproblems' optimal solutions is the global optimum.

Algorithm 7 Branch-and-bound algorithm

Input: The optimization problem P_{X^0} with the objective function $f : X^0 \to \mathbb{R}$, the subset I of the variable index set, and the subset X^0 of $\mathbb{R}^n$.

Output: The optimal solution x^* with the objective value θ^*, or conclusion that P_{X^0} has no solutions, indicated by $\theta^* = \infty$

1. Initialize $\Omega := \{0\}$, $\bar{\theta} := \infty$, $i := 0$.
2. If $\Omega = \phi$, stop and return $x^* = \bar{x}$ and $\theta^* = \bar{\theta}$.
3. Choose $k \in \Omega$, and set $\Omega := \Omega \backslash \{k\}$.
4. Solve the relaxation problem $\tilde{P}_{X^k}$. Let $\tilde{x}^k$ be an optimal solution of $\tilde{P}_{X^k}$ and $\tilde{\theta}^k$ its objective value.
5. If $\tilde{\theta}^k \geq \bar{\theta}$, go to Step 2.
6. If $\tilde{x}^k$ is feasible for P_{X^k}, set $\bar{x} := \tilde{x}^k$, $\bar{\theta} := \tilde{\theta}^k$ and go to Step 2.
7. Select variable index $j \in I$ with a fractional $\tilde{x}^k_j$, and set $X^{i+1} := X^k \cap \{x \in \mathbb{R}^n : x_j \leq \lfloor \tilde{x}^k_j \rfloor\}$, $X^{i+2} := X^k \cap \{x \in \mathbb{R}^n : x_j \geq \lceil \tilde{x}^k_j \rceil\}$, $\Omega := \Omega \cup \{i+1, i+2\}$.
8. Set $i := i + 2$, and go to Step 2.

The intention of Step 5 is to avoid a complete enumeration of all potential solutions of P_{X^0}, which are usually exponentially many. Since $\tilde{\theta}^k$ is the lower bound of the optimal value θ^k of the subproblem P_{X^k}, if $\tilde{\theta}^k \geq \bar{\theta}$, then $\theta^k \geq \bar{\theta}$. This implies that P_{X^k} has no better solutions than the current best solution $\bar{x}$ with the objective value $\bar{\theta}$. Then, the subproblem P_{X^k} does not need to be solved, and P_{X^k} is pruned.

Next, we explain Step 6 and Step 7. If $\tilde{x}^k$ is the feasible solution of P_{X^k}, $\tilde{\theta}^k$ is an upper bound of the optimal value θ^k of the subproblem P_{X^k}. Then, θ^k is equal to

---

[1]This is not a precise definition of the optimal value of (1). In fact, some of optimization problems have a finite optimal value, but no optimal solutions. Hence this definition is inappropriate to such optimization problems. However, this definition is valid for optimization problems in this chapter except for Appendix because they always have optimal solutions.

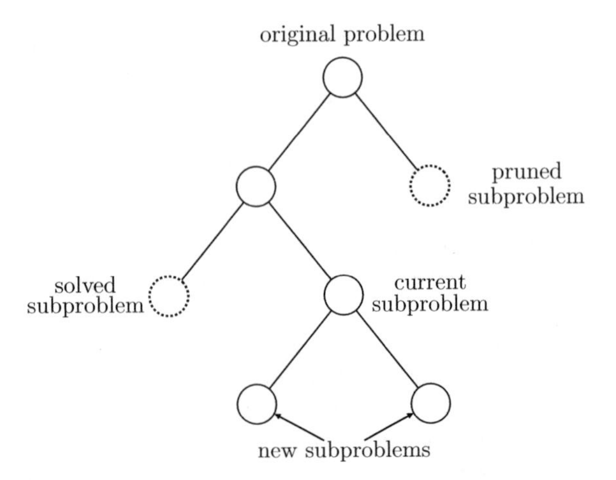

Fig. 1 Branching tree

$\tilde{\theta}^k$, and P_{X^k} is solved. Otherwise, a feasible solution set of P_{X^k} is divided into two sets in Step 7. This is called *branching*, and creates *a branching tree* with each node representing one of the subproblems (see Fig. 1). Let F_X be the feasible solution set of P_X for the subset X of $\mathbb{R}^n$. The branching in Step 7 divides F_{X^k} into $F_{X^{i+1}} \subseteq F_{X^0}$ and $F_{X^{i+2}} \subseteq F_{X^0}$ as follows: $F_{X^k} = F_{X^{i+1}} \cup F_{X^{i+2}}$ and $F_{X^{i+1}} \cap F_{X^{i+2}} = \emptyset$. Therefore, for any $k \in \Omega$, if $\tilde{x}^k$ is feasible for P_{X^k}, $\tilde{x}^k$ is also feasible for the original problem P_{X^0}. Then, the current best solution $\bar{x}$ and value $\bar{\theta}$ are updated in Step 6. Thus the branch-and-bound algorithm proceeds until the individual subproblems are solved or pruned from the branching tree.

If the objective function f in P_X is a quadratic function, and X in P_X is a polyhedron, i.e.,

$$f(x) = \frac{1}{2}x^T Qx + q^T x + r \text{ and } X = \{x \in \mathbb{R}^n : a_i^T x \leq b_i \ (i = 1, \dots, m)\},$$

where $Q \in \mathbb{R}^{n \times n}$ is a symmetric matrix, $q, a_1, \dots, a_m \in \mathbb{R}^n$ are vectors, and $r \in \mathbb{R}$ is a scalar, then P_X is called *a mixed integer quadratic problem* (MIQP). If the symmetric matrix $Q \in \mathbb{R}^{n \times n}$ is positive definite, then f is a strongly convex quadratic function, and the relaxation problem $\tilde{P}_X$ of P_X can be solved by using existing algorithms, e.g., an interior-point method.

We can solve the optimization problem such as (1) by using optimization software based on the branch-and-bound algorithm, e.g., CPLEX [14], Gurobi [16] and SCIP [17]. SCIP is one of the fastest noncommercial optimization solvers and a highly customizable framework for the branch-and-bound algorithm. CPLEX and Gurobi are one of the fastest commercial optimization solvers. We use CPLEX for numerical experiments in Sect. 5.

3 MIQP Formulation

We provide a mixed integer quadratic programming (MIQP) formulation of the shortest vector problem (SVP) in this section. Let $n \in \mathbb{N}$ and $\{b_1, \ldots, b_n\} \subset \mathbb{Z}^n$ be a basis of $\mathbb{R}^n$. In addition, we define the $n \times n$ matrix $B = (b_1, \ldots, b_n)$. We remark that each vector b_j is a column of B. SVP can be mathematically formulated as follows:

$$\theta^* := \min_{\beta_i} \left\{ \|B\beta\|_2^2 : \beta \neq 0_n, \beta = (\beta_1, \ldots, \beta_n)^T \in \mathbb{Z}^n \right\}, \tag{2}$$

where 0_n denotes the n-dimensional zero vector, $\|\cdot\|_2$ is the 2-norm and θ^* stands for the optimal value of (2). As the constraint $\beta \neq 0_n$ means that at least one of the elements of β is nonzero, this is not the form of the MIQP problem. For this, we define the following three optimization problems:

$$\theta_1^* := \min_{\beta_i} \left\{ \|B\beta\|_2^2 : \sum_{i=1}^n \beta_i \geq 1, \beta \in \mathbb{Z}^n \right\}, \tag{3}$$

$$\theta_{-1}^* := \min_{\beta_i} \left\{ \|B\beta\|_2^2 : \sum_{i=1}^n \beta_i \leq -1, \beta \in \mathbb{Z}^n \right\}, \tag{4}$$

$$\theta_0^* := \min_{\beta_i} \left\{ \|B\beta\|_2^2 : \sum_{i=1}^n \beta_i = 0, \beta \in \mathbb{Z}^n, \beta \neq 0_n \right\}. \tag{5}$$

We remark that we remove the constraint $\beta \neq 0_n$ from (3) and (4) because any feasible solution in (3) and (4) is nonzero. In addition, it is clear that $\theta^* = \min\{\theta_1^*, \theta_{-1}^*, \theta_0^*\}$ and that if β is feasible for (3), then $-\beta$ is also feasible for (4) with the same objective value as $\|B\beta\|_2^2$, and vice versa. Thus we have $\theta_1^* = \theta_{-1}^*$ and $\theta^* = \min\{\theta_1^*, \theta_0^*\}$.

The problem (3) is a form of an MIQP problem, while (5) is not a form of the MIQP problem because (5) still has the constraint $\beta \neq 0_n$. However, if we know upper and lower bounds $\ell_i, u_i \in \mathbb{Z}$ $(i = 1, \ldots, n)$ of any feasible solution of (5), i.e., $\ell_i \leq \beta_i \leq u_i$ for $i = 1, \ldots, n$ and any feasible solution β of (5), we can reformulate (5) as an MIQP problem. In fact, we add auxiliary variables $x_{i,v}$ for $i = 1, \ldots, n$, $v = \ell_i, \ell_i + 1, \ldots, u_i - 1, u_i$ into (5) as follows:

$$\theta_0^* = \min_{\beta_i, x_{i,v}} \left\{ \|B\beta\|_2^2 : \begin{array}{l} \sum_{i=1}^n \beta_i = 0, \beta_i = \sum_{v=\ell_i}^{u_i} v x_{i,v}, \sum_{v=\ell_i}^{u_i} x_{i,v} = 1, \\ \beta \in \mathbb{Z}^n, \sum_{i=1}^n x_{i,0} \leq n-1, \\ x_{i,v} \in \{0,1\} \ (i = 1, \ldots, n, v \in [\ell_i, u_i] \cap \mathbb{Z}) \end{array} \right\}. \tag{6}$$

As all the auxiliary variables $x_{i,v}$ in (6) are binary, the constraint $\sum_{v=\ell_i}^{u_i} x_{i,v} = 1$ implies that only one variable $x_{i,v}$ is 1 and the others are 0, and thus $\beta_i = v$ from the constraint $\beta_i = \sum_{v=\ell_i}^{u_i} v x_{i,v}$. In addition, the constraint $\sum_{i=1}^{n} x_{i,0} \leq n-1$ implies that $\beta \neq 0_n$. We see that (6) is the form of an MIQP problem if upper and lower bounds on β_i for $i = 1, \ldots, n$ are obtained in advance. Such bounds ℓ_i and u_i can be found by applying a technique proposed in the next section. In fact, we will discuss this at the end of the next section. On the other hand, we will observe in Sect. 5 that we have $\theta_1^* \leq \theta_0^*$ for almost all problems and that solving (6) spends more computational cost than solving (3).

4 Presolve Based on Second-Order Cone Programs

We introduce a technique to solve the MIQP problems (3) and (5) more efficiently. We first focus on (3) for the sake of simplicity although this can be applied to (5). This technique adds new valid upper/lower bounds of all variables β_i to (3) by using second-order cone programs (SOCPs). As a result, we restrict the search space of (3) and (5). There exist efficient algorithms, e.g., primal–dual interior-point methods, such that they can find an approximate solution to a given SOCP problem in polynomial time to any fixed tolerance. Since the MIQP problem obtained by this technique has the same optimal solution, this technique can be regarded as a preprocessing based on SOCP for (3) and (5).

This technique is one of the presolve techniques introduced in [2] and is called *optimization-based bound tightening*. The presolve is a set of routines to remove redundant variables and constraints, and strengthen the formulation of a given mixed integer program. We apply it to (3) and (6), respectively, although this technique can be also directly applied to (2) after removing the constraint $\beta \neq 0_n$ for (2).

We define the set X and the feasible region F of (3) by

$$X = \left\{ \beta \in \mathbb{R}^n : \sum_{i=1}^{n} \beta_i \geq 1 \right\} \text{ and } F = X \cap \mathbb{Z}^n.$$

We remark that $\theta_1^* = \min_{\beta_i} \{ \|B\beta\|_2^2 : \beta \in F \}$. Let M be the value of $\|B\hat{\beta}\|_2^2$ for a feasible solution $\hat{\beta}$ of (3). For instance, M can be chosen as the minimum over $\|b_1\|_2^2, \ldots, \|b_n\|_2^2$. Then $\hat{\beta} = e_{i^*}$ for some $i^* \in \{1, \ldots, n\}$. Here e_{i^*} stands for the i^* unit vector in $\mathbb{R}^n$. Denote the sets X_M and F_M by

$$X_M = X \cap \left\{ \beta \in \mathbb{R}^n : \|B\beta\|_2^2 \leq M \right\} \text{ and } F_M = X_M \cap \mathbb{Z}^n.$$

Then the following lemma holds:

Lemma 4.1 *We consider the following optimization problem:*

$$\min_{\beta_i} \left\{ \|B\beta\|_2^2 : \beta = (\beta_1, \ldots, \beta_n)^T \in F_M \right\}. \tag{7}$$

Then the optimal value of (7) is equal to θ_1^*.

Proof η^* denotes the optimal value of (7). It is clear that $\theta_1^* \leq \eta^*$ because $F_M \subseteq F$. It is sufficient to prove $\eta^* \leq \theta_1^*$. We assume that an optimal solution $\tilde{\beta}$ of (3) satisfies $\|B\tilde{\beta}\|_2^2 > M$. Otherwise, $\tilde{\beta}$ is also feasible for (7), and thus $\eta^* \leq \theta_1^*$. As $\hat{\beta}$ is feasible for both (3) and (7) and $\|B\hat{\beta}\|_2^2 = M < \|B\tilde{\beta}\|_2^2$, this inequality contradicts the optimality of $\tilde{\beta}$ for (3). Therefore, any optimal solution $\tilde{\beta}$ of (3) satisfies $\|B\tilde{\beta}\|_2^2 \leq M$, and thus $\eta_1^* = \theta_1^*$. □

For any feasible solution $\beta \in F_M$, we have

$$\min_{\beta_i} \{\beta_i : \beta \in F_M\} \leq \beta_i \leq \max_{\beta_i} \{\beta_i : \beta \in F_M\} \tag{8}$$

for all $i = 1, \ldots, n$. Relaxing the optimization problems at the both sides, we have the following inequalities for all feasible solutions $\beta \in F_M$ and $i = 1, \ldots, n$:

$$\min_{\beta_i} \{\beta_i : \beta \in X_M\} \leq \beta_i \leq \max_{\beta_i} \{\beta_i : \beta \in X_M\}. \tag{9}$$

The optimization problems at both sides of (9) are called second-order cone programming (SOCP) problems. See Appendix 1 of this chapter and [5] for the details of SOCP. On the other hand, the optimization problems at the both sides of (8) are called mixed integer second-order cone programming (MISOCP) problems, and CPLEX and Gurobi can handle problems of this type. In our preliminary numerical experiment, we will observe that solving (8) is much more computationally intensive than solving (9).

Similarly, for any feasible solution $\beta \in F_M$, we can refine the coefficients of the linear inequality in (3) by using the following inequalities:

$$\min_{\beta_i} \left\{\sum_{i=1}^n \beta_i : \beta \in X_M\right\} \leq \sum_{i=1}^n \beta_i \leq \max_{\beta_i} \left\{\sum_{i=1}^n \beta_i : \beta \in X_M\right\}. \tag{10}$$

We propose the presolve techniques based on SOCP in Algorithm 8. Here the feasible region of the relaxation of (3) in Algorithm 8 is defined as follows:

$$F(\ell_1, \ldots, \ell_n, u_1, \ldots, u_n, c, d) := \left\{\beta \in \mathbb{R}^n : \begin{array}{c} \beta \in X_M, c \leq \sum_{i=1}^n \beta_i \leq d, \\ \ell_i \leq \beta_i \leq u_i \ (i = 1, \ldots, n) \end{array}\right\}.$$

In Algorithm 8, we apply the rounding up and rounding down to the computed values because the all variables in the original problem (3) are integer.

Algorithm 8 Algorithm to refine bounds of variables and coefficients in constraints of (3)

Input: $B = (b_1, \dots, b_n) \in \mathbb{Z}^{n \times n}$ and $M \in \mathbb{Z}$
Output: Lower and upper bounds ℓ_i, u_i of the variable β_i and coefficients in (3)
$\ell_i \leftarrow -\infty, u_i \leftarrow +\infty \ (i = 1, \dots, n), c \leftarrow 1, d \leftarrow +\infty$ **do**
  /* Computation of upper and lower bounds u_i, ℓ_i */
  for $i \leftarrow 1$ **to** n **do**
    $v \leftarrow \min\{\beta_i : \beta \in F(\ell_1, \dots, \ell_n, u_1, \dots, u_n, c, d)\}$
    if $\lceil v \rceil > \ell_i$ **then**
      $\ell_i \leftarrow \lceil v \rceil$, update $F(\ell_1, \dots, \ell_n, u_1, \dots, u_n, c, d)$
    end
    $v \leftarrow \max\{\beta_i : \beta \in F(\ell_1, \dots, \ell_n, u_1, \dots, u_n, c, d)\}$
    if $\lfloor v \rfloor < u_i$ **then**
      $u_i \leftarrow \lfloor v \rfloor$, update $F(\ell_1, \dots, \ell_n, u_1, \dots, u_n, c, d)$
    end
  end
  /* Computation of coefficients c, d in (3) */
  $v \leftarrow \min\left\{\sum_{i=1}^{n} \beta_i : \beta \in F(\ell_1, \dots, \ell_n, u_1, \dots, u_n, c, d)\right\}$
  if $\lfloor v \rfloor > c$ **then**
    $c \leftarrow \lfloor v \rfloor$ and update $F(\ell_1, \dots, \ell_n, u_1, \dots, u_n, c, d)$
  end
  $v \leftarrow \max\left\{\sum_{i=1}^{n} \beta_i : \beta \in F(\ell_1, \dots, \ell_n, u_1, \dots, u_n, c, d)\right\}$
  if $\lceil v \rceil < d$ **then**
    $d \leftarrow \lceil v \rceil$ and update $F(\ell_1, \dots, \ell_n, u_1, \dots, u_n, c, d)$
  end
while $F(\ell_1, \dots, \ell_n, u_1, \dots, u_n, c, d)$ *is updated*;

We need to solve SOCP in Algorithm 8. In general, as SOCP is a nonlinear programming problem, it may not have any optimal solutions even if the optimal value is finite. Propositions 8.1 and 8.2 in Appendix ensure that SOCP problems in (9) and (10), and their duals have optimal solutions under assumptions. We observe that the assumptions hold for numerical experiment in Sect. 5.

We can apply this type of the presolve technique to (5). We replace the set X with the following set:

$$X = \left\{\beta \in \mathbb{R}^n : \sum_{j=1}^{n} \beta_j = 0\right\}.$$

Then the feasible region F of (5) is included in the set $X \cap \mathbb{Z}^n$. We choose the optimal value or the upper bound of (3) as $M \in \mathbb{Z}$, and consider the following optimization problem:

$$\min_{\beta_i} \left\{\|B\beta\|_2^2 : \beta = (\beta_1, \dots, \beta_n)^T \in X \cap \mathbb{Z}^n, \beta \neq 0_n, \|B\beta\|_2^2 \leq M\right\}. \quad (11)$$

We remark that if (11) is infeasible, i.e., the feasible region is empty, then $\theta_1^* \leq \theta_0^*$ and thus $\theta^* = \theta_1^*$. In addition, if M is the optimal value of (3) and (11) has a feasible solution, then $\theta_0^* \leq \theta_1^*$ and $\theta^* = \theta_0^*$.

From (11) and a similar manner to (9), every feasible solution $\beta = (\beta_1, \ldots, \beta_n)^T$ of (5) satisfies

$$\min_{\beta_i} \left\{\beta_i : \beta \in X, \|B\beta\|_2^2 \leq M\right\} \leq \beta_i \leq \max_{\beta_i} \left\{\beta_i : \beta \in X, \|B\beta\|_2^2 \leq M\right\}. \quad (12)$$

We can use the optimal values of the both sides as lower and upper bounds of β_i in (5), respectively.

5 Preliminary Numerical Experiment

In this section, we report numerical results for some SVP (2) obtained by the generator that is available at [18], and show the performance of Algorithm 8. The computational environment that we used in this numerical experiment is

OS Ubuntu 16.04 1 LTS,
CPU Intel® Xeon® CPU E5-2687W with 3.1 GHz and 32 threads,
Memory 128GB.

We applied LLL algorithm implemented in `fplll` [15] to the generated matrix B to improve computational efficiency. We used CPLEX 12.6.3 [14] to solve (3), (6) and optimization problems appeared in Algorithm 8. The parameters in CPLEX are default except for

- `preprocessing presolve = no`,
- `mip tolerances mipgap = 1e-10`,
- `timelimit = 86400`.

The first parameter indicates whether we execute some of the presolve techniques implemented in CPLEX or not. We switched it off to observe the performance of Algorithm 8. We chose the minimum value over $\|b_1\|_2^2, \ldots, \|b_n\|_2^2$ as M in Algorithm 8.

We solved the following MISOCP problems to obtain tighter bounds ℓ_i and u_i instead of (9) in Algorithm 8:

$$\min_{\beta_i} \{\beta_i : \beta_i \in \mathbb{Z}, \beta \in X_M\} \leq \beta_i \leq \max_{\beta_i} \{\beta_i : \beta_i \in \mathbb{Z}, \beta \in X_M\}. \quad (13)$$

We observe that all problems in (13) can be solved by CPLEX in a few minutes for our SVPs.

Table 1 shows the numerical results of some SVP to which Algorithm 8 is applied. The first column shows the dimension n and seed that we used in the generator. The third column indicates the CPU time to solve the generated MIQP problem in seconds.

Table 1 Numerical results for some benchmark in [18]

(n, seed)	Problem	Time [s]	Gap [%]	$\|B\beta^*\|_2$	Approx. Factor
(40, 0)	(3)	209.30	0.00	**1702.46**	1.03
	(6)	465.41	0.00	1811.98	1.09
(40, 11)	(3)	83.68	0.00	**1587.12**	0.96
	(6)	1410.75	0.00	1819.72	1.10
(40, 12)	(3)	19.66	0.00	**1464.76**	0.89
	(6)	697.76	0.00	1758.36	1.07
(40, 76)	(3)	11.90	0.00	**1434.38**	0.87
	(6)	1340.06	0.00	1854.68	1.13
(41, 31)	(3)	272.94	0.00	1674.03	1.00
	(6)	142.12	0.00	**1561.65**	0.93
(41, 135)	(3)	30.74	0.00	**1480.57**	0.89
	(6)	1630.42	0.00	1830.64	1.10
(42, 47)	(3)	115.03	0.00	**1495.81**	0.89
	(6)	17994.53	0.00	1901.30	1.13
(43, 2)	(3)	328.02	0.00	**1545.39**	0.90
	(6)	11269.10	0.00	1814.72	1.06
(44, 8)	(3)	86.02	0.00	**1573.49**	0.91
	(6)	1275.68	0.00	1754.25	1.01
(45, 79)	(3)	427.39	0.00	1547.23	0.88
	(6)	>86400	19.67	1909.88	1.09
(46, 16)	(3)	1098.46	0.00	1565.88	0.89
	(6)	>86400	10.65	1839.76	1.04
(47, 95)	(3)	9053.29	0.00	1678.65	0.94
	(6)	>86400	19.62	1914.31	1.08
(48, 7)	(3)	45872.33	0.00	1703.24	0.95
	(6)	>86400	30.59	1873.50	1.04
(49, 7)	(3)	>86400	30.12	1885.76	1.05
	(6)	>86400	51.14	2004.51	1.11
(49, 72)	(3)	17778.13	0.00	1707.36	0.94
	(6)	>86400	36.82	1945.93	1.08
(49, 126)	(3)	9429.87	0.00	1659.49	0.91
	(6)	>86400	36.07	1979.29	1.09

">86400" means that the algorithm cannot solve the problem within 86400 s = 1 day. The forth column indicates the gap between the upper and the lower bounds of the optimal value θ^* of (2). This value is defined by

$$\text{Gap} := \frac{\text{upper bound} - \text{lower bound}}{\max\{1, |\text{upper bound}|\}} \times 100.$$

Table 2 Numerical comparisons of Algorithm 8 for some benchmark in [18]

	Without Algorithm 8		With Algorithm 8	
(n, seed)	Time [s]	Gap [%]	Time [s]	Gap [%]
(40, 0)	**204.69**	0.00	209.30	0.00
(40, 11)	**74.18**	0.00	83.68	0.00
(40, 12)	**19.40**	0.00	19.66	0.00
(40, 76)	13.67	0.00	**11.90**	0.00
(41, 31)	**214.01**	0.00	272.94	0.00
(41, 135)	58.48	0.00	**30.74**	0.00
(42, 47)	**110.71**	0.00	115.03	0.00
(43, 2)	**300.60**	0.00	328.02	0.00
(44, 8)	97.30	0.00	**86.02**	0.00
(45, 79)	462.13	0.00	**427.39**	0.00
(46, 16)	4147.39	0.00	**1098.46**	0.00
(47, 95)	**8880.61**	0.00	9053.29	0.00
(48, 7)	61404.46	0.00	**45872.33**	0.00
(49, 7)	>86400	**29.63**	>86400	30.12
(49, 72)	26369.60	0.00	**17778.13**	0.00
(49, 126)	12160.23	0.00	**9429.87**	0.00

It is clear that CPLEX finds the optimal value and solution of the MIQP problem if the gap is zero. The fifth column stands for the optimal value or computed upper bound of the optimal value. In particular, the value with bold font style is the optimal value θ^* of (2). The sixth column is the following value:

$$\text{Approx. Factor} := \frac{\|B\beta^*\|_2}{\Gamma(n/2+1)^{1/n}|\det(B)|^{1/n}/\sqrt{\pi}},$$

where β^* is the computed solution by CPLEX, and $\Gamma(n/2+1)$ stands for the value of the gamma function at $(n/2+1)$. In particular, Approx. Factor is used in the case where CPLEX cannot find any optimal solution, but find some feasible solutions, and is a measure of the quality of the computed solution. In [18], it is required to find a feasible solution whose Approx. Factor is less than 1.05.

We observe from Table 1 that we can solve (3) much faster than (6). In Table 2, we show the numerical performance of Algorithm 8 for only MIQP problems (3). We observe from Table 2 that Algorithm 8 improves the numerical performance for SVPs with $n \geq 44$. A further improvement of the numerical performance is future work.

6 Conclusion

We introduce a general scheme of branch-and-bound algorithm for mixed integer program and propose an MIQP formulation of SVP and one of the presolve techniques to improve the computational efficiency, which is described in Algorithm 8. We observe that the Algorithm 8 has a potential to improve the numerical performance of the state-of-the-art software, e.g., CPLEX and Gurobi, etc.

In this chapter, although we only apply Algorithm 8 to the original problem in the root node of the branching tree, we can apply it to the subproblem of every node in the tree. We plan to solve larger-scale SVPs. For this, we use SCIP [17] for the implementation. In addition, Gleixner et al. [8] propose some techniques to tighten the upper and lower bounds for mixed integer nonlinear programming problems. The application of these techniques is involved in our future work.

Acknowledgements We would like to thank Dr. Masaya Yasuda in Kyushu University for fruitful discussions.

Appendix 1 : A Second-Order Cone Program Problem

We provide a brief introduction of a second-order cone program (SOCP) problem. Denote the set K by

$$K = \left\{ x = \begin{pmatrix} x_1 \\ x_2 \end{pmatrix} \in \mathbb{R}^n : x_1 \in \mathbb{R}, x_2 \in \mathbb{R}^{n-1}, x_1 \geq \|x_2\|_2 \right\},$$

where $\|\cdot\|_2$ stands for the 2-norm and K is called *the second-order cone*. For given $A \in \mathbb{R}^{m\times n}$, $b \in \mathbb{R}^m$ and $c \in \mathbb{R}^n$, the SOCP problem (14) and its dual (15) are formulated as follows:

$$\theta_P^* := \inf_x \left\{ c^T x : Ax = b, x \in K \right\}, \tag{14}$$

$$\theta_D^* := \sup_y \left\{ b^T y : c - A^T y \in K^*, y \in \mathbb{R}^m \right\}. \tag{15}$$

Here K^* stands for the dual cone of K and is defined by $K^* = \{s = (s_1, s_2) \in \mathbb{R}^n : s_1 \geq \|s_2\|_2\}$. We remark that the SOCP problem (14) and its dual (15) may not have any optimal solutions even if their optimal values are finite. Such examples are given in [11]. For this, we use the terminology "inf" and "sup" instead of "min" and "max" in (14) and (15), respectively. We define $\theta_P^* = -\infty$ when (14) has no feasible solutions. Similarly, we define $\theta_D^* = -\infty$ when (15) has no feasible solutions.

For any pair (x, y) of feasible solutions of (14) and (15), we have

$$c^T x \geq b^T y.$$

This inequality is called *the weak duality* and implies $\theta_P^* \geq \theta_D^*$. It follows from the weak duality that if there exist a pair of feasible solutions x of (14) and y of (15) such that $c^T x = b^T y$, $\theta_P^* = \theta_D^*$ holds and thus the pair (x, y) of the feasible solution is a pair of optimal solutions for (14) and (15). *The strong duality theorem* ensures $\theta_P^* = \theta_D^*$ under an assumption to (14) and (15). To this end, we introduce the strong feasibility for (14) and (15). (14) is said to be *strongly feasible or strictly feasible* if there exists a feasible solution $\hat{x}$ of (14) such that $A\hat{x} = b$ and $(\hat{x})_1 > \|\hat{x}_2\|_2$. (15) is said to be *strongly feasible or strictly feasible* if there exists a feasible solution $\hat{y}$ of (15) such that $c - A^T \hat{y} \in \text{int}(K^*)$. Here $\text{int}(K^*)$ stands for the interior of K^*.

Theorem 7.1 (Strong duality; see, e.g., [11]) *If (14) is strongly feasible and (15) is feasible, then $\theta_P^* = \theta_D^*$ and (15) has an optimal solution. Similarly, if (15) is strongly feasible and (14) is feasible, then $\theta_P^* = \theta_D^*$ and (14) has an optimal solution.*

The theoretical convergence of most of algorithms to solve (14) requires the strong feasibility for both (14) and (15). In fact, primal–dual interior-point methods approximately compute $(x, y) \in \mathbb{R}^n \times \mathbb{R}^n$ that satisfies

$$\begin{cases} Ax = b, x \in K, \\ c - A^T y \in K^*, \\ c^T x - b^T y = 0. \end{cases}$$

If (14) and (15) are strongly feasible, it follows from Theorem 7.1 that the above system has an solution, which is optimal for (14) and (15).

Finally, we give a characterization for (14) and (15) to be strongly feasible.

Theorem 7.2 (See [10, 13]) *The exactly one of the following two statements is true:*

1. *(14) is strongly feasible.*
2. *There exists $y \in \mathbb{R}^m \setminus \{0\}$ such that $-A^T y \in K^*$ and $b^T y \geq 0$.*

In particular, if there exists $y \in \mathbb{R}^m \setminus \{0\}$ such that $-A^T y \in K^*$ and $b^T y > 0$, (14) is infeasible. Similarly, the exactly one of the following two statements is true:

1. *(15) is strongly feasible.*
2. *There exists $x \in \mathbb{R}^n \setminus \{0\}$ such that $Ax = 0$, $x \in K$ and $c^T x \leq 0$.*

In particular, if there exists $y \in \mathbb{R}^m \setminus \{0\}$ such that $Ax = 0$, $x \in K$ and $c^T x < 0$, (15) is infeasible.

Appendix 2 : An SOCP Formulation for (9) and (10), and Its Strong Feasibility

In this section, we provide an SOCP formulation for (9) and (10), and prove that both the SOCP problem and its dual are strongly feasible under a mild assumption. The latter implies that both problems have optimal solutions, and that algorithms to solve SOCP problems, e.g., primal–dual interior-point methods converge an optimal solution.

We assume that (3) has a feasible solution $\hat{\beta}$. In addition, we allow $\ell_i = -\infty$ and $u_I = +\infty$, but assume $\ell_i \leq u_i$ for all $i = 1, \ldots, n$.

To restrict the search space of a given SVP, we choose $M = \|B\hat{\beta}\|_2^2$. The SOCP formulation for (9) and (10) are

$$\sup_{y} \left\{ f^T y : \begin{array}{l} \ell_j \leq y_j \ (i \in L), y_j \leq u_j \ (j \in U), e^T y \geq 1, \\ \|By\|_2 \leq \sqrt{M}, y \in \mathbb{R}^n \end{array} \right\}, \tag{16}$$

where we define

$$L = \left\{ j \in \{1, \ldots, n\} : \ell_j \text{ is finite} \right\} \text{ and } U = \left\{ j \in \{1, \ldots, n\} : u_j \text{ is finite} \right\},$$

and e stands for the n-dimensional ones vector and f is $\pm e_i$ for (9) and $\pm e$ for (10). Its dual can be formulated as follows:

$$\inf_{\substack{s_j, t_j, w, \\ x_1, x_2}} \left\{ \begin{array}{l} \sum_{j \in U} u_j s_j - \sum_{j \in L} \ell_j t_j \\ -w + \sqrt{M} x_1 \end{array} : \begin{array}{l} s_j - t_j - w - b_j^T x_2 = f_j \ (j \in U \cap L), \\ s_j - w - b_j^T x_2 = f_j \ (j \in U \setminus L), \\ -t_j - w - b_j^T x_2 = f_j \ (j \in L \setminus U), \\ -w - b_j^T x_2 = f_j \ (j \in \{1, \ldots, n\} \setminus (U \cup L)), \\ s_j \geq 0 \ (j \in U), t_j \geq 0 \ (j \in L), \\ w \geq 0, x_1 \geq \|x_2\|_2 \end{array} \right\}. \tag{17}$$

By applying Theorem 7.2 to (17), we can prove the strong feasibility of (17).

Proposition 8.1 (17) *is strongly feasible.*

Proof We consider the following system

$$\begin{cases} f^T y \geq 0, \\ 0 \leq y_j \ (j \in L), y_j \leq 0 \ (j \in U), e^T y \geq 0, \\ \|By\|_2 \leq 0, y \in \mathbb{R}^n \end{cases} \tag{18}$$

As B is nonsingular, $\|By\|_2 \leq 0$ implies $y = 0$ and (18) does not have any nonzero solutions. It follows from Theorem 7.2 that (17) is strongly feasible. □

It should be noted that the strong feasibility of (17) is independent of the choice of f, L, and U although the feasible region of the dual (17) depends on them.

By applying Theorem 7.2 to (16) for the strong feasibility of (16), it is sufficient to consider the following system:

$$\begin{cases} \sum_{j\in U} u_j s_j - \sum_{j\in L} \ell_j t_j - w + \sqrt{M} x_1 \le 0, \\ s_j - t_j - w - b_j^T x_2 = 0 \ (j \in U \cap L), \\ s_j - w - b_j^T x_2 = 0 \ (j \in U \setminus L), \\ -t_j - w - b_j^T x_2 = 0 \ (j \in L \setminus U), \\ -w - b_j^T x_2 = 0 \ (j \in \{1, \dots, n\} \setminus (U \cup L)), \\ s_j \ge 0 \ (j \in U), t_j \ge 0 \ (j \in L), w \ge 0, x_1 \ge \|x_2\|_2 \end{cases} \tag{19}$$

We prove that (19) has only the zero solution under a mild assumption. This implies (16) is strongly feasible.

Proposition 8.2 *Assume $\ell < \hat{\beta} < u$, i.e., $\ell_i < \hat{\beta}_i < u_i$ for all $i = 1, \dots, n$. Moreover, we assume that there exists $j \in \{1, \dots, n\}$ such that $(B\hat{\beta})^T b_j \ne M$. Then (19) does not have any nonzero solutions.*

Proof From (19), we obtain

$$\hat{\beta}_j w = \begin{cases} \hat{\beta}_j (s_j - t_j - b_j^T x_2) & \text{if } j \in U \cap L, \\ \hat{\beta}_j (s_j - b_j^T x_2) & \text{if } j \in U \setminus L, \\ \hat{\beta}_j (-t_j - b_j^T x_2) & \text{if } j \in L \setminus U, \\ -\hat{\beta}_j b_j^T x_2 & \text{if } j \in \{1, \dots, n\} \setminus (U \cup L), \end{cases} \quad (j = 1, \dots, n).$$

Then we obtain $w(\hat{\beta}^T e) = \sum_{j\in U} \hat{\beta}_j s_j - \sum_{j \in L} \hat{\beta}_j t_j - (B\hat{\beta})^T x_2$, and thus

$$\sum_{j\in U} u_j s_j - \sum_{j\in L} \ell_j t_j + \sqrt{M} x_1 \le w \le w(\hat{\beta}^T e) = \sum_{j\in U} \hat{\beta}_j s_j - \sum_{j\in L} \hat{\beta}_j t_j - (B\hat{\beta})^T x_2.$$

It follows from $x_1 \ge \|x_2\|_2$ that the left hand side can be replaced by

$$\sum_{j\in U} u_j s_j - \sum_{j\in L} \ell_j t_j + \sqrt{M} \|x_2\|_2.$$

We have $\sqrt{M}\|x_2\|_2 = \|B\hat{\beta}\|_2 \|x_2\|_2 \ge |(B\hat{\beta})^T x_2| \ge -(B\hat{\beta})^T x_2$ due to the Cauchy-Schwarz inequality. In addition, the assumption on bounds u and ℓ implies

$$\sum_{j\in U} u_j s_j - \sum_{j\in L} \ell_j t_j \ge \sum_{j\in U} \hat{\beta}_j s_j - \sum_{j\in L} \hat{\beta}_j t_j.$$

As both s and t are nonnegative, we obtain the following equalities from these inequalities:

$$s_j = 0 \ (j \in U), t_j = 0 \ (j \in L), x_1 = \|x_2\|_2, \sqrt{M}\|x_2\|_2 = -(B\hat{\beta})^T x_2.$$

It follows from the last equality that we have $x_2 = -\alpha B\hat{\beta}$ for some $\alpha \geq 0$. Substituting them into (19), we can rewrite (19) into the following system:

$$\begin{cases} -w + |\alpha|M = 0, \alpha \geq 0, \\ w = \alpha b_j^T B\hat{\beta} \ (j = 1, \ldots, n), \\ w \geq 0, x_1 = |\alpha|\sqrt{M}. \end{cases} \tag{20}$$

From $w \geq 0$, the first and second equalities, we have $M = b_j^T B\hat{\beta}$ for all $j = 1, \ldots, n$ if $\alpha \neq 0$. This contradicts the assumption. Hence $\alpha = 0$ and $w = x_1 = 0$ and $x_2 = 0_{n-1}$, and (16) is strongly feasible under the assumption. □

References

1. T. Achterberg, Constraint Integer Programming, Ph.D. thesis, Technische Universität Berlin (2007)
2. T. Achterberg, R.E. Bixby, Z. Gu, E. Rothberg, D. Weninger, Multi-row presolve reductions in mixed integer programming, in *Proceedings of the Twenty-Sixth RAMP Symposium* (2014), pp. 181–196
3. M. Ajtai, The shortest vector problem in L_2 is NP-hard for randomized reductions (extended abstract), in *Proceedings of the thirtieth Annual Symposium on the Theory of Computing* (1998), pp. 10–19
4. M. Ajtai, R. Kumar, D. Sivakumar, A sieve algorithm for the shortest lattice vector problem, in *Proceedings of the Thirty-third Annual ACM Symposium on Theory of Computing* (2001), pp. 601–610
5. F. Alizadeh, D. Goldfarb, Second-order cone programming. Math. Program. **95**, 3–51 (2003)
6. C. Bliek, P. Bonami, A. Lodi, Solving mixed-integer quadratic programming problems with IBM-CPLEX: a progress report, in *Proceedings of the Twenty-Sixth RAMP Symposium* (2014), pp. 171–180
7. M. Fukase, K. Kashiwabara, An accelerated algorithm for solving SVP based on statistical analysis. J. Inf. Process. **23**, 1–15 (2015)
8. A.M. Gleixner, T. Bertholdm, B.Müller, S. Weltge, Three enhancements for optimization-based bound tightening. J. Glob. Optim. (2016). doi:10.1007/s10898-016-0450-4
9. A.K. Lenstra, H.W. Lenstra Jr., L. Lovász, Factoring polynomials with rational coefficients. Math. Ann. **261**, 515–534 (1982)
10. G. Pataki, Strong duality in conic linear programming: facial reduction and extended dual, in *Computational and Analytical Mathematics*, vol. 50, ed. By D. Bailey et al. Springer Proceedings in Mathematics & Statistics (2013), pp. 613–634
11. J. Renegar, A mathematical view of interior-point methods in convex optimization. SIAM (2001)
12. C.P. Schnorr, M. Euchner, Lattice basis reduction: improved practical algorithms and solving subset sum problems. Math. program. **66**, 181–199 (1994)
13. H. Waki, M. Muramatsu, Facial reduction algorithms for conic optimization problems. J. Optim. Theory Appl. **158**, 188–215 (2013)

14. IBM ILOG CPLEX Optimizer 12.6.3, IBM ILOG (2015)
15. The FPLLL development team, fplll, a lattice reduction library (2016), https://github.com/fplll/fplll
16. Gurobi Optimization, Inc., Gurobi Optimizer Reference Manual (2015), http://www.gurobi.com
17. SCIP: Solving Constraint Integer Programs, http://scip.zib.de/
18. SVP CHALLENGE, https://www.latticechallenge.org/svp-challenge/

On Analysis of Recovering Short Generator Problems via Upper and Lower Bounds of Dirichlet L-Functions: Part 1

Shingo Sugiyama

Abstract This article is a survey on upper and lower bounds of Dirichlet L-functions $L(s, \chi)$ associated with Dirichlet characters χ at $s = 1$. We give proofs of well-known upper and lower bounds of $L(1, \chi)$ to let the reader know the difficulty of giving (lower) bounds of $L(1, \chi)$. In the last part, we also review some explicit upper and lower bounds of Dirichlet L-functions which will be applied to security analysis of ideal lattice-based cryptography for cyclotomic fields explained in Part 2 (S. Okumura, On analysis of recovering short generator problems via upper and lower bounds of Dirichlet L-functions : Part 2 [20]).

Keywords Upper and lower bounds of Dirichlet L-functions · Zero-free regions · Siegel zeros · Ideal lattice-based cryptography

1 Introduction

Now, on the earth, there are so many zeta functions and L-functions as generalizations of the Riemann zeta function

$$\zeta(s) = \sum_{n=1}^{\infty} \frac{1}{n^s} \qquad \Re(s) > 1$$

in any aspect beyond number theory. Among such functions, Dirichlet in 1831 first introduced the so-called *Dirichlet L-functions* as follows. Let χ be a Dirichlet character χ modulo q, which is defined as a group homomorphism from $(\mathbb{Z}/q\mathbb{Z})^\times$ to $\mathbb{C}^\times$. Then, χ can be naturally extended to a function $\tilde{\chi} : \mathbb{Z} \to \mathbb{C}$ such that $\tilde{\chi}(a) = 0$ if $\gcd(a, q) \neq 1$. Then the Dirichlet L-function associated with χ is defined by

S. Sugiyama (✉)
Institute of Mathematics for Industry, Kyushu University, 744, Motooka,
Nishi-ku, Fukuoka 819-0395, Japan
e-mail: s-sugiyama@imi.kyushu-u.ac.jp

T. Takagi et al. (eds.), *Mathematical Modelling for Next-Generation Cryptography*,
Mathematics for Industry 29, DOI 10.1007/978-981-10-5065-7_14

$$L(s,\chi)=\sum_{n=1}^{\infty}\frac{\tilde{\chi}(n)}{n^s}\qquad \Re(s)>1. \tag{1}$$

Dirichlet L-functions have been studied by many researchers in the field of analytic number theory for some applications such as infinitude of prime numbers in arithmetic progression, the prime number theorem for arithmetic progressions, the class number formula, the growth of class numbers of imaginary quadratic fields, and so on. Even for Dirichlet L-functions, some problems remain unknown and attract us mysteriously, and in particular a distribution of zeros of Dirichlet L-functions is of great interest in analytic number theory related to the generalized Riemann hypothesis (GRH). Recently, the deep Riemann hypothesis (DRH) was suggested in [9]. GRH and DRH will be our task to resolve in the future.

This article is an introductory survey for nonexperts on upper and lower bounds of the special value of the Dirichlet L-function $L(s,\chi)$ associated with a Dirichlet character χ at $s=1$. This is also a preparatory lecture for application to ideal lattice-based cryptography for cyclotomic fields $\mathbb{Q}(\zeta_{p^n})$ of some prime power conductor (cf. [2, 20–22]). One can verify that, if χ is trivial, then $L(s,\chi)$ is equal to $\zeta(s)$ up to elementary factor. When χ is nontrivial, the abscissa of convergence of $L(s,\chi)$ is $\Re(s)=0$ by a periodic property of χ, in contrast to the Riemann zeta function $\zeta(s)$. Hence, the special value $L(1,\chi)$ makes sense for any nontrivial Dirichlet character χ. This value is important to be studied in number theory; for example, Dirichlet's result $L(1,\chi)\neq 0$ led us to infinitude of prime numbers in arithmetic progression, and estimates of $L(1,\chi)$ are useful to investigate class numbers of number fields. The following upper bound is easily confirmed (cf. [3], [13, Proposition 4] and [18, p. 47]).

Theorem 1.1 *Let χ be a nontrivial Dirichlet character modulo q. Then, we have*

$$|L(1,\chi)|\ll \log q,$$

where the implied constant is independent of χ and q.

In contrast to upper bounds, to give lower bounds of $L(1,\chi)$ uniformly for both χ and q is very difficult and needs fine complex analysis. We call that a Dirichlet character χ is quadratic if χ is nontrivial and real-valued, i.e., the image of χ is $\{\pm 1\}$. Landau [10] first gave the following lower bound (see also [15]).

Theorem 1.2 *Let χ be a non-trivial Dirichlet character modulo q. Suppose that χ is not quadratic. Then, we have*

$$|L(1,\chi)|\gg \frac{1}{\log q},$$

where the implied constant is independent of χ and q.

The case where χ is quadratic was harder by technical reasons. However, Siegel [25] gave a breakthrough for such a case as follows.

Theorem 1.3 (Siegel's theorem) *Let χ be a quadratic Dirichlet character modulo q. Then, for any $\varepsilon > 0$ we have*

$$L(1, \chi) \gg_{\varepsilon} \frac{1}{q^{\varepsilon}},$$

where the implied constant is independent of χ and q (but may depend on ε).

Here we remark $L(1, \chi) > 0$ for any quadratic character χ. We also remark that there are a lot of other proofs as in [1, 5, 6, 13]. Since Theorem 1.3 includes ineffectiveness of the implied constant, an effective version of Siegel's theorem was considered in [26] by removing at most one possible ineffective quadratic Dirichlet character. Some types of Siegel's theorem for generalized L-functions are known (cf. [16, Chap. 2]).

Similarly to the Riemann hypothesis that $\zeta(s)$ is analyticaly continued to the whole complex plane would have only zeros whose real parts are all $1/2$ except for negative even integers, the generalized Riemann hypothesis (GRH) for L-functions has been studied. In particular, GRH for the Dirichlet L-function $L(s, \chi)$ means a conjecture that $L(s, \chi)$ would have only zeros whose real part is $1/2$ on the region $0 < \Re(s) < 1$. For difficulty, we are still far to rearch an answer for GRH, even for the original Riemann hypothesis. As for upper and lower bounds, Littlewood [11] improved them as follows by assuming GRH.

Theorem 1.4 ([11]) *Let χ be a nontrivial Dirichlet character of modulo q. Under GRH for $L(s, \chi)$, we have*

$$\frac{1}{\log\log q} \ll |L(1, \chi)| \ll \log\log q,$$

where the implied constant is independent of χ and q.

Theorem 1.4 is proved in the same way as [17, Corollary 13.16], where we omit to prove it in this article (see also [11] or [17, Exercises 13.2.1, 6]).

This article is organized as follows. We focus on proofs of Theorems 1.1–1.3 to let the reader understand the difficulty of giving bounds of $L(1, \chi)$ in spite of its easy forms. We review notions of Dirichlet characters in Sect. 2, and of Dirichlet L-functions in Sect. 3, respectively. After proving Theorem 1.1 in Sect. 4, we prove Theorem 1.2 in Sects. 5 and 6. In Sect. 7, we prove Theorem 1.3 following Estermann's method [5]. Several lemmas in complex analysis will be explained in Sect. 8 (Appendix 1). In Sect. 9 (Appendix 2), an estimate of logarithmic derivative of $L(s, \chi)$ used thoroughly in Sect. 6 will be explained. In Sect. 10 (Appendix 3), we summarize explicit versions of upper and lower bounds, which will be applied to security analysis of ideal lattice-based cryptography in Part 2 ([20]).

Notation

Let $\mathbb{N}$ be the set of positive integers. For $a, b \in \mathbb{Z}$, let $(a, b) = \gcd(a, b)$ be the positive greatest common divisor of a and b unless $a = b = 0$. The Euler totient function is denoted by φ, i.e., $\varphi(n) = \#(\mathbb{Z}/n\mathbb{Z})^\times$. Here we set $(\mathbb{Z}/\mathbb{Z})^\times := \{1\}$ formally. The formula $\varphi(q) = q \prod_{p|q}(1 - p^{-1})$ is easily confirmed, where p runs over prime numbers such that $p|q$. For $n \in \mathbb{N}$ and a prime p, let $\mathrm{ord}_p(n)$ denote the maximal non-negative integer a satisfying that p^a divides n.

In this paper throughout, we adopt Vinogradov's notation $\ll$ and $\ll_\varepsilon$. For real-valued nonnegative functions $f(x)$ and $g(x)$ on a set X, $f(x) \ll g(x)$ means that there exists a constant $C > 0$ independent of x such that $f(x) \le Cg(x)$ for all $x \in X$. If the constant C may depend on $\varepsilon > 0$, we sometimes write $f(x) \ll_\varepsilon g(x)$ for $f(x) \ll g(x)$. We say that a constant in an inequality is absolute if it is independent of any parameters.

We always represent a prime number by the symbol p without notice.

2 Dirichlet Characters

In this article, a Dirichlet character (or character) modulo q is a homomorphism from $(\mathbb{Z}/q\mathbb{Z})^\times$ to $\mathbb{C}^\times$ for some $q \in \mathbb{N}$. We call a Dirichlet character χ modulo q trivial (or principal) if $\chi(a) = 1$ for all $a \in (\mathbb{Z}/q\mathbb{Z})^\times$. The Refs. [3, 17, 19, 28] are useful to study Dirichlet L-functions. We notice that any Dirichlet character modulo $q \in \{1, 2\}$ is trivial. The character group of $(\mathbb{Z}/q\mathbb{Z})^\times$ is non-canonically isomorphic to $(\mathbb{Z}/q\mathbb{Z})^\times$ itself and thus the number of Dirichlet characters modulo q equals $\varphi(q)$. For a Dirichlet character χ modulo q, the conductor $f_\chi \in \mathbb{N}$ of χ is defined to be the minimal positive divisor d of q such that χ factors through a Dirichlet character χ' modulo d:

$$\chi : (\mathbb{Z}/q\mathbb{Z})^\times \twoheadrightarrow (\mathbb{Z}/d\mathbb{Z})^\times \xrightarrow{\chi'} \mathbb{C}^\times.$$

We say that χ is primitive if $f_\chi = q$. Then, χ is induced from the unique primitive Dirichlet character modulo f_χ. Such a primitive character is denoted by χ^*.

The following is easily confirmed (cf. [17, Corollary 4.6 and Lemma 9.3]).

Proposition 2.1 *Let q_1 and q_2 be relatively prime natural numbers. Then, any Dirichlet character χ modulo q_1q_2 is decomposed into the product of Dirichlet characters χ_1 modulo q_1 and χ_2 modulo q_2, and such χ_1 and χ_2 are unique. Moreover, χ is primitive if and only if both χ_1 and χ_2 are primitive.*

Let $\varphi_2(q)$ denote the number of primitive Dirichlet characters modulo q.

Proposition 2.2 ([17, 9.1.1, Exercises, 6]) *We have*

$$\varphi_2(q) = q \prod_{\substack{p \\ \mathrm{ord}(q)=1}} \left(1 - \frac{2}{p}\right) \prod_{\substack{p \\ \mathrm{ord}_p(q)\geq 2}} \left(1 - \frac{1}{p}\right)^2.$$

In particular, $\varphi_2(q) > 0$ *if and only if* $q \not\equiv 2 (mod\ 4)$.

Proof Since $\varphi_2(q)$ is multiplicative by Proposition 2.1, i.e., $\varphi_2(ab) = \varphi_2(a)\varphi_2(b)$ whenever $(a, b) = 1$, we have only to consider $\varphi_2(p^k)$ for any prime number p and $k \in \mathbb{N}$. By noting that any non-primitive character modulo p^k is induced from a Dirichlet character modulo p^{k-1}, we have $\varphi_2(p^k) = \varphi(p^k) - \varphi(p^{k-1})$. By $\varphi_2(p) = p - 2$ for $k = 1$ and $\varphi_2(p^k) = (p^{k-1} - p^{k-2})(p - 1) = p^k(1 - p^{-1})^2$ for $k \geq 2$, we are done. □

We say that a Dirichlet character χ modulo q is even (resp. odd) if $\chi(-1) = 1$ (resp. $\chi(-1) = -1$), and call χ quadratic if χ^2 is a trivial character but so is not χ. It is well-known that, by class field theory, any quadratic extension F of $\mathbb{Q}$ corresponds to the unique quadratic Dirichlet character χ_F modulo the absolute discriminant D_F satisfying that, for any fixed unramified prime p in F, $\chi_F(p) = 1$ if and only if p splits in F (cf. [3, 19, 28]). Then, even quadratic characters correspond to real quadratic fields and odd quadratic ones to imaginary quadratic fields, respectively.

The following is used in security analysis of ideal lattice-based cryptography in [22].

Proposition 2.3 *Let p be a prime and k a positive integer. Let $X(p^k)$ be the number of even primitive Dirichlet characters modulo p^k. Then, we have*

$$X(2^k) = \begin{cases} 0 & (k = 1, 2), \\ 2^{k-3} & (k \geq 3) \end{cases}$$

for $p = 2$, and

$$X(p^k) = \begin{cases} \frac{p-3}{2} & (k = 1), \\ \frac{p^{k-2}(p-1)^2}{2} & (k \geq 2) \end{cases}$$

for any odd prime p.

Proof Any even Dirichlet character modulo $q \neq 1, 2, 4$ induces a character of $(\mathbb{Z}/q\mathbb{Z})^\times/\{\pm 1\}$. Thus, for odd prime p, $X(p^k)$ is equal to $\frac{1}{2}\varphi(p^k) - \frac{1}{2}\varphi(p^{k-1}) = \frac{p^{k-2}(p-1)^2}{2}$ when $k \geq 2$, and to $\frac{1}{2}\varphi(p) - \varphi(1) = \frac{p-3}{2}$ when $k = 1$, respectively. As for $p = 2$, it follows that $X(2^k) = 0$ for $k = 1, 2$ immediately. For $k \geq 3$, we have $X(2^k) = \frac{1}{2}\varphi(2^k) - \frac{1}{2}\varphi(2^{k-1}) = 2^{k-3}$ similarly to the case $p \neq 2$.

Remark 2.4 If a prime number p satisfies $p \equiv 3 \pmod 4$, there exist no even quadratic Dirichlet characters modulo p^k. Indeed, by the isomorphism $(\mathbb{Z}/p^k\mathbb{Z})^\times \cong \mathbb{Z}/(p-1)\mathbb{Z} \times \mathbb{Z}/p^{k-1}\mathbb{Z}$, the cardinality of $(\mathbb{Z}/p^k\mathbb{Z})^\times/\{\pm 1\}$ is odd.

3 Dirichlet L-functions

Let χ be a Dirichlet character modulo q. We extend χ to a mapping $\tilde{\chi} : \mathbb{Z} \to \mathbb{C}$ so that $\tilde{\chi}(a) = \chi(a \pmod q)$ if $(a, q) = 1$ and 0 otherwise. Then, the Dirichlet L-function $L(s, \chi)$ associated with χ is defined by the series (1) in the introduction. This is absolutely and locally uniformly convergent in the region $\Re(s) > 1$. Furthermore, if χ is non-trivial, then $L(s, \chi)$ is conditionally convergent for $0 < \Re(s) \leq 1$. (cf. the proof of Lemma 5.2 (1)). For $\Re(s) > 1$, the Dirichlet L-function $L(s, \chi)$ is well-known to have the Euler product

$$L(s, \chi) = \prod_p (1 - \tilde{\chi}(p)p^{-s})^{-1}.$$

Since $L(s, \chi) = \{\prod_{p|q}(1 - \tilde{\chi^*}(p)p^{-s})\}L(s, \chi^*)$, analytic properties of $L(s, \chi)$ such as locations of its zeros and poles are inherent in $L(s, \chi^*)$. Here, we notice that in Washington's book [28], χ is identified with χ^* although such an identification does not preserve L-functions. The following shows an analytic continuation and a functional equation of $L(s, \chi)$ as the Riemann zeta function has.

Proposition 3.1 *Let χ be a primitive Dirichlet character modulo q.*

(1) If χ is nontrivial, then $L(s, \chi)$ is continued to an entire function on $\mathbb{C}$. If χ is trivial, then $L(s, \chi) = \{\prod_{p|q}(1 - p^{-s})\}\zeta(s)$ has a meromorphic continuation on $\mathbb{C}$; the only pole is $s = 1$ and simple.

(2) Let δ_χ be the element of $\{0, 1\}$ such that $\chi(-1) = (-1)^{\delta_\chi}$. Set $\Gamma_{\mathbb{R}}(s) = \pi^{-s/2}$ $\Gamma(s/2)$ and $\xi(s, \chi) = \Gamma_{\mathbb{R}}(s + \delta_\chi)L(s, \chi)$. Then, the functional equation

$$\xi(s, \chi) = \varepsilon(s, \chi)\xi(1 - s, \chi^{-1})$$

holds, where $\varepsilon(s, \chi)$, what is called the epsilon factor associated with χ, is a nonzero entire function $\varepsilon(s, \chi)$ uniquely determined by χ.

(3) The function $\varepsilon(s, \chi)$ is a unit of the ring $\mathbb{C}[q^{-s}, q^s]$ and explicitly described as

$$\varepsilon(s, \chi) = \frac{\tau(\chi)}{i^{\delta_\chi} q^{1/2}} \times q^{1/2-s} \quad \text{with} \quad \tau(\chi) = \sum_{a \in (\mathbb{Z}/q\mathbb{Z})^\times} \chi(a) \exp\left(\frac{2\pi i a}{q}\right).$$

Here $\tau(\chi)$ is called the Gauss sum of χ, and $\varepsilon(1/2, \chi)$ is called the root number of χ (note $|\varepsilon(1/2, \chi)| = 1$).

From the Euler product expression and a functional equation, $\xi(s, \chi)$ has no zeros in the region $\Re(s) > 1$ and in $\Re(s) < 0$. Since $L(s, \chi)$ has no zeros on the vertical line $\Re(s) = 1$ (cf. [23, Theorem 7–28]), $\xi(s, \chi)$ has zeros only in the region $0 < \Re(s) < 1$.

As an application of Dirichlet L-functions, there is a beautiful formula, the Dirichlet class number formula. For any quadratic extension $F/\mathbb{Q}$, let χ_F denote the primitive quadratic Dirichlet character modulo D_F corresponding to F by class field theory. Here, D_F is the absolute discriminant of $F/\mathbb{Q}$. Let w_F be the number of roots of unity in F and let ε_F be the fundamental unit of F such that $\varepsilon_F > 1$. Then, we have the Dirichlet class number formula $L(1, \chi_F) = \frac{2h_F \log \varepsilon_F}{\sqrt{D_F}}$ when $F/\mathbb{Q}$ is a real quadratic extension, and $L(1, \chi_F) = \frac{2\pi h_F}{w_F\sqrt{D_F}}$ when $F/\mathbb{Q}$ is an imaginary quadratic extension (cf. [3, Sect. 6], [19, Chap. VII, Corollary 5.11] and [23, Theorem 7–25]).

4 Upper Bounds

In this section we prove the following, which deduces Theorem 1.1.

Theorem 4.1 *Let χ be a nontrivial Dirichlet character modulo q. We have*

$$|L(\sigma, \chi)| \ll \log q, \qquad |L'(\sigma, \chi)| \ll (\log q)^2$$

uniformly for q, χ and σ such that $1 - 1/\log q \le \sigma \le 1$.

Proof First we easily have

$$\left|\sum_{n=1}^{q} \frac{\tilde{\chi}(n)}{n^\sigma}\right| \le e \sum_{n=1}^{q} \frac{1}{n} \le e\left(1 + \int_1^q \frac{dx}{x}\right) = e(\log q + 1).$$

Next set $S(x) = \sum_{1 \le n \le x} \tilde{\chi}(n)$ for $x > 0$. Then we notice $|S(x)| \le q$. By the evaluation

$$\sum_{n=q+1}^{\infty} \frac{\tilde{\chi}(n)}{n^s} = \sum_{n=q+1}^{\infty} \frac{S(n) - S(n-1)}{n^s} = \sum_{n=q+1}^{\infty} S(n) \left\{\frac{1}{n^s} - \frac{1}{(n+1)^s}\right\}$$

for $\Re(s) > 0$, with the aid of the inequality $n^\sigma = e^{\sigma \log n} \ge n/e$ when $1 - 1/\log q \le \sigma \le 1$ and $n \le q$, we have

$$\left|\sum_{n=q+1}^{\infty} \frac{\tilde{\chi}(n)}{n^s}\right| \le \max_{n>q} |S(n)| \times q^{-\sigma} \le q \times q^{-\sigma} \le q \times e/q = e.$$

This completes the proof of $|L(\sigma, \chi)| \ll \log q$. The inequality $|L'(\sigma, \chi)| \ll (\log q)^2$ is similarly proved. □

5 Lower Bounds

In this section, we prove Theorem 1.2. More precisely, we prove the following.

Proposition 5.1 *Let χ be a nontrivial Dirichlet character modulo q. Assume that there exists a constant $C > 0$ independent of χ and q such that $L(s, \chi) \neq 0$ if $s = \sigma + it$ $(\sigma, t \in \mathbb{R})$ satisfies*

$$\sigma > 1 - \frac{C}{\log\{q(3 + |t|)\}}.$$

Then, we have

$$|L(1, \chi)| \gg \frac{1}{\log q}.$$

We will show later the existence of a constant C in Proposition 6.1 whenever χ is non-quadratic (cf. [3, Sect. 14] or [8, Theorem 5.10]). Hence, we obtain Theorem 1.2 by combining this with Proposition 5.1.

From now we show a proof of Proposition 5.1, following [10]. In advance let us view a brief outline of the proof. To give the estimate $|L(1, \chi)| \ll 1/\log q$, it suffices to prove

$$\log \frac{1}{|L(1, \chi)|} \leq \log\log q + C$$

with an absolute constant $C > 0$. As an approach to give such an estimate, we use the fundamental theorem of calculus:

$$\log \frac{1}{|L(1, \chi)|} = -\Re(\log L(1, \chi)) = \Re\left(\log \frac{1}{L(s_0, \chi)}\right) + \int_1^{s_0} \Re\left(\frac{L'(s, \chi)}{L(s, \chi)}\right) ds$$

with $s_0 = 1 + \frac{1}{c \log q}$ for a suitable $c > 0$. Although the value $L(1, \chi)$ is not easy to control, we can control the value $L(s_0, \chi)$ $(s_0 > 1)$ by virtue of the series expression. This transfer of $L(1, \chi)$ to $L(s_0, \chi)$ as above is a key ingredient. Next we consider estimates of "much easier" values $\Re(\log \frac{1}{L(s_0, \chi)})$ and $\Re\left(\frac{L'(s,\chi)}{L(s,\chi)}\right)$, from which the desired estimate of $1/|L(1, \chi)|$ is deduced. We make use of an easy estimate

$$\frac{1}{|L(s_0, \chi)|} \leq \frac{2}{s_0 - 1} = 2c \log q$$

for $s_0 = 1 + \frac{1}{c \log q}$ for a suitable $c > 0$. Another essential ingredient of the proof of Proposition 5.1 is the estimate $\Re\left(\frac{L'(s,\chi)}{L(s,\chi)}\right) < C_1 \log q$ for $1 \leq s \leq s_0$ with an absolute constant $C_1 > 0$ by using Proposition 8.4 in Sect. 8 (Appendix 1). The nonvanishing assumption of $L(s, \chi)$ in Proposition 5.1 is essentially used here.

Here are fundamental estimates.

Lemma 5.2 ([3, p. 82 (14)] and [10]) *Let χ be a Dirichlet character modulo q.*

(1) We have

$$|L(s, \chi)| \leq 2q|s|, \qquad (s = \sigma + it, \ \sigma \geq 1/2, \ t \in \mathbb{R}).$$

(2) For $\sigma > 1$, we have

$$\frac{1}{\sigma - 1} < \zeta(\sigma) < 1 + \frac{1}{\sigma - 1}, \qquad \sum_{n=2}^{\infty} \frac{\log n}{n^{\sigma}} < 1 + \frac{1}{(\sigma - 1)^2}.$$

(3) For $1 < \sigma < 2$, we have

$$\left| \frac{1}{L(s, \chi)} \right| < \frac{2}{\sigma - 1}, \qquad \left| \frac{L'(\sigma, \chi)}{L(\sigma, \chi)} \right| < 1 + \frac{1}{\sigma - 1}.$$

Proof Set $S(x) = \sum_{0 \leq n \leq x} \tilde{\chi}(n)$ for $x > 0$. In the same way as in the proof of Theorem 4.1, for $\Re(s) > 0$, we have

$$\begin{aligned} L(s, \chi) &= \sum_{n=1}^{\infty} S(n) \left\{ \frac{1}{n^s} - \frac{1}{(n+1)^s} \right\} = \sum_{n=1}^{\infty} S(n) \times s \int_n^{n+1} x^{-s-1} dx \\ &= s \sum_{n=1}^{\infty} \int_n^{n+1} S(x) x^{-s-1} dx = s \int_1^{\infty} S(x) x^{-s-1} dx. \end{aligned}$$

Since $S(x)$ is bounded, the last integral is holomorphic on $\Re(s) > 0$. For $\Re(s) > 0$, we obtain $|L(s, \chi)| \leq |s| \int_1^{\infty} q x^{-\Re(s)-1} dx = |s| q / \Re(s)$. Thus, the assertion (1) follows. The assertion (2) follows from the estimations

$$\frac{1}{\sigma - 1} = \int_1^{\infty} x^{-\sigma} dx < \zeta(\sigma) < 1 + \int_1^{\infty} x^{-\sigma} dx = 1 + \frac{1}{\sigma - 1}$$

and

$$\sum_{n=2}^{\infty} \frac{\log n}{n^{\sigma}} < \frac{\log 2}{2} + \frac{\log 3}{3} + \int_1^{\infty} \frac{\log x}{x^{\sigma}} dx = \frac{\log 2}{2} + \frac{\log 3}{3} + \frac{1}{(\sigma - 1)^2} < 1 + \frac{1}{(\sigma - 1)^2}.$$

The assertion (3) follows from

$$\left| \frac{1}{L(s, \chi)} \right| = \left| \sum_{n=1}^{\infty} \frac{\mu(n) \tilde{\chi}(n)}{n^{\sigma}} \right| \leq \sum_{n=1}^{\infty} \frac{1}{n^{\sigma}} < \frac{2}{\sigma - 1}$$

and

$$\left|\frac{L'(\sigma,\chi)}{L(\sigma,\chi)}\right| \le -\frac{\zeta'(\sigma)}{\zeta(\sigma)} < \frac{1+\frac{1}{(\sigma-1)^2}}{\frac{1}{\sigma-1}} < 1+\frac{1}{\sigma-1}.$$

Here μ is the Möbius function. From the consideration as above, we are done. □

Proof of Proposition 5.1: Set $s_0 = 1 + \frac{1}{c\log q}$ for a sufficientlly large $c > 0$. If $|s - s_0| < \frac{1}{2}$, we have

$$\left|\frac{L(s,\chi)}{L(s_0,\chi)}\right| \ll \frac{q}{s_0-1} = cq\log q \ll q^2 = e^{2\log q}$$

by Lemma 5.2 (1) and (3), and

$$\left|\frac{L'(s_0,\chi)}{L(s_0,\chi)}\right| < 1+\frac{1}{s_0-1} = 1 + c\log q \ll \log q$$

by Lemma 5.2 (3). By applying Lemma 8.4 for $f(s) = L(s,\chi)$, $\sigma_0 = s_0$, $M = C_0\log q$, $r = 1/2$ and $r' = \frac{1}{8c\log q}$ for a sufficiently large $C_0 > 0$ and by virtue of the assumption and two inequalities as above, we obtain

$$\left|\frac{L'(s,\chi)}{L(s,\chi)}\right| < C_1\log q \qquad |s-s_0| < r' = \frac{1}{8c\log q}$$

with an absolute constant $C_1 > 0$. Hence, combining this with the argument explained after Proposion 5.1, we finally obtain $\log(|L(1,\chi)|^{-1}) < A\log\log q$ with a sufficiently large constant $A > 0$ independent of χ and q, and hence $1/|L(1,\chi)| < e^A\log q$. This completes the proof.

6 Zero-Free Regions and Siegel Zeros

In this section, lower bounds of $L(s,\chi)$ is given when χ is not quadratic. We introduce a zero-free region of $L(s,\chi)$ and explain why Theorem 1.2 needs the assumption that χ is non-quadratic. In the case where χ is quadratic, we need more technical analysis. A difficulty arises from the possible existence of a real zero of $L(s,\chi)$ for any quadratic character χ (cf. [3, 8]).

Proposition 6.1 ([3, p.93] or the special case of [8, Theorem 5.10]) *Let χ be a Dirichlet character modulo q. Then, there exists $C > 0$ independent of q and χ such that $L(s,\chi) \neq 0$ for all $s = \sigma + it$ $(\sigma, t \in \mathbb{R})$ satisfying*

$$\sigma > 1 - \frac{C}{\log\{q(3+|t|)\}}$$

except for at most one simple real zero $\beta \in (1/2, 1)$, which may arise if χ is quadratic. In particular, $L(s, \chi)$ has no such a real zero if χ is not quadratic.

The region given by $s = \sigma + it$ and $\sigma > 1 - C/\log\{q(3 + |t|)\}$ is called a zero-free region of $L(s, \chi)$, and the possible real zero β is called a Siegel zero (or a Siegel–Landau zero) of $L(s, \chi)$ (cf. [16, Chap. 2]). We expect that there are no Siegel zeros as we see from the generalized Riemann hypothesis.

We remark that $L(s, (\chi^2)^*)$ is entire if and only if χ is not quadratic.

Lemma 6.2 *Let χ be a Dirichlet character modulo q. For $\sigma > 1$ and $t \in \mathbb{R}$, we have*

$$- 3\Re\left(\frac{\zeta'(\sigma)}{\zeta(\sigma)}\right) - 4\Re\left(\frac{L'(\sigma + it, \chi^*)}{L(\sigma + it, \chi^*)}\right) - \Re\left(\frac{L'(\sigma + 2it, (\chi^2)^*)}{L(\sigma + 2it, (\chi^2)^*)}\right) \geq 0.$$

Proof One can easily check

$$-\frac{L'(s, \chi)}{L(s, \chi)} = \sum_{n=1,(n,q)=1}^{\infty} \frac{\Lambda(n)\tilde{\chi}(n)}{n^\sigma} e^{-it\log n} \qquad (s = \sigma + it,\ \sigma > 1,\ t \in \mathbb{R})$$

by taking the logatrithmic derivative of the Euler product of $L(s, \chi)$, where $\Lambda(n)$ is the von Mangoldt function defined by $\Lambda(n) = \log p$ if $n = p^k$ for a prime p and $k \in \mathbb{N}$, and $\Lambda(n) = 0$ otherwise. By this formula, the left-hand side of the assertion is evaluated as

$$\sum_{\substack{n=1\\(n,q)=1}}^{\infty} \frac{\Lambda(n)}{n^\sigma}\{3 + 4\Re(\widetilde{\chi^*}(n)e^{-it\log n}) + \Re(\widetilde{(\chi^2)^*}(n)e^{-2it\log n})\} \geq 0,$$

where we use $3 + 4\cos\theta + \cos 2\theta = 2(1 + \cos\theta)^2 \geq 0$ $(\theta \in \mathbb{R})$. □

Proof of Proposition 6.1: Since the inequality $\Re(\frac{1}{\sigma-\rho}) = \frac{\sigma-\Re(\rho)}{|\sigma-\rho|^2} \geq 0$ holds, we can ignore any terms of $\Re(\frac{1}{s-\rho})$ in the inequality in Lemma 9.1.

First, let us consider the case where χ is not quadratic. Let $\beta + i\gamma$ ($\beta \in (0, 1)$, $\gamma \in \mathbb{R}$) be a zero of $L(s, \chi)$. By combining Lemma 6.2 with three estimates

$$-\frac{\zeta'(\sigma)}{\zeta(\sigma)} < \frac{1}{\sigma - 1} + 1,$$

$$\begin{aligned}
-\Re\left(\frac{L'(\sigma + i\gamma, \chi^*)}{L(\sigma + i\gamma, \chi^*)}\right) &< C\log\{q(3 + |\gamma|)\} - \sum_{\rho}\Re\left(\frac{1}{\sigma + i\gamma - \rho}\right)\\
&< C\log\{q(3 + |\gamma|)\} - \frac{1}{\sigma - \beta}
\end{aligned}$$

and

$$-\Re\left(\frac{L'(\sigma+2i\gamma,(\chi^2)^*)}{L(\sigma+2i\gamma,(\chi^2)^*)}\right) < C\log\{q(3+|\gamma|)\}$$

from Lemma 9.1, we obtain

$$0 < \frac{3}{\sigma-1} + c_0\log\{q(3+|\gamma|)\} - \frac{4}{\sigma-\beta}$$

for $1 < \sigma < 2$ with an absolute constant $c_0 > 0$. By taking $\sigma = 1 + c_1/\log\{q(3+|\gamma|)\}$ for a sufficiently large $c_1 > 0$, we obtain $\beta < 1 - \frac{c_2}{\log\{q(3+|\gamma|)\}}$. Thus, we are done with the proof when χ is not quadratic.

The case where χ is quadratic is more difficult than the non-quadratic case. In such a case $L(s,(\chi^2)^*) = \zeta(s)$ has a pole at $s = 1$. In the argument as above, we replace the inequality for $-\Re\left(\frac{L'(s+2i\gamma,(\chi^2)^*)}{L(s+2i\gamma,(\chi^2)^*)}\right)$ with

$$-\Re\left(\frac{L'(\sigma+2i\gamma,(\chi^2)^*)}{L(\sigma+2i\gamma,(\chi^2)^*)}\right) = -\Re\left(\frac{\zeta'(\sigma+2i\gamma)}{\zeta(\sigma+2i\gamma)}\right) < \Re\left(\frac{1}{\sigma+2i\gamma-1}\right) + c_3\log(3+|\gamma|)$$

for $\sigma > 1$ with $c_3 > 0$, which is given by Lemma 9.1. Therefore, in the same way as the argument above, we have

$$\frac{4}{\sigma-\beta} < \frac{3}{\sigma-1} + \Re\left(\frac{1}{\sigma-1+2i\gamma}\right) + c_4\log\{q(3+|\gamma|)\}$$

with an absolute constant c_4. By taking $\sigma = 1 + \delta/\log\{q(3+|\gamma|)\}$ for some $\delta > 0$ and assuming $|\gamma| \geq \delta/\log\{q(3+|\gamma|)\}$, we have

$$\frac{4}{\sigma-\beta} < \frac{3\log\{q(3+|\gamma|)\}}{\delta} + \frac{\log\{q(3+|\gamma|)\}}{5\delta} + c_4\log\{q(3+|\gamma|)\}.$$

Hence, by taking any sufficiently small $\delta > 0$ so that $\delta < 2/15c_4$, we obtain

$$\beta < 1 - \frac{4-5c_4\delta}{16+5c_4\delta}\frac{\delta}{\log\{q(3+|\gamma|)\}} < 1 - \frac{\delta}{5\log\{q(3+|\gamma|)\}}$$

under the assumption $|\gamma| \geq \delta/\log q$.

The rest of the proof is to show that there exists at most one zero $\beta + i\gamma$ ($\beta \in (0,1), \gamma \in \mathbb{R}$) of $L(s,\chi)$ for any quadratic Dirichlet character χ such that $\beta > 1 - \delta/\log\{q(3+|\gamma|)\}$ and $|\gamma| < \delta/\log q$. Indeed, if we prove this claim, then the possible zero $\beta + i\gamma$ must satisfy $\gamma = 0$ and that β is simple since $\beta - i\gamma$ is also a zero by a symmetry of a functional equation and $\chi = \chi^{-1}$. For $\sigma > 1$, the inequality

$$-\frac{L'(\sigma,\chi^*)}{L(\sigma,\chi^*)} < c_5 \log q - \sum_\rho \frac{1}{\sigma-\rho}$$

follows from Lemma 9.1, where we use $\sum_\rho \frac{1}{\sigma-\rho} \in \mathbb{R}$ by a symmetry of a functional equation. If $\rho = \beta + i\gamma$ is a zero of $L(s,\chi)$ with $0 < \beta < 1$ and $0 \neq |\gamma| < \delta/\log q$, then we have

$$-\frac{L'(\sigma,\chi^*)}{L(\sigma,\chi^*)} < c_5 \log q - \left(\frac{1}{\sigma-\rho} + \frac{1}{\sigma-\overline{\rho}}\right) = c_5 \log q - \frac{2(\sigma-\beta)}{(\sigma-\beta)^2+\gamma^2}.$$

Since the image of χ equals $\{\pm 1\}$, we have also

$$-\frac{L'(\sigma,\chi^*)}{L(\sigma,\chi^*)} \geq -\sum_{n=1}^{\infty} \frac{\Lambda(n)}{n^\sigma} = \frac{\zeta'(\sigma)}{\zeta(\sigma)} > -\frac{1}{\sigma-1} - 1.$$

Combining these, we obtain $-\frac{1}{\sigma-1} < c_6 \log q - \frac{2(\sigma-\beta)}{(\sigma-\beta)^2+\gamma^2}$. By taking $\sigma = 1 + 2\delta/\log q$, we obtain $|\gamma| < \frac{\delta}{\log q} = \frac{\sigma-1}{2} < \frac{\sigma-\beta}{2}$ and hence $-\frac{1}{\sigma-1} < c_6 \log q - \frac{8}{5(\sigma-\beta)}$. Thus we obtain $\sigma - \beta > \frac{16\delta}{5(2\delta c_6+1)} \frac{1}{\log q}$. By taking $\delta > 0$ so that δc_6 is sufficiently small, we have $\beta < 1 - \delta'/\log q$ with an absolute constant $\delta' > 0$.

The case $\gamma = 0$ is similarly treated; In such a case, we have only to consider two distinct real zeros or a double real zero instead of two imaginary zeros ρ and $\overline{\rho}$. Hence we are done.

7 Estermann's Proof of Siegel's Theorem

We give a proof of Theorem 1.3, following Estermann's elementary method in [5] and [3, Sect. 21]. Throughout in this section, we use symbols $C_1, C_2, \cdots$ as positive absolute constants without notice, and we identify any Dirichlet character χ with $\tilde{\chi}$.

Let χ_1 and χ_2 be primitive quadratic Dirichlet characters of conductor q_1 and q_2, respectively. Suppose $q_1 \neq q_2$. Set $F(s) = \zeta(s)L(s,\chi_1)L(s,\chi_2)L(s,\chi_1\chi_2)$ for $s \in \mathbb{C} - \{1\}$. Then, $F(s)$ is holomorphic on $\mathbb{C} - \{1\}$ and has a simple pole at $s = 1$ with residue $\lambda = \mathrm{Res}_{s=1} F(s) = L(1,\chi_1)L(1,\chi_2)L(1,\chi_1\chi_2)$. Here we note $\chi_1\chi_2$ is not always primitive. The following is a key inequality.

Proposition 7.1 *There exists a constant $C > 0$ independent of q_1, q_2, χ_1 and χ_2 such that*

$$F(s) > \frac{1}{2} - \frac{C\lambda}{1-s}(q_1 q_2)^{8(1-s)}$$

holds for $7/8 < s < 1$.

We prepare a lemma below to prove Proposition 7.1. By the multiplication of Euler products, we can define a sequence $(a_n)_{n\in\mathbb{N}}$ by $F(s) = \sum_{n=1}^{\infty} \frac{a_n}{n^s}$, $\Re(s) > 1$.

Lemma 7.2 *For any $n \in \mathbb{N}$, we have $a_n \in \mathbb{R}$ and $a_n \geq 0$ for all n.*

Proof By the expression

$$F(s) = \prod_p (1-p^{-s})^{-1}(1-\chi_1(p)p^{-s})^{-1}(1-\chi_2(p)p^{-s})^{-1}(1-\chi_1(p)\chi_2(p)p^{-s})^{-1}$$

for $\Re(s) > 1$ and $\log(1-x)^{-1} = \sum_{m=1}^{\infty} \frac{x^m}{m}$ $(|x| < 1)$, we have

$$\begin{aligned}\log F(s) =& \sum_p \{\log(1-p^{-s})^{-1} + \log(1-\chi_1(p)p^{-s})^{-1} + \log(1-\chi_2(p)p^{-s})^{-1} \\ &+ \log(1-\chi_1(p)\chi_2(p)p^{-s})^{-1}\} \\ =& \sum_p \sum_{m=1}^{\infty} \frac{(1+\chi_1(p^m))(1+\chi_2(p^m))}{mp^{ms}}.\end{aligned}$$

By noting $(1+\chi_1(p^m))(1+\chi_2(p^m)) \geq 0$ and $F(s) = \exp(\log F(s)) = \sum_{n=0}^{\infty} \frac{\{\log F(s)\}^n}{n!}$, we have the lemma. □

Proof of Proposition 7.1: We consider the Taylor expansion $F(s) = \sum_{m=0}^{\infty} b_m (2-s)^m$ for $|2-s| < 1$. Here, we note $b_m = \sum_{n=1}^{\infty} \frac{a_n(\log n)^m}{n^2} \geq 0$ and $b_0 \geq 1$ by Lemma 7.2. From $(s-1)^{-1} = \sum_{m=0}^{\infty}(2-s)^m$ for $|2-s| < 1$, we have

$$F(s) - \frac{\lambda}{s-1} = \sum_{m=0}^{\infty}(b_m - \lambda)(2-s)^m.$$

Since the left-hand side of the equality above is entire on $\mathbb{C}$, the series $\sum_{m=0}^{\infty}(b_m - \lambda)(2-s)^m$ makes sense even for $|2-s| < 2$. On $|2-s| = 3/2$, Lemma 5.2 (1) yields

$$|\zeta(s)| < C_1, \quad |L(s,\chi_1)| < C_1 q_1, \quad |L(s,\chi_2)| < C_1 q_2, \quad |L(s,\chi_1\chi_2)| < C_1 q_1 q_2.$$

Thus $F(s)$ is evaluated as $|F(s)| < C_2 q_1^2 q_2^2$ for $|s-2| = 3/2$. Similarly, $|\frac{\lambda}{s-1}| < C_2 q_1^2 q_2^2$ is also deduced on the circle $|s-2| = 3/2$. Then, with the aid of Cauchy's integral theorem, we have

$$b_m - \lambda = (-1)^m \frac{1}{2\pi i} \oint_{|s-2|=3/2} \frac{F(s) - \lambda(s-1)^{-1}}{(s-2)^{m+1}} ds$$

and hence $|b_m - \lambda| < 2C_3 q_1^2 q_2^2 (2/3)^m$.

Let $M > 1$ be a number chosen suitably later. When $7/8 < s < 1$, we obtain

$$\sum_{m=M}^{\infty} |b_m - \lambda|(2-s)^m \leq \sum_{m=M}^{\infty} 2C_6 q_1^2 q_2^2 |(2-s)2/3|^m \leq 2C_3 q_1^2 q_2^2 \sum_{m=M}^{\infty} (3/4)^m$$
$$< C_4 q_1^2 q_2^2 (3/4)^M < C_4 q_1^2 q_2^2 e^{-M/4},$$

where we use $e^{1/4} < 4/3$. Therefore, $F(s) - \frac{\lambda}{s-1}$ is evaluated as

$$\begin{aligned} F(s) - \frac{\lambda}{s-1} &= \sum_{m=0}^{M} (b_m - \lambda)(2-s)^m - C_4 q_1^2 q_2^2 e^{-M/4} \\ &\geq (1-\lambda) + \sum_{m=1}^{M-1} (-\lambda)(2-s)^m - C_4 q_1^2 q_2^2 e^{-M/4} \\ &= 1 - \lambda \frac{(2-s)^M - 1}{(1-s)} - C_4 q_1^2 q_2^2 e^{-M/4} \end{aligned}$$

for $7/8 < s < 1$. By taking $M = M_{q_1,q_2} \in \mathbb{N}$ such that $4\log(2C_4 q_1^2 q_2^2) \leq M < 4\log(2C_4 q_1^2 q_2^2) + 1$, we have $e^{-1/4}/2 < C_4 q_1^2 q_2^2 e^{-M/4} < 1/2$, and hence we obtain

$$F(s) > \frac{1}{2} - \frac{\lambda}{1-s}(2-s)^M.$$

Since the inequality $M \geq 8\log(q_1 q_2) + C_5$ with $C_5 = 4(4\log C_4 + 1)$ gives us $(2-s)^M = \exp(M\log(1+1-s)) < \exp(M(1-s)) < e^{C_5}(q_1 q_2)^{8(1-s)}$, we finally obtain

$$F(s) > 1/2 - \frac{e^{C_5}\lambda}{1-s}(q_1 q_2)^{8(1-s)}.$$

This completes the proof.

Lemma 7.3 *For any $\varepsilon > 0$, there exists a quadratic primitive character χ_ε and $\beta_\varepsilon \in (1-\varepsilon/16, 1)$ such that $F(\beta_\varepsilon) \leq 0$ for $\chi_1 = \chi_\varepsilon$ and all primitive quadratic $\chi_2 \neq \chi_\varepsilon$.*

Proof First assume that there exist no real zeros of $L(s, \chi)$ in $(1-\varepsilon/16, 1)$ for all quadratic characters χ. We fix any quadratic primitive character χ_1 and $\beta \in (1-\varepsilon/16, 1)$. Then, $L(s, \chi_1)$, $L(s, \chi_2)$ and $L(s, \chi_1\chi_2)$ are all positive on $(1-\varepsilon/16, 1)$. Thus $F(\beta) < 0$ holds with the aid of $\zeta(s) < 0$ for $s \in (0, 1)$. This negativity is independent of χ_1 and χ_2. Second, assume that there exists a real zero of $L(s, \chi)$ for some quadratic character χ in $(1-\varepsilon/16, 1)$. We fix such a character χ_ε and a real zero $\beta_\varepsilon \in (1-\varepsilon/16, 1)$ of $L(s, \chi_\varepsilon)$, and set $\chi_1 = \chi_\varepsilon$. Then, $F(\beta_\varepsilon) = 0$ for all $\chi_2 \neq \chi_\varepsilon$. □

Proof of Theorem 1.3: We may assume that χ is primitive by $L(1,\chi)=\{\prod_{p|q}(1-\tilde{\chi^*}(p)p^{-1})\}L(1,\chi^*)$ and $\frac{\varphi(q)}{q}\gg\frac{1}{\log\log q}\gg_\varepsilon\frac{1}{q^\varepsilon}$ (cf. [17, Theorem 2.9]). For $\varepsilon>0$, we fix $\chi_1=\chi_\varepsilon$ and β_ε same as in Lemma 7.3. The conductor of χ_ε is denoted by q_ε. We put $\chi_2=\chi$ and assume $q>q_\varepsilon$. From $F(\beta_\varepsilon)\le 0$, Proposition 7.1 gives us

$$C_6\lambda>\frac{1-\beta_\varepsilon}{2}(q_\varepsilon q)^{-8(1-\beta_\varepsilon)}.$$

By combing this with $\lambda=L(1,\chi_\varepsilon)L(1,\chi)L(1,\chi_\varepsilon\chi)<C_7(\log q_\varepsilon)L(1,\chi)(\log(q_\varepsilon q))$ (Theorem 4.1), we obtain

$$L(1,\chi)>\frac{1-\beta_\varepsilon}{2C_6C_7\log q_\varepsilon}q_\varepsilon^{-8(1-\beta_\varepsilon)}\times\frac{q^{-8(1-\beta_\varepsilon)}}{\log(q_\varepsilon q)}.$$

Putting $C_{1,\varepsilon}=\frac{1-\beta_\varepsilon}{2C_6C_7\log q_\varepsilon}q_\varepsilon^{-8(1-\beta_\varepsilon)}$ and using $8(1-\beta_\varepsilon)<\varepsilon/2$, we conclude

$$L(1,\chi)>C_{1,\varepsilon}\times\frac{q^{-\varepsilon/2}}{\log(q_\varepsilon q)}>C_{1,\varepsilon}C_{2,\varepsilon}^{-1}\frac{1}{q^\varepsilon}.$$

Here we use the inequality $\log(q_\varepsilon q)<C_{2,\varepsilon}q^{\varepsilon/2}$ with a constant $C_{2,\varepsilon}>0$ depending on $\varepsilon>0$ but not on $q>q_\varepsilon$. Thus we are done.

Remark 7.4 The implied constant in Theorem 1.3 depends on a possible Siegel zero β_ε. Thus, C_ε is not effective.

As an application of Siegel's theorem, we can measure the size of Siegel zeros.

Corollary 7.5 *Let χ be any quadratic character modulo q and suppose that $L(s,\chi)$ has a Siegel zero β. Then, for any $\varepsilon>0$, there exists $C_\varepsilon>0$ independent of q and χ such that $\beta<1-\frac{C_\varepsilon}{q^\varepsilon}$.*

Proof We may assume $1-1/\log q<\beta<1$. By the mean value theorem, there exists $\sigma\in(\beta,1)$ such that $L(1,\chi)=L(1,\chi)-L(\beta,\chi)=(1-\beta)L'(\sigma,\chi)$. Since $|L'(\sigma,\chi)|\ll(\log q)^2$ holds by Theorem 4.1, we have

$$1-\beta=\frac{L(1,\chi)}{L'(\sigma,\chi)}\ge C\frac{L(1,\chi)}{(\log q)^2}$$

with an absolute constant $C>0$, and hence we are done by virtue of Theorem 1.3. □

With the aid of the class number formula, Siegel's theorem is also applied to the Gauss conjecture $\lim_{D\to\infty}h_{\mathbb{Q}(\sqrt{-D})}=\infty$, which was first proved by Heilbronn [7].

8 Appendix 1

We prepare several lemmas on complex analysis for Theorem 1.2.

Lemma 8.1 (The Schwartz lemma) *Set* $D = \{z \in \mathbb{C} \mid |z| < 1\}$. *Let* $f : D \to D$ *be a holomorphic map such that* $f(0) = 0$. *Then,* $|f(z)| \leq |z|$ *for all* $z \in D$ *and* $f'(0) \leq 1$.

Moreover, suppose either that there exists $z_0 \in D$ *such that* $|f(z_0)| = |z_0|$, *or that* $f'(0) = 1$ *holds. Then we have* $f(z) = az$ *for some* $a \in \partial D$.

Proof Set $g(z) = f(z)/z$ if $z \in D - \{0\}$ and $g(z) = f'(0)$ if $z = 0$. Then, g is holomorphic on D. For each $r \in (0, 1)$ and the closed disc $D_r = \{z \in \mathbb{C} \mid |z| \leq r\}$, by the maximum modulus principle, there exists $z_r \in \partial D_r$ such that $\max_{z \in D_r} |g(z)| = |g(z_r)| = \frac{|f(z_r)|}{|z_r|} \leq 1/r$. As we take the limit $r \to 1 - 0$, we have the first assertion. As for the second assertion, we have only to verify that g is a constant, which is proved by the maximum modulus principle similarly. □

Lemma 8.2 (The Borel–Carathéodory theorem) *Let* $f(s)$ *be a holomorphic function on* $|s| \leq R$ *with* $R > 0$. *For any* $r \in (0, R)$, *we have*

$$\max_{|s| \leq r} |f(s)| \; (= \max_{|s|=r} |f(s)|) \leq \frac{2r}{R - r} \max_{|s| \leq R} \Re(f(s)) + \frac{R + r}{R - r} |f(0)|.$$

Proof First consider the case $f(0) = 0$. Set $A = \max_{|s| \leq R} \Re(f(s))$. Then $A \geq \Re(f(0)) = 0$. In what follows, we may assume $A > 0$; indeed, if $A = 0$, the function $\Re(f(x + iy))$ in (x, y) with $x + iy \in D_R$ is a harmonic function and attains a maximum at the interior point 0 of D_R. By the maximum modulus principle, $\Re(f(s))$ must be a constant $\Re(f(0)) = 0$ on D_R. Thus $f(s)$ is identically equal to zero on D_R.

Set $H_a = \{z \in \mathbb{C} \mid \Re(z) \leq a\}$ for $a \in \mathbb{R}$. Let us define the functions g and h by $g(z) = z/A - 1$ and $h(z) = \frac{z+1}{z-1}$, respectively. By applying the Schwartz lemma for the composite

$$D_R \xrightarrow{f} H_A \xrightarrow{g} H_0 \xrightarrow{h} D_R,$$

we have $|h \circ g \circ f(s)| \leq |s/R|$ for all $s \in D_R$, that is,

$$\frac{|R\, f(s)|}{|f(s) - 2A|} \leq |s|,$$

by which we have $R|f(s)| \leq r|f(s) - 2A| \leq r|f(s)| + 2Ar$ on D_r. Hence we obtain $|f(s)| \leq \frac{2r}{R-r} A$, which proves the assertion for f such that $f(0) = 0$. For the case $f(0) \neq 0$, by applying the estimate above to $f(s) - f(0)$, we obtain

$$|f(s)| \leq |f(s) - f(0)| + |f(0)| \leq \frac{2r}{R - r} \max_{|s| \leq R} \Re(f(s) - f(0)) + |f(0)|.$$

for $s \in D_r$. This completes the proof. □

Lemma 8.3 ([27, p. 56]) *Suppose $r > 0$ and let $f(s)$ be a holomorphic function on $|s - s_0| \leq r$ with $f(s_0) \neq 0$. For a constant $M > 0$, suppose that $\left|\frac{f(s)}{f(s_0)}\right| < e^M$ on $|s - s_0| \leq r$. Then, there exists an absolute constant $A > 0$ such that*

$$\left|\frac{f'(s)}{f(s)} - \sum_{\rho} \frac{1}{s - \rho}\right| < \frac{AM}{r},$$

where ρ runs over the zeros of $f(s)$ such that $|\rho - s_0| \leq r/2$ (the set of zeros is regarded as a multi-set).

Furthermore, if we suppose $f(s) \neq 0$ for all $s \in \mathbb{C}$ such that $|s - s_0| \leq r$ and $\Re(s_0) < \Re(s)$, then we have

$$-\Re\left(\frac{f'(s_0)}{f(s_0)}\right) < \frac{AM}{r}.$$

Proof Set $g(s) = f(s) \prod_{\rho}(s - \rho)^{-1}$ and $h(s) = \log \frac{g(s)}{g(s_0)}$, where ρ runs over the zeros of $f(s)$ such that $|\rho - s_0| \leq r/2$. Then, $g(s)$ is holomorphic on $|s - s_0| \leq r$ and non-zero for $|s - s_0| \leq r/2$. For $s \in \mathbb{C}$ such that $|s - s_0| = r$, we have

$$\left|\frac{g(s)}{g(s_0)}\right| = \left|\frac{f(s)}{f(s_0)} \prod_{\rho} \frac{s_0 - \rho}{s - \rho}\right| \leq \left|\frac{f(s)}{f(s_0)}\right| < e^M$$

by $|s_0 - \rho| \leq r/2 \leq |s - \rho|$. This inequality as above is still valid for $|s - s_0| \leq r$ by the maximum modulus principle. From this, we have $\Re(h(s)) < M$ for $|s - s_0| \leq r$. Then, by the Borel–Carathéodory theorem, the estimate

$$|h(s)| < \frac{2 \times 3r/8}{r/2 - 3r/8} M \leq A_0 M$$

holds for $|s - s_0| \leq 3r/8$ with $A_0 > 0$ being an absolute constant ($A_0 = 6$ is sufficient). Hence, for $|s - s_0| \leq r/4$ we obtain

$$\left|\frac{f'(s)}{f(s)} - \sum_{\rho} \frac{1}{s - \rho}\right| = |h'(s)| = \left|\frac{1}{2\pi i} \oint_{|z-s|=r/4} \frac{h(z)}{(z - s)^2} dz\right| < \frac{4A_0 M}{r}.$$

This completes the proof of the first assertion. The inequality as above gives us

$$-\Re\left(\frac{f'(s_0)}{f(s_0)}\right) < \frac{AM}{r} - \sum_{\rho} \Re\left(\frac{1}{s_0 - \rho}\right).$$

This completes the proof of the second assertion since $\Re\left(\frac{1}{s_0 - \rho}\right) \geq 0$ holds under the assumption. □

Proposition 8.4 ([27, p. 57] *Let $f(s)$ be a holomorphic function on $|s - s_0| \leq r$ such that $f(s_0) \neq 0$*

$$\left|\frac{f(s)}{f(s_0)}\right| < e^M \quad \text{and} \quad \left|\frac{f'(s_0)}{f(s_0)}\right| < \frac{M}{r} \quad \text{on} \quad |s - s_0| \leq r$$

for a constant $M > 0$. Furthermore, suppose that $f(s) \neq 0$ for all $s = \sigma + it$ such that $\sigma \geq \sigma_0 - 2r'$ and $|s - s_0| \leq r$, where we set $\sigma_0 = \Re(s_0)$ and r' is some real number such that $0 < r' < r/4$. Then, there exists an absolute constant $A' > 0$ such that

$$\left|\frac{f'(s)}{f(s)}\right| < \frac{A'M}{r} \qquad |s - s_0| \leq r'.$$

Proof As in Lemma 8.3, we have $-\Re\left(\frac{f'(s)}{f(s)}\right) < \frac{AM}{r}$ for all $s = \sigma + it$ such that $|s - s_0| \leq r/4$ and $\sigma \geq \sigma_0 - 2r'$. Therefore, the Borel–Carathéodory theorem yields the inequalities

$$\max_{|s-s_0|\leq r'} \left|-\frac{f'(s)}{f(s)}\right| \leq \frac{2r'}{2r' - r'} \max_{|s-s_0|\leq 2r'} \Re\left(\frac{-f'(s)}{f(s)}\right) + \frac{2r' + r'}{2r' - r'} \left|\frac{-f'(s_0)}{f(s_0)}\right| < \frac{(2A+3)M}{r}.$$

Thus we are done. □

9 Appendix 2

Let χ be a primitive Dirichlet character modulo q. With the aid of a functional equation of $L(s, \chi)$ in Proposition 3.1 and the Hadamard factorization theorem of entire functions of finite order (cf. [3, Sects. 11 and 12] or [17, Sect. 10.2]), and by taking the logarithmic derivative, there exists a constant $B(\chi) \in \mathbb{C}$ such that

$$\frac{L'(s, \chi)}{L(s, \chi)} = -\frac{a_\chi}{s - 1} - \frac{1}{2}\log\frac{q}{\pi} - \frac{1}{2}\frac{\Gamma'(s/2 + \delta_\chi/2)}{\Gamma(s/2 + \delta_\chi/2)} + B(\chi) + \sum_\rho \left(\frac{1}{s - \rho} - \frac{1}{\rho}\right),$$

where ρ runs over all zeros of $L(s, \chi)$ in the region $0 < \Re(s) < 1$, and we put $a_\chi = 1$ if χ is trivial and $a_\chi = 0$ otherwise. Furthermore, taking the real part yields $-\Re\left(\frac{L'(s,\chi)}{L(s,\chi)}\right) = \frac{1}{2}\log\frac{q}{\pi} + \frac{1}{2}\Re\left(\frac{\Gamma'(s/2+\delta_\chi/2)}{\Gamma(s/2+\delta_\chi/2)}\right) - \Re(B(\chi)) - \Re\sum_\rho\left(\frac{1}{s-\rho} - \frac{1}{\rho}\right)$. With the aid of $\Re(B(\chi)) = -\sum_\rho \Re(\frac{1}{\rho})$ (cf. [3, p. 83, (18)]) and $\Re\left(\frac{\Gamma'(s/2+\delta_\chi/2)}{\Gamma(s/2+\delta_\chi/2)}\right) = \mathcal{O}(\log(3 + |t|))$ from Stirling's formula, we have the following.

Lemma 9.1 *There exists an absolute constant $C > 0$ such that*

$$-\Re\left(\frac{L'(s,\chi)}{L(s,\chi)}\right) < \Re\left(\frac{a_\chi}{s-1}\right) + C\log\{q(3+|t|)\} - \sum_{\rho}\Re\left(\frac{1}{s-\rho}\right) \quad \Re(s) > 1.$$

This inequality is often used in the proof of Proposition 6.1 on zero-free regions. Here we notice that the inequality in Proposition 8.4 holds on a closed disc but Lemma 9.1 holds on a half plane.

10 Appendix 3

For application to security analysis of ideal lattice-based cryptography for cyclotomic fields, we review explicit upper and lower bounds of $L(1,\chi)$ (cf. [20, 22]). Such bounds of Dirichlet L-functions are related to the size of a dual basis of a log-unit lattice for a cyclotomic field, and moreover the size of such a dual basis is concerned with the RSG attack[1] as in [2].

Let $\delta_{a,b}$ denote the Kronecker delta.

Theorem 10.1 ([24, Corollaries 1 and 3] (see also [12, Corollary 1.2]) *For any non-trivial primitive Dirichlet character modulo q with $\chi(-1) = (-1)^\varepsilon$ for $\varepsilon \in \{0, 1\}$, we have*

$$|L(1,\chi)| \le \frac{1}{2}\log q + \delta_{\varepsilon,1}\left(\frac{5}{2} - \log 6\right).$$

Furthermore, if $2|q$, then we have

$$|L(1,\chi)| \le \frac{1}{4}\log q + \delta_{\varepsilon,0}\left(\frac{1}{2}\log 2\right) + \delta_{\varepsilon,1}\left(\frac{5}{4} - \frac{1}{2}\log 3\right).$$

Theorem 10.2 ([4, Theorem 1.1]) *For any primitive Dirichlet character modulo q such that $3|q$ and $\chi(-1) = (-1)^\varepsilon$ for $\varepsilon \in \{0, 1\}$, we have*

$$|L(1,\chi)| \le \frac{1}{3}\log q + \delta_{\varepsilon,0}\, 0.368296 + \delta_{\varepsilon,1}\, 0.838374.$$

By combining Theorems 10.1 and 10.2, we have the following estimates.

Corollary 10.1 *Let χ be a non-trivial even primitive Dirichlet character modulo $q = p^k$, where p is prime and $k \in \mathbb{N}$. Let k_χ be the integer such that $f_\chi = p^{k_\chi}$. Then, we have*

---

[1] RSG is an abbreviation for *Recovering Short Generator*.

$$|L(1,\chi)| \le \begin{cases} \frac{1}{4}(\log f_\chi + 2\log 2) & (p=2), \\ \delta_{k_\chi,2}\left(\frac{1}{2}\log f_\chi\right) + \left(\sum_{m=3}^{\infty}\delta_{k_\chi,m}\right)\frac{1}{3}(\log f_\chi + 1.104888) & (p=3), \\ \frac{1}{2}\log f_\chi & (p\ge 5). \end{cases}$$

Here we note $k_\chi \ge 2$ if $p = 3$ by the fact that there exists no even non-trivial Dirichlet characters modulo 3.

Theorem 10.3 ([14, Corollary 2]) *For any non-quadratic primitive Dirichlet character modulo $q > 1$, we have*

$$|L(1,\chi)| \ge \frac{1}{10\log(q/\pi)}.$$

References

1. S. Chowla, A new proof of a theorem of Siegel. Ann. Math. **2**(51), 120–122 (1950)
2. R. Cramer, L. Ducas, C. Peikert, O. Regev, Recovering short generators of principal ideals in cyclotomic rings, in *Advances in Cryptology - EUROCRYPT 2016*, Part II, vol. 9666 (LNCS, 2016), pp. 559–585
3. H. Davenport, *Graduate Texts in Math*, vol. 74, 3rd edn., Multiplicative Number Theory (Springer, New York, 2000)
4. S.S. Eddin, D.J. Platt, Explicit upper bounds for $|L(1,\chi)|$ when $\chi(3) = 0$. Colloq. Math. **133**(1), 23–34 (2013)
5. T. Estermann, On Dirichlet's L-functions. J. Lond. Math. Soc. **23**, 275–279 (1948)
6. D. Goldfeld, A simple proof of Siegel's theorem. Proc. Nat. Acad. Sci. USA **71**(4), 1055–1055 (1974)
7. H. Heilbronn, On the class number in imaginary quadratic fields. Quarterly J. Math. **5**, 150–160 (1934)
8. H. Iwaniec, E. Kowalski, *American Mathematical Society Colloquium Publications*, vol. 53, Analytic number theory (American Mathematical socity, Providence, 2004)
9. T. Kimura, S. Koyama, N. Kurokawa, Euler products beyond the boundary. Lett. Math. Phys. **104**, 1–19 (2014)
10. E. Landau, Über Dirichletsche Reihen mit komplexen Charakteren. J. Reine Angew. Math. **157**, 26–32 (1927)
11. J.E. Littlewood, On the class number of the corpus $P(\sqrt{-k})$. Proc. Lond. Math. Soc. **27**, 358–372 (1928)
12. S. Louboutin, Majorations explicites de $|L(1,\chi)|$ (quatrième partie). C. R. Acad. Sci. Paris **334**, 625–628 (2002)
13. S. Louboutin, Simple proofs of the Siegel–Tatuzawa and Brauer–Siegel theorems. Colloq. Math. **108**, 277–283 (2007)
14. S. Louboutin, An explicit lower bound on moduli of Dirichlet L-functions at $s = 1$. J. Ramanujan Math Soc. **101–113** (2015)
15. T. Metsänkylä, Estimations for L-functions and the class numbers of certain imaginary cyclic fields. Ann. Univ. Turku. Ser. A **I**(140), 1–11 (1970)
16. G. Molteni, L-functions: Siegel-type theorems and structure theorems, Ph.D. thesis, University of Milan, Milan (1999)
17. H.L. Montgomery, R.C. Vaughan, *Multiplicative Number Theory I* (Cambridge University Press, Cambridge, 2006)

18. W. Narkiewicz, *Rational Number Theory in the 20th Century : From PNT to FLT*, Springer monographs in mathematics (Springer, Berlin, 2012)
19. J. Neukirch, *Algebraic Number Theory*, Grundlehren Math. Wiss. (Springer, Berlin, 1999)
20. S. Okumura, On analysis of recovering short generator problems via upper and lower bounds of Dirichlet L-functions : Part 2, in this book
21. S. Okumura, M. Yasuda, T. Takagi, An improvement on the Recovering Short Generator Attack over Ideal Lattices and its Countermeasure, preprint
22. S. Okumura, S. Sugiyama, M. Yasuda, T. Takagi, Security Analysis of Cryptosystems Using Short Generators Over Ideal Lattices, preprint, https://eprint.iacr.org/2015/1004
23. D. Ramakrishnan, R.J. Valenza, *Fourier Analysis on Number Fields*, vol. 186, Graduate texts in math (Springer, New York, 1999)
24. O. Ramaré, Approximate formulae for $L(1, \chi)$. Acta Arith. **100**, 245–266 (2001)
25. C.L. Siegel, Über die Classenzahl quadratischer Zahlkörper. Acta Arith. **1**, 83–86 (1935)
26. T. Tatuzawa, On a theorem of Siegel. Jpn. J. Math. **21**, 163–178 (1951)
27. E.C. Titchmarsh, *The Theory of Riemann Zeta-Function*, 2nd edn. (Oxford science publications, Clarendon Press, Oxford, 1986). revised by H. Brown
28. L. Washington, *Introduction to Cyclotomic Fields*, vol. 83, 2nd edn., Graduate texts in math (Springer, New York, 1997)

On Analysis of Recovering Short Generator Problems via Upper and Lower Bounds of Dirichlet L-functions: Part 2

Shinya Okumura

Abstract In recent years, some fully homomorphic encryption schemes and cryptographic multilinear maps have been constructed by using short generators and ideal lattices arising from 2^kth cyclotomic fields. Moreover, these systems are expected to have resistance to the attacks by quantum computers. The security of some of such cryptosystems depends on the principal ideal problem (PIP) and the recovering short generator problem (RSGP). Biasse and Song showed a quantum algorithm solving PIP on arbitrary number fields in polynomial time under GRH. On the other hand, Campbell et al. explain an algorithm solving RSGP on 2^kth cyclotomic fields. Their algorithm is analyzed independently by Cramer, Ducas, Peikert and Regev/Okumura, Sugiyama, Yasuda and Takagi. Their analyses suggest that RSGP on 2^kth cyclotomic fields is solved easily for practical parameters, and that cryptosystems of which the security is based on PIP and RSGP may not be post-quantum cryptosystems. Important tools in their analyses are upper and lower bounds of special values of Dirichlet L-functions at 1. In this paper, we give a survey on their analyses and explain some cryptographic and number theoretic open problems on RSGP.

Keywords Post-quantum cryptography · Recovering short generator problem · Cyclotomic fields · Dirichlet L-functions

1 Introduction

Constructing fully homomorphic encryption (FHE) schemes and cryptographic multilinear maps have been strongly desired in cryptography. An FHE scheme allows us to compute addition, subtraction and multiplication of plaintexts (numerical data)

S. Okumura (✉)
Information Security Laboratory, Institute of Systems, Information Technologies and Nanotechnologies, 2-1-22, Momochihama, Sawara-ku, Fukuoka 814-0001, Japan
e-mail: okumura@isit.or.jp

T. Takagi et al. (eds.), *Mathematical Modelling for Next-Generation Cryptography*, Mathematics for Industry 29, DOI 10.1007/978-981-10-5065-7_15

any number of times without decryption. Such a scheme has many application, e.g., cloud computing. A cryptographic multilinear map allows us to construct a multiparty Diffie-Hellman key exchange protocol and an efficient broadcast encryption scheme [4].

No FHE scheme or multilinear map was proposed for a long time, but Gentry [12] and Garg et al. [11] first proposed a FHE scheme and a candidate of multilinear map called GGH map, respectively, at last. Gentry's FHE scheme and GGH map are based on the difficulty of problems on ideal lattices arising from 2^kth cyclotomic fields. After these breakthrough, some improvements of them, e.g., SV-FHE [25] and GGHLite [15], were proposed and they are also based on such ideal lattices.

These encryption schemes [12, 25] and multilinear maps [11, 15] have two common characteristics. One is that an integral basis of a principal ideal of the 2^kth cyclotomic field is used as a public key or can be obtained from public data. The other is that a generator satisfying certain conditions, called a short generator, is used as a secret key. Therefore one can recover short generators (secret keys) of [11, 12, 15, 25] by solving the principal ideal problem (PIP) and the recovering short generator problem (RSGP). Recall that PIP is a problem to find a generator of a principal ideal for a given integral basis of the principal ideal of a number field, and that RSGP is a problem to find a short generator generating a principal ideal generated by a given generator of the principal ideal of a number field.

Biasse and Song [3] showed a quantum algorithm solving PIP in polynomial time under GRH for the Dedekind zeta function of an underlying number field. Moreover, recently, Espitau et al. [10] announced that PIP on the 2^kth cyclotomic fields can be solved in sub-exponential time with classical computers. They demonstrated in their paper how to recover secret keys (short generators) of SV-FHE scheme for the smallest recommended parameters.

On the other hand, independently, Bernstein [2] and Campbell et al. [6] showed algorithms solving RSGP on 2^kth cyclotomc fields for reasonable k. Their idea is to reduce solving RSGP to solving the bounded distance decoding on the log-unit lattices arising from the 2^kth cyclotomc fields. In particular, Campbell et al. suggest using Babai's round-off algorithm [1] and the canonical generators of the groups of cyclotomic units for such cyclotomic fields. We refer to Campbell et al.'s method as the RSG attack.

However, Bernstein and Campbell et al.'s papers do not contain enough theoretical evidence and any experiments to support their algorithms. Theoretical and experimental analysis on their approaches is first given by Cramer, Ducas, Peikert and Regev [8]. Note that they analyzed RSGP over the qth cyclotomic fields with prime power integers q. They study the geometry of the canonical generators of the group of cyclotomic units by using Dirichlet L-functions associated with Dirichlet characters modulo q. More precisely, they showed a relation between Dirichlet L-functions and the Euclidean norm of a dual basis vector of the log-unit lattice, which is a vector necessary for the RSG attack. They used a well-known "implicit" lower bound of a special value of Dirichlet L-function at 1 to give an upper bound of the Euclidean norm of the dual basis vector. Moreover, they also used a statistic technique, "Tail Bounds", to estimate the success probability of the RSG attack from the upper bound of the Euclidean norm of the dual basis vector.

The second analysis on RSGP is given by Okumura, Sugiyama, Yasuda and Takagi [21]. Okumura et al. gave explicit upper and lower bounds of special values of Dirichlet L-functions at 1 by works [9, 16, 17, 23]. Okumura et al. also gave explicit upper and lower bounds of the Euclidean norm of the dual basis vector by using their explicit upper and lower bounds of special values of Dirichlet L-functions at 1 and by counting the number of even Dirichlet characters of given conductors exactly.

The third analysis is given by Okumura, Yasuda and Takagi [22]. They showed that a dual basis of the log-unit lattice is computable efficiently by using a fast Fourier transformation (FFT) [7]. They also showed that the RSG attack over cyclotomic fields of 2-power conductors may be avoided by proving the fact that the short generators resisting the RSG attack over the 2^kth cyclotomic field can resist the RSG attack over the 2^{k_0}th cyclotomic field for $k \leq k_0$.

In this paper, we give a survey on the analyses of Cramer et al. [8] and of Okumura et al. [21, 22] on RSGP over the qth cyclotomic fields with prime power integers q. Moreover, we explain some cryptographic and number theoretic open problems on RSGP.

The rest of this paper is organized as follows: Sect. 2 describes ideal lattices, short generators and key generation algorithms of SV-FHE and GGH/GGHLite. Section 3 gives preliminaries to explain the RSG attack and describes the RSG attack, its success condition and some remarks. Section 4 describes the first part of Cramer et al.'s analysis. Section 5 describes Okumura et al.'s second and third analyses. Section 6 gives some open problems on RSGP.

2 Ideal Lattices and Short Generators in Cryptography

In this section, we describe ideal lattices, short generators and key generation algorithms of SV-FHE and GGH/GGHLite.

2.1 Definitions of Ideal Lattices and Short Generators

We consider a ring $R := \mathbb{Z}[x]/(f(x))$ for a polynomial $f(x) \in \mathbb{Z}[x]$. For any $\alpha \in R$, there exists a unique polynomial $f_\alpha(x) = \sum_{0 \leq j \leq n-1} f_{\alpha,j} x^j \in \mathbb{Z}[x]$ such that $\alpha = f_\alpha(x) \pmod{f(x)}$ when $\deg f(x) = n$. Then we can identify α with $\mathbf{f}_\alpha := (f_{\alpha,n-1}, \ldots, f_{\alpha,0}) \in \mathbb{Z}^n$ and consider arbitrary norms of α, such as the Euclidean norm $\|\alpha\| := \|\mathbf{f}_\alpha\|$ and the ∞-norm $\|\alpha\|_\infty := \|\mathbf{f}_\alpha\|_\infty$. An *ideal lattice* is defined as a lattice corresponding to an ideal of R by the above correspondence between

α and $\mathbf{f}_\alpha$ (see Sect. 3.1 for a short description of lattices). We can choose vectors corresponding to $\{g, xg, \ldots, x^{n-1}g\}$ as a basis of the lattice corresponding to a principal ideal $(g) \subset R$ for $g \in R$.

We can easily perform arithmetic on R and compute a basis of an ideal lattice when $f(x) = x^n + 1$ for a 2-power integer n (note that $\mathbb{Z}[x]/(x^n + 1)$ is isomorphic to the ring of integers of the $2n$th cyclotomic field). This is a reason why the 2^kth cyclotomic fields are important in cryptography (note that arithmetic on cyclotomic fields other than the 2^kth cyclotomic fields can be also performed [19] with only little loss of efficiency).

A generator of a principal ideal is called a *short generator* of the principal ideal if the generator is "short" in the sense of a given norm. Note that the meaning of "short" depends on cryptosystems described below.

2.2 Key Generation of Cryptosystems Using Short Generators

Next, we describe procedures for generating public keys pk and secret keys sk of SV-FHE and GGH/GGHLite.

2.2.1 Fully Homomorphic Encryption Scheme [25]

Algorithm 1 below shows the process of key generation of a homomorphic encryption scheme by Smart et al. [25]. The homomorphic encryption scheme is extended to a fully homomorphic encryption scheme in [25, Sect. 5].

In the case of SV-FHE, a short generator $g := G(x) \pmod{(F(x))}$ is not in sk but is used to generate one of elements in sk. On the other hand, a $\mathbb{Z}$-basis of the principal ideal $(g) \subset \mathbb{Z}[x]/(F(x))$ is obtained from a public key pk (see [25]).

2.2.2 Cryptographic Multilinear Maps [11, 15]

Algorithm 2 below shows the process of key generation of cryptographic multilinear maps [11, 15].

Note that a $\mathbb{Z}$-basis of (g) is not public and does not seem to be obtained immediately (we omit to describe a public key). However, the *zeroizing attack* may enable us to obtain a $\mathbb{Z}$-basis of (g) (see [11] for more details).

Algorithm 1 Key generation of SV-FHE [25]

Parameters
N: Dimension of an ideal lattice;
η: Parameter on the length of a short generator;
Notation
$\mathbb{Z}[x]_{\deg=n} := \{f(x) \in \mathbb{Z}[x] \mid \deg f(x) = n\}$;
$\mathbb{Z}_{\leq m}[x] := \{f(x) \in \mathbb{Z}[x] \mid \|f(x)\|_\infty \leq m\}$, where $\|\cdot\|_\infty$ is defined as in Sect. 2;
$a \leftarrow b$ means to substitute an element b for a;
$a \leftarrow_R B$ means to substitute an element chosen uniformly at random from a set B for a;
Input: $(N, \eta) = (N, 2^{\sqrt{N}})$;
Output: $\mathsf{sk} = (p, B)$, $\mathsf{pk} = (p, \alpha)$;
$F(x) \leftarrow_R \mathbb{Z}[x]_{\deg=N}$ ($F(x)$ is monic);
while $F(x)$ is reducible over $\mathbb{Z}$ **do**
$F(x) \leftarrow_R \mathbb{Z}[x]_{\deg=N}$ ($F(x)$ is monic);
end while
$S(x) \leftarrow_R \mathbb{Z}_{\leq \frac{\eta}{2}}[x] \cap \mathbb{Z}[x]_{\deg=N}$;
$G(x) \leftarrow 1 + 2S(x)$;
$p \leftarrow \mathrm{Res}(F(x), G(x))$;
while p is not a prime number **do**
$S(x) \leftarrow_R \mathbb{Z}_{\leq \frac{\eta}{2}}[x] \cap \mathbb{Z}[x]_{\deg=N}$;
$G(x) \leftarrow 1 + 2S(x)$ ($G(x) \pmod{(F(x))}$ is a short generator);
$p \leftarrow \mathrm{Res}(F(x), G(x))$;
end while
$D(x) \leftarrow \gcd(F(x), G(x)) \pmod{p} \in \mathbb{F}_p[x]$;
$\alpha \leftarrow$ root of $D(x)$ in $\mathbb{F}_p$;
$Z(x) \leftarrow \mathbb{Z}[x]$ satisfying $Z(x)G(x) \equiv p \pmod{(F(x))}$;
$z_0 \leftarrow$ constant term of $Z(x)$;
$B \leftarrow z_0 \pmod{2p}$;

Algorithm 2 Key generation of GGH, GGHLite [11, 15]

Parameters
n: Dimension of an ideal lattice;
q: Modulus parameter;
$\ell_{g^{-1}}$: Upper bound of the Euclidean norm of inverse of a short generator;
Notation
$D_{\sigma,\mathbb{Z}}$: Discrete Gaussian distribution over $\mathbb{Z}$ with mean 0 and variance σ^2;
$R_n := \mathbb{Z}[x]/(x^n + 1)$;
$R_{n,q} := (\mathbb{Z}/q\mathbb{Z})[x]/(x^n + 1)$;
$a \leftarrow_{D_{\sigma,\mathbb{Z}}} \mathbb{Z}$ means to substitute an integer chosen from $D_{\sigma,\mathbb{Z}}$ for a;
Input: $(n, q, \sigma, \ell_{g^{-1}})$ ($n = 2^k$ and $\sigma > 0$);
Output: $\mathsf{sk} = (g, z)$ (g: short generator);
$g_i \leftarrow_{D_{\sigma,\mathbb{Z}}} \mathbb{Z}$ $(i = 0, \ldots, n-1)$;
$g \leftarrow \sum_{0 \leq i \leq n-1} g_i x^i \pmod{(x^n + 1)}$;
while $\|g\| > \sigma\sqrt{n}$, $\|g^{-1}\| > \ell_{g^{-1}}$ or $(g) \subset R_n$ is not a prime ideal **do**
$g_i \leftarrow_{D_{\sigma,\mathbb{Z}}} \mathbb{Z}$ $(i = 0, \ldots, n-1)$;
$g \leftarrow \sum_{0 \leq i \leq n-1} g_i x^i \pmod{(x^n + 1)}$;
end while
$z \leftarrow_R R_{n,q}$;

3 Preliminary

In this section, we give preliminaries to describe a method to recover short generators (RSG attack) and the success condition of the RSG attack.

3.1 Lattices

In general, an n-dimensional lattice is a discrete additive subgroup of an n-dimensional $\mathbb{R}$-vector space V. For cryptographic application, we set $V = \mathbb{R}^n$. For a lattice $\mathscr{L} \subset \mathbb{R}^n$, there exist $\mathbb{R}$-linearly independent vectors $\mathbf{b}_1, \ldots, \mathbf{b}_m$ generating $\mathscr{L}$, i.e., $\mathscr{L} = \sum_{1 \le i \le m} \mathbb{Z}\mathbf{b}_i$. The rank of $\mathscr{L}$, denoted by $\mathrm{rank}(\mathscr{L})$, is the rank as a $\mathbb{Z}$-module. We denote by $\mathrm{span}(\mathscr{L})$ the vector space $\mathbb{R} \otimes_{\mathbb{Z}} \mathscr{L} \subset \mathbb{R}^n$. The dual basis $\mathbf{B}^\vee = \{\mathbf{b}_1^\vee, \ldots, \mathbf{b}_m^\vee\} \subset \mathrm{span}(\mathscr{L})$ of the basis $\mathbf{B} = \{\mathbf{b}_1, \ldots, \mathbf{b}_m\}$ of $\mathscr{L}$ is defined as a set of vectors satisfying $\langle \mathbf{b}_i, \mathbf{b}_j^\vee \rangle = \delta_{ij}$ for all $1 \le i, j \le m$, where $\langle \cdot, \cdot \rangle$ and δ_{ij} denote the natural inner product on $\mathbb{R}^n$ and the Kronecker delta, respectively.

The bounded distance decoding (BDD) is a problem that for a given lattice L and $\mathbf{t}' \in \mathrm{span}(L)$ satisfying $\min_{\mathbf{v} \in L} \|\mathbf{t}' - \mathbf{v}\| \le r$ for some $r < \frac{\lambda_1(L)}{2}$, find the unique vector $\mathbf{v}$ satisfying $\|\mathbf{t}' - \mathbf{v}\| \le r$, where $\lambda_1(L)$ denotes the length of shortest vectors in L. BDD is one of well-known computationally hard problems on lattices, and one tries to solve BDD on log-unit lattices in the RSG attack. Babai's round-off algorithm is a standard algorithm for solving BDD and is used in the RSG attack (see Sect. 3.4). For solving BDD on the above $\mathscr{L}$ and $\mathbf{t} \in \mathrm{span}(\mathscr{L})$, Babai's round-off algorithm outputs $\mathbf{B}\lfloor {}^t(\mathbf{B}^\vee)\, \mathbf{t} \rceil$. We note that $\mathbf{B}$ is identified with the matrix whose jth column vector is $\mathbf{b}_j$, and that the vector $\lfloor {}^t(\mathbf{B}^\vee)\, \mathbf{t} \rceil$ is obtained by applying round function $\lfloor \cdot \rceil$ to each entry of ${}^t(\mathbf{B}^\vee)\, \mathbf{t}$ ($\lfloor c \rceil := \lfloor c + \frac{1}{2} \rfloor$ for any $c \in \mathbb{R}$). The next lemma is important for the RSG attack.

Lemma 3.1 ([8, Claim in Sect. 2.1]) *Let $\mathscr{L}$ and $\mathbf{t}$ be as above. Suppose that $\mathbf{t} = \mathbf{v} + \mathbf{e}$ ($\mathbf{v} \in \mathscr{L}, \mathbf{e} \in \mathbb{R}^n$). If $\left\langle \mathbf{e}, \mathbf{b}_j^\vee \right\rangle \in \left[-\frac{1}{2}, \frac{1}{2}\right)$ for all $j = 1, \ldots, m$, then Babai's round-off algorithm outputs $\mathbf{v}$.*

3.2 Cyclotomic Fields

In this subsection, we describe cyclotomic fields briefly (see textbooks of number theory, e.g. [20, 27], for more detail). Let $\zeta_q \in \mathbb{C}$ be a qth primitive root of unity. A field $K := \mathbb{Q}(\zeta_q)$ is called the qth cyclotomic field. The field K is an extension of $\mathbb{Q}$ with degree $[K : \mathbb{Q}] = \varphi(q)$, where $\varphi(\cdot)$ is the Euler totient function. The ring of integers O_K of K coincides with $\mathbb{Z}[\zeta_q]$. For any embedding σ of K into $\mathbb{C}$, there exists $j \in \mathbb{Z}$ satisfying $1 \le j \le q - 1$, $\gcd(j, q) = 1$ and $\sigma(\zeta_q) = \zeta_q^j$. Throughout this paper, we denote by σ_j such σ.

Next, we describe the group of cyclotomic units of K. Let A be the subgroup of K^* generated by $\pm\zeta_q$ and $z_j := \zeta_q^j - 1$ for $1 \leq j \leq q-1$ and $\gcd(j, q) = 1$. The group C of cyclotomic units of K is defined as $A \cap O_K^*$. The next lemma shows that certain generators of C, called the canonical generators, are easily computable if q is a prime power integer.

Lemma 3.2 ([27, Lemma 8.1]) *For a prime power integer q and $G := (\mathbb{Z}/q\mathbb{Z})^*/\{\pm 1\}$, let K be the qth cyclotomic field and set $b_j := \frac{z_j}{z_1}$ for each $j \in G \setminus \{1\}$, where z_j is defined as above. Then the group C of cyclotomic units of K is generated by $\pm\zeta_q$ and b_j for $j \in G \setminus \{1\}$.*

The next lemma on C is important in considering RSGP (see Sect. 3.4).

Lemma 3.3 ([27, Exercise 8.5]) *Let K be the qth cyclotomic field and C the group of cyclotomic units of K. We denote by $h^+(q)$ the class number of $K^+ := \mathbb{Q}\left(\zeta_q + \overline{\zeta_q}\right)$. Then we have the following equality:*

$$[O_K^* : C] = h^+(q).$$

3.3 Log-unit Lattices

In this subsection, we describe log-unit lattices. Let G and b_j be as in Lemma 3.2. The logarithmic embedding $\mathrm{Log}(\cdot)$ of the qth cyclotomic field K is defined as a map $\mathrm{Log} : K^* \longrightarrow \mathbb{R}^{\frac{\varphi(q)}{2}}$ by $a \mapsto \left(\log|\sigma_j(a)|\right)_{j\in G}$, where σ_j is defined as in Sect. 3.2. Let $\mu(K)$ be the subgroup of K^* generated by all roots of unity in K. Then we have $\mathrm{Ker}(\mathrm{Log}) = \mu(K)$. A log-unit lattice for K is defined as $\Lambda := \mathrm{Log}(O_K^*) \subset \mathbb{R}^{\frac{\varphi(q)}{2}}$. In fact, Λ becomes a lattice of $\mathrm{rank}(\Lambda) = \frac{\varphi(q)}{2} - 1$ from Dirichlet's Unit Theorem (see [20, Chap. I, Theorem 7.4]). All vectors in Λ are orthogonal to $\mathbf{1} := (1, \ldots, 1) \in \mathbb{R}^{\frac{\varphi(q)}{2}}$ because we have $N_{K/\mathbb{Q}}(\varepsilon) = \pm 1$ for any $\varepsilon \in O_K^*$. Therefore, all vectors in a basis of Λ and $\mathbf{1}$ form a basis of $\mathbb{R}^{\frac{\varphi(q)}{2}}$ as an $\mathbb{R}$-vector space.

Lemma 3.2 shows that a basis $\{\mathbf{b}_j := \mathrm{Log}(b_j) \mid j \in G \setminus \{1\}\}$ of a sublattice $\mathrm{Log}(C)$ of Λ is easily computable if q is a prime power integer. Therefore, when the class number of K^+ is equal to 1, a basis of Λ, arising from the canonical generators of C, is easily computable.

3.4 Algorithm for RSG Attack and Its Success Condition

Now, we can explain the RSG attack and its success condition. Assume that q is a prime power integer, and that the class number of K^+ is equal to 1. As we mentioned in the end of Sect. 3.3, the basis $\mathbf{B} := \{\mathbf{b}_j \mid j \in G \setminus \{1\}\}$ of Λ is easily computable. First, we explain an idea of the RSG attack. Take $g \in O_K$ as a short generator (secret key). Assume that we obtain a generator $g' \in O_K$ of the principal ideal (g) by solving PIP. Then there exists $u \in O_K^*$ such that $g' = ug$. The RSG attack is aimed at computing $u' \in O_K^*$ such that $\mathrm{Log}(u) = \mathrm{Log}(u')$. One can recover the short generator g after computing such u'. In fact, there exists a unique $\ell \in \mathbb{Z}$ satisfying $0 \leq \ell \leq q-1$ and $u' = \zeta_q^\ell u$ because of $\mathrm{Ker}(\mathrm{Log}) = \mu(K)$, and such ℓ can be found by a brute force for practical sizes of q.

Next, we explain a method to find the above u'. We have $\mathrm{Log}(g') = \mathrm{Log}(u) + \mathrm{Log}(g)$ from the definition of Log. Then $\mathrm{Log}(g) = \mathrm{Log}(g') - \mathrm{Log}(u)$ can be expected to be a short vector with respect to the Euclidean norm, i.e., $\mathrm{Log}(u)$ may be a vector in Λ close to $\mathrm{Log}(g')$, because g is a short generator. It is suggested in [6] that Babai's round-off algorithm (see Sect. 3.1), which is an algorithm to solve CVP and BDD, should be applied to solve BDD on Λ and $\mathrm{Log}(g')$. Cramer et al. [8] and Okumura et al. [21] analyzed the RSG attack in the case where Babai's round-off algorithm is applied to the RSG attack. In the following, we show an algorithm of the RSG attack.

Algorithm 3 RSG attack

Parameter
q: Parameter on the underlying cyclotomic field $K := \mathbb{Q}(\zeta_q)$ (q is prime power);
Notation;
$G := (\mathbb{Z}/q\mathbb{Z})^* / \{\pm 1\}$;
$\{b_j\}_{j \in G \setminus \{1\}}$: Canonical generators of the group of cyclotomic units C of K (see Sect. 3.2);
$\mathbf{B} := \{\mathrm{Log}(b_j) \mid j \in G \setminus \{1\}\}$: Basis of the log-unit lattice of K under the assumption below;
$\mathbf{B}^\vee := \{\mathbf{b}_j^\vee \mid j \in G \setminus \{1\}\}$: Dual basis of $\mathbf{B}$, i.e., $\left\langle \mathbf{b}_i, \mathbf{b}_j^\vee \right\rangle = \delta_{ij}$, where δ_{ij} denotes Kronecker's delta;
Assumption: Class number of $\mathbb{Q}\left(\zeta_q + \overline{\zeta_q}\right)$ is equal to 1;
Input: $g' = ug$ ($u \in C = O_K^*, g \in O_K$), $\mathbf{B}^\vee$
Output: Short generator g or "failure"
$(a_j)_{j \in G \setminus \{1\}} \leftarrow \lfloor {}^t(\mathbf{B}^\vee)\, \mathrm{Log}(g') \rceil$
$u' \leftarrow \prod_{j \in G \setminus \{1\}} b_j^{a_j}$
$F \leftarrow \{j \in \mathbb{Z} \mid 0 \leq j < q\}$
$g_{\mathrm{cand}} \leftarrow \frac{g'}{\zeta_q^j u'}$ for some $j \in F$.
$F \leftarrow F \setminus j$
while $\|g_{\mathrm{cand}}\|$ is not sufficiently small and $F \neq \emptyset$ **do**
  $g_{\mathrm{cand}} \leftarrow \frac{g'}{\zeta_q^j u'}$ for some $j \in F$.
  $F \leftarrow F \setminus j$
end while
if $F = \emptyset$ **then**
  Output "failure"
else
  Output g_{cand}
end if

As we mentioned above, the short generator g can be recovered if we can compute $u' \in O_K^*$ satisfying $\mathrm{Log}(u) = \mathrm{Log}(u')$ by Babai's round-off algorithm. From Lemma 3.1 in Sect. 3.1, a success condition for $\mathrm{Log}(u) = \mathrm{Log}(u')$ is as follows:

$$\left\langle \mathrm{Log}(g), \mathbf{b}_j^{\vee} \right\rangle \in \left[-\frac{1}{2}, \frac{1}{2} \right) \quad (\forall j \in G \setminus \{1\}). \tag{1}$$

We expect $C = O_K^*$, i.e., the class number of K^+ is equal to 1, because we need $u \in C$ in Algorithm 3.

Remark 3.4 In general, we have $\mathrm{Log}(g') \notin \mathrm{span}(\Lambda)$. In fact, we have $\mathrm{Log}(u) \in \Lambda$ and $\mathrm{Log}(g) \in \mathbb{R}^{\frac{\varphi(q)}{2}}$. Set $a := \frac{1}{\#G}\mathrm{Log}\sqrt{|\mathrm{N}_{K/\mathbb{Q}}(a)|}$. Then we have

$$\mathrm{Log}(g') = \sum_{j \in G \setminus \{1\}} a_j \mathbf{b}_j + a\mathbf{1} + \sum_{j \in G \setminus \{1\}} a'_j \mathbf{b}_j \quad (a_j \in \mathbb{Z},\ a, a'_j \in \mathbb{R}).$$

Therefore we should consider not "BDD on Λ and $\mathrm{Log}(g')$" but "BDD on Λ and $\mathrm{Log}(g') - a\mathbf{1}$". However, since we have

$$\left\langle \mathrm{Log}(g), \mathbf{b}_j^{\vee} \right\rangle = \left\langle \mathrm{Log}(g) - a\mathbf{1}, \mathbf{b}_j^{\vee} \right\rangle \quad (\forall j \in G \setminus \{1\}),$$

we do not need to compute a.

4 Cramer et al.'s Analysis on RSGP [8]

In this section, we give a survey on Cramer et al.'s first analysis on RSGP over the qth cyclotomic fields with prime power integers q [8, Sects. 3–5]. Throughout this section, we assume that q is prime power.

4.1 *Upper Bounds of* $\|\mathbf{b}_j^{\vee}\|$

One of main Cramer et al.'s results is the following upper bound of $\|\mathbf{b}_j^{\vee}\|$.

Theorem 4.1 ([8, Theorem 3]) *Let p be a prime number and $q = p^k$. Let G, $\{\mathbf{b}_j\}_{j \in G \setminus \{1\}}$ and $\{\mathbf{b}_j^{\vee}\}_{j \in G \setminus \{1\}}$ be as in Sect. 3. Then we have $\|\mathbf{b}_i^{\vee}\| = \|\mathbf{b}_j^{\vee}\|$ for all $i, j \in G \setminus \{1\}$ and*

$$\|\mathbf{b}_j^{\vee}\|^2 = 2k|G|^{-1}\left(\ell(q)^2 + O(1)\right) = O\left(q^{-1}(\log q)^3\right),$$

where $\ell(q)$ is a constant occurring in [8, Theorem 1].

From the success condition of the RSG attack, we need to analyze the values of $\left\langle \mathrm{Log}(g), \mathbf{b}_j^{\vee} \right\rangle$ for all $j \in G \setminus \{1\}$. Since $\langle \mathrm{Log}(g), \mathbf{b}_j^{\vee} \rangle = \|\mathrm{Log}(g)\| \cdot \|\mathbf{b}_j^{\vee}\| \cos\theta_j$ for some $\theta_j \in [0, 2\pi)$, giving upper bounds of $\|\mathbf{b}_j^\vee\|$ seems to be important as Cramer et al. did. For another importance of giving upper bounds of $\|\mathbf{b}_j^\vee\|$, see Sect. 4.2 below.

In Cramer et al.'s proof of Theorem 4.1, they showed the following relation between special values of Dirichlet L-functions for even Dirichlet characters modulo q at 1 and the Euclidean norm of $\mathbf{b}_j^\vee$.

Lemma 4.2 *Let $\hat{G}$ be the character group of G, i.e., the set of even Dirichlet characters modulo q, and $L(s, \chi)$ the Dirichlet L-function associated with $\chi \in \hat{G}$. We denote by f_χ the conductor of $\chi \in \hat{G}$. Then the following equality holds:*

$$\|\mathbf{b}_j^\vee\|^2 = 4|G|^{-1} \sum_{\chi \in \hat{G}} \frac{1}{f_\chi |L(1, \chi)|}.$$

For definitions and fundamental properties of Dirichlet characters and Dirichlet L-functions, see [26]. In addition, Cramer et al. used a well-known lower bound of $|L(1, \chi)|$ proved by Landau [14] to give their upper bound of $\|\mathbf{b}_j^\vee\|$ (see also [26]).

Finally, we give two lemmas which explain the reason why $|L(1, \chi)|$ occurs in Lemma 4.2.

Lemma 4.3 ([8, Corollary 1]) *For each even Dirichlet character χ modulo q of conductor $f_\chi > 1$, we have*

$$\left| \sum_{a \in (\mathbb{Z}/q\mathbb{Z})^*} \overline{\chi(a)} \cdot \log|1 - \zeta_q^a| \right| = \sqrt{f_\chi} |L(1, \chi)|. \tag{2}$$

Lemma 4.4 ([8, Lemma 3]) *The following equality holds for all $j \in G \setminus \{1\}$:*

$$\|\mathbf{b}_j^\vee\|^2 = |G|^{-1} \sum_{\chi \in \hat{G} \setminus \{1\}} |\lambda_\chi|^{-2},$$

where $\lambda_\chi = \frac{1}{2} \sum_{a \in (\mathbb{Z}/q\mathbb{Z})^*} \overline{\chi(a)} \cdot \log|1 - \zeta_q^a|$ for $\chi \in \hat{G}$ (see Cramer et al.'s proof of Theorem 4.1). In particular, the value of $\|\mathbf{b}_j^\vee\|$ is independent of $j \in G \setminus \{1\}$.

4.2 Statistical Analysis

In this subsection, we describe Cramer et al.'s analysis by the statistical technique, which also shows why estimating upper bounds of $\|b_j^\vee\|$ is important. Cramer et al. pointed out that if short generators are chosen from a continuous or discrete Gaussian distribution with some conditions, then the success condition of the RSG attack is satisfied with non-negligible probability. In GGH and GGHLite, a secret short generator is chosen from a discrete Gaussian distribution of mean 0 and variance σ^2 for some $\sigma \in \mathbb{R}_{>0}$. Thus, if we have $h^+(q) = 1$ and Cramer et al.'s result is true, then we obtain $\pm\zeta_q^j g$ with non-negligible probability. Cramer et al.'s result on the continuous Gaussian distribution comes from the next theorem:

Theorem 4.5 ([8, Lemma 5]) *Let $X_1, \ldots, X_n, X'_1, \ldots, X'_n$ be i.i.d. $N(0, r)$ variables for some $r \in \mathbb{R}_{>0}$. Put $\hat{X}_i := \left(X_i^2 + X_i'^2\right)^{\frac{1}{2}}$ for $i = 1, \ldots, n$. Then for any unit vectors $\mathbf{a}^{(1)}, \ldots, \mathbf{a}^{(\ell)} \in \mathbb{R}^n$ satisfying $\langle \mathbf{a}^{(i)}, \mathbf{1} \rangle = 0$ $(i = 1, \ldots, \ell)$, and every $t \geq c$ for some universal constant c, we have*

$$\Pr\left[\exists j, \left|\sum_{1 \leq i \leq n} a_i \log(\hat{X}_i)\right| \geq t\right] \leq 2\ell \exp(-t/2). \tag{3}$$

Note that the inequality (3) does not involve the variance r. It implies that the success probability does not rely on the variance of continuous Gaussian distribution. We also note that $\langle \mathbf{b}_j^\vee, \mathbf{1} \rangle = 0$ for all $j \in G \setminus \{1\}$.

Cramer et al.'s proof of Theorem 4.5 is based on theory of sub-exponential random variables defined below and the tail bound (see Lemma 4.7). Before we define sub-exponential random variables and present the tail bound, we describe why Theorem 4.5 is related to the success probability of the RSG attack if a result similar to Theorem 4.5 is true for discrete Gaussian distribution. Let $n := \varphi(q)/2$ and $\mathbf{a}^{(j)} := (a_1^{(j)}, \ldots, a_n^{(j)}) = \mathbf{b}_j^\vee / \|\mathbf{b}_j^\vee\|$ for $j \in G \setminus \{1\}$. For each σ_i, let $Y_i + Y_i'\sqrt{-1} := \sigma_i(g)$ and $\hat{Y}_i := \left(Y_i^2 + Y_i'^2\right)^{\frac{1}{2}}$. Then we have $\mathrm{Log}(g) = (\log(\hat{Y}_i))_{i \in G}$ and thus the condition $\langle \mathrm{Log}(g), \mathbf{b}_i^\vee \rangle \in [-\frac{1}{2}, \frac{1}{2})$ is equivalent to

$$\sum_{j \in G} a_j^{(i)} \log(\hat{Y}_i) \in \left[-\frac{1}{2||\mathbf{b}_j^\vee||}, \frac{1}{2||\mathbf{b}_j^\vee||}\right).$$

It implies that if $Y_1, \ldots, Y_n, Y'_1, \ldots, Y'_n$ are i.i.d. $N(0, \sigma^2)$ variables for some $\sigma \in \mathbb{R}_{>0}$, then we can apply (the result similar to) Theorem 4.5 to those variables and the unit vectors by taking $t = 1/2||\mathbf{b}_j^\vee||$.

Now, we define the sub-exponential random variables.

Definition 4.6 For any $\alpha, \beta > 0$, a random variable X is said to be (α, β)-sub-exponential if an inequality $E[\cosh(\alpha X)] \leq \beta$ holds, where $E[Y]$ is the expected value for a random variable Y.

Observe that if we have $E[Y] < \infty$ and $E[Y^{-1}] < \infty$ for a random variable Y, then a variable $\log(Y)$ is $(1, \beta)$-sub-exponential random variable for some $\beta \in \mathbb{R}$. This fact is also useful for proving Theorem 4.5 (see Lemma 4.7). Finally, we present the next lemma which is essential to estimating the success probability.

Lemma 4.7 ([8, Lemma 4]) *Let $X_1, \ldots, X_n$ be independent centered (α, β)-sub-exponential random variables for $\alpha, \beta \in \mathbb{R}$. Then, for each real vector $\mathbf{a} := (a_1, \ldots, a_n$ satisfying $\langle \mathbf{a}, \mathbf{1} \rangle = 0$, and for every $t \in \mathbb{R}_{\geq 0}$, we have*

$$\Pr\left[\left|\sum_{1 \leq i \leq n} a_i X_i\right| \geq t\right] \leq 2\exp\left(-\min\left(\frac{\alpha^2 t^2}{8\beta||\mathbf{a}||^2}, \frac{\alpha t}{2||\mathbf{a}||_\infty}\right)\right).$$

5 Okumura et al.'s Analysis on RSGP [21, 22]

In this section, we give a survey on Okumura et al.'s first analysis [21] and second analysis [22] on RSGP over the qth cyclotomic fields with prime power integers q.

5.1 Explicit Upper and Lower Bounds of $\|\mathbf{b}_j^\vee\|$

In this subsection, we give Okumura et al.'s explicit upper and lower bounds of $\|\mathbf{b}_j^\vee\|$. In Cramer et al.'s analysis, they give only implicit upper bounds of $\|\mathbf{b}_j^\vee\|$, and therefore one cannot use their upper bounds to analyze the RSG attack easily for each q. (Cramer et al. give experimental results on $\|\mathbf{b}_j^\vee\|$ for some q. However, it may be necessary to analyze RSGP for more large q, and therefore explicit bounds are important). Note that although upper bounds are important for analyzing the RSG attack, Okumura et al.'s lower bound is a better approximation of $\|\mathbf{b}_j^\vee\|$ when $q = 2^k$ [21, Table 1].

Now, we give Okumura et al.'s explicit bounds for $q = 2^k$.

Theorem 5.1 *For $q := 2^k$ and $k \geq 3$, we have*

$$\|\mathbf{b}_j^\vee\|^2 \leq \frac{400}{2^{k+1}}\left[\frac{k(k+1)(2k+1) - 84}{6}(\log 2)^2 - (\log 2)(\log \pi)\{k(k+1) - 12\} + (\log \pi)^2(k-3)\right] + \frac{1}{2^{k-2}\{\log(1+\sqrt{2})\}^2}.$$

$$\|\mathbf{b}_j^\vee\|^2 \geq \frac{8}{2^{k-2}(\log 2)^2}\sum_{j=1}^{k-2}\frac{1}{(j+4)^2}.$$

For general prime power integers q, see [21, Theorem 6]. The above explicit result is obtained from explicit upper and lower bounds of special values of Dirichlet L-functions and from counting the exact number of even Dirichlet characters modulo q of given conductors (cf. [26]).

5.2 Experimental Verification of RSGP over $\mathbb{Q}(\zeta_{2^k})$ ($k = 6, 8, 10$)

In this subsection, we describe Okumura et al.'s experiments on RSGP over $\mathbb{Q}(\zeta_{2^k})$ for $k = 6, 8, 10$ [21, Sect. 8]. In Okumura et al.'s experiments, 1000 short generators were generated, and they checked whether for each short generator, the success condition of RSG attack is satisfied or not. In GGH and GGHLite, short generators are chosen randomly from discrete Gaussian distributions and generate prime principal ideals. On the other hand, in SV-FHE, short generators are chosen randomly from uniform distribution and generate prime principal ideals. Therefore Okumura et al. executed many experiments in various situations as above. Moreover, Okumura et al. also executed many experiments to check whether the success of RSG attack depends on the above conditions on short generators.

As a result, the success of RSG attack seems to be independent of the above conditions on short generators, and the RSG attack succeeds in recovering short generators with probability being about 50% ($k = 6$), 85% ($k = 8$) and 100% ($k = 10$).

5.3 Efficient Computation of $\mathbf{b}_j^\vee$

A method of the previous work [21] to compute $\mathbf{b}_j^\vee$ is roughly divided into two strategies. First, one needs to compute the basis of the log-unit lattice, which is obtained from the canonical generators of the group of cyclotomic units for the qth cyclotomic field. Second, one can compute the dual basis by the basis obtained in the first strategy and by the internal function "DualBasisLattice" of Magma [5]. The rank and dimension of the log-unit lattice of the qth cyclotomic field are equal to $\frac{\varphi(q)}{2} - 1$ and $\frac{\varphi(q)}{2}$, respectively. It implies that it takes a lot of time to compute the dual basis for very large q, e.g., it took more than 310 h to compute the dual basis on the log-unit lattice of the 2^{15}th cyclotomic field (see [21, Table 1]).

On the other hand, Okumura et al. [22] showed that one should compute a certain dual basis vector for the RSG attack because other dual basis vectors can be obtained by rearranging the entries of the dual basis vector. Moreover, Okumura et al. also showed that one can compute a dual basis vector by solving a certain linear equation system with the circulant coefficient matrix. Such a linear equation system is solvable efficiently by using FFT.

Algorithm 4 shows an efficient algorithm for computing $\mathbf{b}_j^\vee$ by Okumura et al.

Algorithm 4 Efficient method for computing the dual basis vector $\mathbf{b}_j^\vee$ [22]

Parameter

q: Parameter on the underlying cyclotomic field $K := \mathbb{Q}(\zeta_q)$;

Notation

$a \leftarrow_{\mathrm{FFT},\mathbf{c}} b$ means to substitute b, computed by FFT for a vector $\mathbf{c}$, for a;

Input: $q = p^k$ (p: prime number, $k \geq 3$);

Output: Dual basis vector $\mathbf{b}_j^\vee$;

$\gamma \leftarrow$ generator of $(\mathbb{Z}/q\mathbb{Z})^* / \{\pm 1\}$;

$G \leftarrow \{1, \gamma, \ldots, \gamma^{\frac{\varphi(q)}{2}-1}\}$;

$\mathbf{z} \leftarrow \left(\log \left|\zeta_q^i - 1\right|\right)_{i \in G}$;

$\lambda_j \leftarrow_{\mathrm{FFT},\mathbf{z}}$ eigenvalue of the circulant matrix whose first column is $\mathbf{z}$ ($j = 0, \ldots, \#G - 1$);

$\lambda \leftarrow \left(\lambda_1^{-1}, \ldots, \lambda_{\#G-1}^{-1}\right)$;

$\mathbf{z}_1^\vee \leftarrow_{\mathrm{InverseFFT},\lambda}$ first column vector of $\mathbf{Z}^{-1}$;

$\mathbf{z}_j^\vee \leftarrow$ a vector obtained by rearranging the entries of $\mathbf{z}_1^\vee$ ($j \in G \setminus \{1\}$);

$\mathbf{b}_j^\vee \leftarrow$ the dual basis vector obtained from $\mathbf{z}_j^\vee$ (see a proof of [8, Lemma 3]);

5.4 Countermeasure of RSG Attack

Okumura et al. showed a countermeasure of RSG attack over 2^kth cyclotomic fields for $k \geq 10$ [22, Sect. 5]. In their experiments of the previous work [21, Sect. 8], although the success condition of RSG attack over such cyclotomic fields is satisfied with probability being 100%, the success condition of RSG attack over 2^kth cyclotomic fields for $k = 6, 8$ is not satisfied with probability being about $15 - 50\%$. From this, Okumura et al. analyzed behavior of the inner product in the success condition of the RSG attack when short generators in the 2^kth cyclotomic field are considered as short generators in the 2^{k_0}th cyclotomic field for $k_0 \geq k$. As a result, Okumura et al. showed that short generators which can resist the RSG attack over the 2^kth cyclotomic can also resist the RSG attack over the 2^{k_0}th cyclotomic field.

In the following, we denote Log, z_j, b_j, $\mathbf{1}$, $\mathbf{b}_j$ and $\mathbf{b}_j^\vee$ associated with $K^{(k)} := \mathbb{Q}(\zeta_{2^k})$ by $\mathrm{Log}^{(k)}$, $z_j^{(k)}$, $b_j^{(k)}$, $\mathbf{1}^{(k)}$, $\mathbf{b}_j^{(k)}$ and $\mathbf{b}_j^{(k)\vee}$, respectively. We set $G_k := \{j \in \mathbb{Z} \mid 3 \leq j \leq 2^{k-1} - 1, \gcd(j, 2) = 1\}$ for each $k \geq 3$. Recall that for a short generator g, the success condition of the RSG attack over $K^{(k)}$ is written as $\left\langle \mathrm{Log}^{(k)}(g), \mathbf{b}_j^{(k)\vee} \right\rangle \in \left[-\frac{1}{2}, \frac{1}{2}\right)$ $(\forall j \in G_k)$. The next theorem [22, Theorem 1] is Okumura et al.'s main result on countermeasure of RSG attack.

Theorem 5.2 *For any $k_0 \geq k$ and $g_k \in O_{K^{(k)}}$, the following inequality holds:*

$$\max_{j \in G_k} \left|\left\langle \mathrm{Log}^{(k)}(g_k), \mathbf{b}_j^{(k)\vee} \right\rangle\right| \leq \max_{j \in G_{k_0}} \left|\left\langle \mathrm{Log}^{(k_0)}(g_k), \mathbf{b}_j^{(k_0)\vee} \right\rangle\right|.$$

From this theorem, we clearly have a method to generate short generators which can resist the RSG attack (see [22, Sect. 5.2]).

Remark 5.3 The RSG attack over $K^{(k)}$ computes a generator $\tilde{g} \in O_{K^{(k)}}$ satisfying $\left\langle \mathrm{Log}^{(k)}(\tilde{g}), \mathbf{b}_j^{(k)\vee} \right\rangle \in \left[-\frac{1}{2}, \frac{1}{2}\right) (\forall j \in G_k)$. Such $\tilde{g}$ may be sufficiently short to break cryptosystems, and therefore Theorem 5.2 may not show that the RSG attack can be avoided completely. Moreover, there may be another attack which can recover a short generator chosen from a smaller cyclotomic field. However, we expect that Theorem 5.2 suggests that there are some important factors to theoretically analyze the RSG attack over the 2^kth cyclotomic fields.

6 Open Problems on RSGP

In this section, we explain some cryptographic and number theoretic open problems. Although RSGP over cyclotomic fields of prime power conductors, which can be applied to cryptography, i.e., their conductors have reasonable sizes, is believed to be solvable in many cryptographic communities, we believe that the following problems are important in cryptography and must be solved:

1. Find an algorithm with reasonable assumptions to solve RSGP over $K = \mathbb{Q}(\zeta_q)$ with any prime power integer q. In the RSG attack, we need to assume that the class number h^+ of $\mathbb{Q}\left(\zeta_q + \overline{\zeta_q}\right)$ is equal to 1. However, justifying this assumption has been an open problem for a long time. Moreover, although it is required that h^+ is small because of $[O_K^* : C_K] = h^+$, where C_K is the group of cyclotomic units, h^+ is extremely large if $h^+ > 1$ [21, Appendix].
2. Find an algorithm with reasonable assumptions to solve RSGP over arbitrary cyclotomic fields. As we noticed in Sect. 2.1, arithmetic on arbitrary cyclotomic fields (with reasonable conductors) is efficient. It suggests that cyclotomic fields of reasonable conductors can be applied to cryptography, and that we need to analyze RSGP over such cyclotomic fields.
3. Analyze the RSG attack over number fields other than cyclotomic fields and clarify number fields that RSGP becomes easy. We can consider RSGP and the RSG attack over arbitrary number fields because there are embeddings from arbitrary number fields to $\mathbb{C}$ (this means that logarithmic embeddings can be defined for arbitrary number fields). If there is a number field which enables us to perform any (or some) arithmetic on it, then such the number field may be used to construct lattice based-cryptosystems.
4. Analyze whether there is a similar algorithm (attack) which can be applied to lattice based-cryptosystems, e.g., NTRU [13], LWE [24] and Ring-LWE [18], other than GGH, GGHLite, SV-FHE and their variants. These three cryptosystems use polynomials with small integer coefficients or vectors with small integer entries as secret keys. Almost all problems on lattices, which are used in cryptography, get more difficult as the dimensions of lattices increase, and this property is a basis of the security of almost all lattice based-cryptosystems. On the other hand, the experimental results [21, Sect. 8] suggest that the success probability of the

RSG attack is getting higher as the dimensions of (log-unit) lattices increase. Thus we expect that the RSG attack is a very effective attack on lattice based-cryptosystems.

7 Conclusion

In this paper, we gave a survey on the recovering short generator problem (RSGP) over the qth cyclotomic fields, which is a basis of the security of some fully homomorphic encryption schemes and cryptographic multilinear maps. These cryptosystems have been expected to have resistance to the attacks by quantum computers. We also gave four cryptographic and number theoretic open problems on RSGP. To the best of our knowledge, RSGP is the first problem that Dirichlet characters and Dirichlet L-functions are applied to the security analysis of cryptosystems. Although the analyses on RSGP by Cramer et al. and by Okumura et al. suggest that RSGP over the 2^kth cyclotomic fields with reasonable sizes of k can be solved in practical time under some assumptions, we believe that more theoretical analysis on RSGP is needed, and that RSGP is still an interesting and important open problem in algorithmic number theory and cryptography.

References

1. L. Babai, On Lovász' lattice reduction and the nearest lattice point problem. Combinatorica **6**(1), 1–13 (1986) (Preliminary version in STACS 1985)
2. D. Bernstein, A subfield-logarithm attack against ideal lattices (2014), http://blog.cr.yp.to/20140213-ideal.html
3. J.-F. Biasse, F. Song, Efficient quantum algorithms for computing class groups and solving the principal ideal problem in arbitrary degree number fields, in *Proceedings of the Twenty-Seventh Annual ACM-SIAM Symposium on Discrete Algorithms, SODA '16* (2016), pp. 893–902
4. D. Boneh, A. Silverberg, Applications of multilinear forms to cryptography, in *Contemporary Mathematics*, vol. 324 (American Mathematical Society, Providence, 2003), pp. 71–90
5. W. Bosma, J. Cannon, C. Playoust, The Magma algebra system. I. The user language. J. Symb. Comput. **24**(3–4), 235–265 (1997)
6. P. Campbell, M. Groves, D. Shepherd, Soliloquy: a cautionary tale, in *ETSI 2nd Quantum-Safe Crypto Workshop* (2014)
7. J.W. Cooley, J.W. Tukey, An algorithm for the machine calculation of complex Fourier series. Math. Comput. **19**, 297–301 (1965)
8. R. Cramer, L. Ducas, C. Peikert, O. Regev, Recovering short generators of principal ideals in cyclotomic rings, in *EUROCRYPT 2016*. LNCS, vol. 9666 (Springer, Berlin, 2016), pp. 559–585
9. S.S. Eddin, D.J. Platt, Explicit upper bounds for $|L(1, \chi)|$ when $\chi(3) = 0$. Colloq. Math. **133**(1), 23–34 (2013)
10. T. Espitau, P.-A. Fouque, A. Gélin, P. Kirchner, Computing generator in cyclotomic integer rings, in *IACR Cryptology ePrint Archive, 2016/957* (2016)
11. S. Garg, C. Gentry, S. Halevi, Candidate multilinear maps from ideal lattices, in *EUROCRYPT 2013*. LNCS, vol. 7881 (Springer, Berlin, 2013), pp. 1–17

12. C. Gentry, Fully homomorphic encryption using ideal lattices, in *Proceedings STOC 2009* (ACM, 2009), pp. 169–178
13. J. Hoffstein, J. Pipher, J.H. Silverman, NTRU: a ring-based public key cryptosystem, in *Proceedings of ANTS-III*. Lecture Notes in Computer Science, vol. 1423 (1998), pp. 267–288
14. E. Landau, Über Dirichletsche Reihen mit komplexen Charakteren. Journal für die reine und angewandte Mathematik **157**, 26–32 (1927)
15. A. Langlois, D. Stehlé, R. Steinfeld, GGHLite: more efficient multilinear maps from ideal lattices, in *EUROCRYPT 2014*. LNCS, vol. 8441 (Springer, Berlin, 2014), pp. 239–256
16. S. Louboutin, Majorations explicites de $|L(1, \chi)|$ (quatrième partie). C. R. Acad. Sci. Paris **334**, 625–628 (2002)
17. S. Louboutin, An explicit lower bound on moduli of Dirichlet L-functions at $s = 1$. J. Ramanujan Math. Soc. **30**(1), 101–113 (2015)
18. V. Lyubashevsky, C. Peikert, O. Regev, On ideal lattices and learning with errors over rings. J. ACM **60**(3), 43 (2013)
19. V. Lyubashevsky, C. Peikert, O. Regev, A toolkit for ring-LWE cryptography, in *IACR Cryptology ePrint Archive, 2013/293* (2013)
20. J. Neukirch, in *Algebraic Number Theory*. Grundlehren der mathematischen Wissenschaften, vol. 322 (Springer, Berlin, 1999)
21. S. Okumura, S. Sugiyama, M. Yasuda, T. Takagi, Security analysis of cryptosystems using short generators over ideal lattices, in *IACR Cryptology ePrint Archive, 2015/1004* (2015)
22. S. Okumura, M. Yasuda, T. Takagi, An improvement on the recovering short generator attack over ideal lattices and its countermeasure, Preprint (2016)
23. O. Ramaré, Approximate formulae for $L(1, \chi)$. Acta Arith. **100**, 245–266 (2001)
24. O. Regev, On lattices, learning with errors, random linear codes, and cryptography, in *Proceedings of the Thirty-seventh Annual ACM Symposium on Theory of Computing, STOC '05* (2005), pp. 84–93
25. N.P. Smart, F. Vercauteren, Fully homomorphic encryption with relatively small key and ciphertext sizes, in *Public Key Cryptography-PKC 2010*. LNCS, vol. 6056 (Springer, Berlin, 2010), pp. 420–443
26. S. Sugiyama, On analysis of recovering short generator problems via upper and lower bounds of Dirichlet L-functions: part 1 (in this proceeding)
27. L. Washington, *Introduction to Cyclotomic Fields*, 2nd edn. Graduate Texts in Mathematics, vol. 83 (Springer, New York, 1997)

Recent Progress on Coppersmith's Lattice-Based Method: A Survey

Yao Lu, Liqiang Peng and Noboru Kunihiro

Abstract In 1996, Coppersmith proposed a lattice-based method to solve the small roots of a univariate modular equation in polynomial time. Since its invention, Coppersmith's method has become an important tool in the cryptanalysis of RSA crypto algorithm and its variants. In 2006, Jochemsz and May introduced a general strategy to solve small roots of any form of multivariate modular equations in polynomial time. Based on Jochemsz–May's strategy, for any given multivariate equations one can easily construct the desired lattices with triangular matrix basis. However, for some attacks, Jochemsz–May's general strategy could not fully capture the algebraic structure of the target polynomials. Thus, some sophisticated techniques that can deeply exploit the algebraic relations have been proposed. In this paper, we give a survey of these recent approaches for lattice constructions, and also give small examples to show how these approaches work.

Keywords Coppersmith's method · Unraveled linearization technique · Exponent trick · Two-step lattice-based method · Small roots of modular equations

1 Introduction

The famous RSA public key cryptosystem invented by Rivest, Shamir, and Adleman [45] has been widely used as it is put forward. A brief description on the key generation algorithm of the basic RSA scheme is given as follows:

Y. Lu (✉) · N. Kunihiro
School of Frontier Sciences, University of Tokyo, 5-1-5 Kashiwanoha, Kashiwa-shi, Chiba 277-8561, Japan
e-mail: luyao@it.k.u-tokyo.ac.jp

N. Kunihiro
e-mail: kunihiro@k.u-tokyo.ac.jp

L. Peng
State Key Laboratory of Information Security, Institute of Information Engineering, Chinese Academy of Sciences, Beijing, China
e-mail: pengliqiang@iie.ac.cn

T. Takagi et al. (eds.), *Mathematical Modelling for Next-Generation Cryptography*, Mathematics for Industry 29, DOI 10.1007/978-981-10-5065-7_16

Key Generation of RSA: Let $N = pq$ be an RSA modulus, where p and q are primes of the same bitlength. Randomly choose an integer e such that $\gcd(e, \varphi(N)) = 1$, where $\varphi(N) = (p-1)(q-1)$ and calculate d such that $ed \equiv 1 \pmod{\varphi(N)}$ by the Extended Euclidean Algorithm. The public keys are N and e and the private keys are p, q, and d.

Since $ed \equiv 1 \pmod{\varphi(N)}$, one can easily recover d when $\varphi(N)$ is known, or equivalently, the prime factorization of N is known. Hence, the security of the RSA scheme is based on the difficulty of the integer factorization problem. Thus, the bitlength of modulus N should be chosen prudently. In 2009, Kleinjung et al. [25] successfully factored a 768-bit RSA modulus by the number field sieve factoring method and due to the work of Lenstra et al. [31], 1024-bit RSA moduli seem to be secure now but it is the tendency to choose larger moduli such as 2048 or 4096 bits. So far, except the Shor's quantum algorithm [50], no algorithm has come close to factoring the RSA modulus in polynomial time and it is still unclear whether quantum computers with a large number of quantum bit can be constructed. However, although the RSA scheme is secure when the bitlength of RSA modulus is large enough, in some cases it also can be broken in polynomial time, such as the private exponent d is chosen to be small for efficient modular exponentiation in the decryption process, or partial bits of the private key p or q are exposed by side-channel attack and so on.

In 1990, Wiener [65] showed that when $d \leq N^{0.25}$, one can factor the modulus N in polynomial time by a continued fraction method. Later, by utilizing a lattice-based method, commonly known as Coppersmith's method [7], Boneh and Durfee [3] significantly improved the bound to $N^{0.292}$. Until now, $N^{0.292}$ is still the best bound for small private exponent attack on the RSA scheme. Let us give a brief explanation of Boneh and Durfee's approach. From the equation $ed \equiv 1 \pmod{\varphi(N)}$, the problem of factoring N can be transformed into solving the unknowns $\varphi(N)$ that equals $N - p - q + 1$. This can be modeled by the following modular polynomial equation

$$f(x, y) = x(N + y) + 1 = 0 \mod e,$$

and the desired root is $(k, -(p+q)+1)$, where $k = \frac{ed-1}{\varphi(N)}$. Then Boneh and Durfee fixed an integer m and selected polynomials that have the same root $(k, -(p+q)+1)$ modulo e^m. They constructed a lattice whose row basis vectors are corresponding to the coefficients vectors of the selected polynomials, and applied the L^3 lattice basis reduction algorithm [30] to the lattice basis. Note that when the private exponent d is small enough, there are two polynomials $h_1(x, y)$ and $h_2(x, y)$ which are corresponding to the reduced lattice basis vectors have that both $h_1(k, -(p+q)+1) = 0$ and $h_2(k, -(p+q)+1) = 0$ hold over the integers. Then if $h_1(x, y)$ and $h_2(x, y)$ are algebraic independent, one can easily recover the root $(k, -(p+q)+1)$. In next section, we will describe the sketch of Coppersmith's method in detail.

Actually, since Coppersmith's method [7] was invented, the lattice-based Coppersmith's method has become an important tool in the cryptanalysis of RSA scheme and its variants. The most important step of Coppersmith's method is how to construct the desired lattice. In 2006, Jochemsz and May [23] introduced a general strategy to solve small roots of any form of multivariate modular equations in polynomial time. Based on Jochemsz–May's strategy, one can construct a triangular matrix basis with calculation the attack condition in an easy way. However, for some hidden algebraic relations among the unknowns of the modular equations, one may not obtain the best lattice constructions only with Jochemsz–May's strategy.

For example, it is easy to obtain a selection of shift polynomials based on Jochemsz–May's strategy and factor the RSA modulus $N = pq$ when $d < N^{0.284}$, but Boneh and Durfee [3] showed that one can use the method of geometrically progressive matrices to construct a non-triangular matrix basis, though this method is quite complex, they successfully obtained a better result, namely $d < N^{0.292}$. In 2010, Herrmann and May [19] used unraveled linearization technique to solve the same problem and gave a simple proof to obtain the same result as Boneh and Durfee's. That means in addition to Jochemsz–May's general strategy of selecting polynomials, there exists several other methods of lattice constructions. Admittedly, Jochemsz–May's strategy provided a general method to choose polynomials to construct a triangular matrix basis which allows an easy derivation of the determinant as the product of the diagonal entries. However in the lattice constructions, one should fully utilize the relations between the monomials or multiple equations to select appropriate polynomials to construct lattices which may obtain better results.

In this paper, we give a survey of recent approaches of lattice constructions to solve small roots of modular equations. For some certain cases, these methods yield better results than purely using Jochemsz–May's general strategy. Specifically, we list three types of lattice-based methods:

- The first method is unraveled linearization technique, as we discussed above, one can give a simple proof of small private exponent attack on RSA scheme by unraveled linearization technique;
- The second method is exponent trick, to make better use of the relations of exponents, one can introduce multiple parameters to select appropriate polynomials to construct lattices, and this trick is very useful for the analyses on Prime Power RSA;
- The last method is two-step lattice-based method, this method is suitable for the multiple relations among the private keys, one can first use one of the relations to construct a low-dimensional lattice, and then from the reduced basis of the lattice one can use other relations to solve the desired vector. This method works very well for the analyses on Dual RSA scheme and implicit hint factorization problem of RSA scheme.

Note that Coppersmith's method also includes the lattice-based method of finding small roots of integer equations [6, 23], in this survey we only focus on the methods of lattice constructions which are used to solve small roots of modular equations.

2 Preliminaries

To describe the sketch of Coppersmith's method, we first give a brief review on the definition of lattices. Let $\mathscr{L}$ be a lattice which is spanned by k linearly independent vectors $\boldsymbol{w}_1, \boldsymbol{w}_2, \ldots, \boldsymbol{w}_k \in \mathbb{Z}^n$. Namely, lattice $\mathscr{L}$ is composed by all integer linear combinations, $c_1\boldsymbol{w}_1 + \cdots + c_k\boldsymbol{w}_k$, of $\boldsymbol{w}_1, \boldsymbol{w}_2, \cdots, \boldsymbol{w}_k$, where $c_1, \ldots, c_k \in \mathbb{Z}$. Then the set of vectors $\boldsymbol{w}_1, \ldots, \boldsymbol{w}_k$ is called a lattice basis of $\mathscr{L}$ and k is the lattice dimension of $\mathscr{L}$. Moreover, for any lattice $\mathscr{L}$ with dimension greater than 1, we can transform the lattice basis $\boldsymbol{w}_1, \boldsymbol{w}_2, \ldots, \boldsymbol{w}_k$ to another equivalent lattice basis of $\mathscr{L}$ by a multiplication with some integer matrix with determinant ± 1. More details about lattices can be referred to [41].

The study on lattices has a long history, and there is a wide range application of lattices in fundamental courses, like mathematics. Moreover, since Lenstra, A.K., Lenstra, H.W., and Lovász, L. introduced the famous L^3 lattice basis reduction algorithm to find a lattice basis with good properties, lattices are increasingly being used in analyses of knapsack-based public-key cryptosystems [11, 29], and solving simultaneous Diophantine equations [40] and so on. More specifically, for any given lattice, one can use L^3 lattice basis reduction algorithm [30] to find out relatively short vectors with following lemma in polynomial time.

Lemma 2.1 (L^3, *[30, 36]*) *Let $\mathscr{L}$ be a lattice of dimension k. Applying the L^3 algorithm to the basis of $\mathscr{L}$, the output reduced basis vectors $\boldsymbol{v}_1, \ldots, \boldsymbol{v}_k$ satisfy that*

$$\|\boldsymbol{v}_i\| \leq 2^{\frac{k(k-i)}{4(k+1-i)}} \det(\mathscr{L})^{\frac{1}{k+1-i}}, \text{ for any } 1 \leq i \leq k.$$

In 1996, Coppersmith [7] successfully applied the L^3 lattice basis reduction algorithm to find small roots of univariate modular equations, typically called Coppersmith's method. Since then, based on Coppersmith's method, many cryptanalysis [3, 8, 12, 24, 27, 36, 38, 46, 47, 58, 59, 61] have been proposed to attack the RSA scheme and its variants. There are also some works which are combined with side-channel techniques [2, 4, 13, 17, 32–34, 48, 54, 55, 60]. In 2006, Jochemsz and May [23] introduced a general strategy to solve small roots of any form of multivariate modular equations in polynomial time.

The following lemma due to Howgrave-Graham [20] gives a sufficient condition which can transform a modular equation into an integer equation. For the convenience of describing the lemma, we define the norm of a polynomial $g(x_1, \ldots, x_k) = \sum_{(i_1,\ldots,i_k)} a_{i_1,\ldots,i_k} x_1^{i_1} \ldots x_k^{i_k}$ as

$$\|g(x_1, \ldots, x_k)\| = \left(\sum_{(i_1,\ldots,i_k)} a_{i_1,\ldots,i_k}^2 \right)^{\frac{1}{2}}.$$

Lemma 2.2 (Howgrave-Graham, *[20]*) *Let $g(x_1, \ldots, x_n) \in \mathbb{Z}[x_1, \ldots, x_n]$ be an integer polynomial with at most k monomials and m be a positive integer. Let*

$p, X_1, \ldots, X_n$ be positive integers. Suppose that

$$g(\widetilde{x}_1, \ldots, \widetilde{x}_n) \equiv 0 \pmod{p^m} \text{ for } |\widetilde{x}_1| \le X_1, \ldots, |\widetilde{x}_n| \le X_n, \text{ and}$$
$$\|g(x_1 X_1, \ldots, x_n X_n)\| < \frac{p^m}{\sqrt{k}}.$$

Then $g(\widetilde{x}_1, \ldots, \widetilde{x}_n) = 0$ holds over the integers.

Then based on the above two lemmas, we give a brief sketch of Coppersmith's method. For a modular equation $f(x_1, \ldots, x_n) = 0$ modulo p, we want to solve the desired roots $(\widetilde{x}_1, \ldots, \widetilde{x}_n)$, the sketch of Coppersmith's method can be divided into 4 steps.

1. Select k polynomials $h_i(x_1, \ldots, x_n)$ which have the same root $(\widetilde{x}_1, \ldots, \widetilde{x}_n)$ modulo p^m, where $i = 1, \ldots, k$ and k is larger than n.
2. Construct a lattice basis whose row vectors correspond to the coefficients of the selected polynomials $h_i(x_1 X_1, \ldots, x_n X_n)$, where $|\widetilde{x}_1| \le X_1, \ldots, |\widetilde{x}_n| \le X_n$.
3. Suppose that by applying L^3 algorithm to the lattice basis, one can obtain n polynomials $v_1(x_1, \ldots, x_n), \ldots, v_n(x_1, \ldots, x_n)$ corresponding to the first n reduced basis vectors whose norms are sufficiently small enough to satisfy Howgrave-Graham's Lemma.
4. Solve the root $\widetilde{x}_1, \ldots, \widetilde{x}_n$ from the polynomials $v_1(x_1, \ldots, x_n)$,...,$v_n(x_1, \ldots, x_n)$ if $v_1(\widetilde{x}_1, \ldots, \widetilde{x}_n) = 0$, ...,$v_n(\widetilde{x}_1, \ldots, \widetilde{x}_n) = 0$.

Note that, due to Lemma 2.1, one have

$$||v_1(x_1 X_1, \ldots, x_n X_n)|| \le \cdots \le ||v_n(x_1 X_1, \ldots, x_n X_n)|| \le 2^{\frac{k(k-1)}{4(k+1-n)}} \det(\mathscr{L})^{\frac{1}{k+1-n}}.$$

Moreover, since the obtained polynomials $v_1(x_1, \ldots, x_n), \ldots, v_n(x_1, \ldots, x_n)$ are some integer combinations of the polynomials $h_i(x_1, \ldots, x_n)$ which are used to construct lattice, $v_1(x_1, \ldots, x_n), \ldots, v_n(x_1, \ldots, x_n)$ have the same root $(\widetilde{x}_1, \ldots, \widetilde{x}_n)$ modulo p^m. Then if the norm of $v_1(x_1, \ldots, x_n), \ldots, v_n(x_1, \ldots, x_n)$ satisfy the second condition of Lemma 2.2, namely if

$$2^{\frac{k(k-1)}{4(k+1-n)}} \det(\mathscr{L})^{\frac{1}{k+1-n}} < \frac{p^m}{\sqrt{k}},$$

we have that $v_1(\widetilde{x}_1, \ldots, \widetilde{x}_n) = 0, \ldots, v_n(\widetilde{x}_1, \ldots, \widetilde{x}_n) = 0$ hold over the integers.

Here, we ignore small terms and only simply check whether $\det(\mathscr{L}) < p^{mk}$ does hold or not. Then based on the following heuristic assumption, we can solve the roots $\widetilde{x}_1, \ldots, \widetilde{x}_n$ from the polynomials $v_1(\widetilde{x}_1, \ldots, \widetilde{x}_n) = 0, \ldots, v_n(\widetilde{x}_1, \ldots, \widetilde{x}_n) = 0$.

Assumption 2.3 *The polynomials $v_1(x_1, \ldots, x_n), \ldots, v_n(x_1, \ldots, x_n)$ derived from L^3 output vectors are algebraically independent. Then the common roots of these polynomials can be efficiently computed by using techniques like calculation of the resultants or finding a Gröbner basis.*

Here, we give a toy example to show how Coppersmith's method implements based on L^3 lattice basis reduction algorithm. Given a univariate modular equation $f(x) = x^2 + ax + b \bmod N$, where a, b are integer coefficients and N is a known modulus with unknown factorization. Based on Step 1 of Coppersmith's method described above, we first choose m as 2, then we can choose a selection of polynomials as follows:

$$g_1(x) = N^2, g_2(x) = N^2x, g_3(x) = Nf(x), g_4(x) = Nxf(x), g_5(x) = f^2(x).$$

Note that, if x_0 is the root of $f(x)$ modulo N, we have that all the selected polynomials have x_0 as the root modulo N^2. Then, we follow Step 2 to construct following lattice $\mathcal{L}$ using the coefficient vectors of $g_1(xX), g_2(xX), g_3(xX), g_4(xX), g_5(xX)$,

$$\begin{array}{r} \\ N^2 \\ N^2x \\ Nf(x) \\ Nxf(x) \\ f^2(x) \end{array} \begin{array}{c} \begin{array}{ccccc} 1 & x & x^2 & x^3 & x^4 \end{array} \\ \begin{bmatrix} N^2 & 0 & 0 & 0 & 0 \\ 0 & N^2X & 0 & 0 & 0 \\ bN & aNX & NX^2 & 0 & 0 \\ 0 & bNX & aNX^2 & NX^3 & 0 \\ b^2 & 2abX & (a^2+2b)X^2 & 2aX^3 & X^4 \end{bmatrix} \end{array}.$$

Then as Step 3, we apply L^3 lattice basis reduction algorithm to the basis of $\mathcal{L}$ and obtain a reduced basis vector $\boldsymbol{v}$ whose length can be roughly upper bounded by $\det(\mathcal{L})^{\frac{1}{5}}$, where $\det(\mathcal{L}) = X^{10}N^6$ is the product of the diagonal entries. Moreover, since the polynomial $h(x)$ which corresponds to the vector $\boldsymbol{v}$ can be expressed as a linear combination of the selected polynomials $g_1(x), g_2(x), g_3(x), g_4(x), g_5(x)$, then we can obtain that x_0 is the root of $h(x)$ modulo N^2. Furthermore, due to Lemma 2.2, once the length of $\boldsymbol{v}$ is smaller than N^2, we have that $h(x_0) = 0$ over the integers. Namely, when

$$\det(\mathcal{L})^{\frac{1}{5}} = (X^{10}N^6)^{\frac{1}{5}} < N^2,$$

or equivalently,

$$X < N^{\frac{2}{5}},$$

we can obtain $h(x_0) = 0$ over the integers. The last step is to use some root finding algorithms to extract the root x_0, which means all the small roots x_0 which are smaller than $N^{\frac{2}{5}}$ could be found out by Coppersmith's method in polynomial time.

The remainder of the paper is organized as follows. In Sect. 2, we describe the sketch of unraveled linearization technique and give an example to explain how to use this method to select polynomials to construct lattices. In Sect. 3, we show that how to exploit the relations of exponents and introduce multiple parameters to choose appropriate polynomials as many as possible. Section 4 is two-step lattice-based method, this method is suitable for multiple relations exist between private keys. Section 5 is the conclusion.

3 Unraveled Linearization Technique

In 2009, Herrmann and May [18] introduced a new technique called unraveled linearization in the lattice constructions to analyze the power generators, which is a hybrid of lattice-based linearization and Coppersmith's method. Simply put, for a modular polynomial equation, linearization makes use of the similarity of coefficients in the equation, whereas Coppersmith's method basically makes use of the structure of the polynomial's monomial set. Based on unraveled linearization technique, Herrmann and May also reproduced the small private attack key on RSA scheme [19] and obtained the same result as Boneh and Durfee's work [3] which applied the method of geometrically progressive matrices. Besides, there are also several works adopt unraveled linearization technique for the better lattice constructions of modular equations [1, 22, 26, 28, 43, 44, 53, 54, 56, 57, 59–62].

To describe unraveled linearization technique specifically, we take the small private exponent attack on RSA scheme as an example and simply introduce the sketch of their method. Recall the key generation of RSA scheme, the problem of factoring $N = pq$ can be transformed into solving the unknowns of $ed \equiv 1 \bmod \varphi(N)$, or equivalently,

$$ed = k(p-1)(q-1) + 1,$$

where k is some unknown integer. This can be modeled by the following modular polynomial equation

$$f(x, y) = x(N + y) + 1 = 0 \mod e,$$

the desired root of this modular polynomial equation is $(-(p+q-1), k)$.

To show the difference between Boneh–Durfee's method of geometrically progressive matrices [3] and Herrmann–May's method of unraveled linearization technique [19], we first recall Boneh–Durfee's work. For the above modular polynomial equation, Boneh and Durfee first fixed an integer m and defined the following selection of polynomials,

$$g_{i,l}(x, y) = x^i f^l(x, y) e^{m-l}, \text{ for } l = 0, \ldots, m, \ i = 0, \ldots, m-l,$$

and

$$h_{j,l}(x, y) = y^j f^l(x, y) e^{m-l}, \text{ for } l = 0, \ldots, m, \ j = 1, \ldots, t,$$

where t is a parameter that can be optimized as $(1-2\delta)l$ and N^δ denotes the upper bound of desired root k. Obviously, all the above polynomials have the same root $(-(p+q-1), k)$ modulo e^m and the solutions can be roughly estimated by $|k| \simeq X(:= N^\delta)$ and $|p+q-1| \simeq Y(:= N^{\frac{1}{2}})$. Then we can construct a lattice which is spanned by the coefficient vectors of $g_{i,l}(xX, yY)$ and $h_{j,l}(xX, yY)$. For $m = 2, t = 1$, we have following example of Boneh–Durfee's basis matrix,

$$\begin{array}{c} \\ e^2 \\ ex^2 \\ fe \\ x^2e^2 \\ xfe \\ f^2 \\ ye^2 \\ yfe \\ yf^2 \end{array} \begin{array}{c} \begin{array}{ccccccccc} 1 & x & xy & x^2 & x^2y & x^2y^2 & y & xy^2 & x^2y^3 \end{array} \\ \left[\begin{array}{ccccccccc} e^2 & 0 & 0 & 0 & 0 & 0 & 0 & 0 & 0 \\ 0 & e^2X & 0 & 0 & 0 & 0 & 0 & 0 & 0 \\ e & eNX & eXY & 0 & 0 & 0 & 0 & 0 & 0 \\ 0 & 0 & 0 & e^2X^2 & 0 & 0 & 0 & 0 & 0 \\ 0 & eX & 0 & eNX^2 & eX^2Y & 0 & 0 & 0 & 0 \\ 1 & 2NX & 2XY & N^2X^2 & 2NX^2Y & X^2Y^2 & 0 & 0 & 0 \\ 0 & 0 & 0 & 0 & 0 & 0 & e^2Y & 0 & 0 \\ 0 & 0 & eNXY & 0 & 0 & 0 & eY & eXY^2 & 0 \\ 0 & 0 & 2NXY & 0 & N^2X^2Y & 2NX^2Y^2 & Y & 2XY^2 & X^2Y^3 \end{array}\right] \end{array}.$$

The sets of $g_{i,l}(x, y)$ and $h_{j,l}(x, y)$ can lead to a triangular matrix basis, hence the determinant of lattice is the product of diagonal entries. To obtain better result, Boneh and Durfee removed some polynomials from the set of $h_{j,l}(x, y)$ which introduce larger diagonal entries. For the above example, based on Boneh–Durfee's method we should exclude the polynomials ye^2 and yfe. Note that if the polynomials yfe and ye^2 are removed, the polynomial yf^2 introduces three new monomials y, xy^2 and x^2y^3. Hence, the resulting lattice basis is not full-rank, and it is hard to calculate the determinant. Boneh and Durfee used the method of geometrically progressive matrices to estimate the determinant of this lattice with non-triangular matrix basis, although the analysis of the determinant is quite complicated.

Then we show how to use unraveled linearization technique to construct a lattice basis which yields the same result as Boneh and Durfee's work. From the modular equation,

$$f(x, y) = x(N + y) + 1 = 0 \mod e,$$

Herrmann and May first transformed it into linearization equation,

$$\widehat{f}(u_1, u_2) = u + Nx = 0 \mod e,$$

where $u = xy + 1$.

Then for an integer m, Herrmann and May chose the following polynomials to construct lattice,

$$g_{i,l}(u, x, y) = x^i \hat{f}^l(x, u)e^{m-l}, \text{ for } l = 0, \ldots, m, \text{ and } i = 0, \ldots, m - l,$$

and

$$h_{j,l}(u, x, y) = y^j \hat{f}^l(x, u)e^{m-l}, \text{ for } j = 1, \ldots, t, \text{ and } l = \lfloor \frac{m}{t} \rfloor j, \ldots, m,$$

where t is a parameter that can be optimized as $(1 - 2\delta)m$ and N^δ denotes the upper bound of desired root k. Each occurrence of monomial xy is replaced by $u - 1$, since $u = xy + 1$. Obviously, all the above polynomials have the same root $(1 - k(p + q - 1), -(p + q - 1), k)$ modulo e^m and the newly introduced solution

u can be roughly estimated by $|1 - k(p + q - 1)| \simeq XY(:= N^{\delta+\frac{1}{2}})$. Then we can construct a lattice which is spanned by the coefficient vectors of $g_{i,l}(uU, xX, yY)$ and $h_{j,l}(uU, xX, yY)$. For $m = 2, t = 1$, we have following example of Herrmann–May's basis matrix.

$$
\begin{array}{c|ccccccc}
 & 1 & x & u & x^2 & ux & u^2 & u^2y \\
\hline
e^2 & e^2 & 0 & 0 & 0 & 0 & 0 & 0 \\
xe^2 & 0 & e^2X & 0 & 0 & 0 & 0 & 0 \\
\hat{f}e & 0 & eNX & eU & 0 & 0 & 0 & 0 \\
x^2e^2 & 0 & 0 & 0 & e^2X^2 & 0 & 0 & 0 \\
x\hat{f}e & 0 & 0 & 0 & eNX^2 & eUX & 0 & 0 \\
\hat{f}^2 & 0 & 0 & 0 & N^2X^2 & 2NUX & U^2 & 0 \\
y\hat{f}^2 & 0 & -NX^2 & -2NU & 0 & N^2UX & 2NU^2 & U^2Y
\end{array}.
$$

For Herrmann–May's work, the polynomial $y\hat{f}^2$ also introduces three new monomials x^2y, uxy and u^2y. However, we can replace each occurrence of monomial xy by $u - 1$, i.e., we replace x^2y by $ux - x$ and uxy by $u^2 - u$. Moreover, the monomials ux, x, u^2 and u are already present in the lattice basis. Thus, $y\hat{f}^2$ only introduces one monomial u^2y, leading to a triangular basis.

In [19], Herrmann and May strictly proved that the basis matrix formed by the coefficient vectors of above unraveled shift polynomials is triangular and therefore it is easy to calculate the determinant. Moreover, they successfully obtained the same bound as Boneh and Durfee's result [3]. Compared with Boneh–Durfee's lattice construction, Herrmann and May's method results in a simple and natural analysis.

To summarize, the unraveled linearization technique not only makes use of the similarity of the polynomials by combining some certain monomials, but also exploits the algebraic relation between the monomials of the original polynomial. Then we list the general strategy of applying Herrmann–May's unraveled linearization technique and more details can be referred to [16]. Here we describe the strategy for a modular polynomial equation $f(x_1, \ldots, x_n) \equiv 0 \bmod N$.

Firstly, one should suitably express the nonlinear polynomial $f(x_1, \ldots, x_n)$ as a linear polynomial $g(u_1, \ldots, u_r)$. Then one can derive relations between the variables $u_1, \ldots, u_r$. And if necessary one may also introduce further variables u_{r+i} which are appropriate expressions in the variables $x_1, \ldots, x_n$. Then select a set of shift polynomials as Coppersmith's method and include appropriate extrashift in the variables u_{r+i}. Note that we should use the relations to perform the both shift and extrashift polynomials, results in a triangular matrix basis or some lattice whose determinant is easily computed. Then we can follow the straightforward computations to solve the small roots.

4 Exponent Trick

Since the Coppersmith's method has been proposed, this approach has been widely applied in the analyses of RSA scheme and its variants. Among them, one of the applications is to solve approximate integer common divisor problem, namely, given two integers that are near-multiples of a hidden integer, output that hidden integer.

This problem was first introduced by Howgrave-Graham [21]. He considered an univariate modular equation and proposed a polynomial time algorithm to find small roots of the following equation,

$$f(x) = x + a = 0 \bmod p,$$

where a is a given integer, and p ($p \geq N^{\beta}$ for some $0 < \beta \leq 1$) is unknown that divides the known modulus N. Note that this type of polynomial can be applied in some RSA-related problems, such as factoring with known bits problem [36].

In 2008, Herrmann and May [17] further extended the univariate linear modular polynomial to polynomial with an arbitrary number of n variables. Due to Coppersmith's method, they presented a polynomial time algorithm to find small roots of linear modular polynomial

$$f(x_1, \ldots, x_n) = a_0 + a_1 x_1 + \cdots + a_n x_n = 0 \bmod p,$$

where p is unknown and divides the known modulus N. Naturally, they applied their results to the problem of factoring with known bits for RSA modulus $N = pq$ where those unknown bits might spread across arbitrary number of blocks of p. Besides, Herrmann–May's [17] algorithm also can be used to cryptanalyze Multi-prime ϕ-Hiding Assumption [15, 63], and attack CRT-RSA signatures [9, 10].

On the other hand, in 2012, Cohn and Heninger [5] generalized Howgrave-Graham's equation to the simultaneous modular univariate linear equations

$$\begin{aligned} f_1(x_1) &= a_1 + x_1 = 0 \bmod p, \\ f_2(x_2) &= a_2 + x_2 = 0 \bmod p, \\ &\vdots \\ f_n(x_n) &= a_n + x_n = 0 \bmod p, \end{aligned} \tag{1}$$

where $a_1, \ldots, a_n$ are given integers, and p ($p \geq N^{\beta}$ for some $0 < \beta < 1$) is an unknown factor of known modulus N. Furthermore, by collecting more helpful polynomials which are selected to construct lattice, Takayasu and Kunihiro [52] further improved previous result and also partially improved Herrmann–May's result [17]. These equations have many applications in public-key cryptanalysis. For example, in 2010, Dijk et al. [64] introduced fully homomorphic encryption over the integers, in which security of their scheme is based on the hardness of solving Eq. (1). In 2011, Sarkar and Maitra [49] investigated implicit factorization problem [39] by solving

Eq. (1). In 2012, Fouque et al. [14] proposed fault attacks on CRT-RSA signatures, which can also be reduced to solving Eq. (1).

In 2015, Lu et al. [35] revisited the above problems and gave a more generalized version. Based on the novel method of exponent trick, Lu et al. selected more appropriate polynomials in constructing desired lattice. Compared with previous results, their method are more flexible and especially suitable for some cases, such as May's result [37] on small private exponent attack on RSA variant with moduli $N = p^r q$ ($r \geq 2$) and Jochemsz-May's attack [23] on Common Prime RSA. (Note that the result of [35] does not contain the result of [52]. In [52], the authors proposed a method which takes into account the sizes of the solution bounds. Here we mainly focus on the method of [35], namely how to introduce parameters to select appropriate polynomials to construct lattice.)

More specifically, Lu et al. considered the following two types of equations, the first equation is that

$$f(x_1, \ldots, x_n) = a_0 + a_1 x_1 + \cdots + a_n x_n = 0 \bmod p^v,$$

for some unknown divisor p^v ($v > 1$) and known composite integer N ($N \equiv 0 \bmod p^u$, $u \geq 1$). Here u, v are positive integers.

The second is that

$$\begin{aligned} f_1(x_1) &= a_1 + x_1 = 0 \bmod p^{r_1}, \\ f_2(x_2) &= a_2 + x_2 = 0 \bmod p^{r_2}, \\ &\vdots \\ f_n(x_n) &= a_n + x_n = 0 \bmod p^{r_n}, \end{aligned}$$

where p ($p \geq N^\beta$ for some $0 < \beta < 1$) is unknown that satisfies $N \equiv 0 \bmod p^r$ and $a_1, \ldots, a_n, r, r_1, \ldots, r_n$ are given integers.

Compared with previous analyses, Lu et al. introduced multiple parameters to select appropriate polynomials to construct lattices and wisely embed the algebraic relations of the exponents in the lattice constructions. Due to this method, they obtained improved analytical results for some attacks on RSA and its variants. Next, we use the following univariate linear equation as an example to show how to use parameters to optimize the selection of polynomials,

$$f(x) = a + x = 0 \bmod p^v,$$

where N is known to be a multiple of p^u for known u and unknown p.

Based on Lu et al.'s method, for an integer m, one can define a collection of polynomials as follows:

$$g_k(x) = f^k(x) N^{\max\{\lceil \frac{v(t-k)}{u} \rceil, 0\}},$$

for $k = 0, \ldots, m$ and integer parameter t will be optimized later. Note that for all k, $g_k(y) \equiv 0 \bmod p^{vt}$, where y is the desired root. Then due to Coppersmith's method, one can construct lattice by using the coefficient vectors of polynomials. In the calculations of determinant, one can obtain the optimized value of the parameter t, which depends on u, v, β. Compared with previous works, this method is more generalized and by introducing multiple parameters, it is much helpful to select the appropriate polynomials to construct lattices.

Moreover, one can also use multiple parameters to choose the selection of polynomials for the case of univariate polynomials with arbitrary degree, the case of univariate linear equations to an arbitrary number of variables, and the case of simultaneous modular univariate linear equations.

5 Two-Step Lattice-Based Method

In this section, we introduce another technique of lattice construction called two-step lattice-based method. In 2014, Peng et al. [42] firstly introduced the two-step lattice-based method and applied it to implicit hint factorization problem of RSA scheme [39]. Due to their work, the implicit hint factorization problem can be successfully tackled with balanced RSA modulus which is a significant improved result on previous results. Later, Peng et al. showed that the two-step lattice-based method can also be utilized to improve small private exponent attacks on Dual RSA scheme and common private exponent RSA scheme [44]. Here we take the analysis on Dual RSA scheme as an example to describe two-step lattice-based method.

In 2007, Sun et al. proposed Dual RSA scheme [51] whose key generation algorithm outputs two distinct RSA moduli $N_1 = p_1 q_1, N_2 = p_2 q_2$ having the same public and private exponents (e, d). This variant of RSA scheme can be used in some applications like blind signatures and authentication/secrecy. Due to security analysis in their design, when the private exponent d is smaller than $N^{\frac{1}{3}}$, the moduli N_1 and N_2 can be factored in polynomial time, where N denotes an integer with the same bitlength as N_1 and N_2. Next, we show that this result can be improved to $d < N^{\frac{9-\sqrt{21}}{12}} \approx N^{0.368}$ by utilizing two-step lattice-based method.

Based on the Dual RSA scheme, one can obtain the following equations,

$$ed = k_1(N_1 - p_1 - q_1 + 1) + 1, \tag{2}$$

$$ed = k_2(N_2 - p_2 - q_2 + 1) + 1, \tag{3}$$

where k_1 and k_2 are integers.

First, we construct a two-dimensional lattice $\mathscr{L}_1$ which is generated by the row vectors of the following matrix

$$\begin{pmatrix} A & e \\ 0 & N_1 \end{pmatrix},$$

where A is an integer.

From Eq. (2), $v = (Ad, 1 - k_1(p_1 + q_1 - 1))$ is a vector in $\mathcal{L}_1$. To balance the target vector v, we set $A \simeq N^{\frac{1}{2}}$ and the spirit of the balanced factor is to increase the determinant $\mathcal{L}_1$ which means the lengths of the reduced basis vector are increased, while we also want to ensure that the length of target vector v is not changed. On the other hand, one can obtain L^3 reduced basis vectors $\lambda_1 = (l_{11}, l_{12})$ and $\lambda_2 = (l_{21}, l_{22})$ by applying L^3 algorithm to the basis of $\mathcal{L}_1$, and the lengths of λ_1, λ_2 can be estimated by $\det(\mathcal{L}_1)$. Thus, the sizes of $l_{11}, l_{12}, l_{21}, l_{22}$ can be also estimated.

Then one can represent v as an integer combination of λ_1 and λ_2, $v = a_1\lambda_1 + a_2\lambda_2$, where a_1 and a_2 are integers, or equivalently,

$$d = a_1 l'_{11} + a_2 l'_{21},$$

where $l'_{11} = \frac{l_{11}}{A}$ and $l'_{21} = \frac{l_{21}}{A}$. The problem of finding vector v has been transformed into solving the unknown coefficients a_1, a_2. Moreover, from Eq. (3), one can obtain that $k_2(N_2 - p_2 - q_2 + 1) + 1 = ed = e(a_1 l'_{11} + a_2 l'_{21})$. Hence, $(k_2, -(p_2 + q_2 - 1), a_1)$ is a solution of following modular equation

$$f(x, y, z) = x(N_2 + y) - el'_{11}z + 1 = 0 \mod el'_{21}.$$

Then due to Coppersmith's method, one can choose a selection of polynomials to construct lattice and solve the small roots when d is smaller than $N^{\frac{9-\sqrt{21}}{12}} \approx N^{0.368}$. Note that, Peng et al. also applied unraveled linearization technique [19] in the lattice construction.

In addition, two-step lattice-based method can be also applied to improve the analysis on implicit hint factorization problem of RSA scheme. Similarly as the previous analysis on Dual RSA scheme, one can also use the implicit hint in different RSA private keys to construct a low-dimensional lattice, where the target vector which includes the private keys is in this lattice. Then one presents the target vector as a linear combination of the L^3 reduced basis vectors. Then after some transformation, one can reduce the problem into solving a modular equation, where the modulus is the product of some prime factors of the RSA moduli. At last one can solve the small roots of the modular equation by utilizing Coppersmith's method.

As it is shown above, two-step lattice-based method can be divided into two steps. The first step is to construct a low-dimensional lattice and by utilizing the reduced basis of the lattice to express the large unknown variable as a linear relation of some unknown and smaller variables. The second step is to derive some modular equations and then solve the smaller variables due to the Coppersmith's method. Note that, to make Assumption 2.3 work, one should use different algebraic relations in the two steps. More specifically, the algebraic relation which results in a modular equation in the second step should be algebraically independent to the algebraic relation which has been used to make the target vector in the low-dimensional lattice in the first step.

6 Conclusion

In this paper, we give a survey of the recent approaches of lattice construction rather than Jochemsz–May's general strategy of solving small roots of modular equations. For some certain cases, these methods yield better results than purely using Jochemsz–May's general strategy, especially for the cases of that finds roots of multiple modular equations or there exists algebraic relations among unknowns. In the lattice constructions, one may obtain better results when he fully utilize the relations among the monomials or multiple equations to select appropriate polynomials to construct lattices.

References

1. A. Bauer, D. Vergnaud, J. Zapalowicz, Inferring sequences produced by nonlinear pseudorandom number generators using Coppersmith's methods, in *PKC 2012* (2012), pp. 609–626
2. J. Blömer, A. May, New partial key exposure attacks on RSA, in *CRYPTO 2003* (2003), pp. 27–43
3. D. Boneh, G. Durfee, Cryptanalysis of RSA with private key d less than $N^{0.292}$. IEEE Trans. Inf. Theory **46**(4), 1339–1349 (2000)
4. D. Boneh, G. Durfee, Y. Frankel, An attack on RSA given a small fraction of the private key bits, in *ASIACRYPT 1998* (1998), pp. 25–34
5. H. Cohn, N. Heninger, Approximate common divisors via lattices, in *ANTS-X* (2012)
6. D. Coppersmith, Finding a small root of a bivariate integer equation; factoring with high bits known, in *EUROCRYPT 1996* (1996), pp. 178–189
7. D. Coppersmith, Finding a small root of a univariate modular equation, in *EUROCRYPT 1996* (1996), pp. 155–165
8. J. Coron, A. May, Deterministic polynomial-time equivalence of computing the RSA secret key and factoring. J. Cryptol. **20**(1), 39–50 (2007)
9. J. Coron, A. Joux, I. Kizhvatov, D. Naccache, P. Paillier, Fault attacks on RSA signatures with partially unknown messages, in *CHES 2009* (2009), pp. 444–456
10. J. Coron, D. Naccache, M. Tibouchi, Fault attacks against EMV signatures, in *CT-RSA 2010* (2010), pp. 208–220
11. M.J. Coster, B.A. LaMacchia, A.M. Odlyzko, An improved low-density subset sum algorithm, in *EUROCRYPT 1991* (1991), pp. 54–67
12. G. Durfee, P.Q. Nguyen, Cryptanalysis of the RSA schemes with short secret exponent from Asiacrypt'99, in *ASIACRYPT 2000* (2000), pp. 14–29
13. M. Ernst, E. Jochemsz, A. May, B. de Weger, Partial key exposure attacks on RSA up to full size exponents, in *EUROCRYPT 2005* (2005), pp. 371–384
14. P.A. Fouque, N. Guillermin, D. Leresteux, M. Tibouchi, J.C. Zapalowicz, Attacking RSA-CRT signatures with faults on montgomery multiplication. J. Cryptogr. Eng. **3**(1), 59–72 (2013)
15. M. Herrmann, Improved cryptanalysis of the multi-prime ϕ-hiding assumption, in *AFRICACRYPT 2011* (2011), pp. 92–99
16. M. Herrmann, Lattice-based cryptanalysis using unravelled linearization. Ph.D. thesis, der Ruhr-Universitat Bochum (2011), http://www-brs.ub.ruhr-uni-bochum.de/netahtml/HSS/Diss/HerrmannMathias/diss.pdf
17. M. Herrmann, A. May, Solving linear equations modulo divisors: on factoring given any bits, in *ASIACRYPT 2008* (2008), pp. 406–424
18. M. Herrmann, A. May, Attacking power generators using unravelled linearization: when do we output too much? in *ASIACRYPT 2009* (2009), pp. 487–504

19. M. Herrmann, A. May, Maximizing small root bounds by linearization and applications to small secret exponent RSA, in *PKC 2010* (2010), pp. 53–69
20. N. Howgrave-Graham, Finding small roots of univariate modular equations revisited, in *Cryptography and Coding 1997* (1997), pp. 131–142
21. N. Howgrave-Graham, Approximate integer common divisors, in *CaLC 2001* (2001), pp. 51–66
22. Z. Huang, L. Hu, J. Xu, Attacking RSA with a composed decryption exponent using unravelled linearization, in *Inscrypt 2014* (2014), pp. 207–219
23. E. Jochemsz, A. May, A strategy for finding roots of multivariate polynomials with new applications in attacking RSA variants, in *ASIACRYPT 2006* (2006), pp. 267–282
24. E. Jochemsz, A. May, A polynomial time attack on RSA with private CRT-exponents smaller than $N^{0.073}$, in *CRYPTO 2007* (2006), pp. 395–411
25. T. Kleinjung, K. Aoki, J. Franke, A.K. Lenstra, E. Thomé, J.W. Bos, P. Gaudry, A. Kruppa, P.L. Montgomery, D.A. Osvik, H.J.J. te Riele, A. Timofeev, P. Zimmermann, Factorization of a 768-bit RSA modulus, in *CRYPTO 2010* (2010), pp. 333–350
26. N. Kunihiro, On optimal bounds of small inverse problems and approximate GCD problmes with higher degree, in *ISC 2012* (2012), pp. 55–69
27. N. Kunihiro, K. Kurosawa, Deterministic polynomial time equivalence between factoring and key-recovery attack on Takagi's RSA, in *PKC 2007* (2007), pp. 412–425
28. N. Kunihiro, N. Shinohara, T. Izu, A unified framework for small secret exponent attack on RSA. IEICE Trans. **97-A**(6), 1285–1295 (2014)
29. J.C. Lagarias, A.M. Odlyzko, Solving low-density subset sum problems. J. ACM **32**(1), 229–246 (1985)
30. A.K. Lenstra, H.W. Lenstra, L. Lovász, Factoring polynomials with rational coefficients. Math. Ann. **261**(4), 515–534 (1982)
31. A.K. Lenstra, E. Tromer, A. Shamir, W. Kortsmit, B. Dodson, J.P. Hughes, P.C. Leyland, Factoring estimates for a 1024-bit RSA modulus, in *ASIACRYPT 2003* (2003), pp. 55–74
32. Y. Lu, R. Zhang, D. Lin, Factoring RSA modulus with known bits from both p and q: a lattice method, in *NSS 2013* (2013), pp. 393–404
33. Y. Lu, R. Zhang, D. Lin, Factoring multi-power RSA modulus $N = p^r q$ with partial known bits, in *ACISP 2013* (2013), pp. 57–71
34. Y. Lu, R. Zhang, D. Lin, New partial key exposure attacks on CRT-RSA with large public exponents, in *ACNS 2014* (2014), pp. 151–162
35. Y. Lu, R. Zhang, L. Peng, D. Lin, Solving linear equations modulo unknown divisors: revisited, in *ASIACRYPT 2015, Part I* (2015), pp. 189–213
36. A. May, New RSA vulnerabilities using lattice reduction methods. Ph.D. thesis, University of Paderborn (2003), http://ubdata.uni-paderborn.de/ediss/17/2003/may/disserta.pdf
37. A. May, Secret exponent attacks on RSA-type schemes with moduli $N = p^r q$, in *PKC 2004* (2004), pp. 218–230
38. A. May, Computing the RSA secret key is deterministic polynomial time equivalent to factoring, in *CRYPTO 2004* (2004), pp. 213–219
39. A. May, M. Ritzenhofen, Implicit factoring: on polynomial time factoring given only an implicit hint, in *Proceedings of the PKC 2009* (2009), pp. 1–14
40. A.J. Menezes, P.C. van Oorschot, S.A. Vanstone, *Handbook of Applied Cryptography* (CRC Press, Boca Raton, 1996), pp. 118–122
41. P.Q. Nguyen, B. Vallée (eds.), *The LLL Algorithm - Survey and Applications*. Information Security and Cryptography (Springer, Heidelberg, 2010)
42. L. Peng, L. Hu, J. Xu, Z. Huang, Y. Xie, Further improvement of factoring RSA moduli with implicit hint, in *AFRICACRYPT 2014* (2014), pp. 165–177
43. L. Peng, L. Hu, Y. Lu, H. Wei, An improved analysis on three variants of the RSA cryptosystem. To appear in Inscrypt (2016)
44. L. Peng, L. Hu, Y. Lu, J. Xu, Z. Huang, Cryptanalysis of dual RSA. Des. Codes Cryptogr. (2016). doi:10.1007/s10623-016-0196-5
45. R.L. Rivest, A. Shamir, L.M. Adleman, A method for obtaining digital signatures and public-key cryptosystems. Commun. ACM **21**(2), 120–126 (1978)

46. S. Sarkar, Small secret exponent attack on RSA variant with modulus $N = p^r q$. Des. Codes Cryptogr. **73**(2), 383–392 (2014)
47. S. Sarkar, Revisiting prime power RSA. Discret. Appl. Math. **203**, 127–133 (2016)
48. S. Sarkar, S. Maitra, Partial key exposure attack on CRT-RSA, in *ACNS 2009* (2009), pp. 473–484
49. S. Sarkar, S. Maitra, Approximate integer common divisor problem relates to implicit factorization. IEEE Trans. Inf. Theory **57**(6), 4002–4013 (2011)
50. P.W. Shor, Algorithms for quantum computation: discrete log and factoring, in *FOCS 1994* (1994), pp. 124–134
51. H. Sun, M. Wu, W. Ting, M.J. Hinek, Dual RSA and its security analysis. IEEE Trans. Inf. Theory **53**(8), 2922–2933 (2007)
52. A. Takayasu, N. Kunihiro, Better lattice constructions for solving multivariate linear equations modulo unknown divisors, in *ACISP 2013* (2013), pp. 118–135
53. A. Takayasu, N. Kunihiro, Cryptanalysis of RSA with multiple small secret exponents, in *ACISP 2014* (2014), pp. 176–191
54. A. Takayasu, N. Kunihiro, Partial key exposure attacks on RSA: achieving the Boneh-Durfee bound, in *SAC 2014* (2014), pp. 345–362
55. A. Takayasu, N. Kunihiro, Partial key exposure attacks on CRT-RSA: better cryptanalysis to full size encryption exponents, in *ACNS 2015* (2015), pp. 518–537
56. A. Takayasu, N. Kunihiro, Partial key exposure attacks on RSA with multiple exponent pairs, in *ACISP 2016* (2016), pp. 243–257
57. A. Takayasu, N. Kunihiro, How to generalize RSA cryptanalysis, in *PKC 2016, Part II* (2016), pp. 67–97
58. A. Takayasu, N. Kunihiro, Partial key exposure attacks on CRT-RSA: general improvement for the exposed least significant bits, in *ISC 2016* (2016), pp. 35–47
59. A. Takayasu, N. Kunihiro, Small secret exponent attacks on RSA with unbalanced prime factors, in *ISITA 2016* (2016), pp. 236–240
60. A. Takayasu, N. Kunihiro, A tool kit for partial key exposure attacks on RSA. To appear in CT-RSA 2017 (2017)
61. A. Takayasu, N. Kunihiro, General bounds for small inverse problems and its applications to multi-prime RSA. IEICE Trans. **100-A**(1), 50–61 (2017)
62. A. Takayasu, Y. Lu, L. Peng, Small CRT-exponent RSA revisited. To appear in EUROCRYPT 2017 (2017)
63. K. Tosu, N. Kunihiro, Optimal bounds for multi-prime ϕ-hiding assumption, in *ACISP 2012* (2012), pp. 1–14
64. M. van Dijk, C. Gentry, S. Halevi, V. Vaikuntanathan, Fully homomorphic encryption over the integers, in *EUROCRYPT 2010* (2010), pp. 24–43
65. M.J. Wiener, Cryptanalysis of short RSA secret exponents. IEEE Trans. Inf. Theory **36**(3), 553–558 (1990)

Part IV
Cryptographic Protocols

How to Strengthen the Security of Signature Schemes in the Leakage Models: A Survey

Yuyu Wang and Keisuke Tanaka

Abstract We give a survey on generic transformations that strengthen the security of signature schemes, which are exploited in most cryptographic protocols, in the leakage models. In ProvSec 2014, Wang and Tanaka proposed a transformation which converts weakly existentially unforgeable signature schemes into strongly existentially unforgeable ones in the bounded leakage model. To obtain the construction, they combined a leakage resilient chameleon hash function with the Generalized Boneh–Shen–Waters (GBSW) transformation proposed by Steinfeld, Pieprzyk, and Wang. In ACISP 2015, Wang and Tanaka proposed another transformation in the continual leakage model. To achieve the goal, they defined a continuous leakage resilient (CLR) chameleon hash function and constructed it based on the CLR signature scheme proposed by Malkin, Teranishi, Vahlis, and Yung. Then they improved the GBSW transformation by making use of the Groth–Sahai proof system and then combine it with CLR chameleon hash functions. In Security and Communication Networks, Wang and Tanaka additionally gave an instantiation of (restricted) fully leakage resilient strong one-time signature based on leakage resilient chameleon hash functions, following the construction of strong one-time signature by Mohassel. They also proved that by combining a (restricted) fully leakage resilient strong one-time signature scheme with the transformation proposed by Huang, Wong, and Zhao, another transformation that can strengthen the security of fully leakage resilient signature schemes without changing signing keys can be obtained.

Keywords Bounded leakage resiliency · Continual leakage resiliency · Signature · Strong existential unforgeability · Chameleon hash function · Generic transformation

Y. Wang (✉) · K. Tanaka
Department of Mathematical and Computing Sciences, Tokyo Institute of Technology, Tokyo 152-8552, Japan
e-mail: wang.y.ar@m.titech.ac.jp

K. Tanaka
e-mail: keisuke@is.titech.ac.jp

Y. Wang
National Institute of Advanced Industrial Science and Technology (AIST), Tokyo 135-0064, Japan

T. Takagi et al. (eds.), *Mathematical Modelling for Next-Generation Cryptography*, Mathematics for Industry 29, DOI 10.1007/978-981-10-5065-7_17

1 Introduction

1.1 Background

Signature schemes are said to be weakly existentially unforgeable (wEUF) if it is hard to forge signatures on messages not signed before. But for some applications, a stronger security called strong existential unforgeability (sEUF) is required which also prevents forgery of new signatures on messages signed before.

If a signature scheme is existentially unforgeable (EUF) while the information of the signing keys and randomness may be leaked, then it is said to be fully leakage resilient (FLR). If only the information of the signing key is leaked, then the scheme is said to be leakage resilient (LR).

There are two common models for leakage resiliency, which are the bounded leakage model and continual leakage model. In the bounded leakage model, the adversary is allowed to learn some bounded leakage on the secret information during the lifetime of the system. In the continual leakage model, the signing key is periodically updated, and there is only bound on leakage between two successive key updates, but no bound on the total leakage during the lifetime of the system. The bounded leakage model was suggested by Akavia, Goldwasser, and Vaikuntanathan [1]. Dodis, Haralambiev, López-Alt, and Wichs [9] and Brakerski, Kalai, Katz, and Vaikuntanathan [8] extended the notion of this model and suggested the continual leakage model. There had been a great deal of research proposed for cryptographic primitives in the bounded leakage model (c.f., [2, 3, 7, 10, 15, 18, 21]) and the continual leakage model (c.f., [8, 9, 11, 17, 19]). However, none of the previously proposed FLR signature schemes satisfies the sEUF property in leakage models till [27], while [27] provided no generic transformation to sEUF-FLR.

Boneh, Shen, and Waters [5] presented a transformation that converts wEUF signature schemes into sEUF ones based on chameleon hash functions. However, this transformation applies to a class of signature schemes called partitioned signature schemes. Later works [4, 24, 25] proposed transformations that can convert any signature scheme into an sEUF one. Another work by Huang, Wong, and Zhao [14] presented such a transformation that keeps key pairs unchanged by making use of strong one-time signature schemes. Note that all the transformations mentioned above need to change key pairs or generate additional randomness in the signing process. If we consider leakage on signing keys or randomness, the security of the signature schemes will not hold after being converted by these methods.

1.2 Generic Transformations in the Leakage Models

In this paper, we give a survey on generic transformations into sEUF signature schemes in the leakage models. Such transformations were firstly studied by Wang and Tanaka [26, 28, 29], who gave the following results:

- They proposed a new definition of LR chameleon hash functions and provided a construction of it. Then they obtained a generic transformation from wEUF-FLR signature schemes to sEUF-FLR ones in the bounded leakage model based on LR chameleon hash functions.
- They proposed a new definition of continuous leakage resilient (CLR) chameleon hash functions and provided a construction of it. Then they presented a generic transformation from wEUF-FLR signature schemes into sEUF-FLR ones in the continual leakage model based on CLR chameleon hash functions.
- They gave a (restricted) FLR strong one-time signature scheme based on LR chameleon hash functions and another transformation in the continual leakage model based on (restricted) FLR strong one-time signature schemes.

The definitions of LR chameleon hash functions and CLR chameleon hash functions are extensions of the notion of chameleon hash functions [16, 23]. For standard chameleon hash functions, collisions can be found efficiently by making use of the secret key, and it is hard to find any collisions without knowing the secret key. In the case of (C)LR chameleon hash functions, it is hard to find collisions even when a part of the information of the secret key is (continuously) leaked. To obtain the construction of LR chameleon hash function, Wang and Tanaka [26, 28] exploited the Okamoto-style second preimage resistant (SPR) relation [22]. Furthermore, to construct CLR chameleon hash functions, they made use of the construction of the signature in [19], in which the tuple of key generation algorithm, updating algorithm, and relation of the public/secret key pair forms a CLR one-way relation (CLR-OWR) defined by Dodis et al. [9]. Their resulting constructions of LR chameleon hash functions and CLR chameleon hash functions inherit the properties of the LR hard relation proposed in [10] and the CLR-OWR scheme, respectively. The former can tolerate any leakage of length $\ell = (1 - o(1))L_1$ bits in total and the latter can tolerate any leakage of length $\ell = (1 - o(1))L_2$ bits between every two successive key updating, where L_1 and L_2 denote the lengths of their secret keys, respectively. Then, they presented generic transformations from wEUF-FLR signature schemes into sEUF-FLR ones based on the Generalized Boneh–Shen–Waters (GBSW) transformation proposed by Steinfeld, Pieprzyk, and Wang [24].

Furthermore, following the technique by Mohassel [20], Wang and Tanaka [29] constructed an instantiation of FLR strong one-time signature based on LR chameleon hash functions. The efficiency of the key generation algorithm is the same as that of generating a verification/signing key pair and a hash value of a LR chameleon function and the efficiency of the signing algorithm is the same as that of the trapdoor collision finder algorithm of a LR chameleon hash function. Then, they proved that by combing the FLR strong one-time signature scheme with the transformation proposed by Huang et al. [14], another transformation in the bounded leakage model can be achieved. Since this transformation does not change the signing key, it can be used in the continual leakage model. What is more, by making use of this transformation, we can obtain an ℓ_s-sEUF-FLR signature scheme where $\ell_s = (1 - o(1))L$ from any wEUF signature scheme that can tolerate leakage of $(1 - o(1))L$ bits, where L is the length of their signing keys. However, comparing with the transformation in [26], this transformation is less efficient and has to hide more secret information from the adversary since it needs to generate a verification/signing key pair of FLR strong one-time signature scheme in every signing procedure.

2 Preliminaries

2.1 Strong Collision-Resistant Hash Function

Now, we recall the strongly collision-resistance property of hash functions.

Definition 2.1 (*Strongly collision-resistant hash function*) A hash function $J : \{0, 1\}^* \rightarrow \mathbb{Z}_p \setminus \{0\}$ is said to be strongly collision-resistant if for any PPT adversary $\mathscr{A}$, we have $\Pr[(x_0, x_1) \leftarrow \mathscr{A}(1^k, J) : x_0 \neq x_1 \wedge J(x_0) = J(x_1)] \leq negl(k)$.

2.2 Chameleon Hash Function

In this section, we describe the definition of chameleon hash functions from [16, 23, 24].

A chameleon hash function scheme consists of the following three PPT algorithms $(\mathsf{KG}_H, \mathsf{TC}_H, H)$.

KG_H is a *key generation* algorithm that takes as input (1^k), and outputs a public/secret key pair (pk_H, sk_H). H is a *hash function evaluation* algorithm that takes as input pk_H, a message m, and a randomizer r, and outputs a hash value $\bar{m} = H_{pk_H}(m; r)$. TC is a *trapdoor collision finder* algorithm that takes as input sk_H, a message/randomizer pair (m, r) and a second message m', and outputs $r' = \mathsf{TC}_H(sk_H, (m, r), m')$ such that $H_{pk_H}(m; r) = H_{pk_H}(m'; r')$.

A chameleon hash function scheme must satisfy two properties, which are *random trapdoor collision* and *collision-resistance*.

The random trapdoor collision property is satisfied if for a secret key sk_H, an arbitrary message pair (m, m') and a randomizer r, $r' = \mathsf{TC}_H(sk_H, (m, r), m')$ has a uniform probability distribution on the randomness space.

The collision-resistance property is satisfied if for any PPT adversary $\mathscr{A}$, we have

$$\Pr[(pk_H, sk_H) \leftarrow \mathsf{KG}_H(1^k), ((m, r), (m', r')) \leftarrow \mathscr{A}(1^k, pk_H) : (m, r) \neq (m', r') \wedge H_{pk_H}(m, r) = H_{pk_H}(m', r')] \leq negl(k).$$

Although the original definition of chameleon hash functions just requires that $m \neq m'$, it is required that $(m, r) \neq (m', r')$ here. As far as we know, all of the chameleon function schemes (except for the continual leakage resilient one in [28]) satisfy this stronger version of collision-resistance.

2.3 Fully Leakage Resilient Signature - Bounded Leakage Model

We now describe the definition of FLR signature schemes in the bounded leakage model from [6, 12, 15].

A signature scheme consists of three PPT algorithms. KG is a randomized algorithm that takes as input 1^k and outputs a public/secret key pair (pk, sk). Sign is a randomized algorithm that takes as input a signing key sk and a message m and returns a signature σ. Verify is a deterministic algorithm that takes as input a verification key pk, a message m, and a signature σ and returns 1 (accept) or 0 (reject). In addition to the correctness, we require the following property.

Definition 2.2 (*wEUF-FLR signature - bounded leakage model* [6, 12, 15]) A signature scheme $(\mathsf{KG}, \mathsf{Sign}, \mathsf{Verify})$ is said to be *wEUF* and *ℓ-FLR* in the bounded leakage model if for any PPT adversary $\mathscr{A}$, we have that $\Pr[\mathscr{A} \text{ wins}] \leq \text{negl}(k)$ in the following game:

1. Compute $(pk, sk) \leftarrow \mathsf{KG}(1^k, \ell)$, set $state = sk$.
2. Run the adversary $\mathscr{A}$ on input tuple $(1^k, pk, \ell)$. The adversary may make adaptive queries to the signing oracle and the leakage oracle, defined as follows:
   - Signing oracle: On receiving a query m_i, the signing oracle samples $r_i \leftarrow \{0, 1\}^*$, and computes $\sigma_i \leftarrow \mathsf{Sign}_{sk}(m_i; r_i)$. It updates $state = state||r_i$ and outputs σ_i.
   - Leakage oracle: On receiving a polynomial-time computable function $f_j : \{0, 1\}^* \rightarrow \{0, 1\}^{\ell_j}$, the leakage oracle outputs $f_j(state)$.
3. $\mathscr{A}$ outputs (m^*, σ^*) and wins if: (a) $\mathsf{Verify}_{pk}(m^*, \sigma^*) = 1$, (b) m^* was not queried to the signing oracle, and (c) $\sum_j \ell_j \leq \ell$.

The definition of *sEUF-FLR signature schemes* in the bounded leakage model is the same as the above one, except the winning condition (b) being set as follows.

- the pair (m^*, σ^*) is new, that is, either m^* was not queried to the signing oracle, or it was and σ^* is not the one(s) generated as a signature of m^* by the signing oracle.

Without loss of generality, we can assume that the adversary makes a leakage query every time after making a signing query.

The definition of *FLR strong one-time signature schemes* was given by Katz and Vaikuntanathan [15], which is the same as the definition of sEUF-FLR signature schemes except that the adversary is allowed to make the signing query for only once instead of polynomial times and $state$ is initially set as r instead of sk where r is the randomizer used to generate (pk, sk) ($(pk, sk) = \mathsf{KG}(1^k, \ell; r)$). Here, we give a restriction that the adversary makes the leakage query after making the signing query. Note that our second transformation only requires an FLR strong one-time signature scheme satisfying this weakened security.

2.4 *Fully Leakage Resilient Signature - Continual Leakage Model*

A signature scheme in the continual leakage model consists of four PPT algorithms (KG, Update, Sign, Verify). KG takes as input 1^k and outputs a signing/verification key pair (pk, sk). Update takes as input a verification/signing key pair (pk, sk) and outputs a refreshed signing key sk'. Sign takes as input a signing key sk and a message m and returns a signature σ. Verify takes as input a verification key pk, a message m, and a signature σ and outputs 1 if the signature is valid, outputs 0 otherwise. In addition to the correctness, we require the following property.

Definition 2.3 (*wEUF-FLR signature - continual leakage model* [6]) A signature scheme (KG, Update, Sign, Verify) is said to be *wEUF* and *ℓ-FLR* in the continual leakage model if for any PPT adversary $\mathscr{A}$, we have $\Pr[\mathscr{A} \text{ wins}] \leq \text{negl}(k)$ in the following game:

1. Compute $(pk, sk) \leftarrow \mathsf{KG}(1^k, \ell)$, set $state = sk$ and $L = 0$.
2. Run the adversary $\mathscr{A}$ on input $(1^k, pk)$. The adversary may make adaptive queries to the signing oracle, the leakage oracle and the updating oracle, defined as follows:
   - Signing oracle: On receiving a query m_i, the signing oracle samples $r_i \leftarrow \{0, 1\}^*$, and computes $\sigma_i \leftarrow \mathsf{Sign}_{sk}(m_i; r_i)$. It updates $state = state||r_i$ and outputs σ_i.
   - Leakage oracle: On receiving a polynomial-time computable function $f_j : \{0, 1\}^* \rightarrow \{0, 1\}^{\ell_j}$, if $L + |f_j(state)| \leq \ell$, then outputs $f_j(state)$ and sets $L = L + |f_j(state)|$. Otherwise it aborts.
   - Updating oracle: On receiving an updating query, the updating oracle computes $sk' \leftarrow \mathsf{Update}(pk, sk)$. It resets $sk = sk'$, $state = sk$, and $L = 0$.
3. $\mathscr{A}$ outputs (m^*, σ^*) and wins if: (a) $\mathsf{Verify}_{pk}(m^*, \sigma^*) = 1$, (b) m^* was not queried to the signing oracle.

The definition of *sEUF-FLR signature schemes* is the same as the above one, except the winning condition (b) being set as follows.

- the pair (m^*, σ^*) is new, that is, either m^* was not queried to the signing oracle or it was, σ^* is not the one(s) generated as a signature of m^* by the signing oracle.

Without loss of generality, we can assume that the adversary makes a leakage query every time after making a signing query.

The definition described above does not consider leakage on key generation and key updating since generating key and updating key may be conducted "off-line", according to [6], as argued in [28].

3 Generic Transformation to sEUF Fully Leakage Resilient Signature - Bounded Leakage Model

In this section, we describe LR chameleon hash functions [26]. The instantiation is constructed by making use of the Okamoto-style SPR relation [22]. By combining the LR chameleon hash function scheme with the GBSW transformation proposed by Steinfeld et al. [24], Wang and Tanaka obtained a transformation which can convert wEUF-FLR signature schemes into sEUF-FLR ones.

3.1 Leakage Resilient Chameleon Hash Function

We now give the definition of *LR chameleon hash functions*, which is an extension of the notion of chameleon hash functions [16, 23].

A chameleon hash function scheme consists of the following three PPT algorithms $(\mathsf{KG}_F, \mathsf{TC}_F, F)$.

KG_F is a *key generation* algorithm that takes as input $(1^k, \ell)$, and outputs (pk_F, sk_F), where pk_F is the public key and sk_F is the secret key. F is a *hash function evaluation* algorithm that takes as input pk_F, a message m, and a randomizer r, and outputs a hash value $h = F_{pk_F}(m; r)$. TC is a *trapdoor collision finder* algorithm that takes as input sk_F, a message/randomizer pair (m, r) and a second message m', and outputs $r' = \mathsf{TC}_F(sk_F, (m, r), m')$ such that $F_{pk_F}(m; r) = F_{pk_F}(m'; r')$.

An ℓ-LR chameleon hash function scheme must satisfy three properties, which are *reversibility*, *random trapdoor collision*, and *collision-resistance*.

The reversibility property is satisfied if $r' = \mathsf{TC}_F(sk_F, (m, r), m')$ is equivalent to $r = \mathsf{TC}_F(sk_F, (m', r'), m)$.

The random trapdoor collision property is satisfied if for a secret key sk_F, an arbitrary message pair (m, m'), and a randomizer r, $r' = \mathsf{TC}_F(sk_F, (m, r), m')$ has a uniform probability distribution on the randomness space.

The LR collision-resistance property is satisfied if for any PPT adversary $\mathscr{A}$, we have

$$\Pr[(pk_F, sk_F) \leftarrow \mathsf{KG}_F(1^k), ((m, r), (m', r')) \leftarrow \mathscr{A}^{\mathcal{O}^{k,\ell}_{sk_F}(\cdot)}(pk_F) : \\ (m, r) \neq (m', r') \wedge F_{pk_F}(m, r) = F_{pk_F}(m', r')] \leq negl(k),$$

where $\mathcal{O}^{k,\ell}_{sk_F}$ is the leakage oracle, where sk_F is a secret value, ℓ is the leakage parameter, and k is the security parameter. The adversary can adaptively access

- Key generation algorithm $\mathsf{KG}_F(1^k)$ outputs $pk_F = (y, \mathbf{h} = (h_1, ..., h_n))$ and $sk_F = \mathbf{x} = (x_1, ..., x_n)$ where $\mathbb{G}$ is a cyclic group of prime order q, $n \geq 2$, $h_1, ..., h_n \overset{R}{\leftarrow} \mathbb{G}$, $x_1, ..., x_n \overset{R}{\leftarrow} \mathbb{Z}_q$ and $y = \prod_{i=1}^{n} h_i^{x_i} \in \mathbb{G}$.
- Hash function evaluation algorithm $F_{pk_F}(m, \mathbf{r}) = (y \prod_{i=1}^{n} h_i^{r_i})^{J(m)} \in \mathbb{G}$ where $\mathbf{r} = (r_1, ..., r_n) \overset{R}{\leftarrow} \mathbb{Z}_q^n$ and J denotes a strongly collision-resistant hash function from $\{0,1\}^*$ to $\mathbb{Z}_q/\{0\}$.
- Trapdoor collision finder algorithm $\mathsf{TC}_F(sk_F, (m, \mathbf{r}), m')$ outputs $\mathbf{r}'$ where $\mathbf{r}' = (r_1', ..., r_n')$ and $r_i' \equiv J(m)(x_i + r_i)/J(m') - x_i \pmod{q}$ for $i = 1, ..., n$.

Fig. 1 LR chameleon hash function

to learn leakage on the secret value. Every time the adversary makes an efficient computable leakage query $f_i : \{0, 1\}^* \rightarrow \{0, 1\}^{\ell_i}$ to the leakage oracle, the oracle answers with $f_i(x)$. It is required that the total number of bits leaked is not more than ℓ.

The construction of LR chameleon hash function in [26] is given in Fig. 1. It is constructed from the Okamoto-style SPR relation [22].

Claim 1 *The scheme described in Fig. 1 is an ℓ-LR chameleon hash function for $\ell = (n - 1)\log(q) - \omega(\log k)$ if the discrete logarithm assumption holds.*

3.2 Generic Transformation

Let $\Sigma = (\mathsf{KG}, \mathsf{Sign}, \mathsf{Verify})$ (with randomness space Ω_Σ) be an ℓ_w-FLR signature scheme that satisfies the wEUF property. $\mathscr{F} = (\mathsf{KG}_F, F, \mathsf{TC}_F)$ (with randomness space $\mathscr{R}_F$) denotes the ℓ-LR chameleon hash function. Wang and Tanaka [26] made use of the GBSW transformation [24] to convert Σ into another signature scheme Σ'. The resulting signature scheme is shown in Fig. 2. What they did is just substituting the standard chameleon hash function used in [24] with an ℓ-LR chameleon hash function.

Theorem 1 *The scheme described in Fig. 2 is an ℓ_s-sEUF-FLR signature scheme for $\ell_s = \min\{\ell, \ell_w\}$ if $\mathscr{F} = (\mathsf{KG}_F, F, \mathsf{TC}_F)$ is an ℓ-LR chameleon hash function scheme.*

4 Generic Transformation to sEUF Fully Leakage Resilient Signature - Continual Leakage Model

In this section, we describe CLR chameleon hash function [28] and give a construction of it. It is derived from the CLR signature scheme proposed by Malkin et al. [19]. Then we describe the transformation which can transform wEUF-FLR signature schemes into sEUF-FLR ones in the continual leakage model [28], based

- $\mathsf{KG}'(1^k)$:
  1. Run $(pk, sk) \leftarrow \mathsf{KG}(1^k)$.
  2. Run $(pk_F, sk_F) \leftarrow \mathsf{KG}_F(1^k)$.
  3. Run $(pk_H, sk_H) \leftarrow \mathsf{KG}_F(1^k)$.
  4. Output verification key $pk_s = (pk, pk_F, pk_H, m', \sigma')$ and signing key $sk_s = (sk, sk_F)$, where m' and σ' are arbitrary fixed strings.
- $\mathsf{Sign}'_{sk_s}(m)$:
  1. Parse $sk_s = (sk, sk_F)$.
  2. Randomly choose $\omega \stackrel{R}{\leftarrow} \Omega_\Sigma$, $s \stackrel{R}{\leftarrow} \mathcal{R}_F$, and $r' \stackrel{R}{\leftarrow} \mathcal{R}_F$.
  3. Compute $h = F_{pk_F}(m'||\sigma'; r')$.
  4. Compute $\bar{m} = F_{pk_H}(h; s)$.
  5. Compute $\sigma = \mathsf{Sign}_{sk}(\bar{m}; \omega)$.
  6. Compute $r = \mathsf{TC}_F(sk_F, (m'||\sigma', r'), m||\sigma)$.
  7. Output $\sigma' = (\sigma, r, s)$.
- $\mathsf{Verify}'(pk_s, m, \sigma')$:
  1. Parse $pk_s = (pk, pk_F, pk_H, m', \sigma')$, and $\sigma' = (\sigma, r, s)$.
  2. Compute $h = F_{pk_F}(m||\sigma; r)$.
  3. Compute $\bar{m} = F_{pk_H}(h; s)$.
  4. Output 1 if $\mathsf{Verify}(pk, \bar{m}, \sigma) = 1$ and output 0 otherwise.

Fig. 2 GBSW transformation in the bounded leakage model

on CLR chameleon hash functions, the GBSW transformation in [24], and the Groth–Sahai proof system [13]. We refer the reader to [13, 28] to the Groth–Sahai proof system.

4.1 Bilinear Map

Let $\mathcal{G}$ be an algorithm takes as input 1^k and outputs $gk = (p, \mathbb{G}_1, \mathbb{G}_2, \mathbb{G}_T, e)$ such that p is prime. $(\mathbb{G}_1, \mathbb{G}_2, \mathbb{G}_T)$ are descriptions of groups of order p, and $e : \mathbb{G}_1 \times \mathbb{G}_2 \to \mathbb{G}_T$ is a nondegenerate efficiently computable bilinear map. It is required that there is no efficient computable mapping between $\mathbb{G}_1$ and $\mathbb{G}_2$.

We use additive notation for pairings, such as $e((a+b)A, B) = a \cdot e(A, B) + b \cdot e(A, B)$. We denote $e(\mathbf{A}^T, \mathbf{B}) = \sum_{i=1}^{n} e(A_i, B_i)$, where $\mathbf{A} = (A_1, \ldots, A_n)^T \in \mathbb{G}_1$, $\mathbf{B} = (B_1, \ldots, B_n)^T \in \mathbb{G}_2$.

4.2 Continual Leakage Resilient Chameleon Hash Functions

An ℓ-CLR chameleon hash function scheme consists of the following four PPT algorithms $(\mathsf{KG}_F, \mathsf{TC}_F, F, \mathsf{UD}_F)$.

KG_F is a *key generation* algorithm that takes as input $(1^k, \ell)$, and outputs a public/private key pair (pk_F, sk_F). F is a *hash function evaluation* algorithm that takes as input pk_F, a message m, and a randomizer r, and outputs a hash value $h = F_{pk_F}(m; r)$. TC_F is a *trapdoor collision finder* algorithm that takes as input sk_F, an arbitrary message pair (m, m'), and a randomizer r, and outputs $r' = \mathsf{TC}_F(sk_F, (m, r), m')$ such that $F_{pk_F}(m; r) = F_{pk_F}(m'; r')$. UD_F is a *key update* algorithm takes as input (pk_F, sk_F) and samples a refreshed secret key sk'_F for pk_F.

An ℓ-CLR chameleon hash function scheme must satisfy four properties, which are *reversibility*, *correctness*, *random trapdoor collision*, and *ℓ-CLR strong collision-resistance*.

The reversibility property is satisfied if $r' = \mathsf{TC}_F(sk_F, (m, r), m')$ is equivalent to $r = \mathsf{TC}_F(sk_F, (m', r'), m)$.

The correctness property is satisfied if for any polynomial $i = i(k)$, any message pair (m, m'), and a randomizer r, if we compute $(pk_F, sk_F^{[0]}) \leftarrow \mathsf{KG}_F(1^k)$, $sk_F^{[1]} \leftarrow \mathsf{UD}_F(pk_F, sk_F^{[0]})$,..., $sk_F^{[i]} \leftarrow \mathsf{UD}_F(pk_F, sk_F^{[i-1]})$, and $r' = \mathsf{TC}_F(sk_F^{[i]}, (m, r), m')$, we have $F_{pk_F}(m; r) = F_{pk_F}(m'; r')$. If the r' has a uniform probability distribution on the randomness space, then the random trapdoor collision property is said to be satisfied.

The ℓ-CLR strong collision-resistance property is satisfied if for any PPT adversary $\mathscr{A}$, we have $\Pr[\mathscr{A} \text{ wins}] \leq \text{negl}(k)$ in the following game:

1. the challenger computes $(pk_F, sk_F^{[0]}) \leftarrow \mathsf{KG}_F(1^k)$ and sets $L = 0$.
2. On input $(1^k, pk_F, F)$, $\mathscr{A}$ runs for arbitrarily many leakage rounds $i = 0, \ldots, q$. In each round i:

   a. $\mathscr{A}$ makes adaptive queries to the challenger. Every time on receiving the description of a polynomial-time computable function f_j, if $L + |f_j(sk_F^{[i]})| \leq \ell$, then the challenger sends $f_j(sk_F^{[i]})$ back to $\mathscr{A}$ and updates $L = L + |f_j(sk_F^{[i]})|$. Otherwise it aborts.
   b. the challenger refreshes the secret key by taking a sample $sk_F^{[i+1]}$ from $\mathsf{Update}(pk_F, sk_F^{[i]})$ and resets $L = 0$ at the end of the round.

3. $\mathscr{A}$ wins if it outputs (m, r) and (m', r') such that $m \neq m'$ and $F_{pk_F}(m, r) = F_{pk_F}(m', r')$.

Notice that in the above definition, we do not prevent an adversary from finding a collision for the same m as opposed to the definition of LR chameleon hash functions.

We describe the construction of CLR chameleon hash function by Wang and Tanaka [28] in Fig. 3.

- Key generation algorithm $\mathsf{KG}_F(1^k)$:
  1. Run $gk = (p, \mathbb{G}_1, \mathbb{G}_2, \mathbb{G}_T, e) \leftarrow \mathcal{G}(1^k)$.
  2. Choose $A \leftarrow \mathbb{G}_1$, $Q \leftarrow \mathbb{G}_2$, $\mathbf{a}, \mathbf{q} \leftarrow \mathbb{Z}_p^n$ satisfying $\langle \mathbf{a}, \mathbf{q} \rangle = 0$. $\mathbf{A} = \mathbf{a}A, \mathbf{Q} = \mathbf{q}Q$.
  3. Choose $\mathbf{W}^{[0]} \leftarrow \mathbb{G}_2^n$.
  4. Compute $T = e(\mathbf{A}^T, \mathbf{W}^{[0]})$.
  5. Return $pk_F = (gk, \mathbf{A}, T, \mathbf{Q})$ and $sk_F^{[0]} = \mathbf{W}^{[0]}$.
- Hash function evaluation algorithm $F_{pk_F}(m, \mathbf{R})$ where $\mathbf{R} \leftarrow \mathbb{G}_2^n$:
  1. Return $h = J(m)(T + e(\mathbf{A}^T, \mathbf{R}))$, where J denotes a strongly collision resistant hash function from $\{0,1\}^*$ to $\mathbb{Z}_p \setminus \{0\}$.
- Trapdoor collision finder algorithm $\mathsf{TC}_F(sk_F^{[i]}, (m, \mathbf{R}), m')$:
  1. Parse $sk_F^{[i]} = \mathbf{W}^{[i]}$.
  2. Return $\mathbf{R}' = \frac{J(m)}{J(m')}(\mathbf{W}^{[i]} + \mathbf{R}) - \mathbf{W}^{[i]}$.
- Updating algorithm $\mathsf{UD}_F(pk_F, sk_F^{[i]})$:
  1. Parse $pk = (gk, \mathbf{A}, T, \mathbf{Q})$ and $sk_F^{[i]} = \mathbf{W}^{[i]}$.
  2. Choose $s \leftarrow \mathbb{Z}_p$.
  3. Return $sk_F^{[i+1]} = \mathbf{W}^{[i+1]} = \mathbf{W}^{[i]} + s\mathbf{Q}$.

Fig. 3 CLR chameleon hash function scheme

Claim 2 *The scheme described in Fig. 3 is an ℓ-CLR chameleon hash function for $\ell = n \log p - (2 + \Theta(1/\sqrt{k})) \log p$ if the Symmetric External Diffie–Hellman (SXDH) assumption holds.*

It is easy to find a collision $((m, \mathbf{R}_1), (m, \mathbf{R}_2))$ for this CLR chameleon hash function such that $F_{pk_F}(m, \mathbf{R}_1) = F_{pk_F}(m, \mathbf{R}_2)$ and $\mathbf{R}_1 \neq \mathbf{R}_2$, since for any m, $s \in \mathbb{Z}_p$, and $\mathbf{R}_1 \in \mathbb{G}_2^n$, we have $F_{pk_F}(m, \mathbf{R}_1) = J(m)(T + e(\mathbf{A}^T, \mathbf{R}_1)) = J(m)(T + e(\mathbf{A}^T, \mathbf{R}_2)) = F_{pk_F}(m, \mathbf{R}_2)$ where $\mathbf{R}_2 = \mathbf{R}_1 + s\mathbf{Q}$.

4.3 Generic Transformation

We now describe the construction of the generic transformation that can convert any wEUF-FLR signature schemes into sEUF-FLR ones in the continual leakage model.

Let $\Sigma = (\mathsf{KG}, \mathsf{Sign}, \mathsf{Verify}, \mathsf{Update})$ be an arbitrary ℓ_w-wEUF-FLR signature scheme with randomness space Ω_Σ in the continual leakage model, $\mathcal{F} = (\mathsf{KG}_F, F, \mathsf{TC}_F, \mathsf{UD}_F)$ the ℓ-CLR chameleon hash function scheme with randomness space $\mathcal{R}_F$, where $\ell = n \log p - (2 + \Theta(1/\sqrt{k})) \log p$, and $\mathcal{H} = (\mathsf{KG}_H, H, \mathsf{TC}_H)$ a standard chameleon hash function with randomness space $\mathcal{R}_H$. The ℓ_s-sEUF-FLR signature scheme Σ' as shown in Fig. 4.

- $\mathsf{KG}'(1^k)$:
  1. Run $gk = (p, \mathbb{G}_1, \mathbb{G}_2, \mathbb{G}_T, e) \leftarrow \mathscr{G}(1^k)$.
  2. Run $(pk, sk) \leftarrow \mathsf{KG}(1^k)$.
  3. Run $(pk_F, sk_F) \leftarrow \mathsf{KG}_F(1^k)$.
  4. Run $(pk_H, sk_H) \leftarrow \mathsf{KG}_H(1^k)$.
  5. Randomly choose $\mathbf{G} \leftarrow \mathbb{G}_2^2$, $\mathbf{H} = (\mathbf{H}_0, \mathbf{H}_1, ..., \mathbf{H}_t) \leftarrow (\mathbb{G}_2^2)^{t+1}$, $\mathbf{B} \leftarrow \mathbb{G}_1^n$.
  6. Output public key $pk_s = (gk, pk, pk_F, pk_H, \mathbf{G}, \mathbf{H}, \mathbf{B}, m', \sigma')$ and secret key $sk_s = (sk, sk_F)$, where m' and σ' are arbitrary fixed strings.
- $\mathsf{Sign}'_{sk_s}(m)$:
  1. Parse $sk_s = (sk, sk_F)$.
  2. Randomly choose $\omega \leftarrow \Omega_\Sigma$, $s \leftarrow \mathscr{R}_H$, $r' \leftarrow \mathscr{R}_F$, $R \leftarrow \mathbb{Z}_p^{n\times 2}$, and $\mathbf{M} \leftarrow \mathbb{G}_2^n$.
  3. Compute $h = F_{pk_F}(m'||\sigma'; r')$.
  4. Compute $\bar{m} = H_{pk_H}(h; s)$.
  5. Compute $\sigma = \mathsf{Sign}_{sk}(\bar{m}; \omega)$.
  6. Compute $\mathbf{Pi} = R^T\mathbf{B}$, $T = e(\mathbf{B}^T, \mathbf{M})$.
  7. Compute $r = \mathsf{TC}_F(sk_F, (m'||\sigma', r'), m||\sigma||\mathbf{Pi}||T)$.
  8. Compute $\mathbf{H}_r = h_{gk}(\mathbf{H}, J'(r))$, where J' denotes a strongly collision resistant hash function from $\{0,1\}^*$ to $\{0,1\}^t$.
  9. Compute $(\mathbf{C}, \mathbf{D}) = (R\mathbf{G}, \mathbf{M} + R\mathbf{H_r})$, $\pi = (\mathbf{Pi}, \mathbf{C}, \mathbf{D})$.
  10. Output $\sigma' = (\sigma, r, s, \pi, T)$.
- $\mathsf{Verify}'(pk_s, m, \sigma')$:
  1. Parse $pk_s = (gk, pk, pk_F, pk_H, \mathbf{G}, \mathbf{H}, \mathbf{B}, m', \sigma')$, $\sigma' = (\sigma, r, s, \pi, T)$, and $\pi = (\mathbf{Pi}, \mathbf{C}, \mathbf{D})$.
  2. Compute $h = F_{pk_F}(m||\sigma||\mathbf{Pi}||T; r)$.
  3. Compute $\bar{m} = H_{pk_H}(h; s)$.
  4. Compute $\mathbf{H}_r = h_{gk}(\mathbf{H}, J'(r))$.
  5. Output 1 if $\mathsf{Verify}(pk, \bar{m}, \sigma) = 1$ and $(e(\mathbf{B}^T, \mathbf{C}), e(\mathbf{B}^T, \mathbf{D})) = (e(\mathbf{Pi}^T, \mathbf{G}), T + e(\mathbf{Pi}^T, \mathbf{H}_r))$, and output 0 otherwise.
- $\mathsf{Update}'(pk_s, sk_s)$:
  1. Parse $pk_s = (gk, pk, pk_F, pk_H, \mathbf{G}, \mathbf{H}, \mathbf{B}, m', \sigma')$ and $sk_s = (sk, sk_F)$.
  2. Compute $sk' \leftarrow \mathsf{Update}(pk, sk)$ and $sk'_F \leftarrow \mathsf{UD}_F(pk_F, sk_F)$.
  3. Output $sk'_s = (sk', sk'_F)$.

Fig. 4 GBSW transformation in the continuous leakage model

Theorem 2 *The signature scheme described in Fig. 4 is an ℓ_s-sEUF-FLR signature scheme for $\ell_s = \min\{\ell, \ell_w\}$ if $\mathscr{F} = (\mathsf{KG}_F, F, \mathsf{TC}_F, \mathsf{UD}_F)$ is an ℓ-CLR chameleon hash function scheme, $\mathscr{H} = (\mathsf{KG}_H, H, \mathsf{TC}_H)$ is a standard chameleon hash function scheme, and the SXDH assumption holds.*

5 Generic Transformation Without Changing Keys

In this section, we describe another transformation by Wang and Tanaka [29] that converts wEUF-FLR signature schemes into sEUF-FLR ones, which works in the continual leakage model. To achieve the goal, they improve the transformation of

Huang et al. [14] by substituting the standard strong one-time signature scheme with an FLR one. They also follow the technique of Mohassel [20] to provide an efficient instantiation of (restricted) FLR strong one-time signature based on LR chameleon hash functions.

5.1 (Restricted) FLR Strong One-Time Signature Scheme Based on LR Chameleon Hash Functions

Let $\mathscr{F} = (\mathsf{KG}_F, F, \mathsf{TC}_F)$ (with randomness space $\mathscr{R}_F$) be an ℓ-LR chameleon hash function scheme and $\mathscr{H} = (\mathsf{KG}_H, H, \mathsf{TC}_H)$ (with randomness space $\mathscr{R}_F$) a standard chameleon hash function scheme. The construction of FLR strong one-time signature scheme is given in Fig. 5.

Theorem 3 *The signature scheme* $(\mathsf{KG}_{\mathsf{ot}}, \mathsf{Sign}_{\mathsf{ot}}, \mathsf{Verify}_{\mathsf{ot}})$ *given in Fig. 5 is an ℓ-FLR strong one-time signature scheme if $\mathscr{F}$=$(\mathsf{KG}_F, F, \mathsf{TC}_F)$ is an ℓ-LR chameleon hash function scheme and $\mathscr{H} = (\mathsf{KG}_H, H, \mathsf{TC}_H)$ is a standard chameleon hash function scheme.*

- $\mathsf{KG}_{\mathsf{ot}}(1^k)$:
  1. Run $(pk_0, sk_0) \leftarrow \mathsf{KG}_H(1^k)$.
  2. Run $(pk_1, sk_1) \leftarrow \mathsf{KG}_F(1^k)$.
  3. Randomly choose $r_0 \leftarrow \mathscr{R}_H$, and $r'_1 \leftarrow \mathscr{R}_F$.
  4. Computes $h_1 = F_{pk_1}(m', r'_1)$, $h_0 = H_{pk_0}(h_1, r_0)$, where m' is some arbitrary fixed string.
  5. Output verification key $pk = (pk_0, pk_1, h_0, m')$ and signing key $sk = (sk_1, r_0, r'_1)$.
- $\mathsf{Sign}_{\mathsf{ot}\,sk}(m)$:
  1. Parse $sk = (sk_1, r_0, r'_1)$.
  2. Compute $r_1 = \mathsf{TC}_F(sk_1, (m', r'_1), m)$.
  3. Output $\sigma = (r_0, r_1)$.
- $\mathsf{Verify}_{\mathsf{ot}}(pk, m, \sigma)$:
  1. Parse $pk = (pk_0, pk_1, h_0, m')$ and $\sigma = (r_0, r_1)$.
  2. Compute $\bar{m} = F_{pk_1}(m; r_1)$.
  3. Output 1 if $h_0 = H_{pk_0}(\bar{m}, r_0)$ and output 0 otherwise.

Fig. 5 FLR strong one-time signature scheme

- $\mathsf{KG}'(1^k)$:
  1. Run $(pk,sk) \leftarrow \mathsf{KG}(1^k)$.
  2. Output verification key $pk_s = pk$ and signing key $sk_s = sk$.
- $\mathsf{Sign}'_{sk_s}(m)$:
  1. Randomly choose $\omega \leftarrow \Omega_\Sigma$, $r \leftarrow \Omega_r$, $r_{ot} \leftarrow \Omega_{ot}$.
  2. Compute $(pk_{ot}, sk_{ot}) = \mathsf{KG}_{\mathsf{ot}}(1^k; r)$.
  3. Compute $\sigma = \mathsf{Sign}_{sk_s}(pk_{ot}, \omega)$.
  4. Compute $s = \mathsf{Sign}_{\mathsf{ot}\, sk_{ot}}(m||\sigma, r_{ot})$.
  5. $\sigma' = (\sigma, s, pk_{ot})$.
- $\mathsf{Verify}'(pk_s, m, \sigma')$:
  1. Parse $\sigma' = (\sigma, s, pk_{ot})$.
  2. Output 1 if $\mathsf{Verify}_{\mathsf{ot}}(pk_{ot}, m||\sigma, s) = 1$ and $\mathsf{Verify}(pk_s, pk_{ot}, \sigma) = 1$. Output 0 otherwise.

Fig. 6 Transformation without changing the signing key in the bounded leakage model

5.2 *Generic Transformation*

Let $\Sigma = (\mathsf{KG}, \mathsf{Sign}, \mathsf{Verify})$ (with randomness space Ω_Σ) be an ℓ_w-FLR signature scheme that satisfies the wEUF property and $\Sigma_{ot} = (\mathsf{KG}_{\mathsf{ot}}, \mathsf{Sign}_{\mathsf{ot}}, \mathsf{Verify}_{\mathsf{ot}})$ (with randomness space Ω_{ot}) the ℓ-FLR strong one-time signature scheme. We describe the transformation by Wang and Tanaka [29] in Fig. 6. What they did is substituting the standard strong one-time signature scheme with the FLR one for the transformation in [14].

Theorem 4 *The signature scheme* $(\mathsf{KG}', \mathsf{Sign}', \mathsf{Verify}')$ *given in Fig. 6 is an* ℓ_s*-sEUF-FLR signature scheme for* $\ell_s = \min\{\ell, \ell_w\}$ *if* $\Sigma = (\mathsf{KG}, \mathsf{Sign}, \mathsf{Verify})$ *is an* ℓ_w*-wEUF-FLR signature scheme and* $\Sigma_{ot} = (\mathsf{KG}_{\mathsf{ot}}, \mathsf{Sign}_{\mathsf{ot}}, \mathsf{Verify}_{\mathsf{ot}})$ *is an* ℓ*-FLR strong one-time signature scheme.*

6 Open Problem

There are several problems left open. (1) Since the security proof of the transformation in Sect. 5.2 is not tight, the first question is how to obtain a transformation which does not change keys or loose reduction tightness. (2) The second question is how to obtain a transformation based on one-way functions rather than the discrete logarithm assumption or the SXDH assumption, without loosing efficiency or exploiting more secret information (in key generation or signing procedures).

Acknowledgements The first author is supported by a JSPS Fellowship for Young Scientists and JSPS KAKENHI 16J10697. The second is supported by Input Output Hong Kong, I-System, Nomura Research Institute, NTT Secure Platform Laboratories and JSPS KAKENHI 16H01705.

References

1. A. Akavia, S. Goldwasser, V. Vaikuntanathan, Simultaneous hardcore bits and cryptography against memory attacks, in *Theory of Cryptography*, ed. by O. Reingold. Lecture Notes in Computer Science, vol. 5444 (Springer, Berlin, 2009), pp. 474–495
2. J. Alwen, Y. Dodis, D. Wichs, Leakage-resilient public-key cryptography in the bounded-retrieval model, in *Advances in Cryptology CRYPTO 2009*, ed. by S. Halevi. Lecture Notes in Computer Science, vol. 5677 (Springer, Berlin, 2009), pp. 36–54
3. J. Alwen, Y. Dodis, M. Naor, G. Segev, S. Walfish, D. Wichs, Public-key encryption in the bounded-retrieval model, in *Advances in Cryptology EUROCRYPT 2010*, ed. by H. Gilbert. Lecture Notes in Computer Science, vol. 6110 (Springer, Berlin, 2010), pp. 113–134
4. M. Bellare, S. Shoup, Two-tier signatures, strongly unforgeable signatures, and Fiat-Shamir without random oracles, in *Public Key Cryptography PKC 2007*, ed. by T. Okamoto, X. Wang. Lecture Notes in Computer Science, vol. 4450 (Springer, Berlin, 2007), pp. 201–216
5. D. Boneh, E. Shen, B. Waters, Strongly unforgeable signatures based on computational Diffie-Hellman, in *Public Key Cryptography PKC 2006*, ed. by M. Yung, Y. Dodis, A. Kiayias, T. Malkin. Lecture Notes in Computer Science, vol. 3958 (Springer, Berlin, 2006), pp. 229–240
6. E. Boyle, G. Segev, D. Wichs, Fully leakage-resilient signatures, in *Advances in Cryptology EUROCRYPT 2011*, ed. by K.G. Paterson. Lecture Notes in Computer Science, vol. 6632 (Springer, Berlin, 2011), pp. 89–108
7. Z. Brakerski, S. Goldwasser, Circular and leakage resilient public-key encryption under subgroup indistinguishability, in *Advances in Cryptology CRYPTO 2010*, ed. by T. Rabin. Lecture Notes in Computer Science, vol. 6223 (Springer, Berlin, 2010), pp. 1–20
8. Z. Brakerski, Y. Kalai, J. Katz, V. Vaikuntanathan, Overcoming the hole in the bucket: public-key cryptography resilient to continual memory leakage, in *2010 51st Annual IEEE Symposium on Foundations of Computer Science (FOCS)* (2010), pp. 501–510
9. Y. Dodis, K. Haralambiev, A. López-Alt, D. Wichs, Cryptography against continuous memory attacks, in *Proceedings of the 2010 IEEE 51st Annual Symposium on Foundations of Computer Science, FOCS'10, Washington, DC, USA* (IEEE Computer Society, 2010), pp. 511–520
10. Y. Dodis, K. Haralambiev, A. López-Alt, D. Wichs, Efficient public-key cryptography in the presence of key leakage, in *Advances in Cryptology ASIACRYPT 2010*, ed. by M. Abe. Lecture Notes in Computer Science, vol. 6477 (Springer, Berlin, 2010), pp. 613–631
11. Y. Dodis, A. Lewko, B. Waters, D. Wichs, Storing secrets on continually leaky devices, in *2011 IEEE 52nd Annual Symposium on Foundations of Computer Science (FOCS)* (2011), pp. 688–697
12. S. Garg, A. Jain, A. Sahai, Leakage-resilient zero knowledge, in *Advances in Cryptology CRYPTO 2011*, ed. by P. Rogaway. Lecture Notes in Computer Science, vol. 6841 (Springer, Berlin, 2011), pp. 297–315
13. J. Groth, A. Sahai, Efficient noninteractive proof systems for bilinear groups. SIAM J. Comput. **41**(5), 1193–1232 (2012)
14. Q. Huang, D.S. Wong, Y. Zhao, Generic transformation to strongly unforgeable signatures, in *Applied Cryptography and Network Security ACNS 2007*, ed. by J. Katz, M. Yung. Lecture Notes in Computer Science, vol. 4521 (Springer, Berlin, 2007), pp. 1–17
15. J. Katz, V. Vaikuntanathan, Signature schemes with bounded leakage resilience, in *Advances in Cryptology ASIACRYPT 2009*, ed. by M. Matsui. Lecture Notes in Computer Science, vol. 5912 (Springer, Berlin, 2009), pp. 703–720
16. H. Krawczyk, T. Rabin, Chameleon signatures, in *NDSS* (The Internet Society, 2000)
17. A. Lewko, M. Lewko, B. Waters, How to leak on key updates, in *Proceedings of the Forty-Third Annual ACM Symposium on Theory of Computing, STOC'11, New York, NY, USA* (ACM, 2011), pp. 725–734
18. V. Lyubashevsky, A. Palacio, G. Segev, Public-key cryptographic primitives provably as secure as subset sum, in *Theory of Cryptography*, ed. by D. Micciancio. Lecture Notes in Computer Science, vol. 5978 (Springer, Berlin, 2010), pp. 382–400

19. T. Malkin, I. Teranishi, Y. Vahlis, M. Yung, Signatures resilient to continual leakage on memory and computation, in *Theory of Cryptography*, ed. by Y. Ishai. Lecture Notes in Computer Science, vol. 6597 (Springer, Berlin, 2011), pp. 89–106
20. P. Mohassel, One-time signatures and chameleon hash functions, in *Selected Areas in Cryptography*, ed. by A. Biryukov, G. Gong, D. Stinson. Lecture Notes in Computer Science, vol. 6544 (Springer, Berlin, 2011), pp. 302–319
21. M. Naor, G. Segev, Public-key cryptosystems resilient to key leakage, in *Advances in Cryptology CRYPTO 2009*, ed. by S. Halevi. Lecture Notes in Computer Science, vol. 5677 (Springer, Berlin, 2009), pp. 18–35
22. T. Okamoto, Provably secure and practical identification schemes and corresponding signature schemes, in *Advances in Cryptology CRYPTO'92*, ed. by E. Brickell. Lecture Notes in Computer Science, vol. 740 (Springer, Berlin, 1993), pp. 31–53
23. A. Shamir, Y. Tauman, Improved online/offline signature schemes, in *Proceedings of the 21st Annual International Cryptology Conference on Advances in Cryptology, CRYPTO'01, London, UK* (Springer, 2001), pp. 355–367
24. R. Steinfeld, J. Pieprzyk, H. Wang, How to strengthen any weakly unforgeable signature into a strongly unforgeable signature, in *Topics in Cryptology CT-RSA 2007*, ed. by M. Abe. Lecture Notes in Computer Science, vol. 4377 (Springer, Berlin, 2006), pp. 357–371
25. I. Teranishi, T. Oyama, W. Ogata, General conversion for obtaining strongly existentially unforgeable signatures, in *Progress in Cryptology INDOCRYPT 2006*, ed. by R. Barua, T. Lange. Lecture Notes in Computer Science, vol. 4329 (Springer, Berlin, 2006), pp. 191–205
26. Y. Wang, K. Tanaka, Generic transformation to strongly existentially unforgeable signature schemes with leakage resiliency, in *Provable Security*, ed. by S.S. Chow, J.K. Liu, L.C. Hui, S.M. Yiu. Lecture Notes in Computer Science, vol. 8782 (Springer International Publishing, New York, 2014), pp. 117–129
27. Y. Wang, K. Tanaka, Strongly simulation-extractable leakage-resilient NIZK, in *Information Security and Privacy*, ed. by W. Susilo, Y. Mu. Lecture Notes in Computer Science, vol. 8544 (Springer International Publishing, New York, 2014), pp. 66–81
28. Y. Wang, K. Tanaka, Generic transformation to strongly existentially unforgeable signature schemes with continuous leakage resiliency, in *Information Security and Privacy*, ed. by E. Foo, D. Stebila. Lecture Notes in Computer Science, vol. 9144 (Springer International Publishing, New York, 2015), pp. 213–229
29. Y. Wang, K. Tanaka, Generic transformations for existentially unforgeable signature schemes in the bounded leakage model. Secur. Commun. Netw. **9**(12), 1829–1842 (2016)

Constructions for the IND-CCA1 Secure Fully Homomorphic Encryption

Satoshi Yasuda, Fuyuki Kitagawa and Keisuke Tanaka

Abstract Homomorphic encryption allows a user to receive encrypted data and to perform arbitrary computation on that data without decrypting it. The homomorphic encryption scheme which supports only a bounded number of homomorphic operations is called "somewhat homomorphic encryption". The scheme which supports arbitrary number of homomorphic operations is called "fully homomorphic encryption". We need to construct an fully homomorphic encryption scheme which satisfies strong security for practical use to use a homomorphic encryption scheme practically, but essentially, we cannot construct a scheme which satisfies IND-CCA2 security Thus, one of the strongest security notions for homomorphic encryption is IND-CCA1 security. In this paper, we construct an fully homomorphic encryption scheme which satisfies IND-CCA1 security. Our construction has a restriction that our scheme can compute an arbitrary number of operations, but the arity of circuits is bounded. Our construction is based on the leakage-resilient bounded arity fully homomorphic encryption scheme proposed by Berkoff and Liu (TCC 2014). We show that their general construction can work for our construction.

Keywords Fully homomorphic encryption · Somewhat homomorphic encryption · Multi-key · IND-CCA1

S. Yasuda (✉) · F. Kitagawa · K. Tanaka
Department of Mathematical and Computing Sciences, Tokyo Institute of Technology, Ookayama 2-12-1, Meguro-ku, Tokyo 152-8552, Japan
e-mail: saith0326@gmail.com

F. Kitagawa
e-mail: kitagaw1@is.titech.ac.jp

K. Tanaka
e-mail: keisuke@is.titech.ac.jp

T. Takagi et al. (eds.), *Mathematical Modelling for Next-Generation Cryptography*, Mathematics for Industry 29, DOI 10.1007/978-981-10-5065-7_18

1 Introduction

1.1 Backgrounds and Motivations

Homomorphic encryption (HE), introduced by Rivest, Adelman and Detouzos [12], allows a user to receive encrypted data and perform arbitrary computation on that data without decrypting it. Recently, many concrete constructions of homomorphic encryption schemes have been proposed, because it is useful to construct cryptographic protocols such as electrical voting, cloud services, and so on.

We can categorize homomorphic encryption schemes to two types depending on the capacity of homomorphic operations. The homomorphic encryption scheme which supports only a bounded number of homomorphic operations is called "somewhat homomorphic encryption (SHE)". A scheme which supports an arbitrary number of homomorphic operations is called "fully homomorphic encryption".

Constructing an FHE scheme is a big goal of the research for homomorphic encryption, because it allows many applications. Gentry [8] proposed the first FHE scheme. He constructed the SHE scheme first. Then, he converted it to the FHE scheme by using the method which is called bootstrapping. This is a general method to convert an SHE scheme to an FHE scheme. After that, FHE schemes based on various computational assumptions have been proposed [4, 7, 9, 13]

In order to use an FHE scheme practically, it is necessary to construct a scheme which satisfies a strong security notion. Indistinguishability against chosen plaintext attack (IND-CPA) and indistinguishability against chosen ciphertext attack/adaptive chosen ciphertext attack (IND-CCA1/IND-CCA2) are the general security notions for encryption schemes, and IND-CCA2 is the strongest notion. However, essentially, we cannot construct a homomorphic encryption scheme that satisfies the IND-CCA2 security. Thus, one of the strongest security notions for homomorphic encryption schemes is the IND-CCA1 security. However, almost all of the FHE schemes which have been proposed do not satisfy the IND-CCA1 security.

For SHE, there is a concrete scheme which satisfies the IND-CCA1 security. Loftus, May, Smart, and Vercauteren [10] proposed the SHE scheme which satisfies the IND-CCA1 security. They constructed it by modifying the decryption algorithm of the homomorphic encryption scheme proposed by Smart and Vercauteren [13]. In particular, they added the process that checks the validity of the ciphertext to the decryption algorithm.

If we apply the bootstrapping to their SHE scheme, we might have an IND-CCA1 secure FHE scheme. However, the resulting FHE scheme will have public keys that contains a ciphertext of the secret key. Thus, the chosen ciphertext attacker can get the partial information about secret keys from public keys. For the above reason, we cannot apply the bootstrapping to their SHE scheme in order to achieve an IND-CCA1 secure FHE scheme.

Berkoff and Liu [1] faced to the similar problem that they could not apply the bootstrapping to their scheme. They proposed the homomorphic encryption scheme which has a key-leakage resilience, but they could not apply the bootstrapping to the scheme. This is because the resulting FHE scheme also has public keys that contains a ciphertext of secret key, and they could not prove the key-leakage resilience of their scheme. Therefore, they proposed the method that converted an SHE scheme to an FHE scheme. They also proved that the method kept the key-leakage resilience. However, the resulting FHE scheme converted by their method has a restriction that the scheme can evaluate homomorphically bounded number of ciphertext. They called it an N-ary FHE scheme.

1.2 Our Contribution

In this paper, we construct an N-ary FHE scheme which satisfies the IND-CCA1 security. Our idea is based on the one of Berkoff and Liu [1]. Their general construction can work for our construction. In particular, we apply their method to an SHE scheme which satisfies the IND-CCA1 security. The reason their method can be applied to an IND-CCA1 secure homomorphic encryption is the similarity between the IND-CCA1 security and the key-leakage resilience. In the both security models, an attacker can query and attempt to get the partial information about the secret key. Therefore, in this paper, we construct an IND-CCA1 secure N-ary FHE scheme by using their method.

1.3 Overview of Our Techniques

We employ two building blocks in order to construct an N-ary FHE scheme. One is an IND-CCA1 secure SHE scheme and the other an IND-CPA secure multikey FHE scheme. In this section, we describe an multikey homomorphic encryption, an N-ary FHE scheme and an overview of our techniques.

Multikey Homomorphic Encryption

Homomorphic encryption proposed by López-Alt, Tromer and Vaikuntanathan [11] has a special function that we can evaluate homomorphically the ciphertexts which are encrypted under different public keys. However the scheme proposed by López-Alt et al. has a restriction that the number of public and secret keys for the homomorphic operations on the ciphertexts is bounded in advance. Recently, the scheme which does not have the above restriction was proposed by Brakerski and Perlman [3].

Generally, a multikey homomorphic encryption scheme uses the special secret key called the "joint secret key" in the decryption part. The calculation methods of the joint secret key are different by each concrete scheme. For example, the scheme proposed by López-Alt et al. calculates it by multiplications of all secret keys each of which corresponds to the ciphertexts for the homomorphic operation. The calculation algorithm of the joint secret key is not defined formally, but this is the general technique for decryption of multikey homomorphic encryption schemes. Thus, we describe this algorithm as $\mathsf{CombineSK}$.

N-ary Fully Homomorphic Encryption

A fully homomorphic encryption scheme has a property that we can evaluate homomorphically the bounded number of ciphertexts. If we consider an evaluations as a circuit, we can evaluate circuits whose depth is an arbitrary but the number of inputs is bounded. In our construction, this restriction depends on the number of public keys of the multikey homomorphic encryption scheme and the number of homomorphic operations which is supported by the SHE scheme. The multikey homomorphic encryption and the somewhat homomorphic encryption scheme are building blocks of our scheme.

Construction of Our Scheme

In order to explain our idea, we first review the idea of the construction of Berkoff and Liu [1].

They used a KEM-DEM like construction with a leakage resilient homomorphic encryption scheme (SHE) and a multikey fully homomorphic encryption scheme (MHE). As a result, they constructed an N-ary fully homomorphic encryption scheme (FHE). We describe the encryption algorithm of SHE as sEnc, the one of MHE as mEnc, the public and secret key of SHE as $(pk_{\mathsf{SHE}}, sk_{\mathsf{SHE}})$ and those of MHE as $(pk_{\mathsf{MHE}}, sk_{\mathsf{MHE}})$.

First, the encryption algorithm of FHE is

$$\begin{aligned}\mathsf{Enc}(pk_{\mathsf{SHE}}, m) &\to (c_1, c_2)\\ &= (\mathsf{mEnc}(pk_{\mathsf{MHE}}, m), \mathsf{sEnc}(pk_{\mathsf{SHE}}, sk_{\mathsf{MHE}})).\end{aligned}$$

In their scheme, it encrypts the plaintext m by using MHE and the sk_{MHE} by using MHE. $(pk_{\mathsf{MHE}}, sk_{\mathsf{MHE}})$ are generated in this Enc algorithm.

Next, we describe the evaluation algorithm of SHE as sEval, and the one of MHE as mEval. The evaluation algorithm of FHE is

$$\begin{aligned}\mathsf{Eval}(pk_{\mathsf{SHE}}, f, (c_1, c_2)) &\to (c_{f,1}, c_{f,2})\\ &= (\mathsf{mEval}(pk_{\mathsf{MHE}}, f, c_1), \mathsf{sEval}(pk_{\mathsf{SHE}}, f_{CSK}, c_2)).\end{aligned}$$

Here, f is an arbitrary circuit, and f_{CSK} is a circuit which represents the algorithm $\mathsf{CombineSK}$. In this algorithm, it evaluates homomorphically an arbitrary circuit by using mEval, and the circuit f_{CSK} by using sEval.

In this paper, we employ two building blocks. One is an IND-CCA1 secure SHE scheme (SHE) and the other an IND-CPA secure multikey fully homomorphic encryption scheme (MHE). By using the above construction and building blocks, we construct an N-ary FHE scheme, and prove this scheme satisfies the IND-CCA1 security.

1.4 Candidate for Building Blocks

We discuss the concrete schemes which we can use as building blocks MHE and SHE.

As far as we know, three multikey fully homomorphic encryption scheme were proposed. First one is proposed by López-Alt et al. [11]. However, this scheme is based on nonstandard assumption. They called it the Decisional Small Polynomial Ratio assumption. Second one is proposed by Clear and McGoldrick [6]. This scheme is based on the Learning with Errors (LWE) assumption. This scheme is the first concrete scheme which is based on a standard assumption. However, their scheme is 1-hop homomorphic. In other words, after an evaluation is completed, no further homomorphic evaluation can be carried out. Third one is proposed by Brakerski and Perlman [3]. This scheme is also based on the LWE assumption. This scheme gets over the restriction of the scheme proposed by Clear and McGoldrick. In other words, their scheme is multi-hop homomorphic.

Next, as far as we know, the SHE scheme proposed by Loftus et al. [10] is the only SHE scheme which satisfies the IND-CCA1 security. However, this scheme is based on the knowledge assumption for ideal lattices and this assumption is not a standard one. Therefore, in this paper, we construct an IND-CCA1 secure homomorphic encryption scheme which is equivalent to an SHE scheme. In particular, in order to construct this scheme, we use an identity based homomorphic encryption scheme and a technique like CHK transformation. CHK transformation is proposed by Boneh et al. [2, 5]. By using their transformation, we can construct an IND-CCA secure public key encryption scheme from an identity based encryption scheme. Therefore, we apply the part of their technique to an identity based homomorphic encryption scheme. As a result, we can construct an IND-CCA1 secure SHE scheme. This resulting scheme does not completely satisfies the compactness, however it has enough compactness used as a building block of our N-ary FHE scheme.

2 Preliminaries

2.1 Notations

In this paper, $x \xleftarrow{r} X$ denotes selecting an element from a finite set X uniformly at random, and $y \leftarrow \mathsf{A}(x)$ denotes assigning to y the output of an algorithm A on an input x. λ denotes a security parameter. We write $f(\lambda) = \mathsf{negl}(\lambda)$ to denote $f(\lambda)$ being a negligible function where $f(\lambda)$ tends to 0 faster than $\frac{1}{\lambda^c}$ for every constant

$c > 0$. PPT stands for probabilistic polynomial time. $[\ell]$ denotes the set of integers $\{1, \cdots, \ell\}$. ϕ denotes an empty set.

2.2 Homomorphic Encryption

In this section we review homomorphic encryption. This definition follows those in [1, 11]

Definition 2.1 *($\mathscr{F}$-Homomorphic Encryption)* Let $\mathscr{F}$ be a class of circuits. A tuple of PPT algorithms $\mathsf{HE} = (\mathsf{KG}, \mathsf{Enc}, \mathsf{Dec}, \mathsf{Eval})$ is an $\mathscr{F}$-homomorphic encryption scheme.

- The key generation algorithm KG, given a security parameter 1^λ, outputs a public key pk and a secret key sk.
- The encryption algorithm Enc, given a public key pk and a message $m \in \mathscr{M}$, outputs a ciphertext c, where $\mathscr{M}$ is the plaintext space of HE.
- The decryption algorithm Dec, given a secret key sk and a ciphertext c, outputs a message $\tilde{m} \in \mathscr{M}$.
- The homomorphic evaluation algorithm Eval, given a public key pk, a circuit $f \in \mathscr{F} : \mathscr{M}^t \rightarrow \mathscr{M}$, and a set of t ciphertexts $c_1, \ldots, c_t$, outputs a ciphertext c_f.

Correctness For any circuit $f \in \mathscr{F}$, and respective inputs $m_1, \ldots, m_t$, it holds that

$$\Pr[\mathsf{Dec}(sk, \mathsf{Eval}(pk, f, c_1, \ldots, c_t)) \neq f(m_1, \ldots, m_t)] = \mathsf{negl}(\lambda)$$

where $(pk, sk) \leftarrow \mathsf{KG}(1^\lambda)$ and $c_i \leftarrow \mathsf{Enc}(pk, m_i)$.

Compactness Let $c^* = \mathsf{Eval}(pk, f, c_1, \ldots, c_t)$. There exists a polynomial $P = \mathsf{poly}(\lambda)$ such that $|c^*| \leq P$. In other words, the size of c^* is independent of t and the size of f.

Definition 2.2 *(Fully Homomorphic Encryption)* A homomorphic encryption scheme is fully homomorphic if it is $\mathscr{F}$-homomorphic for the class $\mathscr{F}$ of all circuits.

Berkoff and Liu [1] defined the N-ary homomorphic encryption in their paper. This is the homomorphic encryption that we can perform homomorphic operations to the bounded number of ciphertexts.

Definition 2.3 *(N-ary Homomorphic Encryption)* Let $N = \mathsf{poly}(\lambda)$. A homomorphic encryption scheme is bounded arity homomorphic if it is $\mathscr{F}$-homomorphic for the class $\mathscr{F}$ of all circuits with less than N where integer $N < 0$.

Next, we review the multikey homomorphic encryption. This is the homomorphic encryption that allows us to perform homomorphic operations to ciphertext which are encrypted under different public keys. This definition follows those in [11]

Definition 2.4 (*Multikey $\mathscr{F}$-Homomorphic Encryption*) Let $\mathscr{F}$ be a class of circuits. A tuple of PPT algorithms $\mathsf{MHE} = (\mathsf{mKG}, \mathsf{mEnc}, \mathsf{mDec}, \mathsf{mEval})$ is an $\mathscr{F}$-homomorphic encryption scheme.

- The key generation algorithm mKG, given a security parameter 1^λ, outputs a public key pk and a secret key sk.
- The encryption algorithm mEnc, given a public key pk and a message $m \in \mathscr{M}$, outputs a ciphertext c, where $\mathscr{M}$ is the plaintext space of MHE.
- The decryption algorithm mDec, given N secret keys sk_i and a ciphertext c, outputs a message $\tilde{m} \in \mathscr{M}$.
- The homomorphic evaluation algorithm mEval, given t pairs (pk_i, c_i) of public key and ciphertext, and a circuit $f \in \mathscr{F} : \mathscr{M}^t \to \mathscr{M}$, outputs a ciphertext c_f and a corresponding public key pk^*.

Correctness For any circuit $f \in \mathscr{F}$, and respective inputs $m_1, \ldots, m_t$, it holds that

$$\Pr[\mathsf{Dec}(sk_1, \ldots, sk_N, \mathsf{Eval}(f, (pk_1, c_1), \ldots, (pk_t, c_t))) \neq f(m_1, \ldots, m_t)] = \mathsf{negl}(\lambda)$$

where $i \in [t]$, $(pk_i, sk_i) \leftarrow \mathsf{KG}(1^\lambda)$ and $c_i \leftarrow \mathsf{Enc}(pk_i, m_i)$.

Compactness Let $c^* = \mathsf{Eval}(f, (pk_1, c_1), \ldots, (pk_t, c_t))$. There exists a polynomial $P = \mathsf{poly}(\lambda, N)$ such that $|c^*| \leq P$. In other words, the size of c^* is independent of t and the size of f. However, we allow the evaluated ciphertext to depend on the number of keys.

The identity-based homomorphic encryption is the identity-based encryption that we can perform homomorphic evaluations. The "multi-identity" means that we can perform homomorphic operations to ciphertext which encrypted different identities. This definition follows those in [6]

Definition 2.5 (*Multi-identity Identity Based (Leveled) Fully Homomorphic Encryption*) A Multi-Identity (Leveled) IBFHE scheme is defined with respect to a message space $\mathscr{M}$, an identity space $\mathscr{I}$ and a class of circuits $\mathscr{F} : \mathscr{M}^* \to \mathscr{M}$. A Multi-Identity IBFHE scheme is a tuple of PPT algorithms $(\mathsf{Setup}, \mathsf{KG}, \mathsf{Enc}, \mathsf{Dec}, \mathsf{Eval})$ defined as follow.

- The setup algorithm Setup, given a security parameter 1^λ, a number of levels $\mathscr{L}$ and the number of distinct identities $\mathscr{D}$ that can be tolerated in an evaluation, outputs public parameters PP and a master secret key MSK.
- The key generation algorithm KG, given a master secret key MSK and an identity id, derives and outputs a secret key sk_{id} for identity id.
- The encryption algorithm Enc, given public parameters PP, an identity id, and a message $m \in \mathscr{M}$, outputs a ciphertext c that encrypts m under the identity id.
- The decryption algorithm Dec, given $d \leq \mathscr{D}$ secret keys $sk_{id_1}, \ldots, sk_{id_d}$ for (resp.) identities $id_1, \ldots, id_d$ and a ciphertext c, outputs $\tilde{m}$ if c is a valid encryption under identities $id_1, \ldots, id_d$; outputs a failure symbol $\perp$ otherwise.

- The evaluation algorithm Eval, given public parameters PP, a circuit $f \in \mathcal{F}$ and ciphertexts $c_1, \ldots, c_l$, outputs an evaluated ciphertext c'.

More precisely, the scheme is required to satisfy the following properties:
Over all choices of $(PP, MSK) \leftarrow \mathsf{Setup}(1^\lambda)$, $d \leq \mathcal{D}$, $id_1, \ldots, id_d \in \mathcal{I}$, $f : \mathcal{M}^l \rightarrow \mathcal{M} \in \mathcal{F}$, $j_1, \ldots, j_l \in [d]$, $\mu_1, \ldots, \mu_l \in \mathcal{M}$, $c_i \leftarrow \mathsf{Enc}(PP, id_{j_i}, \mu_i)$ for $i \in [l]$, and $c' \leftarrow \mathsf{Eval}(PP, f, c_1, \ldots, c_l)$:

Correctness

$$\mathsf{Dec}(sk_1, \ldots, sk_d, c') = f(\mu_1, \ldots, \mu_l)$$

for any $sk_i \leftarrow \mathsf{KG}(MSK, id_i)$ for $i \in [l]$

Compactness

$$|c'| \leq \mathsf{poly}(\lambda, \mathcal{L}, d)$$

where $d \leq \mathcal{D}$ is the number of distinct identities; that is, $d = |\{j_1, \ldots, j_l\}|$.

2.3 Security Notions

We review the IND-CPA and the IND-CCA1 security for homomorphic encryption schemes. This is the same as the IND-CPA and the IND-CCA1 security for general public key encryption schemes.

Definition 2.6 (IND-CPA *Security*) Let $\mathsf{HE} = (\mathsf{KG}, \mathsf{Enc}, \mathsf{Dec}, \mathsf{Eval})$ be an homomorphic encryption scheme. We define the IND-CPA game between a challenger and an adversary $\mathcal{A}$ as follows.

Initialization First the challenger chooses a challenge bit $b \xleftarrow{r} \{0, 1\}$. Next the challenger generates a key pair $(pk, sk) \leftarrow \mathsf{KG}(1^\lambda)$ and sends pk to $\mathcal{A}$.
Challenge query $\mathcal{A}$ can send (m_0, m_1), where m_0, m_1 are messages. The challenger returns $c^* \leftarrow \mathsf{Enc}(pk, m_b)$.
Final phase $\mathcal{A}$ outputs $b' \in \{0, 1\}$.

We say that a homomorphic encryption scheme HE is IND-CPA secure if for any PPT adversary $\mathcal{A}$, we have $\mathsf{Adv}^{indcpa}_{\mathsf{HE},\mathcal{A}}(\lambda) := |\Pr[b = b'] - \frac{1}{2}| = \mathsf{negl}(\lambda)$.

Definition 2.7 (IND-CCA1 *Security*) Let HE be a homomorphic encryption scheme. We define the IND-CCA1 game between a challenger and an adversary $\mathcal{A}$ as follows.

Initialization First the challenger chooses a challenge bit $b \xleftarrow{r} \{0, 1\}$. Next the challenger generates a key pair $(pk, sk) \leftarrow \mathsf{KG}(1^\lambda)$ and sends pk to $\mathcal{A}$.
$\mathcal{A}$ may make polynomially many decryption queries.
Decryption queries $\mathcal{A}$ can send (c), where c is a ciphertext. The challenger returns $m \leftarrow \mathsf{Dec}(sk, c)$.

Challenge query $\mathscr{A}$ can send (m_0, m_1), where m_0, m_1 are messages. The challenger returns $c^* \leftarrow \mathsf{Enc}(pk, m_b)$.
Final phase $\mathscr{A}$ outputs $b' \in \{0, 1\}$.

We say that an homomorphic encryption scheme HE is IND-CCA1 secure if for any PPT adversary $\mathscr{A}$, we have $\mathsf{Adv}^{indcca1}_{\mathsf{HE},\mathscr{A}}(\lambda) := |\Pr[b = b'] - \frac{1}{2}| = \mathsf{negl}(\lambda)$.

Next we review an IND-sID-CPA security for IBHE schemes. This is the same as the IND-sID-CPA security for general identity based encryption schemes.

Definition 2.8 (*IND-sID-CPA Security*) Let IBHE be a IBHE scheme. We define the IND-sID-CPA game between a challenger and an adversary $\mathscr{A}$ as follows.

Initialization First, the adversary chooses a identity id^* and sends it to the challenger. Next, the challenger chooses a challenge bit $b \xleftarrow{r} \{0, 1\}$. The challenger generates public parameters and a master secret key $(PP, MSK) \leftarrow \mathsf{Setup}(\lambda)$ and sends PP to $\mathscr{A}$.
Key derivation queries $\mathscr{A}$ can send (id), where id is an identity. The challenger returns $sk_{id} \leftarrow \mathsf{KG}(MSK, id)$. The adversary can query any identities expect the identity id^*.
Challenge query $\mathscr{A}$ can send (m_0, m_1), where m_0, m_1 are messages. The challenger returns $c^* \leftarrow \mathsf{Enc}(PP, id^*, m_b)$.
Final phase $\mathscr{A}$ outputs $b' \in \{0, 1\}$.

We say that an IBHE is IND-sID-CPA secure if for any PPT adversary $\mathscr{A}$, we have $\mathsf{Adv}^{ind\text{-}sid\text{-}cpa}_{\mathsf{IBHE},\mathscr{A}}(\lambda) := |\Pr[b = b'] - \frac{1}{2}| = \mathsf{negl}(\lambda)$.

3 Construction

In this section, we show our main technical result: how to construct an IND-CCA1 secure N-ary FHE scheme. Our construction is based on the idea of the construction by Berkoff and Liu [1]. Their general construction can work for our construction. In our construction, we employ two building blocks. One is an IND-CCA1 secure SHE scheme and the other an IND-CPA secure multikey FHE scheme. For the multikey FHE scheme, we require a condition that we can divide the decryption algorithm into two parts. The first part is the calculation part of the "joint secret key". The second part is decrypting the ciphertext by using "joint secret key". We represent the algorithm of the first part as CombineSK, and the circuit that calculate CombineSK as f_{CSK}.

Let SHE = (sKG, sEnc, sDec, sEval) be an SHE scheme, and MHE = (mKG, mEnc, mDec, mEval) an MHE scheme. Then using SHE and MHE as the building blocks, we construct an N-ary FHE scheme HE = (KG, Enc, Dec, Eval) as follows.

$\mathsf{KG}(1^\lambda)$:

$(pk_{\mathsf{SHE}}, sk_{\mathsf{SHE}}) \leftarrow \mathsf{sKG}(1^\lambda)$
Output : $(pk_{\mathsf{SHE}}, sk_{\mathsf{SHE}})$

$\mathsf{Enc}(pk_{\mathsf{SHE}}, m)$:

$(pk_{\mathsf{MHE}}, sk_{\mathsf{MHE}}) \leftarrow \mathsf{mKG}(1^\lambda)$
$c_1 \leftarrow \mathsf{mEnc}(pk_{\mathsf{MHE}}, m)$
$c_2 \leftarrow \mathsf{sEnc}(pk_{\mathsf{SHE}}, sk_{\mathsf{MHE}})$
Output : $C = (pk_{\mathsf{MHE}}, c_1, c_2)$

$\mathsf{Dec}(sk_{\mathsf{SHE}}, C)$:

Parse c to $(pk_{\mathsf{MHE}}, c_1, c_2)$
$sk_{\mathsf{MHE}} \leftarrow \mathsf{sDec}(sk_{\mathsf{SHE}}, c_2)$
$\tilde{m} \leftarrow \mathsf{mDec}(sk_{\mathsf{MHE}}, c_1)$
Output : $\tilde{m}$

$\mathsf{Eval}(pk_{\mathsf{SHE}}, f, C_1, \ldots, C_t)$:

Parse C_i to $(pk_{\mathsf{MHE}_i}, c_{i.1}, c_{i.2})$
$(pk^*_{\mathsf{MHE}}, c^*_1) \leftarrow \mathsf{mEval}(f, (pk_{\mathsf{MHE}_i}, c_{1.1}), \ldots, (pk_{\mathsf{MHE}_t}, c_{t.1}))$
$c^*_2 \leftarrow \mathsf{sEval}(pk_{\mathsf{SHE}}, f_{CSK}, c_{1.2}, \ldots, c_{t.2})$
Output : $C = (pk^*_{\mathsf{MHE}}, c^*_1, c^*_2)$

In general MHE, mDec is given some secret keys. It calculates "joint secret key" from given secret keys, and then decrypts the ciphertext by using the joint secret key. Berkoff and Liu divided mDec to the above two steps, and constructed HE by putting the first step to Eval and the second step to Dec.

In particular, Eval and Dec are as follows. In Eval, we run sEval to obtain a ciphertext of the joint secret key with the input ciphertext of the secret key sk_{MHE} and f_{CSK} i.e. $c^*_2 \leftarrow \mathsf{sEval}(pk_{\mathsf{SHE}}, f_{CSK}, c_{1.2}, \ldots, c_{t.2})$. f_{CSK} represents the circuit to calculate the joint secret key. In Dec, we first decrypt the ciphertext of the joint secret key and then run mDec with its joint secret key.

As the homomorphism and the compactness of this construction, the following theorem holds.

Theorem 3.1 $N = \mathsf{poly}(\lambda)$. *Let* SHE *be an SHE scheme which can evaluate* f_{CSK} *and* MHE *a N-multikey FHE scheme. Then,* HE *is an N-ary fully homomorphic encryption scheme.*

Proof First we discuss correctness. In HE, we perform homomorphic evaluation to the message with MHE because the messages is encrypted with MHE. MHE can evaluate the circuit with an arbitrary depth, then HE can also evaluate the circuit with an arbitrary depth. However the restriction that HE can only evaluate circuits with at most N inputs depends on N that the number of key pairs for the MHE operations. The ciphertext c_2 is encrypted with SHE. We consider SHE can evaluate f_{CSK}, so we can calculate the correct ciphertexts of joint secret key from the ciphertext c_2. From the above, HE can evaluate the circuit with an arbitrary depth and at most N inputs.

For compactness, both MHE and SHE satisfy compactness. Thus HE satisfies compactness.

From the above discussion, HE is an N-ary fully homomorphic encryption scheme. □

4 Security of the Construction

In this section, we prove the IND-CCA1 security of our construction.

Theorem 4.1 *Let* SHE *satisfy the IND-CCA1 security, and* MHE *satisfy the IND-CPA security. Then,* HE *satisfies the IND-CCA1 security.*

Proof We prove this theorem via a sequence of games. Let $\mathcal{A}$ be an adversary that attacks the IND-CCA1 security of our scheme HE. Now consider the following sequence of games.

Game 0 This is the IND-CCA1 game regarding our scheme HE.
Game 1 Same as Game 0 except that in challenge query the challenger makes c_2^* as follows: $c_2^* \leftarrow \mathsf{sEnc}(pk_{\mathsf{SHE}}, 0^{|m_b|})$.
Game 2 Same as Game 1 except that in challenge query the challenger makes c_1^* as follows: $c_1^* \leftarrow \mathsf{mEnc}(pk_{\mathsf{MHE}}, 0^{|m_b|})$.

For $i = 0, 1, 2$, we define the following events in Game i:

S_i: $\mathcal{A}$ succeeds in guessing the challenge bit, that is $b = b'$ occurs.

Then, we can estimate $\mathsf{Adv}^{indcca1}_{\mathsf{HE},\mathcal{A}}(\lambda)$ as follows:

$$\begin{aligned}\mathsf{Adv}^{indcca1}_{\mathsf{HE},\mathcal{A}}(\lambda) &= |\Pr[S_0] - \frac{1}{2}| \\ &\leq \sum_{i=0,1} |\Pr[S_i] - \Pr[S_{i+1}]| + |\Pr[S_2] - \frac{1}{2}|. \end{aligned} \tag{1}$$

Below, we show that each term of the right side of inequality is negligible.

Lemma 4.2 *Let* SHE *be IND-CCA1 secure. Then* $|\Pr[S_0] - \Pr[S_1]| = \mathsf{negl}(\lambda)$.

Proof Using the adversary $\mathcal{A}$ that attacks HE, we construct an adversary $\mathcal{B}$ that attacks the IND-CCA1 security of SHE. We describe the adversary $\mathcal{B}$ as follows.

Initialization On input public key pk_{SHE}, $\mathcal{B}$ sends pk_{SHE} to $\mathcal{A}$.

Decryption queries For a decryption query (C), $\mathscr{B}$ parses C to $(pk_{\mathsf{MHE}}, c_1, c_2)$. Then $\mathscr{B}$ makes a decryption query (c_2) to the challenger, gets sk_{MHE}, and computes $m \leftarrow \mathsf{mDec}(sk_{\mathsf{MHE}}, c_1)$. Finally, $\mathscr{B}$ returns m to $\mathscr{A}$.

Challenge query For a challenge query (m_0, m_1), $\mathscr{B}$ computes $(pk^*_{\mathsf{MHE}}, sk^*_{\mathsf{MHE}}) \leftarrow \mathsf{mKG}(1^\lambda)$. Then, $\mathscr{B}$ makes a challenge query $(m'_0 = sk^*_{\mathsf{MHE}}, m'_1 = 0^{|SK|})$ to the challenger, and gets c^*_2. Next, $\mathscr{B}$ flips random coin $b \leftarrow \{0, 1\}$, then computes $c^*_1 \leftarrow \mathsf{mEnc}(pk^*_{\mathsf{MHE}}, m_b)$. Finally, $\mathscr{B}$ returns $C = (pk^*_{\mathsf{MHE}}, c^*_1, c^*_2)$ to $\mathscr{A}$.

Final phase When $\mathscr{A}$ outputs $b' \in \{0, 1\}$, $\mathscr{B}$ outputs $\beta' = 1$ if $b = b'$, otherwise outputs $\beta' = 0$.

Let β be a challenge bit in the game between the challenger and $\mathscr{B}$. When $\beta = 0$, $\mathscr{B}$ perfectly simulates Game 0. When $\beta = 1$, $\mathscr{B}$ perfectly simulates Game 1. Therefore. we have

$$\begin{aligned}\mathsf{Adv}^{indcca1}_{\mathsf{SHE},\mathscr{B}}(\lambda) &= \frac{1}{2}|\Pr[\beta' = 1|\beta = 0] - \Pr[\beta' = 1|\beta = 1]| \\ &= \frac{1}{2}|\Pr[S_0] - \Pr[S_1]|.\end{aligned}$$

Since SHE is IND-CCA1 secure, we see that $|\Pr[S_0] - \Pr[S_1]| = \mathsf{negl}(\lambda)$. □

Lemma 4.3 *Let* MHE *be IND-CPA secure. Then* $|\Pr[S_1] - \Pr[S_2]| = \mathsf{negl}(\lambda)$.

Proof Using the adversary $\mathscr{A}$ that attacks HE, we construct an adversary $\mathscr{B}$ that attacks the IND-CPA security of MHE. We describe the adversary $\mathscr{B}$ as follows.

Initialization On input public key pk_{MHE}, $\mathscr{B}$ computes $(pk_{\mathsf{SHE}}, sk_{\mathsf{SHE}}) \leftarrow \mathsf{sKG}(1^\lambda)$. Then, $\mathscr{B}$ sends pk_{SHE} to $\mathscr{A}$.

Decryption queries For a decryption query (C), $\mathscr{B}$ pareses C to $(pk_{\mathsf{MHE}}, c_1, c_2)$. Then, $\mathscr{B}$ computes $sk_{\mathsf{MHE}} \leftarrow \mathsf{sDec}(sk_{\mathsf{SHE}}, c_2)$, and $m \leftarrow \mathsf{mDec}(sk_{\mathsf{MHE}}, c_1)$. Finally, $\mathscr{B}$ returns m to $\mathscr{A}$.

Challenge query For a challenge query, $\mathscr{B}$ flips random coin $b \leftarrow \{0, 1\}$. Then, $\mathscr{B}$ makes a challenge query $(m'_0 = m_b, m'_1 = 0^\lambda)$ to the challenger, and get c^*_1. Next, $\mathscr{B}$ computes $c^*_2 \leftarrow \mathsf{sEnc}(pk_{\mathsf{SHE}}, 0^\lambda)$. Finally, $\mathscr{B}$ returns $C=(pk^*_{\mathsf{MHE}}, c^*_1, c^*_2)$ to $\mathscr{A}$.

Final phase When $\mathscr{A}$ outputs $b' \in \{0, 1\}$, $\mathscr{B}$ outputs $\beta' = 1$ if $b = b'$, otherwise outputs $\beta' = 0$.

Let β be a challenge bit in the game between the challenger and $\mathscr{B}$. When $\beta = 0$, $\mathscr{B}$ perfectly simulates Game 1. When $\beta = 1$, $\mathscr{B}$ perfectly simulates Game 2. Therefore, we have

$$\begin{aligned}\mathsf{Adv}^{indcpa}_{\mathsf{MHE},\mathscr{B}}(\lambda) &= \frac{1}{2}|\Pr[\beta' = 1|\beta = 0] - \Pr[\beta' = 1|\beta = 1]| \\ &= \frac{1}{2}|\Pr[S_1] - \Pr[S_2]|.\end{aligned}$$

Since MHE is IND-CPA secure, we see that $|\Pr[S_1] - \Pr[S_2]| = \mathsf{negl}(\lambda)$. □

Finally, in Game 2, the challenge ciphertext does not contain information of challenge bit. Thus $|\Pr[S_2] - 1/2| = 0$. From the inequality 1 and Lemmas 4.2 and 4.3, we have $\mathsf{Adv}^{indcca1}_{\mathsf{HE},\mathscr{A}}(\lambda) = negl(\lambda)$. Thus we see that HE satisfies the IND-CCA1 security. □

5 Candidates of the CCA1 Secure SHE

In our construction, we employ two building blocks. One is an IND-CCA1 secure SHE scheme and the other an IND-CPA secure multikey FHE scheme. In this section, we discuss concrete schemes for SHE.

As far as we know, the SHE scheme which Loftus et al. proposed is the only scheme which satisfies the IND-CCA1 security. However the security of their scheme is based on the knowledge assumption and this assumption is not standard. We construct an IND-CCA1 secure SHE scheme which is based on standard assumptions. To construct this, we use a multi-identity IBHE and a transformation like CHK transformation. Let Multi-Identity IBFHE = (Setup, KG, Enc, Dec, Eval) be a multi-identity IBHE. We assume that the identity space $\mathscr{I}$ of IBFHE is super polynomially large. Then we construct an SHE scheme SHE = (sKG, sEnc, sDec, sEval) as follows.

$\mathsf{sKG}(1^\lambda)$:

$(PP, MSK) \leftarrow \mathsf{Setup}(1^\lambda)$
pk := PP, sk := MSK
Output : (pk, sk)

$\mathsf{sEnc}(pk, m)$:

$id \leftarrow \mathscr{I}$
$c \leftarrow \mathsf{Enc}(pk, id, m)$
Output : $C := (id, c)$

$\mathsf{sDec}(sk, C)$:

Parse C to $(\{id_i, \ldots, id_d\}, c)$
For all $i \in [d]$, $sk_{id_i} \leftarrow \mathsf{KG}(id_i, sk)$
$\tilde{m} \leftarrow \mathsf{Dec}(sk_{id_1}, \ldots, sk_{id_d}, c)$
Output : $\tilde{m}$

$\mathsf{Eval}(pk, f, C_1, \ldots, C_t)$:

Parse C_i to (id_i, c_i)
$c^* \leftarrow \mathsf{Eval}(f, (id_1, c_1), \ldots, (id_t, c_t))$
Output : $C := (\{id_1, \ldots, id_t\}, c^*)$

We can prove that the resulting SHE satisfies the IND-CCA1 security and prove this later. However, the above scheme does not completely satisfy the compactness because the output of sEval contains the set of id thus the length of output ciphertext depends on the number of input ciphertexts. Though it does not satisfy the compactness, we can use it as a building block of our N-ary FHE scheme. The maximum number of the input ciphertexts of Eval of our N-ary FHE scheme is N. This N is determined in advance. Thus, the maximum number of the input ciphertexts of sEval is also N. As a result, we can bound the length of output ciphertexts of sEval by the polynomial of N.

Next, we prove this scheme satisfies the IND-CCA1 security.

Theorem 5.1 *Let* IBFHE *be IND-sID-CPA secure. Then* SHE *satisfies the IND-CCA1 security.*

Proof We prove this theorem via a sequence of games. Let $\mathscr{A}$ be an adversary that attacks the IND-CCA1 security of our scheme SHE. Now consider the following sequence of games.

Game 0 This is the IND-CCA1 game regarding our scheme SHE
Game 1 Same as Game 0 expect that at the beginning of the game the challenger chooses the identity id^*.
Game 2 Same as Game 1 expect that if $\mathscr{A}$ makes a decryption query (id^*, C), then the challenger returns $\perp$.
Game 3 Same as Game 2 expect that the challenger makes c^* as follows: $c^* \leftarrow \mathsf{Enc}(pk, id^*, 0^{|m_b|})$

For $i = 0, \ldots, 3$, we define the following events in Game i:

S_i: $\mathscr{A}$ succeeds in guessing the challenge bit, that is $b = b'$ occurs.

Then, we can estimate $\mathsf{Adv}^{indcca1}_{\mathsf{HE},\mathscr{A}}(\lambda)$ as follows.

$$\begin{aligned}\mathsf{Adv}^{indcca1}_{\mathsf{SHE},\mathscr{A}}(\lambda) &= |\Pr[S_0] - \frac{1}{2}| \\ &\leq \sum_{i=0}^{2} |\Pr[S_i] - \Pr[S_{i+1}]| + |\Pr[S_3] - \frac{1}{2}|.\end{aligned}$$

It is clear that $|\Pr[S_0] - \Pr[S_1]| = 0$ because the adversary cannot know when the challenger chooses id^*.

Below, we show that each term of the right side of inequality is negligible.

Lemma 5.2 *Let* IBFHE *have exponential identity space* $\mathscr{I}$*. Then* $|\Pr[S_1] - \Pr[S_2]| = \mathsf{negl}(1^\lambda)$.

Proof we define the following events in Game 1 and Game 2.

B_i: $\mathscr{A}$ makes decryption query (id^*, c).

Now Game 1 and Game 2 are the same game until the event B_1 and B_2 happen in respective game. Thus following equality holds:

$$|\Pr[S_1] - \Pr[S_2]| \leq \Pr[B_1] = \Pr[B_2].$$

Let the number of decryption queries made by $\mathscr{A}$ be q, where q is a polynomial. Then, the probability that the event B_i occurs is $\Pr[B_i] = \frac{q}{|\mathscr{I}|}$. Since IBFHE has super polynomially large identity space $\mathscr{I}$ and q is a polynomial, we have $\frac{q}{|\mathscr{I}|} = \mathsf{negl}(1^\lambda)$. From above, we see that $|\Pr[S_1] - \Pr[S_2]| = \mathsf{negl}(1^\lambda)$. □

Lemma 5.3 *Let* IBFHE *be IND-sID-CPA secure. Then* $|\Pr[S_2] - \Pr[S_3]| = \mathsf{negl}(\lambda)$.

Proof Using the adversary $\mathscr{A}$ that attacks HE, we construct an adversary $\mathscr{B}$ that attacks the IND-sID-CPA security of IBFHE. We describe the adversary $\mathscr{B}$ as follows.

Initialization $\mathscr{B}$ randomly chooses $id^* \leftarrow \mathscr{I}$, and send it to the challenger. On input public parameter PP, $\mathscr{B}$ sends $pk := PP$ to $\mathscr{A}$.

Decryption queries For a decryption query (C) from $\mathscr{A}$, $\mathscr{B}$ parses C to (id, c). Then $\mathscr{B}$ makes a key generation query (id) to the challenger, gets SK_id, and computes $m \leftarrow \mathsf{mDec}(SK_{id}, c)$. Finally, $\mathscr{B}$ returns m to $\mathscr{A}$.

Challenge query For a challenge query (m_0, m_1) from $\mathscr{A}$, $\mathscr{B}$ flips random coin $b \leftarrow \{0, 1\}$. Then $\mathscr{B}$ makes a challenge query $(m'_0 = m_b, m'_1 = 0^{|m_b|})$ and gets c^*. Finally, $\mathscr{B}$ returns $C = (id^*, c^*)$ to $\mathscr{A}$.

Final phase When $\mathscr{A}$ outputs $b' \in \{0, 1\}$, $\mathscr{B}$ outputs $\beta' = 1$ if $b = B'$, otherwise outputs $\beta' = 0$.

Let β be a challenge bit in the game between the challenger and $\mathscr{B}$. The ciphertext c^* is the output of $\mathsf{Enc}(PP, ID^*, m'_\beta)$ When $\beta = 0$, $\mathscr{B}$ perfectly simulates Game 2. On the other hand, when $\beta = 1$, $\mathscr{B}$ perfectly simulates Game 3. Therefore, we have

$$\begin{aligned}\mathsf{Adv}^{ind\text{-}sid\text{-}cpa}_{\mathsf{IBFHE},\mathscr{B}}(\lambda) &= \frac{1}{2}|\Pr[\beta' = 1|\beta = 0] - \Pr[\beta' = 1|\beta = 1]| \\ &= \frac{1}{2}|\Pr[S_2] - \Pr[S_3]|.\end{aligned}$$

Since IBFHE is IND-sID-CPA secure, we see that $|\Pr[S_2] - \Pr[S_3]| = \mathsf{negl}(\lambda)$. □

Finally, in Game 3, the challenge ciphertext does not contain information of challenge bit. So $|\Pr[S_3] - \frac{1}{2}| = 0$. From this equality and above lemma, we have $\mathsf{Adv}^{indcca1}_{\mathsf{SHE},\mathscr{A}}(\lambda) = \mathsf{negl}(\lambda)$. Thus we see that SHE satisfies the IND-CCA1 security. □

We discuss the building block IBFHE of our SHE. The concrete multi-identity IBFHE scheme is proposed by Clear and McGoldrick [6]. Their scheme is based on the LWE assumption and satisfies the IND-sID-CPA security in the random oracle model. Thus our SHE also satisfies the IND-CCA1 security under the same assumption.

6 Conclusion

In this paper, we have constructed an N-ary FHE scheme which satisfies the IND-CCA1 security. To construct our scheme, we have employed two building blocks. One is an IND-CCA1 secure SHE scheme and the other an IND-CPA secure multikey FHE scheme. In particular, we have proved the following two theorems.

Theorem 1 $N = \mathsf{poly}(\lambda)$. *Let* SHE *be an SHE scheme which can evaluate* f_{CSK} *and* MHE *a* N*-multikey FHE scheme. Then,* HE *is an N-ary fully homomorphic encryption scheme.*

Theorem 2 *Let* SHE *satisfy the IND-CCA1 security, and* MHE *satisfy the IND-CPA security. Then,* HE *satisfies the IND-CCA1 security.*

We have also discussed the building block SHE. We have constructed an IND-CCA1 secure SHE scheme by using general construction which converted a multi-identity IBHE to a IND-CCA1 secure SHE scheme. Formally, we have showed following theorem.

Theorem 3 *Let* IBFHE *be IND-sID-CPA secure. Then* SHE *satisfies the IND-CCA1 security.*

Our FHE scheme satisfies the IND-CCA1 security. However, the scheme has a restriction that it can only evaluate homomorphically a bounded number of ciphertexts. Thus, as a future work, it needs to construct "pure" FHE scheme which satisfies the IND-CCA1 security.

Acknowledgements The second author was supported by Grant-in-Aid for JSPS Research Fellow and JSPS KAKENHI Grant Number JP16J10322. The third author was supported by Input Output Hong Kong, I-System, Nomura Research Institute, NTT Secure Platform Laboratories, JST OPERA, and JSPS KAKENHI 16H01705.

References

1. A. Berkoff, F.-H. Liu, Leakage resilient fully homomorphic encryption, in *TCC*. Lecture Notes in Computer Science, vol. 8349 (Springer, Berlin, 2014), pp. 515–539
2. D. Boneh, R. Canetti, S. Halevi, J. Katz, Chosen-ciphertext security from identity-based encryption. SIAM J. Comput. **36**(5), 1301–1328 (2007)
3. Z. Brakerski, R. Perlman, Lattice-based fully dynamic multi-key FHE with short ciphertexts, in *CRYPTO (1)*. Lecture Notes in Computer Science, vol. 9814 (Springer, Berlin, 2016), pp. 190–213

4. Z. Brakerski, V. Vaikuntanathan, Efficient fully homomorphic encryption from (standard) LWE, in *FOCS* (IEEE Computer Society, New Jersey, 2011), pp. 97–106
5. R. Canetti, S. Halevi, J. Katz, Chosen-ciphertext security from identity-based encryption, in *EUROCRYPT*. Lecture Notes in Computer Science, vol. 3027 (Springer, Berlin, 2004), pp. 207–222
6. M. Clear, C. McGoldrick, Multi-identity and multi-key leveled FHE from learning with errors, in *CRYPTO (2)*. Lecture Notes in Computer Science, vol. 9216 (Springer, Berlin, 2015), pp. 630–656
7. J.-S. Coron, A. Mandal, D. Naccache, M. Tibouchi, Fully homomorphic encryption over the integers with shorter public keys, in *CRYPTO*. Lecture Notes in Computer Science, vol. 6841 (Springer, Berlin, 2011)
8. C. Gentry, Fully homomorphic encryption using ideal lattices, in *STOC* (ACM, 2009), pp. 169–178
9. C. Gentry, A. Sahai, B. Waters, Homomorphic encryption from learning with errors: Conceptually-simpler, asymptotically-faster, attribute-based, in *CRYPTO (1)*. Lecture Notes in Computer Science, vol. 8042 (Springer, Berlin, 2013), pp. 75–92
10. J. Loftus, A. May, N.P. Smart, F. Vercauteren, On CCA-secure somewhat homomorphic encryption, in *Selected Areas in Cryptography*. Lecture Notes in Computer Science, vol. 7118 (Springer, Berlin, 2011), pp. 55–72
11. A. López-Alt, E. Tromer, V. Vaikuntanathan, On-the-fly multiparty computation on the cloud via multikey fully homomorphic encryption, in *STOC* (ACM, 2012), pp. 1219–1234
12. R.L. Rivest, L. Adleman, M.L. Dertouzos, On data banks and privacy homomorphisms. Found. Secur. Comput. **4**(11), 169–180 (1978)
13. N.P. Smart, F. Vercauteren, Fully homomorphic encryption with relatively small key and ciphertext sizes, in *Public Key Cryptography*. Lecture Notes in Computer Science, vol. 6056 (Springer, Berlin, 2010), pp. 420–443

A Survey on Identity-Based Encryption from Lattices

Goichiro Hanaoka and Shota Yamada

Abstract Lattice-based cryptography is one of the most important topics in the area of cryptography, because of its (asymptotic) efficiency, post-quantum security, and expressiveness. In this survey, we provide an overview of lattice-based *identity-based encryption* (IBE), which is also an important topic in the area. In more details, we first introduce dual Regev public key encryption. Then, we change it to obtain Gentry–Peikert–Vaikuntanathan IBE, which is secure in the random oracle model. We then provide a framework for capturing constructions in the standard model. Then, by instantiating the framework, we show that we can capture the Cash–Hofheinz–Kiltz–Peikert and Agrawal–Boneh–Boyen scheme. Finally, we mention several recent works aiming at reducing parameters or tight security reductions.

Keywords Lattice-based cryptography · Identity-based encryption · Public key encryption

1 Preliminaries

The notion of identity-based encryption (IBE) is introduced by Shamir [42]. In IBE, a ciphertext and a private key are associated with arbitrary strings, such as an e-mail address, and the decryption is possible if and only if these strings are the same. It took nearly two decades for the first realizations of IBE [9, 22, 41] to appear. The first realization (with security against unbounded collusion) was given by Sakai, Ohgishi, and Kasahara [41] and Boneh and Franklin [9] independently. Both constructions are based on groups equipped with bilinear maps. The construction based on the composite-order groups, which is completely different algebraic structure

G. Hanaoka
National Institute of Advanced Industrial Science and Technology (AIST), Tokyo Waterfront Bio-IT Research Building, 2-4-7 Aomi, Koto-ku, Tokyo 135-0064, Japan
e-mail: hanaoka-goichiro@aist.go.jp

S. Yamada (✉)
Tokyo Waterfront Bio-IT Research Building, 2-4-7 Aomi, Koto-ku, Tokyo 135-0064, Japan
e-mail: yamada-shota@aist.go.jp

T. Takagi et al. (eds.), *Mathematical Modelling for Next-Generation Cryptography*, Mathematics for Industry 29, DOI 10.1007/978-981-10-5065-7_19

from bilinear groups, was also given by Cocks [22]. Since then, the constructions of identity-based encryption (IBE) has been one of the central topics in the study of cryptography. In the subsequent works, extension to the hierarchical IBE (HIBE) [24], the constructions in the standard model [10, 19, 23], the adaptively secure schemes [11, 44], and other constructions with various features [12, 15] have been proposed.

This survey provides an overview of identity-based encryption from lattices. Lattice-based cryptography is one of the most important topics in the cryptography because of the following reasons. First, it is expected to be secure against the quantum algorithms. Second, it is asymptotically fast and (relatively) simple, because encryption and decryption algorithms typically only involve linear algebraic operations. Finally, it can provide new functionalities that are not possible (e.g., attribute-based encryption for any circuit [13, 26] and fully homomorphic encryption [17]) using other algebraic structures such as elliptic curves or RSA modulus. We refer to [39] for an extensive survey on lattice-based cryptography in general.

The first lattice-based and identity-based encryption scheme was proposed by Gentry, Peikert, and Vaikuntanathan [25]. Then, the first realization in the standard model was given by Cash, Hofheinz, Kiltz, and Peikert [20]. Soon after, Agrawal, Boneh, and Boyen gave a more efficient construction [1, 2, 14]. In this survey, we introduce these schemes and then mention several recent schemes [5, 16, 33, 46, 50].

1.1 Notations

For a finite set S, $s \hookleftarrow S$ denotes an action of sampling a random element s from S. Throughout the paper, λ denotes the security parameter. We say that a function $f(\cdot) : \mathbb{N} \to \mathbb{R}_{\geq}$ is negligible when $\forall c \in \mathbb{N}\ \exists \lambda_0 \in \mathbb{N}\ \forall \lambda > \lambda_0\ f(\lambda) < 1/\lambda^c$.

1.2 Identity-Based Encryption

We adopt the standard definition of (anonymous) identity-based encryption. The following definition is from [33].

Syntax. Let $\mathcal{ID}$ and $\mathcal{M}$ be the ID and the message space of the scheme. If a collision resistant hash function $CRH : \{0,1\}^* \to \mathcal{ID}$ is available, one can use an arbitrary string as an identity. We note that we will set $\mathcal{M} = \{0,1\}$ in the description of the actual schemes. An IBE scheme is defined by the following four algorithms.

$\mathsf{Setup}(1^\lambda) \to (\mathsf{mpk}, \mathsf{msk})$:
The setup algorithm takes as input a security parameter 1^λ and outputs a master public key mpk and a master secret key msk.

$\mathsf{KeyGen}(\mathsf{mpk}, \mathsf{msk}, \mathsf{ID}) \to \mathsf{sk}_{\mathsf{ID}}$:
The key generation algorithm takes as input the master public key mpk, the master secret key msk, and an identity $\mathsf{ID} \in \mathcal{ID}$. It outputs a private key $\mathsf{sk}_{\mathsf{ID}}$.

We assume that ID is implicitly included in $\mathsf{sk}_{\mathsf{ID}}$.

$\mathsf{Encrypt}(\mathsf{mpk}, \mathsf{ID}, \mathsf{M}) \to C$:
The encryption algorithm takes as input a master public key mpk, an identity $\mathsf{ID} \in \mathcal{ID}$, and a message $\mathsf{M} \in \mathcal{M}$. It outputs a ciphertext C.

$\mathsf{Decrypt}(\mathsf{mpk}, \mathsf{sk}_{\mathsf{ID}}, C) \to \mathsf{M}$ or $\perp$:
The decryption algorithm takes as input the master public key mpk, a private key $\mathsf{sk}_{\mathsf{ID}}$, and a ciphertext C. It outputs the message M or $\perp$, which means that the ciphertext is not in a valid form.

Correctness. We require correctness of decryption: that is, for all λ, all $\mathsf{ID} \in \mathcal{ID}$, and all $\mathsf{M} \in \mathcal{M}$,

$$\Pr[\mathsf{Decrypt}(\mathsf{mpk}, \mathsf{sk}_{\mathsf{ID}}, \mathsf{Encrypt}(\mathsf{mpk}, \mathsf{ID}, \mathsf{M})) = \mathsf{M}] = 1 - \mathsf{negl}(\lambda)$$

holds, where the probability is taken over the randomness used in $(\mathsf{mpk}, \mathsf{msk}) \hookleftarrow \mathsf{Setup}(1^\lambda)$, $\mathsf{sk}_{\mathsf{ID}} \hookleftarrow \mathsf{KeyGen}(\mathsf{mpk}, \mathsf{msk}, \mathsf{ID})$, and $\mathsf{Encrypt}(\mathsf{mpk}, \mathsf{ID}, \mathsf{M})$.

Security. We now define the security for an IBE scheme Π. This security notion is defined by the following game between a challenger and an adversary $\mathcal{A}$.

- Setup. At the outset of the game, the challenger runs $\mathsf{Setup}(1^\lambda) \to (\mathsf{mpk}, \mathsf{msk})$ and gives mpk to $\mathcal{A}$.

- Phase 1. $\mathcal{A}$ may adaptively make key-extraction queries. If $\mathcal{A}$ submits $\mathsf{ID} \in \mathcal{ID}$ to the challenger, the challenger returns $\mathsf{sk}_{\mathsf{ID}} \leftarrow \mathsf{KeyGen}(\mathsf{mpk}, \mathsf{msk}, \mathsf{ID})$.

- Challenge Phase. At some point, $\mathcal{A}$ outputs a message $\mathsf{M} \in \mathcal{M}$ and an identity $\mathsf{ID}^\star \in \mathcal{ID}$, on which it wishes to be challenged. Then, the challenger picks a random coin $\mathsf{coin} \hookleftarrow \{0, 1\}$ and a random ciphertext $C \hookleftarrow \mathcal{C}$ from the ciphertext space. If $\mathsf{coin} = 0$, it runs $\mathsf{Encrypt}(\mathsf{mpk}, \mathsf{ID}^\star, \mathsf{M}) \to C^\star$ and gives the challenge ciphertext $C^\star$ to $\mathcal{A}$. If $\mathsf{coin} = 1$, it sets the challenge ciphertext as $C^\star = C$ and gives it to $\mathcal{A}$.

- Phase 2. After the challenge query, $\mathcal{A}$ may continue to make key-extraction queries, with the added restriction that $\mathsf{ID} \neq \mathsf{ID}^\star$.

- Guess. Finally, $\mathcal{A}$ outputs guess a $\widehat{\mathsf{coin}}$ for coin. The advantage of $\mathcal{A}$ is defined as

$$\mathsf{Adv}^{\mathsf{IBE}}_{\mathcal{A},\Pi} = \left| \Pr[\widehat{\mathsf{coin}} = \mathsf{coin}] - \frac{1}{2} \right|.$$

We say that Π is adaptively-anonymous secure, if the advantage of any PPT $\mathcal{A}$ is negligible. The term anonymous captures the fact the the ciphertext does not reveal the identity for which it was sent to.

We also consider a weaker security notion that we call selective security. To define the notion, we consider a modified version of the above game in which $\mathcal{A}$ should declare its target $\mathsf{ID}^\star$ at the beginning of the game, even before seeing mpk. We say that Π is selectively secure if the advantage of the adversary $\mathcal{A}$ is negligible in this modified game.

1.3 Preliminaries on Lattices

For positive integers q, m, n, a matrix $\mathbf{A} \in \mathbb{Z}_q^{n \times m}$, and a vector $\mathbf{u} \in \mathbb{Z}_q^m$, the m-dimensional integer lattices $\Lambda_q^{\perp}(\mathbf{A})$ and $\Lambda_q^{\mathbf{u}}(\mathbf{A})$ are defined as

$$\Lambda_q^{\perp}(\mathbf{A}) = \{\mathbf{e} \in \mathbb{Z}^m : \mathbf{A}\mathbf{e} = \mathbf{0} \mod q\}$$
$$\Lambda_q^{\mathbf{u}}(\mathbf{A}) = \{\mathbf{e} \in \mathbb{Z}^m : \mathbf{A}\mathbf{e} = \mathbf{u} \mod q\}.$$

Let $D_{\Lambda,\mathbf{c},\sigma}$ denote the discrete Gaussian distribution over Λ with center $\mathbf{c}$ and parameter σ. When $\mathbf{c}$ is omitted, we set $\mathbf{c} = \mathbf{0}$.

Matrix Norms. For a vector $\mathbf{u}$, we let $\|\mathbf{u}\|$ and $\|\mathbf{u}\|_\infty$ denote its ℓ_2 and ℓ_∞ norm respectively. For a matrix $\mathbf{R} \in \mathbb{Z}^{k \times m}$ we denote three matrix norms:

- $\|\mathbf{R}\|$ denotes the ℓ_2 length of the longest column of $\mathbf{R}$.
- $\|\mathbf{R}\|_{\text{GS}}$ denotes $\|\tilde{\mathbf{R}}\|$ where $\tilde{\mathbf{R}}$ is the result of applying Gram-Schmidt to the columns of $\mathbf{R}$.
- $\|\mathbf{R}\|_2$ is the operator norm of $\mathbf{R}$ defined as $\|\mathbf{R}\|_2 = \sup_{\|\mathbf{x}\|=1} \|\mathbf{R}\mathbf{x}\|$.

We have that the following lemma holds [1].

Lemma 1.1 *Let m, n, q be positive integers with $m > n$, $\mathbf{A} \in \mathbb{Z}_q^{n \times m}$ be a matrix, $\mathbf{u} \in \mathbb{Z}_q^n$ be a vector, $\mathbf{T_A}$ be a basis for $\Lambda_q^{\perp}(\mathbf{A})$, and $\sigma > \|\mathbf{T_A}\|_{\text{GS}} \cdot \omega(\sqrt{\log m})$. Then we have* $\Pr[\mathbf{x} \hookleftarrow D_{\Lambda_q^{\mathbf{u}}(\mathbf{A}),\sigma} : \|\mathbf{x}\| > \sqrt{m}\sigma] < \mathsf{negl}(n)$.

Learning with Errors. The learning with errors (LWE) problem was introduced by Regev [40] who showed that solving it on the average is as hard as (quantumly) solving several standard lattice problems in the worst case.

Definition 1.2 (*LWE*) For an integers n, $m = m(n)$, a prime integer $q = q(n) > 2$, an error distribution $\chi = \chi(n)$ over $\mathbb{Z}_q$, and an PPT algorithm $\mathscr{A}$, an advantage for the learning with errors problem $\mathsf{dLWE}_{n,m,q,\chi}$ of $\mathscr{A}$ is defined as follows:

$$\mathsf{Adv}_{\mathscr{A}}^{\mathsf{dLWE}_{n,m,q,\chi}} = |\Pr[\mathscr{A}(\mathbf{A}, \mathbf{s}^\top\mathbf{A} + \mathbf{x}^\top) \rightarrow 1] - \Pr[\mathscr{A}(\mathbf{A}, \mathbf{v}^\top) \rightarrow 1]|$$

where $\mathbf{A} \hookleftarrow \mathbb{Z}_q^{n \times m}$, $\mathbf{s} \hookleftarrow \mathbb{Z}_q^n$, $\mathbf{x} \hookleftarrow \chi^m$, $\mathbf{v} \hookleftarrow \mathbb{Z}_q^m$. We say that $\mathsf{dLWE}_{n,m,q,\chi}$ assumption holds if $\mathsf{Adv}_{\mathscr{A}}^{\mathsf{dLWE}_{n,m,q,\chi}}$ is negligible for all PPT $\mathscr{A}$.

Let $B = B(n) \in \mathbb{N}$. A family of distributions $\chi = \{\chi_n\}$ is called B-bounded if $\Pr[\chi \in [-B, B]] = 1$. For any constant $d > 0$ and sufficiently large q, Regev [40] through a quantum reduction showed that taking χ as a q/n^d-bounded (truncated) discretized Gaussian distribution, the $\mathsf{dLWE}_{n,m,q,\chi}$ problem is as hard as approximating the worst-case GapSVP to $n^{O(d)}$ factors. Here, "quantumly reduce" means that the quantum algorithm is used in a part of the reduction. If approximating the worst-case GapSVP is hard even for quantum algorithms, which is a widely believed assumption, the $\mathsf{dLWE}_{n,m,q,\chi}$ is also hard for quantum quantum algorithms. The latter trivially implies that $\mathsf{dLWE}_{n,m,q,\chi}$ is hard for classical algorithms. In subsequent works, (partial) dequantization of the Regev's reduction were achieved [18, 38]. More generally, let $\chi_{\max} < q$ be the bound on the noise distribution. The difficulty of the problem is measured by the ratio $q/\chi_{\max}$. This ratio is always bigger than 1 and the smaller it is the harder the problem. The problem appears to remain hard even when $q/\chi_{\max} < 2^{n^{\varepsilon}}$ for some fixed ε that is $0 < \varepsilon < 1/2$.

Leftover Hash Lemma. As observed in [1, 40], the following lemma is obtained as a corollary to the (general) leftover hash lemma.

Lemma 1.3 (Leftover Hash Lemma) *Let $q \in \mathbb{N}$ be an odd prime and let $m > (n+1)\log q + \omega(\log n)$. Let $\mathbf{R} \hookleftarrow \{-1, 1\}^{m\times m}$ and $\mathbf{A}, \mathbf{A}' \hookleftarrow \mathbb{Z}_q^{n\times m}$ be uniformly random matrices. Then the following distributions are* $\mathsf{negl}(n)$*-close:*

$$(\mathbf{A}, \mathbf{AR}) \approx (\mathbf{A}, \mathbf{A}').$$

2 Trapdoor Mechanism for LWE and Construction of Public Key Encryption

Let us start by recalling the definition of the LWE assumption. The LWE assumption says that the following distributions are computationally indistinguishable:

$$(\mathbf{A}, \mathbf{s}^\top\mathbf{A} + \mathbf{x}^\top) \approx (\mathbf{A}, \mathbf{v}^\top)$$

where $\mathbf{A} \hookleftarrow \mathbb{Z}_q^{n\times m}$, $\mathbf{s} \hookleftarrow \mathbb{Z}_q^n$, $\mathbf{v} \hookleftarrow \mathbb{Z}_q^m$, and $\mathbf{x}$ is sampled from the Gaussian distribution $D_{\mathbb{Z}^m,\alpha}$. Throughout this survey, we assume that $\mathbf{x}$ is a short vector and $\|\mathbf{x}\| \ll q$ holds. The LWE assumption implies that it is hard to recover $\mathbf{s}$ given $(\mathbf{A}, \mathbf{s}^\top\mathbf{A} + \mathbf{x}^\top)$. In other words, the function $f_{\mathbf{A}}(\mathbf{s}, \mathbf{x}) = \mathbf{s}^\top\mathbf{A} + \mathbf{x}^\top$ is a one-way function.

Our first goal is to obtain a public-key encryption scheme from the LWE assumption. Toward this goal, we would like to find a trapdoor $t_{\mathbf{A}}$ that allows us to efficiently invert $f_{\mathbf{A}}$. Actually, a matrix $\mathbf{T}_{\mathbf{A}} \in \mathbb{Z}^{m\times m}$ that is $\mathbf{A}\mathbf{T}_{\mathbf{A}} = \mathbf{0}$ *mod* q and $\|\mathbf{T}_{\mathbf{A}}\| \ll q/\alpha$ serves as such a trapdoor. (In other words, $\mathbf{T}_{\mathbf{A}}$ is a short basis of the lattice $\Lambda_q^{\perp}(\mathbf{A})$.) Concretely, we prove the following claim:

Claim *If $\|\mathbf{T}_{\mathbf{A}}\|$ is sufficiently short, there is an efficient algorithm that takes $\mathbf{T}_{\mathbf{A}}$ and $\mathbf{v}^\top = \mathbf{s}^\top\mathbf{A} + \mathbf{x}^\top$ as inputs and computes $\mathbf{s}$ and $\mathbf{x}$ efficiently.*

Proof We first compute $\mathbf{v}^\top\mathbf{T}_{\mathbf{A}}$ *mod* q. We have

$$\mathbf{v}^\top\mathbf{T}_{\mathbf{A}} \equiv (\mathbf{s}^\top\mathbf{A} + \mathbf{x}^\top)\mathbf{T}_{\mathbf{A}} \equiv \mathbf{s}^\top \underbrace{\mathbf{A}\mathbf{T}_{\mathbf{A}}}_{\equiv \mathbf{0} \ mod\ q} + \underbrace{\mathbf{x}^\top\mathbf{T}_{\mathbf{A}}}_{short} \equiv \mathbf{x}^\top\mathbf{T}_{\mathbf{A}} \quad mod\ q.$$

In the above, each coefficient of $\mathbf{x}^\top\mathbf{T}_{\mathbf{A}}$ is small because of our assumption that $\|\mathbf{x}\|$ and $\|\mathbf{T}_{\mathbf{A}}\|$ are small. In particular, it holds that $\mathbf{x}^\top\mathbf{T}_{\mathbf{A}}$ over $\mathbb{Z}$ corresponds to $\mathbf{x}^\top\mathbf{T}_{\mathbf{A}}$ *mod* q, when we regard an element in $\mathbb{Z}_q$ as an element in $\{-(q-1)/2, -(q-1)/2+1, \ldots, -1, 0, 1, \ldots, (q-1)/2\}$. Namely, we have $(\mathbf{v}^\top \mathbf{T}_\mathbf{A}\ mod\ q) = \mathbf{x}^\top \mathbf{T}_\mathbf{A}$ over $\mathbb{Z}$. Then, since $\mathbf{T}_A$ is invertible over $\mathbb{Z}$, we can recover $\mathbf{x}$ by computing

$$(\mathbf{v}^\top \mathbf{T}_\mathbf{A}\ mod\ q) \cdot \mathbf{T}_\mathbf{A}^{-1} = \mathbf{x}^\top.$$

Given $\mathbf{x}$, we can compute $\mathbf{s}$ from $\mathbf{v}$ by solving linear equations over $mod\ q$. This completes the proof of the claim. □

2.1 Public-Key Encryption from LWE

Ajtai [3] proved that there is an efficient way to generate uniformly random matrix $\mathbf{A} \in \mathbb{Z}_q^{n \times m}$ along with trapdoor $\mathbf{T}_\mathbf{A}$. The lemma is quite useful for constructing lattice PKE and beyond, and is improved by several subsequent works [4, 36].

Lemma 2.1 *There is an efficient randomized algorithm* $\mathsf{TrapGen}(1^n, 1^m, q) \to (\mathbf{A}, \mathbf{T}_\mathbf{A})$ *that, when* $m \geq 6n\lceil \log q \rceil$*, outputs a full rank matrix* $\mathbf{A} \in \mathbb{Z}_q^{n \times m}$ *and a basis* $\mathbf{T}_\mathbf{A} \in \mathbb{Z}^{m \times m}$ *for* $\Lambda_q^\perp(\mathbf{A})$ *such that* $\mathbf{A}$ *is* $\mathsf{negl}(n)$*-close to uniform and* $\|\mathbf{T}_\mathbf{A}\|_{\mathrm{GS}} = O(\sqrt{n \log q})$ *with all but negligible probability in* n.

Given the above algorithm, we can construct PKE from the LWE assumption as follows:

DUAL REGEV ENCRYPTION WITH FULL TRAPDOOR SECRET KEY.

$\mathsf{Setup}(1^\lambda) \to (\mathsf{pk}, \mathsf{sk})$:
The setup algorithm runs $\mathsf{TrapGen}(1^n, 1^m, q)$ to obtain $(\mathbf{A}, \mathbf{T}_\mathbf{A})$. It then also samples $\mathbf{u} \hookleftarrow \mathbb{Z}_q^n$. Finally it outputs the public key $\mathsf{pk} = (\mathbf{A}, \mathbf{u})$ and the secret key $\mathsf{sk} = \mathbf{T}_\mathbf{A}$. Here, n, m, and q would be chosen depending on λ. Typical chose would be $n = \Theta(\lambda)$, $m = \Theta(n \log n)$, $q = n^c$ for some constant c.

$\mathsf{Encrypt}(\mathsf{pk}, \mathsf{M}) \to C$:
To encrypt the message $\mathsf{M} \in \{0, 1\}$, it first samples $\mathbf{s} \hookleftarrow \mathbb{Z}_q^n$, $x_0 \hookleftarrow D_{\mathbb{Z},\alpha}$, and $\mathbf{x} \hookleftarrow D_{\mathbb{Z}^m,\alpha}$. The final output is the ciphertext

$$C = \left(c_0 = \mathbf{s}^\top \mathbf{u} + x_0 + \mathsf{M} \cdot \lceil q/2 \rceil, \mathbf{c}_1^\top = \mathbf{s}^\top \mathbf{A} + \mathbf{x}^\top\right).$$

$\mathsf{Decrypt}(\mathsf{pk}, \mathsf{sk}, C) \to \mathsf{M}$ or $\perp$:
To decrypt the ciphertext, it first recovers $\mathbf{s}$ from $\mathbf{c}_1$ using $\mathbf{T}_\mathbf{A}$. Then, it computes $c_0 - \mathbf{c}_1^\top \mathbf{u} = x_0 + \lceil \mathsf{M} \cdot q/2 \rceil$. If the value is closer to $q/2$ than 0 (over $\mathbb{Z}_q$), it output 1 and otherwise 0.

The correctness of the scheme follows when x_0 is sufficiently small. We next explain that the scheme is semantically secure. Due to the LWE assumption, we have the following distributions are computationally indistinguishable:

$$\big([\mathbf{u}\|\mathbf{A}], \mathbf{s}^\top[\mathbf{u}\|\mathbf{A} + \mathbf{x}^\top]\big) \approx \big(\mathbf{A}, [v_0\|\mathbf{v}^\top]\big)$$

where $\mathbf{A} \hookleftarrow \mathbb{Z}_q^{n\times m}$, $\mathbf{s} \hookleftarrow \mathbb{Z}_q^n$, $\mathbf{x} \hookleftarrow D_{\mathbb{Z}^m,\alpha}$, $v_0 \hookleftarrow \mathbb{Z}_q$, and $\mathbf{v} \hookleftarrow \mathbb{Z}_q^m$. The above holds even if we replace the distribution of $\mathbf{A}$ to be chosen by $(\mathbf{A}, \mathbf{T_A}) \hookleftarrow \mathsf{TrapGen}(1^n, 1^m, q)$ by Lemma 2.1. This implies that the following distributions are computationally indistinguishable:

$$\begin{aligned}\big(c_0 = \mathbf{s}^\top\mathbf{u} + x_0 + \mathsf{M}\cdot\lceil q/2\rceil,\quad \mathbf{c}_1^\top = \mathbf{s}^\top\mathbf{A} + \mathbf{x}^\top\big) &\approx (v_0 + \lceil \mathsf{M}q/2\rceil, \mathbf{v}) \\ &\approx (v_0, \mathbf{v}),\end{aligned}$$

where the first and second distributions are indistinguishable from the LWE assumption, and the second and the third are the same distribution. We can observe that the last distribution is independent of the value of the message M. This implies that the scheme is secure.

3 Identity-Based Encryption Scheme from LWE in the Random Oracle Model

The first IBE scheme from lattices is proposed by Gentry, Peikert, and Vaikuntanathan [25] in the random oracle model. In this section, we introduce their result. To do so, we first introduce the dual Regev PKE [25] and then modify it to obtain IBE.

3.1 Dual Regev Public-Key Encryption

We first state the following lemma [25].

Lemma 3.1 *For all but negligible fraction of* $\mathbf{A} \in \mathbb{Z}_q^{n\times m}$, *the following distributions are statistically close if* $\sigma \geq \omega(\sqrt{\log m})$*:*

$$\mathbf{Ae} \approx \mathbf{u} \tag{1}$$

where $\mathbf{e} \hookleftarrow D_{\mathbb{Z}^m,\sigma}$ *and* $\mathbf{u} \hookleftarrow \mathbb{Z}_q^n$.

The lemma can be seen as an analogue of the leftover hash lemma (Lemma 1.3), where the input distribution to the pairwise independent hash function $\mathbf{x} \mapsto \mathbf{Ax}$ is replaced by Gaussian distribution with (typically small, but) sufficiently large σ.

We now describe the dual Regev encryption scheme proposed in [25].

DUAL REGEV ENCRYPTION.

$\mathsf{Setup}(1^\lambda) \to (\mathsf{pk}, \mathsf{sk})$:

The setup algorithm samples $\mathbf{A} \hookleftarrow \mathbb{Z}_q^{n\times m}$ and $\mathbf{e} \hookleftarrow D_{\mathbb{Z},\sigma}$. Then it sets $\mathbf{u} = \mathbf{Ae} \bmod q$. Finally it outputs the public key $\mathsf{pk} = (\mathbf{A}, \mathbf{u})$ and the secret key $\mathsf{sk} = \mathbf{e}$.

$\mathsf{Encrypt}(\mathsf{pk}, \mathsf{M}) \to C$:

To encrypt the message $\mathsf{M} \in \{0, 1\}$, it first samples $\mathbf{s} \hookleftarrow \mathbb{Z}_q^n$, $x_0 \hookleftarrow D_{\mathbb{Z},\alpha}$, and $\mathbf{x} \hookleftarrow D_{\mathbb{Z}^m,\alpha}$. The final output is the ciphertext

$$C = \left(c_0 = \mathbf{s}^\top \mathbf{u} + x_0 + \mathsf{M} \cdot \lceil q/2 \rceil, \mathbf{c}_1^\top = \mathbf{s}^\top \mathbf{A} + \mathbf{x}^\top\right).$$

$\mathsf{Decrypt}(\mathsf{pk}, \mathsf{sk}, C) \to \mathsf{M}$ or $\perp$:

To decrypt the ciphertext, it first computes $w = c_0 - \mathbf{c}_1^\top \mathbf{e}$ *mod* q. If the value is closer to $q/2$ than 0 (over $\mathbb{Z}_q$), it output 1 and otherwise 0. (We note that the decryption algorithm outputs $\perp$ when the inputs are not in the specified form, e.g., $\mathsf{pk} \notin \mathbb{Z}_q^{n\times m} \times \mathbb{Z}_q^n$.)

Then, we show the correctness of the scheme. When the cryptosystem is operated as specified, we have during decryption,

$$w = c_0 - \mathbf{c}_1^\top \cdot \mathbf{e} = \mathsf{M} \cdot \lceil q/2 \rceil + \underbrace{x_0 - \mathbf{x}^\top \cdot \mathbf{e}}_{\text{error term}}.$$

Since $\mathbf{x}$ and $\mathbf{e}$ are short vectors, their inner product is also a small value. The error term is much smaller than q, if we choose q sufficiently large. Concretely, we can correctly recover the message from w if the error term is less than $q/4$ with overwhelming probability.

It is straightforward to see that the security of the above scheme follows that of the scheme in Sect. 2.1.

3.2 Gentry–Peikert–Vaikuntanathan Identity-Based Encryption Scheme

The following lemma is at the heart of the construction of the GPV IBE scheme.

Lemma 3.2 (GPV sampling) *There is a PPT algorithm that given* $\mathbf{A} \in \mathbb{Z}_q^{n\times m}$, $\mathbf{T_A} \in \mathbb{Z}^{m\times m}$, $\sigma > \|\mathbf{T_A}\|_{\mathrm{GS}} \cdot \omega(\sqrt{\log n})$, *and* $\mathbf{u} \in \mathbb{Z}_q^n$ *and outputs a sample from a distribution statistically close to* $D_{\Lambda^{\mathbf{u}}(\mathbf{A}),\sigma}$. *We denote the algorithm by* $\mathsf{GPVSample}$.

The lemma intuitively says that given short enough trapdoor $\mathbf{T_A}$ for $\mathbf{A}$, we can sample short $\mathbf{e} \in \mathbb{Z}^m$ satisfying $\mathbf{Ae} = \mathbf{u}$ *mod* q. Furthermore, the distribution of $\mathbf{e}$ does not depend on the trapdoor $\mathbf{T_A}$.

Then we proceed to the construction of the IBE scheme.

GPV IBE SCHEME.

$\mathsf{Setup}(1^\lambda) \to (\mathsf{mpk}, \mathsf{sk})$:

The setup algorithm runs $\mathsf{TrapGen}(1^n, 1^m, q) \to (\mathbf{A}, \mathbf{T_A})$. It also chooses some hash function $\mathcal{H} : \mathcal{ID} \to \mathbb{Z}_q^n$. The hash function will be modeled as the random oracle in the security proof. Finally it outputs the master public key $\mathsf{mpk} = (\mathbf{A}, \mathcal{H})$ and the master secret key $\mathsf{msk} = \mathbf{T_A}$.

$\mathsf{KeyGen}(\mathsf{mpk}, \mathsf{msk}, \mathsf{ID}) \to \mathsf{sk}_{\mathsf{ID}}$:

To generate a secret key for ID, it first computes $\mathbf{u}_{\mathsf{ID}} = \mathscr{H}(\mathsf{ID}) \in \mathbb{Z}_q^n$. Then, it runs

$$\mathbf{e} \hookleftarrow \mathsf{GPVSample}(\mathbf{A}, \mathbf{T_A}, \sigma, \mathbf{u}). \tag{2}$$

The secret key is $\mathsf{sk}_{\mathsf{ID}} = \mathbf{e}$. Note that $\mathbf{e}$ is a short vector satisfying $\mathbf{Ae} = \mathbf{u}$.

$\mathsf{Encrypt}(\mathsf{mpk}, \mathsf{ID}, \mathsf{M}) \to C$:

To encrypt the message $\mathsf{M} \in \{0, 1\}$, it first samples $\mathbf{s} \hookleftarrow \mathbb{Z}_q^n$, $x_0 \hookleftarrow D_{\mathbb{Z},\alpha}$, and $\mathbf{x} \hookleftarrow D_{\mathbb{Z}^m,\alpha}$. It then also computes $\mathbf{u}_{\mathsf{ID}} = \mathscr{H}(\mathsf{ID}) \in \mathbb{Z}_q^n$. The final output is the ciphertext

$$C = \left(c_0 = \mathbf{s}^\top \mathbf{u}_{\mathsf{ID}} + x_0 + \mathsf{M} \cdot \lceil q/2 \rceil, \mathbf{c}_1^\top = \mathbf{s}^\top \mathbf{A} + \mathbf{x}^\top\right).$$

$\mathsf{Decrypt}(\mathsf{mpk}, \mathsf{sk}_{\mathsf{ID}}, C) \to \mathsf{M}$ or $\perp$:

To decrypt the ciphertext, it first computes $w = c_0 - \mathbf{c}_1^\top \mathbf{e} \mod q$. If the value is closer to $q/2$ than 0 (over $\mathbb{Z}_q$), it output 1 and otherwise 0.

The correctness of the scheme can be shown similarly to that of the security of the dual regev encryption scheme in Sect. 3.1. We prove adaptive security for the scheme in the random oracle model.

Theorem 3.3 *The above scheme is adaptively secure under the LWE assumption in the random oracle model.*

Proof Let us assume an adversary $\mathscr{A}$ who breaks the security of the above IBE scheme. We construct an adversary $\mathscr{B}$ against the dual Regev encryption scheme as follows. Without loss of generality, we assume that the adversary makes a hash query for an identity ID before making a key extraction query for the (same) identity ID. We also assume that the adversary $\mathscr{A}$ makes a hash query for $\mathsf{ID}^\star$. Otherwise, $\mathbf{u}_{\mathsf{ID}^\star}$ is information theoretically hidden from $\mathscr{A}$ and thus the message to be encrypted is also information theoretically hidden. We denote the number of the hash queries and the key extraction queries that $\mathscr{A}$ makes by Q_H and Q_K respectively. We also denote the identity for which the i-th hash query is made by $\mathsf{ID}^{(i)}$.

Master Public Key. First, the adversary $\mathscr{B}$ is given the master public key of the dual Regev encryption scheme $(\mathbf{A}, \mathbf{u})$. Then, $\mathscr{B}$ sets $\mathsf{mpk} = \mathbf{A}$ and returns it to $\mathscr{A}$. $\mathscr{B}$ also picks $i^\star \hookleftarrow Q_H$ and keeps it secretly.

Answering the Hash Queries. For the i-th query $\mathsf{ID}^{(i)}$ that $\mathscr{A}$ makes, $\mathscr{B}$ proceeds as follows.

- If $i = i^\star$, $\mathscr{A}$ sets $\mathscr{H}(\mathsf{ID}^{(i^\star)}) = \mathbf{u}$. Namely, we embed the problem instance and expect that $\mathsf{ID}^\star = \mathsf{ID}^{(i^\star)}$. Since $\mathbf{u} \hookleftarrow \mathbb{Z}_q^n$, $\mathscr{B}$ correctly simulates the hash value.
- If $i \neq i^\star$, $\mathscr{A}$ picks $\mathbf{e}^{(i)} \hookleftarrow D_{\mathbb{Z}^m,\sigma}$ and set $\mathscr{H}(\mathsf{ID}) = \mathbf{Ae}^{(i)}$. By Lemma 3.1, the distribution of $\mathscr{H}(\mathsf{ID})$ is statistically close to the uniform distribution over $\mathbb{Z}_q^n$ and thus the distribution is correct as desired.

Answering the Key Extraction Queries. When $\mathscr{A}$ makes a key extraction query for the identity ID, $\mathscr{B}$ first searches for an index i such that $\mathsf{ID}^{(i)} = \mathsf{ID}$. There exists such i because of our assumption. If $i = i^\star$, $\mathscr{B}$ aborts. Otherwise, it returns $\mathbf{e}^{(i)}$.

Answering the Challenge Queries. When $\mathscr{A}$ makes the challenge query for $\mathsf{ID}^\star$ and $\mathsf{M}_0, \mathsf{M}_1$, it queries $(\mathsf{M}_0, \mathsf{M}_1)$ for its challenger. Then, $\mathscr{B}$ is given the challenge ciphertext $C^\star$. Then, $\mathscr{B}$ passes $C^\star$ to $\mathscr{A}$. If $\mathscr{H}(\mathsf{ID}^\star) = \mathbf{u}$ (namely, if $i^\star = i$), $\mathscr{B}$ correctly simulates the challenge ciphertext.

Guess. Finally, $\mathscr{A}$ outputs $\widehat{\mathsf{coin}}$ as its guess for coin. $\mathscr{B}$ outputs the same bit.

The simulation succeeds when $\mathscr{B}$ correctly guesses $i^\star$ such that $\mathsf{ID}^{(i^\star)} = \mathsf{ID}^\star$. Therefore, the simulation succeeds with probability $1/Q_H$. By the standard analysis, we can show that the advantage of $\mathscr{B}$ is $1/Q_H$ of $\mathscr{A}$. □

4 Identity-Based Encryption Scheme from LWE in the Standard Model

The GPV IBE scheme is only proven secure in the random oracle model. The first realization of the lattice IBE scheme in the standard model was given by Cash et al. [20]. Soon after the work, more efficient constructions were given by Agrawal et al. [1]. In this section, we provide a template for lattice IBE schemes that encompasses these schemes. The template was first described in [46] in an informal manner. Soon after, the template was made formal and rigorous by Zhang et al. [50] who introduced the notion of programmable hash functions [30] in the lattice settings.

To describe the template, we introduce the following lemmas.

Lemma 4.1 ([36]) *Let $m > n\lceil \log q \rceil$. Then there is a fixed full-rank matrix $\mathbf{G} \in \mathbb{Z}_q^{n\times m}$ such that the lattice $\Lambda_q^\perp(\mathbf{G})$ has a publicly known basis $\mathbf{T_G} \in \mathbb{Z}^{m\times m}$ with $\|\mathbf{T_G}\|_{\mathrm{GS}} \le \sqrt{5}$.*

Lemma 4.2 *There are two efficient algorithms* SampleLeft *and* SampleRight *with the following properties.*

- *([20]):* $\mathsf{SampleLeft}(\mathbf{A}, \mathbf{F}, \mathbf{u}, \mathbf{T_A}, \sigma) \to \mathbf{e}$
  a randomized algorithm that, given a full rank matrix $\mathbf{A} \in \mathbb{Z}_q^{n\times m}$, a matrix $\mathbf{F} \in \mathbb{Z}_q^{n\times m}$, a vector $\mathbf{u} \in \mathbb{Z}_q^n$, a basis $\mathbf{T_A}$ for $\Lambda_q^\perp(\mathbf{A})$, and a Gaussian parameter $\sigma > \|\mathbf{T_A}\|_{\mathrm{GS}} \cdot \omega(\sqrt{\log m})$, outputs a vector $\mathbf{e} \in \mathbb{Z}^{2m}$ sampled from a distribution which is $\mathsf{negl}(n)$-close to $D_{\Lambda_q^{\mathbf{u}}([\mathbf{A}\|\mathbf{F}]),\sigma}$.
- *([1]):* $\mathsf{SampleRight}(\mathbf{A}, \mathbf{G}, \mathbf{R}, \mathbf{Y}, \mathbf{u}, \mathbf{T_G}, \sigma) \to \mathbf{e}$ *where $\mathbf{F} = \mathbf{AR} + y\mathbf{G}$*
  a randomized algorithm that, given a full rank matrix $\mathbf{A}, \mathbf{G} \in \mathbb{Z}_q^{n\times m}$, an invertible matrix $\mathbf{Y} \in \mathbb{Z}_q^{n\times n}$, a matrix $\mathbf{R} \in \mathbb{Z}^{m\times m}$, a vector $\mathbf{u} \in \mathbb{Z}_q^n$, a basis $\mathbf{T_G}$ for $\Lambda_q^\perp(\mathbf{G})$, and a Gaussian parameter $\sigma > \|\mathbf{T_G}\|_{\mathrm{GS}} \cdot \|\mathbf{R}\|_2 \cdot \omega(\sqrt{\log m})$ outputs a vector $\mathbf{e} \in \mathbb{Z}^{2m}$ sampled from a distribution which is $\mathsf{negl}(n)$-close to $D_{\Lambda_q^{\mathbf{u}}([\mathbf{A}\|\mathbf{F}]),\sigma}$.

The above lemma says that there are two kinds of trapdoor for a matrix in the form of $[\mathbf{A}\|\mathbf{AR}+\mathbf{XG}]$ when $\mathbf{X}$ is invertible. Namely, $\mathbf{T_A}$ and $\mathbf{R}$. These trapdoors allows us to sample a short vector $\mathbf{e}$ such that $[\mathbf{A}\|\mathbf{AR}+\mathbf{XG}]\mathbf{e}=\mathbf{u}$ for any $\mathbf{u}$. Notably, the distribution of $\mathbf{e}$ does not depend on whether we use $\mathbf{T_A}$ or $\mathbf{R}$.

4.1 Template IBE Scheme

Now we provide the template for the scheme in the following.

$\mathsf{Setup}(1^\lambda) \to (\mathsf{mpk}, \mathsf{sk})$:

The setup algorithm runs $\mathsf{TrapGen}(1^n, 1^m, q) \to (\mathbf{A}, \mathbf{T_A})$. It also chooses several other matrices $\mathbf{B}_1, \ldots, \mathbf{B}_\ell \hookleftarrow \mathbb{Z}_q^{n\times m}$. These matrices define a hash function $\mathscr{H}: \mathscr{I}\mathscr{D} \to \mathbb{Z}_q^{n\times m'}$, which can be efficiently computed from $\mathbf{B}_1, \ldots, \mathbf{B}_\ell$. How to define the hash function varies in each schemes. It also samples $\mathbf{u} \hookleftarrow \mathbb{Z}_q^n$ and outputs the master public key $\mathsf{mpk} = (\mathbf{A}, \mathbf{B}_1, \ldots, \mathbf{B}_\ell, \mathbf{u})$ and the master secret key $\mathsf{msk} = \mathbf{T_A}$.

$\mathsf{KeyGen}(\mathsf{mpk}, \mathsf{msk}, \mathsf{ID}) \to \mathsf{sk_{ID}}$:

To generate a secret key for ID, it first computes $\mathbf{A}_{\mathsf{ID}} = \mathscr{H}(\mathsf{ID}) \in \mathbb{Z}_q^{n\times(m+m')}$. Then, it runs

$$\mathsf{SampleLeft}(\mathbf{A}, \mathscr{H}(\mathsf{ID}), \mathbf{u}, \mathbf{T_A}, \sigma) \to \mathbf{e} \tag{3}$$

to obtain an short vector $\mathbf{e}$ that satisfies $[\mathbf{A}\|\mathscr{H}(\mathsf{ID})]\mathbf{e} = \mathbf{u}$. The secret key is $\mathsf{sk_{ID}} = \mathbf{e}$.

$\mathsf{Encrypt}(\mathsf{mpk}, \mathsf{ID}, \mathsf{M}) \to C$:

To encrypt the message $\mathsf{M} \in \{0, 1\}$, it first samples $\mathbf{s} \hookleftarrow \mathbb{Z}_q^n$, $x_0 \hookleftarrow D_{\mathbb{Z},\alpha}$, and $\mathbf{x}_1, \mathbf{x}_2 \hookleftarrow D_{\mathbb{Z}^m,\alpha}$. It then also computes $\mathscr{H}(\mathsf{ID}) \in \mathbb{Z}_q^{n\times m}$. The final outputs are the ciphertext

$$C = \left(c_0 = \mathbf{s}^\top\mathbf{u} + x_0 + \mathsf{M}\cdot\lceil q/2\rceil, \mathbf{c}_1^\top = \mathbf{s}^\top[\mathbf{A}\|\mathscr{H}(\mathsf{ID})] + [\mathbf{x}_1^\top\|\mathbf{x}_2^\top]\right).$$

$\mathsf{Decrypt}(\mathsf{mpk}, \mathsf{sk_{ID}}, C) \to \mathsf{M}$ or $\bot$:

To decrypt the ciphertext, it first computes $w = c_0 - \mathbf{c}_1^\top\mathbf{e} \bmod q$. If the value is closer to $q/2$ than 0 (over $\mathbb{Z}_q$), it output 1 and otherwise 0.

Note that the critical difference from the GPV scheme in the above scheme is that we choose the matrix $[\mathbf{A}\|\mathscr{H}(\mathsf{ID})]$ depending on the value of the identity ID instead of the vector $\mathbf{u}$.

The correctness of the scheme can be shown as in the dual Regev scheme, if we choose the parameters appropriately.

SECURITY. We next argue the security of the template scheme. We would like to prove the security from the LWE assumption. We are given the problem instance of the LWE $\mathbf{A}', \mathbf{v}$ where $\mathbf{A}' \hookleftarrow \mathbb{Z}_q^{n\times(m+1)}$ and $\mathbf{v} \hookleftarrow \mathbb{Z}_q^{m+1}$ or $\mathbf{v}^\top = \mathbf{s}^\top \mathbf{A}' + \mathbf{x}'$ for small Gaussian noise $\mathbf{x}'$.

To prove the security, we first parse $\mathbf{A}'$ to be $\mathbf{A}' \to [\mathbf{u}|\mathbf{A}]$. We embed $\mathbf{A}$ and $\mathbf{u}$ into the master public key. To simulate the other parts of the master public key, we pick random $\mathbf{Y}_0, \mathbf{Y}_1, \ldots, \mathbf{Y}_\ell \in \mathbb{Z}_q^{n\times n}$ from certain distribution. Then, the reduction algorithm picks $\mathbf{R}_0, \mathbf{R}_i \hookleftarrow \{-1, 1\}^{m\times m}$ and embeds these values into the public parameters as

$$\mathbf{B}_i = \mathbf{A}\mathbf{R}_i + \mathbf{Y}_i\mathbf{G} \tag{4}$$

for $i \in [1, \ell]$. It can be seen that the distribution of each $\mathbf{B}_i$ is negligibly close to uniform (and thus the distribution in the real world) by Lemma 1.3.

We require that $\mathscr{H}(\mathsf{ID})$ can be represented as

$$\mathscr{H}(\mathsf{ID}) = \mathbf{A}\mathbf{R}_{\mathsf{ID}} + \mathbf{X}_{\mathsf{ID}}\mathbf{G} \tag{5}$$

where $\mathbf{R}_{\mathsf{ID}}$ and $\mathbf{X}_{\mathsf{ID}}$ can be efficiently computed from $\mathsf{ID}, \mathbf{R}_1, \ldots, \mathbf{R}_\ell$, and $\mathbf{Y}_1, \ldots, \mathbf{Y}_\ell$. Furthermore, we have to choose $\mathbf{Y}_1, \ldots, \mathbf{Y}_\ell$ so that we have

$$\mathbf{X}_{\mathsf{ID}} = \begin{cases} \mathbf{0} & \text{If } \mathsf{ID} = \mathsf{ID}^\star \\ \text{invertible} & \text{If } \mathsf{ID} = \mathsf{ID}^{(j)} \text{ for some } j \in [Q] \end{cases} \tag{6}$$

with noticeable probability,[1] where $\mathsf{ID}^\star$ is the challenge identity and $\mathsf{ID}_1, \ldots, \mathsf{ID}_Q$ are the identities for which key extraction query was made.

We claim that if the Eq. 6 is satisfied, the simulation can be successful. First, key extraction queries for $\mathsf{ID} = \mathsf{ID}^{(j)}$ for $j \in [Q]$ can be handled using SampleRight. The simulation of the challenge ciphertext is a bit more complicated. We first parse $\mathbf{v}$ as $\mathbf{v} \to [v_0 \| \mathbf{v}_1]$. Recall that our task is to distinguish whether $v_0 \hookleftarrow \mathbb{Z}_q$ and $\mathbf{v}_1 \hookleftarrow \mathbb{Z}_q^m$ or $v_0 = \mathbf{s}^\top \mathbf{u} + x_0'$ and $\mathbf{v}_1 = \mathbf{s}^\top \mathbf{A} + \mathbf{x}_1'$, where we parse $\mathbf{x}'$ as $\mathbf{x}' = [x_0' \| \mathbf{x}_1']$. We set $c_0 = v_0 + \mathsf{M}_{\mathsf{coin}} \cdot \lceil q/2 \rceil$. We can see that $c_0 \hookleftarrow \mathbb{Z}_q$ if $\mathbf{v} \hookleftarrow \mathbb{Z}_q^{m+1}$ and the distribution of c_0 is the same as the real world if $\mathbf{A}', \mathbf{v}$ is the LWE sample. The more difficult part is to simulate $\mathbf{c}_1$. Let us set $\mathbf{c}_1^\top = \mathbf{v}^\top [\mathbf{I}_m \| \mathbf{R}_{\mathsf{ID}}]$. If $\mathbf{v}$ is random, $\mathbf{c}_1$ is random by the leftover hash lemma. On the other hand, if $\mathbf{v}$ is taken from the LWE sample, we have

$$\begin{aligned} \mathbf{c}_1^\top &= \mathbf{v}^\top [\mathbf{I}_m \| \mathbf{R}_{\mathsf{ID}^\star}] \\ &= (\mathbf{s}^\top \mathbf{A} + \mathbf{x}_1') \cdot [\mathbf{I}_m \| \mathbf{R}_{\mathsf{ID}^\star}] \\ &= \mathbf{s}^\top [\mathbf{A} \| \mathbf{A}\mathbf{R}_{\mathsf{ID}^\star}] + \mathbf{x'}_1^\top [\mathbf{I}_m \| \mathbf{R}_{\mathsf{ID}}] \\ &= \mathbf{s}^\top [\mathbf{A} \| \mathbf{B}_{\mathsf{ID}^\star}] + \underbrace{\mathbf{x'}_1^\top [\mathbf{I}_m \| \mathbf{R}_{\mathsf{ID}^\star}]}_{\text{short}}. \end{aligned}$$

[1] We note that even if this probability is noticeable, the security proof requires additional complication in the simulation and the computation of the advantage [7, 44].

Here, ${\mathbf{x}'_1}^\top[\mathbf{I}_m \| \mathbf{R}_{\mathsf{ID}}]$ is a short vector, since $\mathbf{x}'_1$ is short and the coefficients of $[\mathbf{I}_m \| \mathbf{R}_{\mathsf{ID}}]$ are small, and the form of $\mathbf{c}_1$ is similar to that in the real world. A subtle problem is that the distribution of the error term ${\mathbf{x}'_1}^\top[\mathbf{I}_m \| \mathbf{R}_{\mathsf{ID}}]$ depends on $\mathbf{R}_{\mathsf{ID}^\star}$, which is different from the real world. This problem can be resolved in several different ways. We do not explain how to resolve this problem in this survey. We refer to [1, 33, 50] for the interested readers.

4.2 Instantiations

Here, we instantiate the framework and recover several existing schemes. Specifically, we have to set $\mathcal{H}(\mathsf{ID})$ and $\mathbf{Y}_1, \ldots, \mathbf{Y}_\ell$ so that Eq. 6 holds with sufficient high probability.

(Selectively Secure Variant of) Cash–Hogheinz–Kiltz–Peikert Scheme [20]. Let us set mpk as $(\mathbf{A}, \mathbf{B}_{1,0}, \mathbf{B}_{1,1}, \ldots, \mathbf{B}_{\ell,0}, \mathbf{B}_{\ell,1}, \mathbf{u})$, where ℓ is the length of the identities. We set $\mathcal{H}(\mathsf{ID})$ as

$$\mathcal{H}(\mathsf{ID}) = [\mathbf{A} \| \mathbf{B}_{1,\mathsf{ID}_1} \| \cdots \| \mathbf{B}_{i,\mathsf{ID}_i} \| \cdots \| \mathbf{B}_{\ell,\mathsf{ID}_\ell}]$$

where ID_i is the i-th bit of ID. We skip the discussion on the security of the scheme, because it slightly deviates from our template. The scheme described above is only selectively secure. As shown in [20], we can boost it to be adaptively secure by the semi-generic technique of using the admissible hash function [11].

(Selectively Secure Variant of) Agrawal–Boneh–Boyen Scheme [1]. In [1], the authors introduce a function F that maps an identity ID to a matrix $F(\mathsf{ID}) \in \mathbb{Z}_q^{n \times n}$ such that $F(\mathsf{ID}) - F(\mathsf{ID}')$ is invertible if and only if $\mathsf{ID} \neq \mathsf{ID}'$. They set mpk be $(\mathbf{A}, \mathbf{B}_1, \mathbf{u})$ and $\mathcal{H}(\mathsf{ID})$ be

$$\mathcal{H}(\mathsf{ID}) = \mathbf{B}_1 + F(\mathsf{ID}) \cdot \mathbf{G}.$$

We can prove the selective security of the scheme. In the simulation, we set $\mathbf{B}_1 = \mathbf{A}\mathbf{R}_1 - F(\mathsf{ID}^\star)\mathbf{G}$. Note that we can embed the value of $\mathsf{ID}^\star$ into the master public key, because $\mathsf{ID}^\star$ is provided by the adversary at the outset of the (selective) security game. It can be seen that the condition (6) always holds by the property of F.

The above scheme is the most efficient scheme in the standard model. However, it only achieves the selective security. The adaptively secure variant has also been proposed.

(Adaptively Secure Variant of) Agrawal–Boneh–Boyen Scheme [1, 14]. Let us assume that the identity space is $\{0, 1\}^\ell$. The master public key in the scheme is $(\mathbf{A}, \mathbf{B}_0, \mathbf{B}_1, \ldots, \mathbf{B}_\ell, \mathbf{u})$. In their scheme, they set $\mathcal{H}(\mathsf{ID})$ as

$$\mathcal{H}(\mathsf{ID}) = \mathbf{B}_0 + \sum_{\{i \in [\ell] | \mathsf{ID}_i = 1\}} \mathbf{B}_i$$

where ID_i is the i-th bit of ID. In the simulation, we set

$$\mathbf{B}_0 = \mathbf{A}\mathbf{R}_0 + y_0 \cdot \mathbf{G}, \qquad \mathbf{B}_i = \mathbf{A}\mathbf{R}_i + y_i \cdot \mathbf{G} \qquad \text{for } i \in [\ell].$$

where $y_0 \hookleftarrow [-O(\ell Q)]$ and $y_i \hookleftarrow O(Q)$. Here, constant terms hidden in the big-O notation should be adjusted appropriately. By the similar analysis to [44], we have $y_0 + \sum_{\{i \in [\ell] | \mathsf{ID}^*_i = 1\}} y_i = 0$ and $\forall j \in [Q]\ y_0 + \sum_{\{i \in [\ell] | \mathsf{ID}^{(j)}_i = 1\}} y_i \neq 0$. If this happens, the simulation will be successful. (We will replace the scalars y_i with the matrix $y_i \mathbf{I}_n$ to apply SampleRight algorithm.) This security proof only works when $Q = o(q)$. Namely, when $Q = \mathsf{poly}(\lambda)$, it can only achieve weaker notion of the security. Boyen [14] refines the above analysis and proves the security without this restriction.

RECENT SCHEMES. The main drawback of the above scheme is that the size of the master public key is very long. That is, it requires $O(\ell)$ basic matrices in the master public key, where ℓ is the length of the identities. Several works try to reduce the size of the master public key. Using the technique of Naccache IBE [37], we can reduce the number of basic matrices to $O(\ell / \log \lambda)$ [43]. Yamada asymptotically reduces the number of basic matrices to be $\tilde{O}(\ell^{1/d})$, where d is arbitrary constant, by generating ℓ matrices from smaller number of matrices using fully homomorphic computation [13]. The technique is somewhat reminiscent of that used in the previous works on (bounded) IBE and CCA-secure PKE in the bilinear map settings [28, 47–49]. One of the drawback of the scheme is that it requires the mudulus q to be super-polynomial. The subsequent work by Katsumata et al. [33] essentially proved that the ring variant [34, 35] of the scheme in [46] is secure even if we use the polynomial-size modulus. Zhang et al. [50] proposed a scheme with Q-bounded security (which is also referred to as the Q-resilient security [29]) with very short master public keys. In their scheme the master public key only contains $O(\log Q)$ matrices. Here, Q-bounded security means that the security of the scheme is guaranteed only when the number of key extraction queries that the adversary makes in the security game is bounded by some *predetermined* polynomial Q (e.g., $Q = \lambda^3$). The security notion is also called k-resilient security [29]. Apon et al. [5] reduced the number of the basic matrices in the master public key to be $O(\ell / \log^2 \lambda)$, and observed that this can be further shrunk to $O(1)$ if we use *exponentially secure* collision-resistant hash function. Another interesting direction is to achieve tight security. Recently, Boyen and Li gave the first tightly secure lattice-based IBE scheme [16]. All the above schemes except for [16] roughly follow the above template.

OPEN PROBLEMS. There are several interesting problems. The first one is whether we can further reduce the number of matrices in the master public key from the adaptively secure IBE schemes (from lattices).

This is achieved in the pairing settings, the adaptively secure schemes are as efficient as selective schemes up to small constant factors [21, 32]. Such schemes are constructed using the powerful machinery of the dual system encryption methodology [45], for which we do not have lattice analogue.

Regarding tight security for lattice IBE schemes, we have several interesting problems. Can we achieve tight security from the LWE assumption with polynomial approximation factors? Can we achieve tight security under the multi-challenge and multi-user settings? Note that the scheme proposed by Boyen and Li [16], which is tightly secure under single-challenge settings, requires super-polynomial approximation factors (if we stick to post-quantum security) and long public parameters. Regarding these questions, it seems that we have better solutions for them in the pairing settings. For the first question, we have tightly secure IBE schemes (in the single challenge setting) from the (standard) DLIN assumption and weaker variants [8, 21]. These schemes have been extended to the multi-challenge setting in the subsequent works, under the same assumptions [6, 27, 31]. The reason why we have better solutions in the pairing settings would be, again, because we can use the powerful machinery of the dual system encryption methodology.

References

1. S. Agrawal, D. Boneh, X. Boyen, Efficient lattice (H)IBE in the standard model, in *EUROCRYPT* (2010), pp. 553–572
2. S. Agrawal, D. Boneh, X. Boyen, Lattice basis delegation in fixed dimension and shorter-ciphertext hierarchical IBE, in *CRYPTO* (2010), pp. 98–115
3. M. Ajtai, Generating hard instances of the short basis problem, in *ICALP* (1999), pp. 1–9
4. J. Alwen, C. Peikert, Generating shorter bases for hard random lattices, in *STACS* (2009), pp. 75–86
5. D. Apon, X. Fan, F. Liu, Fully-secure lattice-based IBE as compact as PKE, in *IACR Cryptology ePrint Archive* 2016:125 (2016)
6. N. Attrapadung, G. Hanaoka, S. Yamada, A framework for identity-based encryption with almost tight security, in *ASIACRYPT (1)* (2015), pp. 521–549
7. M. Bellare, T. Ristenpart, Simulation without the artificial abort: simplified proof and improved concrete security for waters' IBE scheme, in *EUROCRYPT* (2009), pp. 407–424
8. O. Blazy, E. Kiltz, J. Pan, (hierarchical) identity-based encryption from affine message authentication, in *CRYPTO* (2014), pp. 408–425
9. D. Boneh, M. Franklin, Identity-based encryption from the weil pairing, in *CRYPTO* (2001), pp. 213–229
10. D. Boneh, X. Boyen, Efficient selective-id secure identity-based encryption without random oracles, in *EUROCRYPT* (2004), pp. 223–238
11. D. Boneh, X. Boyen, Secure identity based encryption without random oracles, in *CRYPTO* (2004), pp. 443–459
12. D. Boneh, X. Boyen, E.J. Goh, Hierarchical identity based encryption with constant size ciphertext, in *EUROCRYPT* (2005), pp. 440–456
13. D. Boneh, C. Gentry, S. Gorbunov, S. Halevi, V. Nikolaenko, G. Segev, V. Vaikuntanathan, D. Vinayagamurthy, Fully key-homomorphic encryption, arithmetic circuit ABE and compact garbled circuits, in *EUROCRYPT* (2014), pp. 533–556
14. X. Boyen, Lattice mixing and vanishing trapdoors: a framework for fully secure short signatures and more, in *PKC* (2010), pp. 499–517
15. X. Boyen, B. Waters, Anonymous hierarchical identity-based encryption (Without Random Oracles), in *CRYPTO* (2006), pp. 290–307
16. X. Boyen, Q. Li, Towards tightly secure lattice short signature and ID-based encryption, in *ASIACRYPT* (to appear) (2016)
17. Z. Brakerski, V. Vaikuntanathan, Lattice-based FHE as secure as PKE, in *ITCS* (2014), pp. 1–12

18. Z. Brakerski, A. Langlois, C. Peikert, O. Regev, D. Stehlé, Classical hardness of learning with errors, in *STOC* (2013), pp. 575–584
19. R. Canetti, S. Halevi, J. Katz, A forward-secure public-key encryption scheme, in *EUROCRYPT* (2003), pp. 255–271
20. D. Cash, D. Hofheinz, E. Kiltz, C. Peikert, Bonsai trees, or how to delegate a lattice basis, in *EUROCRYPT* (2010), pp. 523–552
21. J. Chen, H. Wee, Fully, (almost) tightly secure IBE and dual system groups, in *CRYPTO* (2013), pp. 435–460
22. C. Cocks, An identity based encryption scheme based on quadratic residues, in *IMA International Conference* (2001), pp. 360–363
23. C. Gentry, Practical identity-based encryption without random oracles, in *EUROCRYPT* (2006), pp. 445–464
24. C. Gentry, A. Silverberg, Hierarchical ID-based cryptography, in *ASIACRYPT* (2002), pp. 548–566
25. C. Gentry, C. Peikert, V. Vaikuntanathan, Trapdoors for hard lattices and new cryptographic constructions, in *STOC* (2008), pp. 197–206
26. S. Gorbunov, V. Vaikuntanathan, H. Wee, Attribute-based encryption for circuits, in *STOC* (2013), pp. 545–554
27. J. Gong, J. Chen, X. Dong, Z. Cao, S. Tang, Extended nested dual system groups, revisited, in *PKC(1)* (2016), pp. 133–163
28. K. Haralambiev, T. Jager, E. Kiltz, V. Shoup, Simple and efficient public-key encryption from computational Diffie-Hellman in the standard model, in *PKC* (2010), pp. 1–18
29. S. Heng, K. Kurosawa, k-Resilient identity-based encryption in the standard model, in *CT-RSA* (2004), pp. 67–80
30. D. Hofheinz, E. Kiltz, Programmable hash functions and their applications, in *CRYPTO* (2008), pp. 21–38
31. D. Hofheinz, J. Koch, C. Striecks, Identity-based encryption with (almost) tight security in the multi-instance, multi-ciphertext setting, in *PKC* (2015), pp. 799–822
32. C. Jutla, A. Roy, Shorter quasi-adaptive NIZK proofs for linear subspaces, in *ASIACRYPT* (2013), pp. 1–20
33. S. Katsumata, S. Yamada, Partitioning via non-linear polynomial functions: more compact IBEs from ideal lattices and bilinear maps, in *ASIACRYPT* (to appear) (2016)
34. V. Lyubashevsky, C. Peikert, O. Regev, On ideal lattices and learning with errors over rings, in *EUROCRYPT* (2010), pp. 1–23
35. V. Lyubashevsky, C. Peikert, O. Regev, A toolkit for ring-LWE cryptography, in *EUROCRYPT* (2013), pp. 35–54
36. D. Micciancio, C. Peikert, Trapdoors for lattices: simpler, tighter, faster, smaller, in *EUROCRYPT* (2012), pp. 700–718
37. D. Naccache, Secure and practical identity-based encryption. IET Inf. Secur. **1**(2), 59–64 (2007)
38. C. Peikert, Public-key cryptosystems from the worst-case shortest vector problem: extended abstract, In *STOC* (2009), pp. 333–342
39. C. Peikert, A decade of lattice cryptography, IACR Cryptology ePrint Archive, Report 2015/939
40. O. Regev, On lattices, learning with errors, random linear codes, and cryptography, in *STOC* (2005), pp. 843–873
41. R. Sakai, K. Ohgishi, M. Kasahara, Cryptosystems based on pairing over elliptic curve, in *The 2000 Symposium on Cryptography and Information Security* (in Japanese) (2000)
42. A. Shamir, Identity-based cryptosystems and signature schemes, in *CRYPTO* (1984), pp. 47–53
43. K. Singh, C. Pandu Rangan, A.K. Banerjee, Adaptively secure efficient lattice (H)IBE in standard model with short public parameters, in *SPACE* (2012), pp. 153–172
44. B. Waters, Efficient identity-based encryption without random oracles, in *UROCRYPT* (2005), pp. 114–127
45. B. Waters, Dual system encryption: realizing fully secure IBE and HIBE under simple assumptions, in *CRYPTO* (2009), pp. 619–636

46. S. Yamada, Adaptively secure identity-based encryption from lattices with asymptotically shorter public parameters, in *EUROCRYPT (2)* (2016), pp. 32–62
47. S. Yamada, Y. Kawai, G. Hanaoka, N. Kunihiro, Public key encryption schemes from the (B)CDH assumption with better efficiency. IEICE Trans. **93–A**(11), 1984–1993 (2010)
48. S. Yamada, G. Hanaoka, N. Kunihiro, Two-dimensional representation of cover free families and its applications: short signatures and more, in *CT-RSA* (2012), pp. 260–277
49. T. Yamakawa, S. Yamada, K. Nuida, G. Hanaoka, N. Kunihiro, Reducing public key sizes in bounded CCA-secure KEMs with optimal ciphertext length, *ISC 2013* (2015), pp. 100–109 (Short Paper)
50. J. Zhang, Y. Chen, Z. Zhang, Programmable hash functions from lattices: short signatures and IBEs with small key sizes, in *CRYPTO(1)*, pp. 214–243

Index

T. Takagi et al. (eds.), *Mathematical Modelling for Next-Generation Cryptography*, Mathematics for Industry 29, DOI 10.1007/978-981-10-5065-7

Printed in the United States
By Bookmasters